# TREATISE



ON

# AMMUNITION

*Printed by Order of the Secretary of State for War*

CORRECTED UP TO DECEMBER, 1877

LONDON:
*Printed under the Superintendence of Her Majesty's Stationery Office*

**The Naval & Military Press Ltd**

Published by
**The Naval & Military Press Ltd**
5 Riverside, Brambleside, Bellbrook
Industrial Estate, Uckfield, East Sussex,
TN22 1QQ England
Tel: +44 (0) 1825 749494
Fax: +44 (0) 1825 765701
www.naval-military-press.com

# TABLE OF CONTENTS.

## LIST OF PLATES.

## LIST OF TABLES.

N.B.—There are several minor tables to be found in the text in connection with the description of the corresponding stores.

The coloured plates in this reprint are placed after page 349

# INTRODUCTION TO FIRST EDITION.

It has become necessary to publish a new work on Ammunition, Parts I and II of Ammunition by Captains Majendie and Browne being out of print.* Smooth bore ordnance is rapidly being discarded, therefore it is unnecessary to treat the subject at length. Part I included not only S. B. Ordnance, but many articles of Ammunition used in conjunction with rifled ordnance. In order to make a complete work, selections have been made from Parts I and II, omitting such portions as have ceased to be of importance owing to changes in material, and making the additions necessary to bring the work up to date.

The book has been arranged so as to embrace in the opening chapters the Ammunition common to the various systems of S. B. and Rifled guns. Each system has then been treated separately; the concluding chapters contain B. L. Small Arm Ammunition and Rockets.

Fuzes have been treated as a whole, some fuzes being common to both Rifled and S. B. guns, and in all cases the similarity is so great that it is convenient to keep them together.

The statements in the text have, as far as possible, been supported by references to Changes in War Stores, where extended information will frequently be found.

The changes are published monthly, Artillery Officers can obtain them through the R. A. Institution, and by means of them can keep this book up to date.

Frequent reference has also been made to Extracts from the Proceedings of the Department of the Director of Artillery. The Extracts are published quarterly, and may also be obtained from the R. A. Institution. Brief accounts of the various experiments carried on will be found in them. They are useful as a guide to the value of the new material which is at present being rapidly introduced into the service.

Many pages have been taken verbatim from Ammunition, Parts I and II, by Captains Majendie and Browne, the greater parts of Chapters XI, XII and XVI have been reprinted from Part II.

---

* Some copies of Ammunition, Part I, may still be obtained, a second thousand having been published by Mr. W. Mitchell, 39, Charing Cross. The official edition is exhausted.

The value of the works above mentioned is well known, and I can hardly hope to equal them in accuracy.

As far as possible detailed information has been given in the form of tables, in order not to encumber the text with a mass of figures useful only for reference.

The proportion of Ammunition issued in connection with ordnance, will be found in Revised Army Regulations, Vol. III, 1870.* Some important changes have been made, which have been notified from time to time in Army Circulars.

In demanding stores the nomenclature of the Priced Vocabulary of Stores should be adhered to, this nomenclature has been followed in the index.

Some important circulars relating to Magazines, Gunpowder, and Laboratory operations are given in the Appendix.

I have received much assistance from Sergeant-Major Macken and Sergeant-Instructors Dickson and Tims in preparing this work for the press.

W. R. BARLOW,
Major R.A.,
Captain-Instructor, Royal Laboratory.

*July* 1874.

# PREFACE TO SECOND EDITION.

The numerous changes which have occurred in Artillery *matériel*, even in the three years which have elapsed since the publication of the first edition, have necessitated a thorough revision of this work. The general arrangement is but little altered, but a considerable amount of new matter has been added, many alterations, rendered necessary by changes made, and a small amount of matter, now rendered obsolete or comparatively unimportant, struck out.

It is difficult to draw the line with regard to experimental stores. As far as possible I have, in this edition, treated in the text only of such stores as are actually in the service, or practically certain to be so very shortly. Stores, still in the earlier experimental stages, have been touched upon in the foot-notes.

The various tables have been corrected up to date.

Sergeant-Major Tims has rendered most valuable assistance, especially in the laborious work of correcting tables, and comparing references.

J. P. CUNDILL,
Captain R.A.,
Assistant Captain-Instructor, Royal Laboratory.

*December* 1877.

* Now to be found in Equipment Regulations, 1876, and in Errata published from time to time with Army Circulars.

# ERRATA AND ADDENDA.

*Corrected up to and including September* 1878.

---

*Page* 3, *line* 11.

*For* "is about 600° F.†," *read* "is about 600° F., but gunpowder will deteriorate at much lower temperatures. It begins to lose sulphur at 212°, and this ingredient passes off rapidly as the temperature rises†."

*Page* 6, 2*nd paragraph.*

"§ 3327 slightly modifies the specification of 'P' powder." The powder is to pass through meshes $\frac{3}{4}''$ in the clear, and be retained on meshes $\frac{3}{8}''$ in the clear. The grains are to be cubical with rounded edges, and to number about 80 to the lb. The absolute density of the finished powder is not to be less than 1·75. The velocity test is slightly altered.

*Page* 6, *line* 3 *from bottom.*

*For* "*" *put* "§."

*Page* 7, *after line* 6 *from bottom.*

"*Service R.L.G.*[2]" is a new powder, approved 8/5/78 and 8/6/78. The grains should pass through a 3-mesh sieve and be retained on a 6-mesh sieve. They should not be flat or flaky in form. The absolute density is not less than 1·65. It should give a velocity from the 9-pr. R.M.L. proof gun, with a 9 lb. 3 oz. shot and a $1\frac{3}{4}$ lb. charge, of not less than 1,380 f.s., or more than 1,430 f.s. The pressure indicated by any one gauge is not to exceed 15 tons. A barrel holds 120 lbs.

*Page* 7, *line* 13 *from bottom.*

*Omit* "fired."

*Page* 7.

*Transpose* "references to footnotes."

*Page* 16, *add to* 1*st paragraph below table.*

§ 3371. The tank above mentioned is to be considered obsolete. Gun-cotton will in future be transported and stored in "Case, wood, packing, gun-cotton, 50 lbs., Mark II." This case measures on the exterior 28″·75 × 12″·5 × 8″·62 over cleats, and contains from 48 to 62 lbs. of wet gun-cotton, the exact amount varying with the form of the cotton and the expansion produced by wetting it. When packed, a label will show the exact contents.

The case is painted red inside and outside. A strip of serge is placed round the inside of the lid and the top edge of the sides. The lid is fixed to the case by brass screws. To allow the gun-cotton to be re-damped, when necessary, without taking off the lid, a G.S. plug is screwed into a brass socket at the lower part of one end, and at the upper part of this end is a small hole closed by a brass screw (which acts as a vent peg). This enables water to be poured into the box when necessary, and the surplus to be drained off, as in the case of the tank.

*Page* 16, *line* 15 *from bottom.*

*For* "present pattern," *read* "Mark I. Case."

*Page* 16, *after* 2*nd paragraph below table.*

§ 3326. Priming charges for submarine mining will in future be issued dry. Each priming charge to consist of four 9 oz. discs (one of them a "primer," *i.e.*, perforated for two detonators), and will be placed in a hermetically closed tin cylinder. Such cylinders will be packed by dozens in skeleton wooden boxes.

*Page* 21, 3*rd column of table.*

*For* "5, 9, or 20 secs. B.L. or M.L. time," *read* "5, 9, or 20 secs. B.L. and 5, 9, 15, or 20 secs. M.L."

*Page* 22, *line* 8.

*After* "p.," *insert* "71, 72."

*Page* 23, *line* 8.

*For* "fuse" *read* "fuze;" *also, for* "if there are no powder channels . . . . . . . . 10 seconds or under," *read* "except in fuzes having two powder channels, when the composition channel is eccentric;" *also, for* "They are essential," *read* "Powder channels are essential."

*Page* 23, *line* 17.

*After* "bored in a fuze," *insert* "such as the 9 seconds."

*Page* 23, *line* 8 *from bottom.*

*For* "last hole in the row," *read* "last hole at least."

*Page* 23, 3*rd foot note.*

*For* "Where there are powder channels," *read* "where there are two powder channels."

*Add to end of note* "In the 15 sec. fuze the hole marked 30, only is bored through and threaded with quickmatch; the powder in the other five channels is supported by connecting quickmatch laid in a circular groove in the base, pressed up into the openings of those channels."

*Page* 24, *line* 16.

*For* "fuse," *read* "fuze."

*Page* 25, *after* 5*th paragraph.*

A special slow-burning composition has been introduced for the 15 sec. fuzes (see Addenda to table, page 319), which burns at the rate of 1 inch in 7½ secs."

*Page* 26, *lines* 6, 15, *and* 33.

*After* "The 5 seconds," *add*, "and 15 seconds."

*Page* 26, *line* 9 *from bottom.*

*After* "II.," *insert* "5, 9, and 20 secs."

*Page* 26, 1*st line of footnote.*

*For* "All fuzes having powder channels," *read* "All fuzes having two powder channels."

*Page* 28, *line* 3.

*Omit* "*," and cancel footnote to which it refers.

*Page* 29, *line* 17.

*For* "5 and 9 seconds," *read* "5, 9, 15, and 20 seconds."

*Page* 29, *line* 11 *from bottom.*

*Add*, "The cylinders containing 15 sec. fuzes are the same as those used for the 5 and 9 secs. fuzes. The top is painted white, and the label printed in black."

*Page* 29, *add to end of second footnote.*

The following method of closing tin cylinders which have been opened for inspection, &c., is recommended by the Superintendent R.L., in the case of cylinders containing quill friction tubes, and will be found generally useful. The lids of the cylinders to be secured by a band of calico coated with equal parts of Swedish pitch, gutta percha, and rosin.

The mixture is thus prepared:—Heat the pitch and stirin the rosin and gutta-percha, the latter being cut into small pieces. Stir the mixture till uniform. Soak the band in it and apply while warm to the cylinder, wrapping it *tightly* round. With the flat part of a knife, &c., apply a little of the mixture round the top and bottom edges of the band after the latter is on the cylinder.

*Page* 30, *line* 21.

*For* "Mark I. B.L. fuze," *read* "Mark I. 9 sec. B.L. fuze."

*Page* 30, *line* 18 *from bottom.*

*After* "p.," *insert* "133."

*Page* 30, *line* 7 *from bottom.*

*For* "exeept," *read* "except."

*Page* 33, *line* 11 *from bottom.*

*For* "arge," *read* "large."

*Page* 36, *line* 13 *and bottom line.*

*For* "Marks of M.L. fuzes," *read* "Marks of the 5, 9, and 20 seconds M.L. fuzes."

*Page* 38, *before description of* 20-*seconds M.L. Fuze.*

§ 3306. 15-*secs. M.L. Fuze, Mark I.*—This fuze in general principles of construction, and in external dimensions resembles the 5 and 9 sec. Mark III., M.L. fuzes. It has, however, the composition channel in the centre, driven with 2″ of slow burning composition (1 inch in 7½ seconds). Above this is a ·25″ pellet of meal powder. There are six powder channels connected at the bottom by quickmatch placed in an annular groove and pressed into the bottom of each channel. The bottom hole of one channel is bored through and threaded with quickmatch. The paper scale gives intervals corresponding to half seconds and quarter seconds of time (the same half second unit as in the other time fuzes).

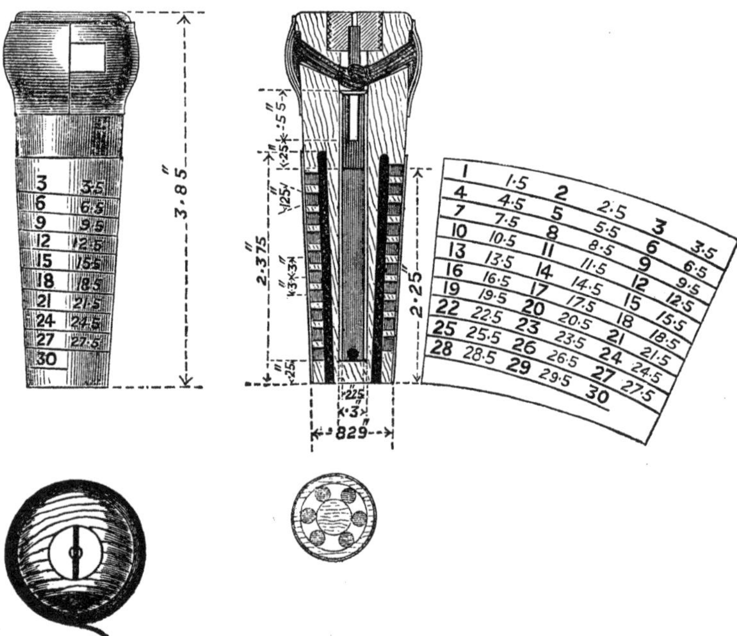

This fuze burnt at rest burns about 15 seconds; when fired in rifled shells, 13 to 14 seconds only. It may be used in place of either 5-seconds or 9-seconds M.L. fuzes, which it is intended eventually to supersede.

*Page* 40, *after 2nd paragraph from bottom.*

*Add,* "The 15-seconds fuze may be used in place of either 5 or 9-seconds M.L. fuzes, which it is intended eventually to supersede. Reckoning the average time of burning in flight at 13·4 seconds (Extracts, Vol. xv., p. 189), it would be available up to about 3,900 yards with the 16-pr., and to about 3,500 yards with the 9-pr. R.M.L. gun."

*Page* 40, *footnote.*

*In margin add* "§ 2360." *Add to note* "By addenda to Army Equipment, 1876, issued 1/2/78, Shrapnel are now issued for heavy guns in L.S. for certain forts and batteries in exceptional positions. Consequently time fuzes will have to be issued in such cases."

*Page* 44, *line* 10.

*For* "fuse," *read* "fuze."

*Page* 57, *at bottom.*

*Add,* "Fuze, percussion, delay action, for base of 64-pr. battering shell."

This fuze consists of a body, slowing chamber, pellet with detonator, and guard with suspending wire and needle. The *body* is of gun-metal with a hexagonal head to take the gas-check spanner, and is tapped on the exterior to screw into the base of the shell, and act as a gas-check plug. It is bored out to various diameters and tapped internally in places to contain the other portions of the fuze.

The *slowing chamber* is of gun-metal, and its lower portion screws into the top of the body. It contains a central channel driven with ·8″ of composition burning at the rate of 1″ in 9 secs. At the top of this composition are two conical side holes, opposite each other, and filled with fine grain powder. They connect the central channel with the outside, and are closed on the exterior with a band of paper shellaced on. There is also a little fine grain powder above the composition. At the bottom of the composition is a small pellet of meal powder, and two pieces of quickmatch projecting slightly downwards.

Below the slowing chamber in the interior of the body is an air space to receive the gas, &c., resulting from the burning composition. Were it not for this air space, the fuze would in all probability burst before the composition was nearly expended. At the bottom of the air space is screwed into the interior of the body a steel plug, pierced with four fire holes, and having a hollow needle projecting downwards. Immediately below the needle is a detonator containing cap composition.

This detonator occupies the upper portion of the gun-metal *pellet*, which is hexagonal in cross section at its lowest part, and slides easily (when free) in the interior of the body.

There is an under-cut neck between the upper and lower portions of this pellet, into which dovetails the hexagonally shaped gun-metal *guard*, when the copper suspending wire is sheared on the shock of discharge

HALF SIZE.

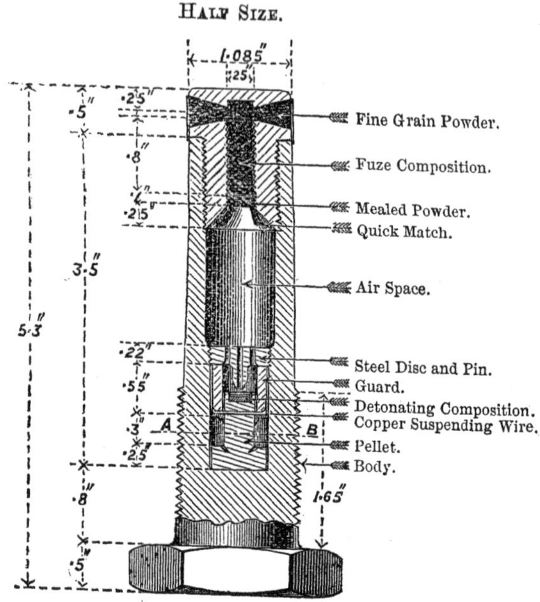


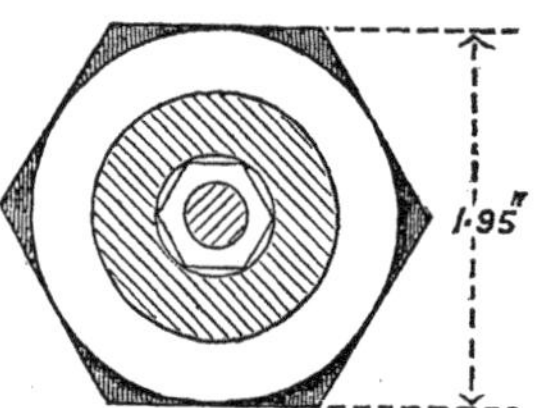


SECTION AT A.B.

*Action.*—On the shock of discharge the suspending wire is sheared, the guard descends and dovetails into the under-cut recess in the pellet, so as to form one mass with it. On impact the pellet and guard fly forward, the cap explodes against the needle, and the flash passes through the hollow needle and fire-holes, and ignites the quickmatch and meal powder pellet. The composition is in turn ignited, and burns till the fine grain powder is reached, when the flash passes through the side holes into the shell and ignites the bursting charge.

The object of the delay-action is to allow the shell plenty of time to penetrate well into masonry, &c., before bursting.

*Page* 58, *table.*

*The second paragraph of second column should read,* "Shrapnel, 7-pr. excepted. Common shell up to 64-pr. gun inclusive, and also 6″·3 and 8″ howitzers. Also 7-pr. double shell."

*The third paragraph of second column should read,* "Double 7-pr., common 25, 40, and 64-prs.; also 6″·3 and 8″ howitzers."

*The fifth paragraph of second column should read,* instead of "40-pr.," "64-pr."

*Add, for L.S. use of sensitive fuze,* "7-pr. common and double, common 6″·3 and 8″ howitzers."

*Add in table,* "Percussion delay-action fuze used with 64-pr. battering shells, L.S."

*Add in table,* "Time, wood, 15-seconds fuze, to eventually supersede 5 and 9-seconds fuzes."

*Page* 59, *at beginning of text.*

*Add,* "N.B.—The following instructions will need slight and obvious modifications, owing to the introduction of gas-checks. Instructions for the mode of fitting these will be found in the addenda to p. 137, and and a description of the new implements required in the addenda to the section on gas-checks."

*Page* 64, *line* 8 *below table.*

*After* "filled with red-lead putty," *add,* "This applies to S.S. only."

*Page* 64, *line* 15 *from bottom.*

*Add,* "also 7″ R.M.L. guns and upwards, when supplied with Shrapnel shell."

*Page* 66, *line* 11.

*For* "fuse," *read* "fuze."

*Page* 67, *under Remarks to table.*

*After* "p.," *add* "61, 64."

*Page* 69, *line* 10.

*Add,* "also for 7″ R.M.L. guns and upwards, when supplied with Shrapnel shell."

*Page* 72, *top line.*

*In margin add,* "3189," *omit* "plain."

*Page* 72, *bottom line.*

*For* "Boxer," *read* "borer."

*Page* 76, *in* 2*nd paragraph of* 1*st footnote.*

*For* "S.," *read* "F."

*Page* 77, *e* 28.

*Add,* "a zinc cylinder, Mark I., to contain the 80-lb. half battering charge for the 12″·5 gun has been introduced. The interior length is 20″·1875, the interior diameter is 12″·7. The introduction of this cylinder renders obsolete the cylinder which took the 130-lb. charge for the 12″·5 gun; but those in store may be utilized for the 110-lb. charge for the 12″ gun of 35 tons." *Alter the table on p.* 79, *accordingly.*

*Page* 77, *line* 13 *from bottom.*

*Add in margin,* "3350."

*Page* 78, *line* 10.

*Omit,* "and 12″·5." *To end of paragraph add,* "A 'C.' cage has been introduced for use with cylinders for 12″·5 guns. It will take two 80-lb. cylinders. Weight of cage 20¾ lbs."

*Page 78, after line 14 from bottom.*

*Add* "§ 3320. *Powder Barrels, Whole, Half, and Quarter. Mark III.*— These are fitted with four copper hoops, two at each end. The whole barrel has 12 ash hoops, the half-barrel 10, and the quarter-barrel 8."

*Page 79, table, 1st column, line 4.*

*For* "1″," *read* "11″."

*Page 82, line 6 from bottom.*

*For* "an," *read* "a."

*Page 83, line 26.*

*Add as note,* "Mark III. barrels are full-bound."

*Page 86, line 10.*

*For* "upport," *read* "support."

*Page 96, line 8.*

*Add,* "This fuze is now obsolete."

*Page 96, line 2 from bottom.*

*Add,* "This fuze is now obsolete."

*Page 97, second footnote.*

*For* "fuzed," *read* "fused."

*Page 98, line 5.*

*Add,* "This detonator will be superseded gradually in submarine service (L.S.) by No. 12.

*Page 98, line 5 from bottom.*

*For* "an," *read* "a."

*Page 98, footnote.*

*For* "Six," *read* "Eight." *Add at end,* "The Bickford fuze is two feet long, and burns about ¾ minute. The free end is protected by a small calico cap which is to be torn or cut off when required for use. The fuze is best fired with a vesuvian.

*Page 99, 2nd footnote.*

*Add,* "Xylonite has, however, proved to be not at all suitable for tubes from the way it holds fire. Reeds have been tried as a material, also tubes made of three quills cemented together. A small ebonite cylinder, grooved longitudinally to take the poles fits into the upper portion, and they are connected by a ·25″ diagonal bridge. The tubes made of three quills are now approved."

*Page 101, after line 11.*

*Add,* "Mark II. differs from the above in having the gun-cotton and meal powder priming instead of gun-cotton yarn. The priming is separated from the fulminate by a calico disc."

*Page 102, after line 5.*

§ 3388. *Add,* "*Detonator, No. 15, for Instantaneous Fuze, Mark I.*— This detonator generally resembles the No. 8 detonator, but the portion into which the fuze fits is much shorter, and is coned, the wider portion of the cone being at the top. The socket should be tightly nipped down

*Page* 80.—*For table given in text substitute the following:—*

PACKAGES OF

| PACKAGE. | | | Rifle, B.L. Martini-Henry, Mark III. Charge, 85 grs. R.F.G.2 Bullet, 480 grs. | | Carbine, B.L. Martini-Henry, Mark II. Charge, 70 grs. R.F.G.2 Bullet, 410 grs. | |
|---|---|---|---|---|---|---|
| Name. | Dimensions, External, for Stowage. | Tare, Average | Rounds. | Gross. | Rounds. | Gross. |
| | | lb. oz. | | lb. oz. | | lb. oz. |
| Bundle ... ... ... ... ... ... ... ... | ... ... | ... | 10 | 1 $1\frac{14}{16}$ | 10 | 1 0 |
| Barrel, Cartridge, Half ... ... ... ... ... | L. $16\frac{9}{10}$ D. $13\frac{4}{10}$ | 12 2 | ... | ... | ... | ... |
| Barrel, Cartridge, Quarter ... ... ... | L. $14\frac{1}{4}$ D. $11\frac{6}{10}$ | 7 12 | 700 | 86 0 | 700 | 77 12 |
| Box, Wood, Ammunition, Small Arm, with Tin Lining, Adams Revolver ... ... II. | $11\frac{3}{4}$ by 8 by $4\frac{1}{4}$ | 4 7 | ... | ... | ... | ... |
| do. do. ... ... ... III. | $8\frac{1}{2}$ ,, $5\frac{1}{2}$ ,, $4\frac{1}{4}$ | 2 4 | ... | ... | ... | ... |
| Bullock ... ... ... ... I. | $20\frac{1}{2}$ ,, $10\frac{1}{4}$ ,, $9\frac{3}{8}$ | 16 8 | 840 | 110 5 | ... | ... |
| do. India Pattern ... ... | $19\frac{1}{4}$ ,, $9\frac{1}{4}$ ,, 10 | 14 4 | 600 | 81 4 | ... | ... |
| Camel ... ... ... ... ... I. | $27\frac{1}{4}$ ,, 14 ,, $13\frac{3}{8}$ | 33 1 | 2600 | 323 9 | ... | ... |
| do. India Pattern ... ... | 25 ,, $10\frac{1}{2}$ ,, 18 | 34 8 | 1800 | 235 8 | ... | ... |
| Special ... ... ... ... ... ... | $27\frac{1}{4}$ ,, 14 ,, $12\frac{1}{8}$ | 31 7 | ... | ... | ... | ... |
| Service ... ... ... ... ... I. | $16\frac{1}{4}$ ,, 7 ,, $8\frac{1}{2}$ | 9 0 | 480 | 62 10 | 500 | 59 0 |
| do. ... II., IV. & V. / Special ... ... ... ...VII. | 22 ,, 7 ,, $8\frac{1}{2}$ | 12 4 | 600 | 79 4 | 630 | 75 4 |
| without Tin Lining, Service I. | $16\frac{1}{4}$ ,, 7 ,, $8\frac{1}{2}$ | 7 10 | 480 | 61 4 | 500 | 57 6 |
| without Tin Lining, Service VI. | 22 ,, 7 ,, $8\frac{1}{2}$ | 9 7 | 600 | 76 7 | 630 | 72 7 |
| Case, Wood, Powder, Metal Lined, Half... ... ... ... | $14\frac{1}{4}$ ,, $13\frac{1}{2}$ ,, $16\frac{1}{4}$ | 30 0 | 1540 | 202 0 | 1620 | 192 6 |
| Case, Wood, Powder, Metal Lined, Quarter ... .. ... | 11 ,, $10\frac{1}{4}$ ,, $13\frac{3}{4}$ | 18 0 | 660 | 91 11 | 660 | 84 0 |

| NAME. | Charge. |
|---|---|
| Cartridges, Small Arm, Ball, Rifle, B.L. Martini-Henry, Mark IV. ... ... ... | 80 grains, R.F.G. |
| Cartridges, Small Arm, Ball, Rifle, B.L. Martini-Henry, solid case ... ... ... | 85 grains, R.F.G.2 |
| Cartridges, Small Arm, Ball, Rifle, M.L., 1853 pattern and Lancaster, Mark I. ... ... | 70 grains, R.F.G. |
| Cartridges, Small Arm, Ball, Rifle or Carbine, B.L., Snider Buck Shot, Mark II. ... ... ... | 54 grains, R.F.G. |
| Cartridges, Small Arm, Ball, Carbine, B.L. Westley Richards', Mark I. ... ... ... ... | 2 drams, R F.G. |
| Cartridges, Small Arm, Ball, Carbine, M.L., Artillery and Cavalry, Mark I. ... ... ... | 2 drams, R.F.G. |
| Cartridges, Small Arm, Ball, Pistol, M.L. Mark I.... ... ... ... ... ... ... | 1 dram, R.F.G. |
| Cartridges, Small Arm, Ball, Pistol, Revolver, Colts' Skin, Mark I. ... ... ... ... | 13 grains, Pistol. |
| Cartridges, Small Arm, Blank, Rifle or Carbine, M.L., Mark I. ... ... ... ... ... ... | $3\frac{1}{2}$ drs. Blank, R.F.G. |
| Cartridges, Small Arm, Blank, Carbine, B.L., Westley Richards', Mark I. ... ... ... ... | 75 grains R.F.G. |
| Cartridges, Small Arm, Blank, Pistol, M.L., Mark I. ... ... ... ... ... ... ... ... | 2 drs. Blank, R.F.G. |
| Cartridges, Small Arm, Proof, Rifle or Carbine, B.L., Snider ... ... ... ... ... ... ... | 5 drs. R.F.G., compr |

BULLETS AND EMPTY CARTRIDGE CASES FOR P

FOR PROOF OF B.L. MARTINI-HENRY.

| | | | Tare lb. oz. | | Gross lb. oz. |
|---|---|---|---|---|---|
| Bullets ... ... | 714 grains, Rifle or Carbine | 1st proof ... ... ... | Tare | 3 10 | Gross 55 2 |
| | 714 grains, Rifle or Carbine | 2nd proof, papered and lubricated | ,, | 3 10 | ,, 55 10 |
| | 480 grains, Rifle, Service, for proof ... ... ... ... | | ,, | 3 0 | ,, 37 8 |
| | 410 grains, Carbine, Service, for proof ... ... ... ... | | ,, | 3 0 | ,, 32 8 |
| Cases, Cartridge, S.A., Ball, Rifle or Carbine ... | | Service for proof ... ... | ,, | 7 6 | ,, 21 4 |
| | | strengthened, for proof ... | ,, | 7 8 | ,, 21 2 |

DIRECTIONS FOR ISSUE OF BOXE

Mark I., tin lined,—If iron nailed are for "Land and Home Service only"; if with brass screws and copper nails, m
,, II., tin lined,—Brass screws and copper nails, zinc fittings for straps for "Home Service."
,, III., tin lined,—Tinned iron screws, nails, fittings, and bands, for "Land and Home Service."
,, IV., tin lined,— ,, ,, ,, ,, ,, ,, ,,
,, V., tin lined,—Brass screws, copper nails and bands, but no fittings for straps for "Land and Sea Service."
,, VII., tin lined,—Brass screws, tinned iron bands, no fittings for straps, "Sp cial, for India only."
,, I., not lined,—For "Home Service only."
,, VI. not lined,—Brass screws, no bands or fittings for straps, for "Home Service only."
All service S.A.A. boxes issued for "Foreign Service" must be made of Mahogany or Teak; the boxes if made of Teak
Repaired boxes are re-issued with original numeral, except Mark III., which becomes Mark IV.
No new boxes of Mark I., II., III. or IV. will be manufactured in future.

NOTE.—Old style fig

## Cartridges, Small Arm.

| BALL. | | | | | | | | | | | | | | | | | | BLANK. | | | | | |
|---|---|---|---|---|---|---|---|---|---|---|---|---|---|---|---|---|---|
| Rifle or Carbine, B.L. Snider Service, Mark IX. Charge, 70 grs. R.F.G. Bullet, 480 grs. | | | Gatling Gun, 0·65 inch, Mark I. Charge, 270grs. R.F.G.2 Bullet, 1422 grs. | | | Gatling Gun, 0·45 inch, Mark I. Charge, 85grs. R.F.G. Bullet, 480 grs. | | | Pistol, Revolver, Adams B.L., Mark II. Charge, 13 grs. Pis. Pr. Bullet, 225 grs. | | | Rifle or Carbine, B.L., Mark IV. Charge, 2½ drs. Blank, R.F.G. or F.G. | | | Rifle or Carbine (Converted) B.L., Mark IV. Charge, 2½ drs. Blank, R.F.G or F.G. | | |
| ounds. | Gross. | | Rounds. | Gross. | | Rounds. | Gross. | | Rounds. | Gross. | | Rounds. | Gross. | | Rounds. | Gross. | |
| | lb. | oz. | | lb. | oz. | | lb. | oz. | | lb. | oz. | | lb. | oz. | | lb. | oz. |
| 10 | 1 | 0 10/16 | 10 | 3 | 7 3/16 | 10 | 1 | 2¼ | 12 | 0 | 8 3/16 | 10 | 0 | 4 | 10 | 0 | 5 6/16 |
| ... | ... | | ... | ... | | ... | ... | | ... | ... | | 2000 | 62 | 2 | 2000 | 79 | 5 |
| 700 | 81 | 4 | 230 | 87 | 1 | 750 | 93 | 5 | ... | ... | | 1300 | 40 | 4 | 1300 | 51 | 7 |
| ... | ... | | ... | ... | | ... | ... | | 600 | 30 | 1 | ... | ... | | ... | ... | |
| ... | ... | | ... | ... | | ... | ... | | 240 | 12 | 9 | ... | ... | | ... | ... | |
| 780 | 98 | 0 | ... | ... | | ... | ... | | ... | ... | | 1320 | 49 | 8 | 1320 | 60 | 13 |
| ... | ... | | ... | ... | | ... | ... | | ... | ... | | ... | ... | | ... | ... | |
| 2400 | 284 | 9 | ... | ... | | ... | ... | | ... | ... | | 4130 | 136 | 5 | 4130 | 171 | 13 |
| ... | ... | | ... | ... | | ... | ... | | ... | ... | | ... | ... | | ... | ... | |
| 2000 | 239 | 4 | ... | ... | | ... | ... | | ... | ... | | ... | ... | | ... | ... | |
| 420 | 52 | 8 | 160 | 64 | 3 | 500 | 66 | 0 | 1680 | 80 | 10 | 740 | 27 | 12 | 740 | 33 | 14 |
| 560 | 70 | 4 | 200 | 81 | 4 | 680 | 89 | 13 | 2136 | 103 | 8 | 960 | 36 | 4 | 960 | 44 | 8 |
| 420 | 51 | 2 | 160 | 62 | 13 | 500 | 64 | 10 | 1680 | 79 | 4 | 740 | 26 | 6 | 740 | 32 | 8 |
| 560 | 67 | 7 | 200 | 78 | 7 | 680 | 87 | 0 | ... | ... | | 960 | 33 | 7 | 960 | 41 | 11 |
| 1440 | 181 | 0 | 540 | 215 | 6 | 1680 | 227 | 0 | 5364 | 258 | 12 | 2400 | 90 | 0 | 2400 | 110 | 10 |
| 560 | 76 | 0 | 220 | 93 | 12 | 680 | 95 | 9 | 2220 | 112 | 11 | 1020 | 43 | 8 | 1020 | 52 | 5 |

| | Bullet. | Bundle. | | Rounds. | Caps. | Package. Gross. | | Description. |
|---|---|---|---|---|---|---|---|---|
| | | No. | Weight. | | | lb. | oz. | |
| | | | oz. | | | | | |
| | 410 grains | 10 | 16 3/16 | 630 | Nil. | 76 | 0 | Box, tin-lined, IV. or V. |
| | 480 grains | 10 | 19½ | 580 | Nil. | 82 | 15 | do. do. do. |
| | 535 grains | 10 | 14 11/16 | 480 | 720 | 53 | 4 | do. do. I. |
| | 13 | | | 760 | Nil. | 79 | 0 | Quarter Barrel. |
| | Buck Shot, | 10 | 15 | 420 | Nil. | 48 | 6 | Box, tin-lined, I. |
| | 220 per lb. | | | 560 | Nil. | 70 | 8 | Quarter M.L. Case. |
| | 402 grains | 10 | 11½ | 800 | 1200 | 69 | 2 | Quarter Barrel. |
| | 535 grains | 10 | 14 5/16 | 800 | 1200 | 83 | 3 | do. do. |
| | 390 grains | 10 | 10 5/16 | 900 | 1350 | 69 | 4 | do. do. |
| | 135 grains | 18 | 6 10/16 | 2250 | Nil. | 69 | 12 | Quarter M.L. Case. |
| or F.G. | ... | 10 | 2½ | 680 | 748 | 21 | 5 | Box, tin-lined, I. |
| | ... | 10 | 1 15/16 | 2500 | 2750 | 48 | 12 | Half Barrel. |
| or F.G. | ... | 10 | 1 7/16 | 900 | 990 | 18 | 3 | Quarter Barrel. |
| essed. | ... | 10 | 7 5/16 | 850 | Nil. | 46 | 9 | do. do. |

ROOF, ARE ISSUED IN PACKING CASES CONTAINING 500 EACH.

**For Proof of B.L. Snider.**

| | | | | Tare lb. | oz. | | Gross lb. | oz. |
|---|---|---|---|---|---|---|---|---|
| Bullets ... ... | 724 grains, Rifle or Carbine | 1st proof ... ... ... ... | Tare | 5 | 14 | Gross | 57 | 10 |
| | | 2nd proof, papered and lubricated | ,, | 5 | 14 | ,, | 58 | 2 |
| | 480 grains, Rifle or Carbine, Service, for proof ... ... ... | | ,, | 5 | 8 | ,, | 40 | 8 |
| Cases, Cartridge, S.A., Ball, Rifle or Carbine, Service, for proof ... ... ... ... | | | ,, | 8 | 15 | ,, | 20 | 14 |

, WOOD, AMMUNITION, SMALL ARM, SERVICE.

y be issued for S.S.

may weigh a few ounces heavier than shewn in table.

ires are not regular Service Packages.

on the fuze when the latter is inserted. The top is closed with a wooden stopper covered with paper. This stopper is pulled out and the paper torn off when the detonator is required for use. Paint, red all over."

*Fuze, Electric, No.* 16, *Disconnecting, Submarine, Mark I.*—The head of this fuze generally resembles that of Nos. 9, 12, and 13, except that a hole ·2″ diameter is bored in the centre of the base of the ebonite head. This hole receives a small quill tube, a section of the body of an ordinary quill friction tube, and above the quill is a small quantity of pistol powder. Immediately below the quill is the bridge of iridio-platinum wire. The object of this arrangement is to ensure the breaking of the bridge, and consequent interruption of the current when the fuze is fired.

The head fits into a short brass socket whose base is closed with a disc of paper. The socket is lined with paper to prevent contact with the poles. The use of this fuze is to disconnect each one of a series of electro-contact mines when fired without injuring the current to the other mines yet unfired. The details do not come within the scope of this work. It is of course a "low tension" fuze.

Priming—Gun-cotton and meal powder.
Resistance—0·32 ohms.
Paint—Head white, body blue.
Service—Submarine.

*Page* 102, *line* 10.

*Add*, "The cylinders are painted to correspond with the colours of the electric fuzes, &c., contained in them; thus a cylinder containing No. 12 detonator would be painted white, blue, and red.

*Page* 102, *line* 14.

*After*, "of the detonator," *add*, "The boxwood was found to be liable to warp and break. Mark II. rectifier is made of lignum vitæ or other suitable wood."

*Page* 102, *to end of* "*Caution.*"

*Add*, "On no account is an electric or other detonator to be taken to pieces for examination or any other purpose.

Any detonator that may have missed fire, or that may be found distorted or injured in any way that would appear to render it unfit for use, should, if means are available, be carefully retained for examination by properly appointed persons, or should at once be destroyed by immersion in water."

*Page* 102, *under heading* "*Cautions.*"

*Add*, "No. 12 electric detonators (except when being tested for sensitiveness) are only to be tested with the cells of high resistance, specially supplied for the purpose, and the 10 coil of the 3 coil galvanometer in circuit, the detonator being placed in a strong iron box or guard while being tested. The maximum amount of current to be used in testing is three-tenths of a Weber."

*Page* 108, *at end.*

*Add*, "*Instantaneous Fuze.*"—This fuze acts, as its name implies, instantaneously, or practically so. It generally resembles Bickford's fuze, but contains a sort of quickmatch leader instead of a column of powder. It is coated with tape, gutta percha, &c., and is lightly braided over with yellow twist to rpevent the coils sticking together. It is issued in 100 yard lengths, packed in a tin box.

*Page* 110, *line* 17 *from bottom.*

*After* "page," *add* "120."

*Page* 113, *line* 3 *from bottom.*

*After first* "practice," *add* "; ".

*Page* 118, *line* 5.

*For* "rivited," *read* "riveted."

*Page* 120, *line* 22 *from bottom.*

*For* "shooting," *read* "shotting."

*Page* 122, 2*nd footnote.*

*In first line omit* "probably," *and at end of note, for* "warps," *read* "warp." *Add in margin,* "§ 3301."

*Page* 123, *in table.*

*The* "7½-lb. charge" *should be bracketed with the* "7, 6, and 24-pr. 8 lb. charges."

*Page* 124, 2*nd footnote.*

*After* "p.," *add* "122."

*Page* 126, *line* 11.

*For* "321," *read* "331."

*Page* 126, *line* 21.

*Add as note,* "§ 3277. A phosphor bronze 4″ needle will be made in future. This metal is harder than brass."

*Page* 128, *footnote.*

*For* "12-pr.," *read* "13-pr.," *and omit* "it is highly probable that."

*Page* 131, *after* 2*nd paragraph.*

*Add,* "A.C., cl. 41, 1878. Filled shell for 7″ R.M.L., and upwards, to be placed upright on their bases in the shell stores. Filled shell for R.B.L. and R.M.L. guns, up to 80-pr. inclusive, may be piled, and 7″ R.B.L. shells of 83 lbs. may be also piled, if the store is not large enough to permit of their placed on their bases.

Projectiles fitted with gas-checks, and standing on their bases, must have battens 9″ × 1″ × 1″ underneath the gas-checks, to allow for the projection of the plug and nut.

*Page* 132, *top.*

*In margin, after* "12 inch," *add,* "35 ton." *After* 2*nd paragraph, add,* "§ 3370. A slight alteration has been made in the holes in the metal bearing plate. They are lengthened to preserve the balance when used with projectiles fitted with gas-checks."

"§ 3259. Selvagees are to be used for slinging projectiles, 9″ to 12″ of 25 tons, when loading. Their strength and length vary with the calibre of the projectiles used.

*Page* 137, *line* 4 *from bottom.*

*For* "cap," *read* "nut." *For* "in turning," *in* 1*st foot note*, *read* "of manufacture."

*Page* 137, *line* 3 *from bottom.*

*Omit from* "The following instructions" *down to* "gas-checks on the same principle," *on page* 138.

*Page* 137.—ADDITIONAL REMARKS ON GAS-CHECKS.

Since the matter printed in the text was written, considerable advances have been made in the application of gas-checks in the service. With the exception of the 6″·3 howitzer, all projectiles at present furnished with gas-checks take them of the Lyon pattern, the check proper being as described in the text. The projectiles at present authorized to be fitted with gas-checks will be given in detail below. It is to be distinctly understood that the question is by no means yet definitely settled, and the particulars given below are only correct to the present date, August, 1878.

*Projectiles for Woolwich Guns.*

Palliser projectiles. § 3374.

All existing Palliser shell and shot, 9″ to 12″·5 inclusive, will be altered to take gas-checks with the following exceptions, viz.:—

| | |
|---|---|
| * Shell, Palliser .. | 9″, Marks I. and II. |
| † Shot, Palliser | 9″, Marks I. and II. |
| | 10″, Mark I. |
| | 12″, Mark I. |

which cannot be fitted with the gas-checks now introduced. Palliser shot, 9″, Marks III. and IV., take a special wrought-iron plug (owing to the small size of the original bush in the base), which will be described below.

With the above exceptions all are fitted as shown in the cut, which represents the Shell, Palliser, 9″, Mark IV.*

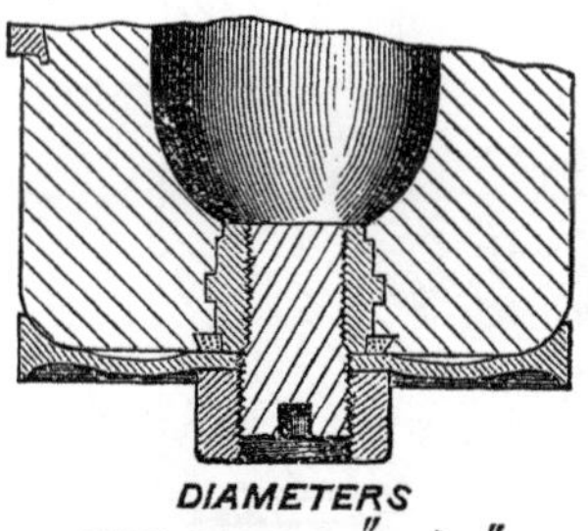


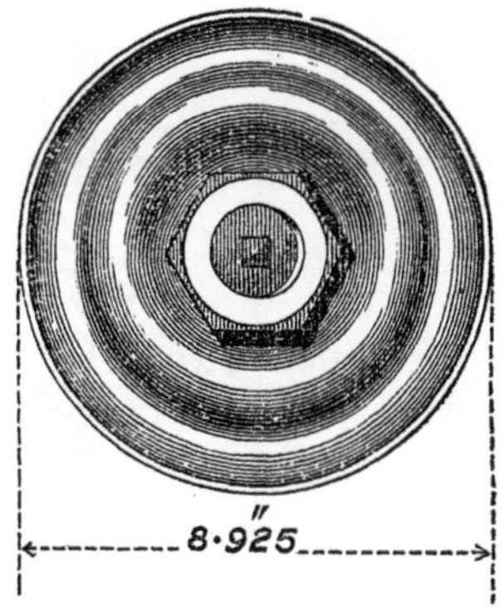


* Because they have flat bases.

† Because they have no screw-plug in base and are chilled all over.

The alteration consists in enlarging the plug hole in the base, and tapping it to take a "plug, metal, for gas-checks," with a left-handed thread. The 9″ Palliser shot, Marks III. and IV., are tapped to take "plug, wrought-iron, for gas-checks." All projectiles so altered are marked with a * in addition to the original numeral.

All new Palliser projectiles will be cast so as to take the gas-check fittings. They differ from the altered projectiles of previous pattern only in having a stronger bush cast in the base. The following projectiles have been approved:—

| | | |
|---|---|---|
| Shell, Palliser | 9″, | Mark V. |
| | 10″, | " III. |
| | 11″, | " III. |
| | 12″, (25-ton) | " III. |
| | 12″, (35-ton) | " III. |

A new pattern (Mark III.) of 12″·5 Palliser shell has also been approved, which has bands (for details see addenda to p. 205). Ere long all Palliser projectiles will have bands.

| | | |
|---|---|---|
| Also Shot, Palliser | 9″, | Mark VI. |
| | 10″, | " V. |
| | 12″, (25-ton) | " V. |

Common shell.

The existing store of 10″, 11″, 12″ (25 tons), and 12″ (35 tons), will be altered to take gas-checks. Also the 7″ double shell. Shell so altered are not to be fired with battering charges.

The 9″ common shells of present pattern are not strong enough to be fired with gas-checks.

The alteration (see cut, which represents a 10″, Mark II* common shell), consists in boring and tapping a hole in the base to take the "plug, metal, for gas-checks." The hole is not bored completely through.

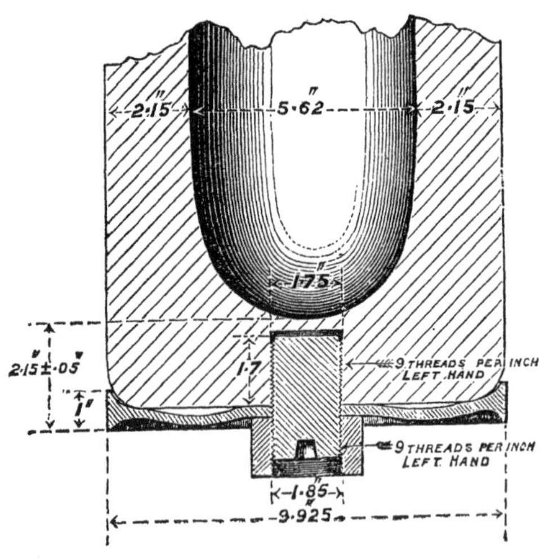

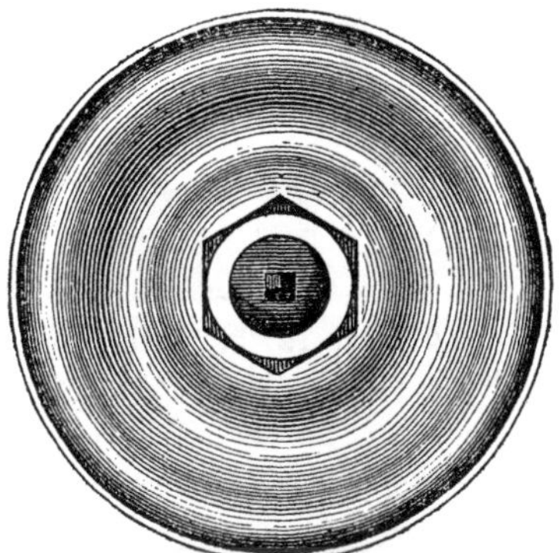

Shells so altered will have a * marked on them in addition to the original numeral.

New shell will be cast with a hole completely through the base, and will be marked with an advanced numeral.

Shrapnel shell.

*At present* only the 12″·5, Mark I., Shrapnel is fitted with a gas-check. For details of this shell, see addenda to p. 194. In future all Shrapnel of 9″ calibre and upwards will have the base rounded to 1″ radius. Efforts are being made to produce a gas-check on the Lyon principle to fit the existing store of Shrapnel, having the base rounded to about 3″ radius.

Case shot.

The 12″·5 case shot, Mark I., has a gas-check. For details of this projectile, see addenda to p. 210.

It will be noted that *at present* gas-checks are not approved for 7″ and 8″ projectiles, with the exception of the 7″ double shell.

Gas-check.

*Gas-check, copper, 12″·5, Mark I.; Gas-check, copper, Palliser and common, Mark I., for 12″, 11″, 10″, and 9″ calibres*; and *Gas-check, copper, Mark I., for 7″ double shell.* All these gas-checks are of the same form (see cuts above) and dimensions (diameters excepted). The depth of the rim is 1″. The mean thickness, except at the rim, is 0″·27. They are cast with a curve to facilitate the radial expansion of the outer rim against the lands, and into the grooves of the gun.

Plug, metal, for gas-checks

*Plug, metal, for Gas-checks, 7″ to 12″·5, Mark I.*, is shown in the cut below, which represents one for 9″ calibre. The diameter of every plug is the same, the length of that part of the plug which enters the base is made to conform to the thickness of the base, or to the depth of the plug hole in the base of the projectile with which it is to be used.

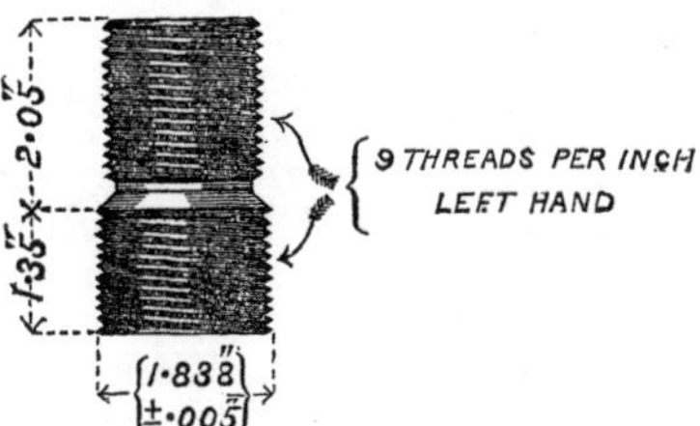


Plug, wrought-iron, for gas-checks.

*Plug, wrought-iron, for Gas-checks, Mark I.*, is used in lieu of the metal plug for 9″ Palliser shot, Marks III. and IV., as explained above. The diameter of the portion which enters the shell is 1″·3. The head is of the same dimensions as that of the metal plug, and takes the same nut.

Nut, wrought-iron, for gas-check plugs.

*Nut, wrought-iron, for Gas-checks, Mark I.*, is for use with all plugs for gas-checks, 7″ to 12·″5. See cut.

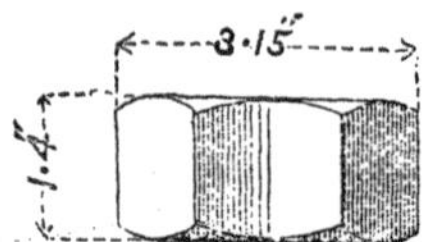


Spanner.

*Spanner, Gas-check Nuts*, 7″ *to* 12″·5, *Mark I.*, comes under the head of fuze and shell implements. It is used to screw and unscrew the nut. See cut.

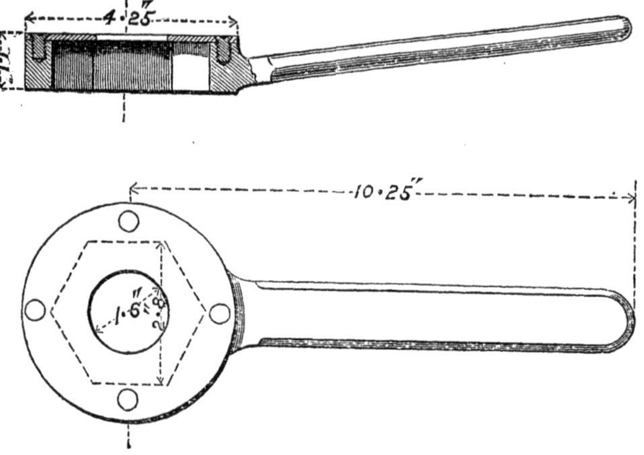


8″ howitzer.

A gas-check is introduced for the 8″ howitzer common shell. It is similar to those for the heavy guns, but is thinner (mean thickness ·15″), has four projections to fit the grooves of the piece, and has eight fire-holes (0″·2 diameter), bored through it close to the rim to allow of the passage of sufficient flash to ignite the R.M.L. time fuze.

It is secured to the shell by a gun-metal plug, with hexagonal head. The same plug, except as regards length, serves for 8″ and 6″·3 howitzers, and for the 64-pr. gun. A spanner to fit this plug is also introduced.

Mark I., Common Shell, can be altered to take the gas-check, and when so altered (similarly to common shell for the heavy guns), is to be marked with a * in addition to original numeral.

Mark II., Common Shell, differs only in being cast with a hole through the base, which is fitted to take the plug.

6″·3 howitzer. § 3285.

The 6″·3 howitzer takes a saucer-shaped gas-check, with projections to fit the grooves of the piece. The shell is cast to gauge, and has no studs. For details, see addenda, p. 226. The gas-check is stamped from 8 B.W.G. sheet copper, and has projections on its convex side to fit into the radial grooves (eight in number) cast in base of shell. It has five fire-holes, 0″·2 diameter. The plug is of the same pattern as that for the 8″ howitzer gas-check, differing in length only.

64-pr. gun.

The 64-pr. Battering Shell (see addenda to p. 230) takes a plain Lyon gas-check, resembling those for heavy guns. Mean thickness ·15″. It is attached to the projectile by the same description of plug as that for the 8″ howitzer.

The 40-pr. R.M.L. Common and Shrapnel Shell take gas-checks of Lyon pattern, with projections to fit the grooves of the gun. It has six fire-holes 0″·2 in diameter. Mean thickness ·125″. They are attached by the same sort of plug as that used for the 8″ howitzer, but of smaller dimensions. Existing shells may be altered, and when so altered take the *. In the common shell the hole is bored completely through the base; but not in the Shrapnel. Mark II., Common Shell, and Mark III., Shrapnel, differ from previous patterns only in being cast to finished dimensions, and in having the hole for the plug cast through the base.

**40-pr. R.M.L. gun.**

---

## Instructions for Fitting Projectiles with Gas-checks, 7 inch to 12·5 inch.

N.B.—The following instructions may not be found to agree *verbatim* with those which will be ultimately issued; but they will be found to be in practical accordance with them:—

*Projectiles that are to be fired empty, or Shells that have already been filled.*

1. Unscrew the wrought-iron nut and remove it, then apply the "wrench, removing plug" to the gas-check plug, and screw it well up in the direction of the arrow,* to ensure its being well home.

N.B.—When unscrewing the nut, if there should be any tendency for the plug to unscrew also, the "wrench, removing plug" should be at once applied to the head of the plug and turned in the direction of the arrow, at the same time as the nut is being turned in the opposite direction.

2. Place the gas-check on the base of the projectile, with the concave, or unpainted side, next the base, and then screw the wrought-iron nut on to the end of the plug with the "spanner, gas-check nut," until the nut binds against the gas-check.

*Shells that require to be filled before fitting with Gas-checks.*

1. Unscrew the wrought-iron nut, and then the gas-check plug, in the case of Palliser shell; in the case of common and Shrapnel shell it is obviously unnecessary to remove the gas-check plug.

N.B.—If difficulty is experienced in removing the plug, or if the threads appear to be clogged or dirty, insert a screw-tap into the plug hole, screwing it up so as to ensure the whole of the screw thread in the plug hole being clear, then remove the screw tap. This operation must only be performed in empty shells.

2. Thoroughly clean the screw thread of the plug-hole (if the plug is removed) with a piece of cotton waste or rag.

3. The shell should now be filled in the usual manner, as described in the Instructions for use of Fuze and Shell Implements.

4. The gas-check plug (if removed) is to be coated with a mixture of cocoanut oil and powdered chalk, or white lead and tallow if the former is not available, and then screwed firmly into the plug-hole.* Care should be taken that no grains of powder or grit adhere to either screw.

5. Place the gas-check on the base of the projectile, with the concave, or unpainted side, next the base, and then screw the wrought-iron

---

* The heads of the gas-check plugs and the wrought-iron nuts will each be stamped with an arrow, to show the direction in which to turn, either when screwing in the gas-check plug, or when screwing on the wrought-iron nut.

nut on to the end of the plug with the "spanner, gas-check nut," till the nut binds against the gas-check.

6. Filled shells which are to be stored in readiness for fitting with gas-checks, will be fitted with the gas-check plug as described in (4), and the wrought-iron nut, well lubricated, screwed on to the plug without the gas-check.

### Instructions for Fitting Gas-checks to Projectiles fitted with Plug with Hexagonal Head.

Unscrew and remove the plug.

Place the gas-check on the base of the projectile, with the concave or unpainted side, next the base (the 6″·3 howitzer gas-check, however, convex side next the base); insert the plug, and screw it well home with the spanner until it binds against the gas-check.

With gas-checks having projections, and used with studded projectiles, see that the projections are in the line of the studs before screwing the plug quite home. In the case of 6″·3 howitzer shell, see that the radial projections on the gas-check fit into the corresponding grooves in the base of the shell.

---

Palliser and other projectiles fitted with plugs and kept in exposed situations, where the plugs are liable to be set fast by corrosion from the action of salt water, &c., should have their plugs unscrewed once at least every six months, cleaned, and freshly lubricated.

Projectiles in charge of the Royal Artillery are to be filled, and the gas-checks fitted on, before being placed in the shell store. In storing them they are to be placed resting on their gas-checks, their stability being ensured by two pieces of wood, 9″ × 1″ × 1″, being placed under them, one on each side of the nut.

*Page* 139, *line* 5.

*Omit from* "The projectiles," *down to* "plain grooves," *and substitute*, "The 64-pr. projectiles have *three* rings of studs; those for the common and Shrapnel are of copper, those for the battering shell of brass. The studs fit both the shunt guns and those with plain grooves."

"The 6″·3 howitzer has no studs."

*Page* 140, *line* 13.

*Add*, "The 12″·5 case is made of three pieces of iron longitudinally riveted together."

*Page* 140, *line* 20.

*For* "is one exception," *read* "are two exceptions," *and after* "six segments" *add*, "and the 12″·5 case, which contains eight."

*Page* 140, *footnote*.

*For* "envelops," *read* "envelopes."

*Page* 141, *table*.

§ 3366.
§ 3382.
§ 3383.

*Add*, "Common shell 7-pr., 25-pr., 40-pr., and 80-pr." *Also*,
"Shrapnel shell 7-pr. Has internal weakening grooves.
" 80-pr. " "
Double shell 7-pr."

This makes all common and Shrapnel shell up to 80-pr. inclusive, cast to gauge, with the exception of the 16-pr. common shell, which will, however, also be cast to gauge in future manufacture. Common and Shrapnel for 12″·5 gun are also cast to gauge, except the bands, which are turned down.

*Page* 141, *line* 6 *from bottom.*

*After* "shells," *add*, "(in S.S.)"

*Page* 146, *line* 3.

*After* "p.," *add* "152."

*Page* 147, *line* 11.

*For* "fuse," *read* "fuze."

*Page* 162, *line* 6 *from bottom.*

*After* "p.," *add* "140."

*Page* 163, *line* 13.

*For* 1*st* "165," *read* "140."

*Page* 165, *table, last line of* 12*th column.*

*For* "3," *read* "12."

*Page* 167, *line* 2.

*Add as note*, "The 40-pr. S.S. cartridge may be also issued for L.S."

*Page* 170, *line* 12.

*After* "cartridges," *add* "†".

*Page* 187.

*Omit* 4th note.

*Page* 187, *line* 5 *from bottom.*

*After* "for the," *add*, "12″·5 gun, and for the."

*Page* 189, *line* 11.

*For* "358," *read* "188." "Use" *in the margin, should be opposite line* 12. *Add, in table at head*—
"12″·5 10″·75 33″·0 6″·0 4″·5 2″·25."
*and in margin* "§ 3297."

*Page* 190, *table.*

Fill in the following details of 12″·5 common shell, Mark I., approved 16/2/78, *see* § 3375.

| | | | |
|---|---|---|---|
| Length, without gas-check | | 36″·93 | ± 0″·3 |
| " with gas-check over nut | | 38″·60 | ± 0″·3 |
| Diameter | Over body | 12″·3 | ± 0″·015 |
| | " bands | 12″·425 | ± 0″·015 |
| | " gas-checks | 12″·425 | ± 0″·015 |
| | " studs | 12″·85 | ± 0″·005 |
| Thickness of metal | Walls .. Top | 2·96″ | |
| | Walls .. Bottom | 3·10″ | |
| | Base | 3·4″ | |

| | | lbs. | oz. |
|---|---|---|---|
| Weight | Shell empty, without gas-check, &c. .. | 780 | 8 |
| | Gas-check .. .. .. .. .. .. .. .. | 12 | 8 |
| | Nut .. .. .. .. .. .. .. .. .. | 1 | 7½ |
| | Plug .. .. .. .. .. .. .. .. .. | 3 | 3½ |
| | Bursting charge.. .. .. .. .. .. .. | 27 | 0 |

Total (mean weight), 825 lbs. ± 1·5 per cent.

This shell, fitted with gas-check, will stand being fixed with battering charges.

§ 3378.—Common shells, 12″, of 35 tons, Mark III., have been approved pending their manufacture with bands. They differ from previous pattern in being cast with hole through base to take gas-check plug. They are not to be fired with battering charges when fitted with gas-check. If battering charges are used the gas-check is to be removed, and the plug and nut screwed well home.

*Page* 192, *line* 1.

*Add*, "12″·5."

*Line* 3.

*Omit*, "These shell are not issued to L.S." *Omit paragraph commencing* "These shells have been withdrawn," &c.

*Omit second foot note.*

*Page* 193, *line* 12.

*For* "If," *read* "When."

*Line* 16.

*After* "available," *read* "The use of 15-sec. fuze would extend the range to about 4,000 yards."

*Page* 194, *table.*

*Fill in for* "12″·5 Shrapnel, Mark I.," from the following, approved 17/2/78, *see* § 3377:—

| | | Inches. | Inches. |
|---|---|---|---|
| Diameter | Over body .. .. .. .. .. | 12·30 | ± 0·01 |
| | „ bands and gas-check.. | 12·425 | ± 0·015 |
| | „ studs .. .. .. .. .. | 12·85 | ± 0·05 |
| Length.. | Without gas-check .. .. | 35·65 | ± 0·3 |
| | With gas-check over nut .. | 37·32 | ± 0·3 |
| Contents | Sand shot .. .. .. | 8 oz. | 284 |
| | Bursting charge .. .. .. | 2 lbs. 7 oz. | |

Total weight, filled and fitted with gas-check, 823 lbs. ± 1·5 per cent.

*In* 1*st paragraph of column of Remarks, for* "of 25 tons," *read* "9″."

*Page* 203, *table.*

*Add to* "12″·5," *in columns left blank*—"10·2 20·2 5·0 2·7."

*Page* 205.

§ 3379.—*Add as note,* "Patterns of 9″, 10″, 11″, 12″ (25 tons), 12″ (35 tons)." Palliser shell with an advanced numeral have been sealed. They differ from previous patterns only in being fitted for gas-checks. The total weights differ somewhat from the preceding patterns, owing to the addition of gas-checks. and fittings. Palliser shells *altered* to take gas-checks have a * in addition to original numeral.

§ 3376.—A Palliser shell, Mark III., has been approved for the 12″·5 gun. It is fitted for a gas-check, and is cast with bands at the base and head, which are ground to a windage of 0″·075 ± 0″·015. The body, from 2″·5 at the base end and 1″·5 from the head, is cast to finished dimensions, and has a windage of 0″·2. The general dimensions are much the same as those of Mark I. The total weight, filled and fitted with gas-check, is 118 lbs. ± 1·5 per cent.

A few only of Mark I. were issued to meet urgent demands. They were not fitted for gas-checks and had no bands. These, when returned and altered to take a gas-check, become Mark I.* A few more shells issued with gas-checks, having a stronger bush than Mark I., were Mark II.

*Page* 207, 1*st footnote.*

*Transfer this note to bottom of page* 206, *and in line* 14 *of that page add* "†."

*Page* 209, *table.*

§ 3379.—*Add as note,* "Patterns of 9″, 10,″ and 12″ (25 tons), have been approved with advanced numerals. They differ from preceding patterns only in being fitted to take gas-checks and in having a stronger bush cast in the base. Shot *altered* to take gas-checks have a * in addition to the original numeral."

*Page* 209, 1*st line of footnote.*

*After* "p.," *add* "205."

*Page* 210, *top line.*

*Add,* "12″·5."

*Page* 210, *line* 16.

*Add,* "The 12″·5 case differs in some particulars from the case used with other calibres, though resembling it in general principles. It has eight segmental linings, and contains a wrought-iron stay bolt, the top end of which is screwed to receive a wrought-iron nut, and the base end is fitted with a head screwed to receive the gas-check and nut for gas-check plugs. The base is rounded at the edge to conform to the shape of other projectiles. For details see addenda to table, p. 212."

N.B.—Mark I., to which the above refers, was approved, but the approval was cancelled 5/9/78.

*Page* 210, *line* 10 *from bottom.*

*Omit from* "As Shrapnel," *to* "against boats."

*Page* 211, *line* 11.

*For* "23," *read* "231."

*Page* 212, *table*.

The following are the details of the 12″·5 case shot, Mark I., approved, 16/2/78; approval cancelled 5/9/78:—

| | | | |
|---|---|---|---|
| Diameter over.. | Body .. .. .. .. .. | 12″·35 | ±0·05 |
| | Gas-check .. .. .. .. | 12″·425 | ±0″·015 |
| Length over .. | Case-body and gas-check . | 33″·5 | ±0″·27 |
| | Nut and handles .. .. | 36″·22 | ±0″·27 |
| | Contents, 834 8-oz. sand shot. | | |
| | Total weight, 798 lbs. ± 12 lbs. | | |

*Page* 217.

*After cut add*, "Mark II. (§ 3307) differs from Mark I., described above, in having a slip of paper pasted on one side, on which is a printed table giving the lengths and diameters of all R.M.L. cartridges. The edge of the scale is graduated to tenths of inches, instead of to inches only. Mark I. is easily converted."

*Page* 219, *table*.

A drill cartridge representing the 80 lb. half charge for the 12″·5 gun has been approved, 20/6/78:—

| | | From | To |
|---|---|---|---|
| Diameter | Body .. | 11″·3 | 11″·6 |
| | Bottom .. | 9″·6 | 9″·9 |
| Length .. | .. .. | 19″·25 | 19″·75 |

*Page* 220, *table*.

§ 3302.—A silk cartridge for the 12″·5 gun to contain 100 lbs. P² has been approved for the Navy. Length, when filled, 24 to 25 inches. Diameter, 12 inches at top, 10·4 inches at bottom. 13 hoops of broad silk braid.

§ 3372. An 80 lb. cartridge has been approved for the half-charge of the 12″·5 gun, two of these cartridges forming the battering charge of 160 lbs. P². The cartridge is made up with a central wooden stick about 1 inch in diameter. One end of this stick is attached to the bottom of the cartridge by means of a strip of silk cloth sewn on the centre of the bottom of the cartridge. The other end of the stick is fixed in the choke. The stick, being of a certain length, ensures a uniform amount of air space in the chamber when the charge and projectile are rammed home. Length filled (over stick) 19″·75, diameter of body 12 inches, of bottom 10·4 inches.

*Page* 220, *last footnote*.

*Omit* "§" *before* "S.S."

*Page* 224, *line* 12.

*Omit from* "in fact," *to end of paragraph*. *At end of* 1*st footnote*, *after* "Chap.," *add* "XVI."

*Page* 225, *before* 3*rd line from bottom*.

§ 3380.—*Add*, "A Mark II. common shell has been approved. It differs from Mark I. only in being fitted with a gas-check, and in being cast with a hole through the base to take the gas-check plug.

Mark I. shell, *altered* to take gas-checks, have a * in addition to original numeral.

*Page* 226, *line* 11.

*Omit from* "This howitzer," *to end of paragraph and insert*, "Common shell, gauge G.S. Fuzes, 9 (or 15) and 20 secs. M.L. time and sensitive percussion."

§ 3285.—The shell has no studs, the rotation being given by the gas-check, and is cast to finished dimensions. Eight radial grooves are cast in the base to receive corresponding projectiles on the saucer shaped gas-check. A hole 1″·295 in diameter through the base receives the gas-check plug. There are two extractor holes in the head. (*See cut.*)

For details see addenda to table, page 234.

*Page* 228, *lines* 22–24.

*After* "P.," *add* "193," *and after* "P.," *add* "231."

*Page* 230, 2*nd footnote.*

*This note to read*, "Fuzes for siege train 64-prs.

Common shell—9 (or 15) and 20 sec. M. L. time R.L. percussion, Mark II.

Shrapnel shell—5 and 9 secs. (or 15 secs.) M.L. time R.L. percussion, Mark II."

*Page* 230.

§ 3381.—*For paragraph about battering shell, substitute*—"A battering shell has been approved for the 64-pr. It is of Palliser form, and the head is chilled to a depth of about 0″·25 inch all round. In general construction it resembles a Palliser shell, but the chill is slighter, and the metal composing the shell is not of quite so hard a quality. It has a hole through the base to take a gas-check plug. It has three rows of brass studs. It is intended for use against masonry, &c., and is fitted with a delayed percussion fuze in the base (see addenda to page 57), which also acts when screwed home as a gas-check plug.

| | | | Inches. | Inches. |
|---|---|---|---|---|
| Diameter over | Body and gas-check | .. | 6·22 | ± 0·01 |
| | Studs | .. | 6·47 | ± 0·005 |
| Length .. | Without gas-check | .. | 16·0 | ± 0·13 |
| | Total over head of plug | .. | 16·65 | ± 0·13 |
| | | | lbs. | oz. |
| Weight .. | Empty, without gas-check | | 85 | 0 |
| | Gas-check.. | .. | 1 | 15 |
| | Plug for gas-check | .. | 1 | 3 |
| | Bursting charge .. | .. | 2 | 1 |
| | Total.. | .. | 90 | 3 ± 1·5 per cent." |

*Page* 234, *table.*

§ 3380.—Common shell for 8″, Mark II., howitzer, differs from Mark I. in being fitted with a gas-check, and in having a hole cast through the base to take the gas-check plug. Shells, Mark I., *altered* to take a gas-check have a * in addition to the original numeral.

Add details to 6·3 howitzer—

| | | Inches. | Inches. | | Inches. | Inches. | Inches. |
|---|---|---|---|---|---|---|---|
| 6/2/78 | 3285 — | 16·0 ± 0·13 | 6·25 | No studs. | ·775 | ·775 | 1·1 |

| lbs. | oz. | lbs. | oz. | lbs. | oz. |
|---|---|---|---|---|---|
| 59 | 7 | 7 | 0 | 69 | 5¼ with plug and gas-check. |

SHELL, R.M.L., 6″·3 HOWITZER. MARK I.

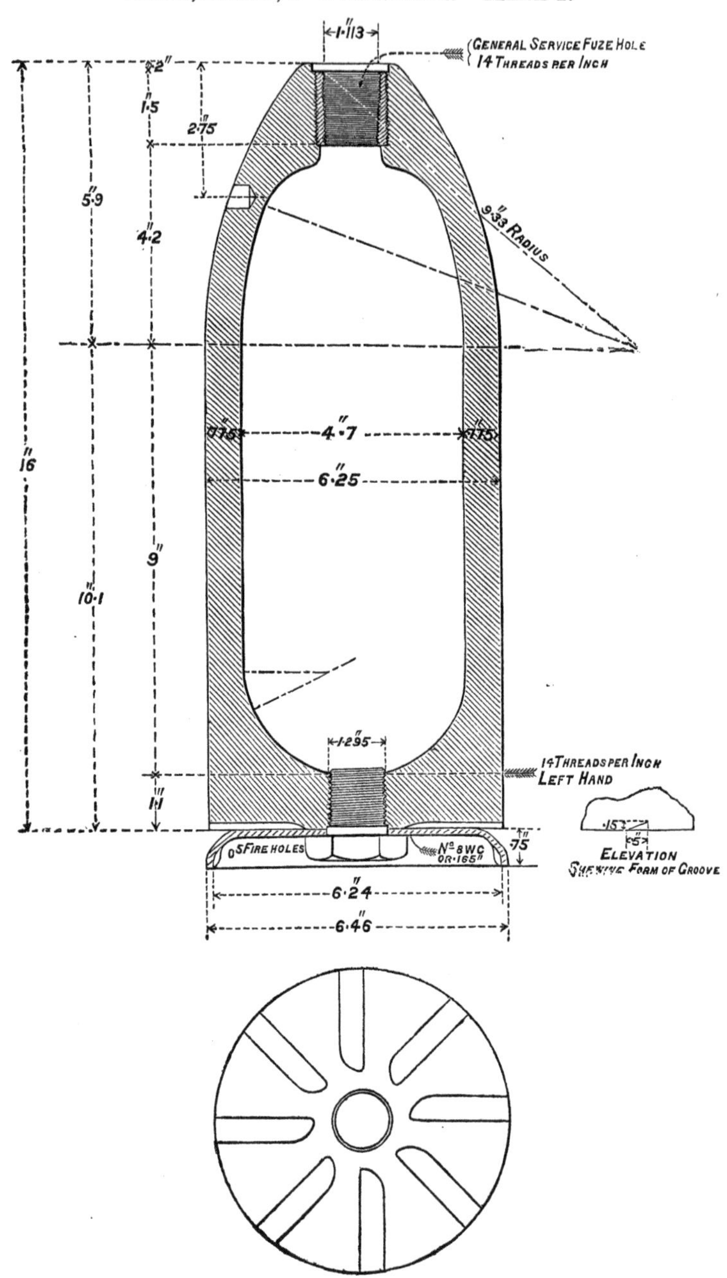

Mark II., 80-pr. common shell, differs from Mark I. in being cast to final dimensions, § 3366.

§ 3382.—Mark II., 40-pr. common shell, differs from Mark I. in being cast to finished dimensions, and in being fitted with a gas-check, a hole being cast through the base to take the gas-check plug. Shells, Mark I., *altered* to take a gas-check, have a * in addition to the original numeral.

Mark II., 25-pr. common shell, differs from Mark I. in being cast to final dimensions, § 3366.

*Page* 235, *table.*

80-pr. Shrapnel shell, Mark IV., differs from Mark III. in being cast to final dimensions, § 3366.

§ 3383.—40-pr. Shrapnel shell, Mark III., differs from Mark II. in being cast to finished dimensions, and in being fitted with a gas-check, a hole being cast through the base to take the gas-check plug. Shells, Mark II., *altered* to take a gas-check, have a * in addition to the original numeral.

*Page* 236, *table.*

The case for 8″ howitzer is the same as that for the 8″ gun. *See* table, p. 212.

*To* " 64-pr. or 80-pr., Mark IV." *add*, " or 6″·3 howitzer, § 3243." The marks on the top of case to be altered accordingly.

*Page* 237, *line* 8.

*Omit*, " for which no cartridges have yet been sealed." *Also* " with the exception at present of the 6″·3 howitzer."

*Page* 237, *line* 13.

" The cartridges for the howitzers," *should read* " The cartridges for the 8″ howitzers."

*Page* 237, *line* 15.

*Before* " 80-prs.," *add* " 6″·3 howitzer."

*Page* 237, *line* 19.

*At end of line*, *add* " 8″."

*Page* 238, *table.*

*Add*, " for 6″·3 howitzer cartridges, date of approval 10/10/77, § 3280.
*Last word of footnote should be* " 220," *instead of* " 200."

*Page* 244, *line* 16.

*For* " 249," *read* " 250."

*Page* 244, *line* 19.

*Omit* " fuze."

*Page* 244, *line* 23.

*Add*, " also 7-pr., Mark VII."

*Page* 245, 1*st footnote.*

*After* " p.," *add* " 157."

*Page* 246, *cut*

*Transpose sections on A A and B B.*

*Page* 248, *margin.*

*For* " A.C. /77, Cl. 137," *read* " A. C. /78, Cl. 72."

*Page* 249, *table.*

" The 7-pr. common shell, Mark IV., and double shell, Mark V., differ from previous patterns in being cast to finished dimensions, § 3366."

*Page* 250, *table.*

*Heading of* 10*th column should read,* " Pistol, F.G., or R.F.G."
7-pr. Shrapnel shell, Mark VII., differs from Mark VI. in being cast to final dimensions, § 3366.

*Page* 251, *table.*

5*th paragraph of column of Remarks should read,* " Iron ring, disc and segments, instead of zinc."

*Page* 255, *line* 13.

*For* " is likely to be soon," *read* " has been," *and add in margin,* " Mark II., § 3303."

*Page* 257, *margin.*

*Under* " Gatling cartridges, § 2644," *add* " § 2324."

*Page* 257, *bottom of page.*

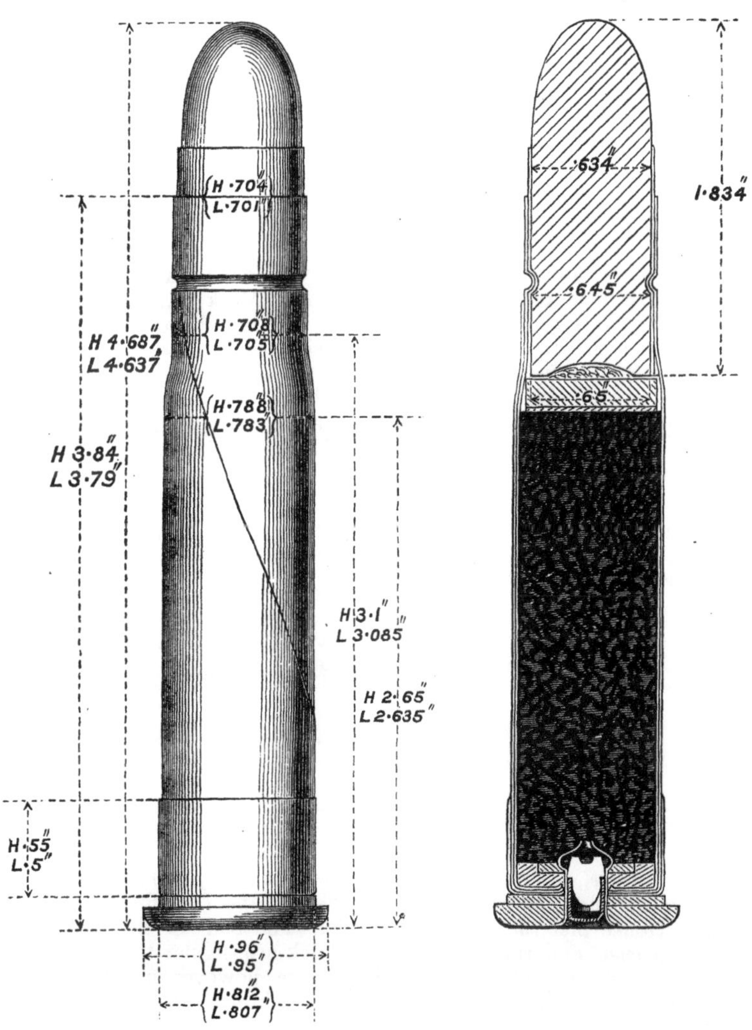


*Page* 258, *top line.*

*Omit from* "The cartridge," *to* "R.L.," *and substitute* "The ·45″ Gatling cartridges have been, till recently, procured by contract, as well as the caps for them, they are now entirely manufactured in the R.L."

*Page* 258, *line* 7.

*For* "lacquered," *read* "with hard brown varnish."

*Page* 259, *margin.*

"§ 1739, Mark I.," refers to the ammunition for Adams' revolver.

*Page* 264, 1*st footnote*, 5*th line.*

*After* "p.," *add* "290."

*Page* 264, 2*nd footnote.*

*After* "p.," *add* "7."

*Page* 266, *line* 12.

*Add*, "Since the above was published it was found that the paper lined rockets were not satisfactory. Accordingly, Mark VI., 24-pr., and Mark VII., 9-pr., were introduced (§ 3284). They have no paper lining, the cases are tinned all over, the inside of the case is covered with three coats of anti-corrosive paint, and the case is corrugated as in earlier patterns. Mark VI., 24-pr., as described above, has also proved unsatisfactory, and the present manufacture is of Mark III., differing only from the original Mark III. in having the canvas protection to the vents."

*Page* 269, *table.*

*Add at end of table*, "See addenda on page 266."

*Page* 272.

*At end of* 1*st paragraph add*, "for L.S. and S.S., § 3298."

*Page* 272, *line* 7 *from bottom.*

*For* "*" *read* "†."

*Page* 278.

*Omit the footnote, and add* "§ 3229. A rocket light, ½ lb., Mark I., and a rocket, sound, gun-cotton, ½ lb., Mark I., have been introduced for Trinity House and the Board of Trade. The body of the rocket proper, paper case, and the fitting of the stick are similar to those of the ordinary ½ lb. signal rocket. They have however a different arrangement for firing. About 4 inches of Bickford fuze is laid up alongside the copper socket for the stick. One end of this fuze passes into the vent round the lower edge of the case, the other end, protected by a paper cap, may be ignited by a vesuvian or other convenient means. The rocket *may* also be fired like an ordinary signal rocket with a port-fire, the paper covering of the vent being broken.

The light rocket has the head filled with a single star of magnesium light (see table, p. 319) contained in a paper case, and matched and primed. It is ignited and blown off when the rocket has reached its maximum height. The star burns about 15 secs.

The sound rocket has instead of a star a 2 oz. disc of dry gun-cotton, coated with paraffin, and a detonator. These are carried separate from the rocket until required for use. The head is a cylindrical paper case, rather larger in diameter than the body of the rocket. It has a piece of calico at the top fitted with a tape. When the gun-cotton is inserted into the head the calico and tape fasten it in, the calico being tied up by the tape like a bag. The detonator is a small tin tube, containing fulminate of mercury, and is ignited by quickmatch passing into the top of the rocket composition.

The stick used with these rockets is the same as that used with the ordinary ½-lb. rocket, but is 12 inches shorter, for more convenient firing in a restricted space, as the gallery of a lighthouse, &c., &c.

Probably no more call rockets of the description given in the text will be made.

*Page* 288, *line* 19.

*After* "page," *add* "291."

*Page* 288, *line* 23.

*After* "page," *add* "264."

*Page* 290, *margin.*

*Omit marginal reference.*

*Page* 314.

The regulations issued with A.C. /78, Cl. 74, have slightly modified those issued with A.C., June 1873. The necessary alterations are given below:—

1*st paragraph: Omit from commencement down to* "shells filled," *and for* "suitable buildings for the purpose," *read* "laboratories."

*At the end of 7th paragraph add,* "and thrown into a bucket or tub, containing water, which, with a stick for the purpose of stirring, is to be kept just outside the Laboratory."

*In* 10*th paragraph, for* "shot," *read* "spherical projectiles."

*In* 11*th paragraph, add* "Care must be taken not to detach the lacquer in this examination; this is often done by the careless use of the copper scraper."

*In* 14*th paragraph, add* "Covered two-wheeled barrows are generally provided for this purpose; if hand-barrows are used the packages must be covered over with wadmiltilts.

*In* 15*th paragraph, after* "hides," *add* "or wadmiltilts."

*For* 17*th paragraph, substitute* "No greater quantity of powder than is absolutely necessary to keep the work going will be in the laboratory, or in transit between the magazine and laboratory, at the same time. The actual quantity must depend on the nature of the cartridges to be made up, or the shells to be filled; the great object in view being, of course, to minimize the damage resulting from a possible explosion. In no case should the quantity exceed two barrels."

*Paragraph* 18, *as given in text, to be numbered* "19." The 18th paragraph runs thus:—"The following regulations are applicable to filling cartridges and shells for all natures; but cartridges over 80 lbs. and shells over 10″ R.M.L. will generally be filled at district laboratories, under special regulations as to the amount of powder and number of shells to be under manipulation at the same time."

*Page* 319.

"Fuze composition for wood fuzes," *insert* "15-second fuzes excepted." The composition for this fuze is—

| | lbs. | oz. |
|---|---|---|
| Saltpetre .. .. .. .. .. | 3 | 12 |
| Sulphur .. .. .. .. .. | 1 | 3 |
| Powder, mealed, pit .. .. | 2 | 0 |

*Page* 324.

*Add in table* "6″·3 howitzer ring gauge. Diameter over body 6″·27, over projections on gas-check 6″·48."

*Page* 332.

Analysis of metals, "No. 8," should be the same as No. 7 (*i.e.*, showing a small proportion of zinc), and should also include 64-pr. projectiles, with the exception of the battering shell, which has brass studs made from S.A.A. cartridge clippings.

"No. 9." *Transpose* "Copper," and "Tin."

*Omit footnote.*

The "copper" gas-checks contain 3 per cent. of zinc, with the exception of that for the 6″·3, which is stamped from sheet copper.

# LIST OF ABBREVIATIONS.

| | |
|---|---|
| F.G. | Fine Grain. |
| R.F.G. | Rifle Fine Grain. |
| R.F.G$^2$. | „ „ (for Martini-Henry Rifles). |
| L.G. | Large Grain. |
| R.L.G. | Rifle Large Grain. |
| P. | Pebble. |
| P$^2$. | Cubical 1½″ powder. |
| L.S. | Land Service |
| S.S. | Sea Service. |
| G.S. | General Service |
| F.S. | Field Service. |
| R.L. | Royal Laboratory. |
| R.G.F. | Royal Gun Factory |
| R.C.D. | Royal Carriage Department. |
| R.G.P.F. | Royal Gunpowder Factory. |
| S.B. | Smooth Bore |
| M.L. | Muzzle Loading. |
| B.L. | Breech Loading. |
| M.L.O. | Muzzle Loading Ordnance. |
| R.M.L.O. | Rifled Muzzle Loading Ordnance. |
| R.B.L.O. | Rifled Breech Loading Ordnance. |
| S.A. | Small Arms. |
| R.A. | Royal Artillery |
| § | Paragraph of Changes in War Stores. |
| A.C. | Army Circular. |
| EXTRACTS. | Extracts from the Proceedings of the Department of the Director of Artillery. |
| O.S.C. | Ordnance Select Committee. |

NOTE.—Stores which are issued for practice only are to be marked with a yellow line. A blue line is painted on obsolete stores in the Royal Laboratory, while red is often used to indicate powder. The colour red is generally used as an indication of danger; thus, a red flag is used at rifle ranges; powder barges, &c., are painted red, and filled shells are marked with red letters. In the case of detonators for gun cotton, red denotes the presence of fulminate of mercury. In demanding stores it is necessary to name them in the same manner as in the Priced Vocabulary, but in this work the ordinary name is generally made use of. Thus, the large mortar fuze would be strictly designated as "fuze, time, wood, Boxer, mortar, large." The nomenclature of the Priced Vocabulary is adhered to in the index.



H6

# CHAPTER I.—GUNPOWDER.

THE PROPERTIES OF ALL GUNPOWDERS USED IN THE SERVICE.—USES OF EACH NATURE.—LIABILITY OF GUNPOWDER TO EXPLODE FROM A BLOW, OR FROM FRICTION.—CONCLUSIONS OF NOBLE AND ABEL.

**Properties of Gunpowder.**

The manufacture of gunpowder will be found fully given in the Handbook on the Manufacture and Proof of Gunpowder, by the late Captain F. M. Smith, R.A., and in the Notes on Gunpowder and Gun-cotton, 1878, by Major Wardell, R.A., Assistant Superintendent R.G.P.F. It is necessary to give a short account of some of the leading properties of gunpowder which bear directly on its employment for cartridges for different natures of ordnance and small arms, as well as other Laboratory uses.

The composition has remained unaltered for many years, having in 100 parts, 75 of saltpetre,* 15 of charcoal, and 10 of sulphur. Its exploding point is about 600° F.†

The qualities which most influence gunpowder with regard to its use in rifled guns, are its density, the size and shape of grain, and the condition of the charcoal. It is essential that gunpowder for rifled guns should be uniform in action, to ensure regularity in range, otherwise the accuracy of shooting due to the guns will be impaired.

The denser gunpowder is, the slower it will burn, even in single grains, and the dense powder will offer a smaller surface to ignition than an equal *weight* of a less dense kind; for a similar reason, a large grain will burn slower than a number of small grains making up the same weight, and a grain of regular shape, such as a cube, or a sphere, will offer less surface than an irregular one of the same mass; a laminated or flakey form is well known to indicate a violent powder. Very much will depend on the degree to which the charring of the wood is carried, as, if it is imperfectly charred, the oxygen and hydrogen retained in the charcoal, cause it to burn much more rapidly than when it is reduced to nearly a pure carbon, and also it is more hygroscopic.‡

---

* In manufacture 76 parts of saltpetre are used, the additional amount being added to cover loss in manufacture.

† This has hitherto been tbe received idea, but a theory has recently been started to the effect that the keeping and damp resisting properties of gunpowder are improved by heating it to a point exceeding that of boiling water. Experiments will be carried on to test the truth of this theory.

‡ The violence of some Madras powder was attributed by the Committee on Explosives (Report, 1872, p. 50) to the low density, flakey form of grain, and quality of charcoal.

There are other minor points influencing the rate of burning which need not be considered in this place.

We may say then, that a large grained, dense gunpowder will burn slower than a small grained powder of low density,* always supposing that the grain is sufficiently large to leave interstices for the passage of the flash, as otherwise, the charge of powder would only be ignited at one place, and would, therefore, burn slowly; this would however, only take place when it is reduced to *meal.*

On the introduction of rifled guns it was considered advisable to use a powder which would burn more gradually, and strain the gun less than that then in use for S.B. guns. More work is thrown on rifled guns, as not only a forward velocity, but also a velocity of rotation is given to the projectile; the weight too of the projectile is much greater in a rifled gun than in a S.B. gun of the same calibre; thus an 8-inch rifled gun throws a projectile of 180 lbs. weight, a S.B. gun of the same calibre, one of 68 lbs. weight.

To attain this end, the density and size of grain were increased, "R.L.G." being introduced for rifled guns. This powder answered for guns of small calibre; but when R.M.L. guns of 7-inch and upwards, burning large quantities of powder, were introduced, it was found advisable to use a slower burning powder than "R.L.G." Hence "P" gunpowder was introduced, a powder of high density and of much larger grain. It burns more slowly than "R.L.G." and consequently strains the gun less. For similar reasons "$P^2$" has been introduced for still heavier guns.

§ 1291.

A considerable stock of "L.G." gunpowder still remains on hand; in order to utilize it, "L.G." is now largely used with rifled guns, except with the 7-pr.; this is, however, only a temporary measure as no more "L.G." powder will be made.†

For small arms a rapidly burning gunpowder is required, therefore a much smaller grain is used. There are five kinds, "F.G." for smooth bore small arms, "R.F.G." and "R.F.$G^2$." for rifled small arms, "Pistol" powder, and Adams' Pistol. "F.G." is used for the charges of the 7-pr. R.M.L. gun, as, owing to the shortness of the bore, a quick burning powder is required. A quick acting powder is also required to burst

---

* The density of powder is measured at Waltham Abbey by means of an instrument called a densimeter. It consists chiefly of a glass globe, which can be exhausted at pleasure by means of an air pump; the globe is furnished with stop-cocks, and can be attached or detached from the air-pump as required. The process of finding the density of a sample of powder is briefly as follows:—The globe is exhausted of air, and mercury is allowed to run in and completely fill up the vacuum thus created. The globe is weighed full of mercury; it is then emptied and a known weight of powder placed in it, usually 100 grammes (1543·235 grs.); it is again attached to the air-pump and exhausted, and mercury allowed to flow in to fill up the space not occupied by the powder. The globe containing mercury and powder is weighed. Then if D represent the density of mercury at the temperature at which the experiment is carried on, $P_1$, the weight of the globe full of mercury, and P, the weight of it when full of mercury and powder, we have

$$\text{Density of powder} = \frac{D \times 100}{(P_1 - P) + 100}.$$

For full details, and full description of the instrument, see Captain Smith's Handbook, pp. 84, 85.

† Experiments carried on by the Committee on Explosives showed that some L.G. powder gave a lower pressure when fired out of an 8-inch S.B. gun than R.L.G. did, it also gave less velocity. This would seem to contradict the statement as to the effect of the size of grain on the rate of burning, but the fact is, that the density of L.G. and also of the earlier manufacture of R.L.G. varies, and thus complicates the question.—Extracts, Vol. IX, pp. 150, 151, 152.

Shrapnel shell, the charge used being small, and the sharper the action the more readily the shell will be opened without scattering the balls, therefore either Pistol or "F.G." are used.*

The size of the gunpowder is determined by the sieve through which it is passed in manufacture, the sieves are distinguished according to the number of sub-divisions in a linear inch, thus an 8-mesh sieve would have 64 holes in a square inch.

The following are the classes into which gunpowder is divided: "Service," "Blank" and "Shell" are *serviceable* powders. The term "Service" is applied only to powders used for firing projectiles.

| Class. | Designation. | Description. |
|---|---|---|
| I. | Service.. .. | 1. All new powder.<br>2. All returned powder (including cannon cartridges) which, on examination, may be found uninjured. |
| II. | Blank .. .. | 1. Powder from broken-up cannon cartridges, too dusty to be placed in Class I.<br>2. Powder from broken-up S.A. ammunition.†<br>3. Service powder found dusty or broken in the grain at periodical inspections, or on return; except such as being only dusty, is firm enough in grain to be re-dusted for service. |
| III. | Shell .. .. | Powder found too dusty for Class II. |
| IV. | Doubtful .. | All powder whatever (except new powder) returned into store and awaiting examination. |
| V. | Condemned for sale | Powder found on examination to be too much deteriorated to be placed in any of the above classes. |
| VI. | Condemned for extraction | Powder obtained from shells, and powder found to have been so much damaged as to be unfit for any purpose but extraction of saltpetre. |

All gunpowder, not made up into cartridges or bursters, is packed in 100 lb. barrels, except "P" and "$P^2$," the higher density of which powders allows of 125 lb. being placed in each barrel.

By the above it will be seen that the only difference between the serviceable classes is in the condition of the powder, and, to put it briefly, we may say that *Service* powder is equal to new, *Blank* is slightly dusty or broken in the grain, while *Shell* is too dusty for *Blank*, and is only to be used as bursting charges for shells.

At first sight it may seem strange that *Blank* should be a better variety than *Shell*, the first being only used for drill, while the latter is used on service; the fact is, however, that a very dusty powder should not be used in cartridges, as it would work its way through.

Two varieties of powder are ordered to be wetted, powder taken out of shells and from B.L.S.A. cartridges, in case of the possible presence of iron or detonating composition. This order includes powder taken from shells in which bags are used.

* R.F.G. will answer as well as F.G. for the charge of the 7-pr. gun, and for the bursting charges of Shrapnel shell, and is to be so used when the present stock of F.G. is exhausted. § 2708.

† Powder obtained by breaking up breech-loading small-arm ammunition should be at once thoroughly wetted, as it may contain small particles of detonating composition. In this state it is useless except for the extraction of saltpetre, and will therefore be placed in Class VI. All powder emptied from shells is also ordered to be wetted.

Use of each kind powder. P². § 3066.

The following gunpowders are used with ordnance and small arms:—

"P²" powder is at present used only with the 12″·5 gun of 38 tons. It will probably be used with all heavy guns from the 10-inch upwards. The grains are approximately cubical in form with rounded edges, the length of each edge being about 1½ inches. The grains run from 5 to 7 to the pound. The density is not less than 1·75. A charge of 200 lb., rammed to a length of 40 inches, in a 38-ton 12″·5 R.M.L. gun, chambered to 14″ must give a muzzle velocity of not less than 1,540 feet to a projectile, fitted with a gas-check, and weighing in all 812 lb. The pressure on any one crusher* is not to exceed 22 tons per square inch, and the mean pressure on the powder chamber is not to exceed 21 tons.†

"P." § 2103.

"P" is used for the battering charges of all rifled guns of 7-inch calibre and over (except the 12″·5 gun), and for all service charges of 40 lb. and upwards. When no "P" is available, "R.L.G." will be used. The grain approaches to a cubical form, the length of the sides being about ⅝ of an inch. The density is between 1·75 and 1·78. When fired from an 8-inch S.B.‡ proof gun, 35 lb. charge, shot 180 lb., the pressure must be under 20 tons on the square inch, and the initial velocity between 1,420 and 1,480 feet per second.

Owing to its high density 125 lbs. of "P" are contained in the barrel, and the same remark applies to "P²."

The charges of "P" powder are considerably larger than the charges of R.L.G. for the same gun, thus the battering charge of the 10-inch gun is either 70 lb. "P" or 60 lb. "R.L.G." In order to enable the cartridges to fit the same packing cases or zinc cylinders, the diameters of the "P" powder cartridges are larger than those of the "R.L.G." (See page 220).

The comparative pressure and velocity given by "R.L.G." and "P" may be seen from the following instance:—

The mean pressure in the powder chamber of the 10-inch gun, charge 70 lb. "P," projectile 400 lb., was 15·7 tons per square inch, and the velocity at 50 yards was 1,412 feet; while the mean pressure given by 60 lb. R.L.G. was 17·9, and the velocity 1,328 feet.*

The reason there is an increase of velocity when "P" powder is used is because, although the initial pressure is lower in the powder

---

* The methods employed for testing the pressure will be found in the Preliminary Report of the Committee on Explosives, p. 4.

Crusher gauges containing small cylinders of cópper, to which the pressure is transmitted by a piston, are inserted into the gun; the compression these cylinders undergo gives an indication of the pressure of the gas. By subjecting similar cylinders to known statical pressures the amount of compression caused by various weights is ascertained and tabulated. By means of these tables the relation is obtained between the amount of compression undergone by the cylinder in the gun, and the pressure to which the compression is due.

† Very good results have been obtained by using "P²" in 10-inch and 11-inch R.M.L. guns. When charges equal in weight to the service charges of "P" powder were used the muzzle velocity was diminished, and the pressure in the bore notably decreased. When, however, the charges of "P²" were increased, the muzzle velocity was considerably greater, while the pressure, though somewhat increased, was still below that given by the service charges of "P" powder. The 10-inch gun was thus placed, with regard to penetrating power, on a par with the 11-inch 25-ton gun, while the latter nearly equalled the service power of the 12-inch gun of 35 tons. For details of the experiments, see Extracts, Vol. XIV, pp. 152, *et seq.*

‡ The gun is identical with the 8-inch R.M.L. gun, except that it is not rifled.

§ A large amount of information on this subject will be found in Extracts, Vol X, p. 80. For comparative pressures and velocities given by "P²" and "P," see Extracts, Vol. XIV, pp. 152, *et seq.*

chamber, it is kept up longer in the bore than is the case when the quicker burning "R.L.G." is used, the velocity depending upon the total pressure exerted before the shot leaves the bore.

It is found that the recoil of the same gun is more violent when "P" powder is used; this incidentally shows that the strain on the gun is less, as the force of the powder is expended in making the mass recoil, and not in exerting a strain upon the particles of the metal of so sudden a nature as to tend to tear them asunder; in fact, the slow burning powder gives sufficient time to allow the gun to recoil, and thus ease the strain. The opposite takes place with a very rapid explosive such as gun cotton, which is stated to occasion a recoil only $\frac{2}{3}$ that of gunpowder.

Extracts, Vol. IX, p. 14.

Bloxam's Chemistry, p. 507.

Service R.L.G.

*Service R.L.G.* is at present used for S.S. for full charges for R.M.L. guns of 7-inch and upwards, under 40 lbs., but when the stock of *L.G.* is exhausted it will be probably used for all charges under 40 lbs. for L.S. as well as S.S. rifled guns, except battering charges, and the charge for the 7-pr. R.M.L. gun.

§ 2838.

According to Changes in War Stores, § 2838, alternative charges of "P" and "R.L.G." powder are marked on cartridges for battering charges up to the 12-inch gun of 25 tons, and for full charges up to that for the 12-inch gun of 35 tons. The Equipment Regulations, 1876, however, only sanction the use of "R.L.G." for the battering charges of 7, 8 and 9 inch R.M.L. guns, for which guns, it is added, "L.G." may be used in case of necessity.

Army Equipment, 1876, p. 61.

Cl. 158, A.C. 1874.

R.L.G. is to be used with R.M.L. field guns in possession of Horse and Field Artillery and the School of Gunnery.

§ 2708.

The size of the grain is determined by its passing through a 4-mesh sieve and resting on an 8-mesh. Its density is 1·68. This applies to all powder made since 1870, but powder of early manufacture varied in density and consequently in regularity.† It is tested by firing from the 9-pr. R.M.L. proof gun at Waltham Abbey, and must give an initial velocity of 1,410 $\pm$ 25 feet a second, using a 9 lb. flatheaded shot and a charge of $1\frac{3}{4}$ lbs.

An instance of the pressure exerted has been given above; for the comparatively small charges made up of R.L.G. the pressure is not so important, as the abnormally high pressures do not take place which are found to occur when a quick burning powder is used in large quantities.

Even in firing blank charges fired from heavy guns, high pressures have been developed with "R.L.G." powder; on firing a blank charge of 30 lbs. in an 8-inch gun, a pressure of $15\frac{1}{4}$ tons was given on a gauge in rear of the powder chamber, $9\frac{3}{4}$ tons on a gauge in the centre, and 9 tons on a gauge in the front of the powder chamber. The Committee considered that the pressure of $15\frac{1}{4}$ tons was local, and that the lower pressures fairly represented the mean pressure due to the explosion of the charge.*

Service L.G. § 2087.

*Service L.G.* is at present used for L.S. for all S.B. and B.L. guns, and for the full charges of R.M.L. guns under the 10-inch (except the R.M.L. Field Artillery (L.S.) and the 7-pr.); and for S.S. for all S.B. and B.L. guns, and for R.M.L. guns, 64-pr. downwards, 7-pr. excepted. The size of grain is determined by its passing through

---

* Preliminary Report, Committee on Explosives, p. 10.

† Extracts, Vol. IX, p. 151.

an 8 and resting on a 16-mesh sieve. It is readily known from "R.L.G." by the absence of glaze,* and the smaller size of the grains.

Extracts, Vol. IX, pp. 150, 151.

Some trials made by the Committee on Explosives showed that L.G. powder gave less pressure and velocity than R.L.G. tried against it, but it must be remembered that powder of early manufacture is far from uniform, and these trials can only be taken as a general indication of the action of the powders. The Committee report that there need be no hesitation in interchanging L.G. with R.L.G. in wrought-iron guns, on the ground of danger to the guns, but the substitution would be attended with a falling off in muzzle velocity, power, and shooting qualities. All cartridges made up with L.G. powder for rifled guns are marked with the letters L.G. in red.†

§ 1998.

Service R.F.G. § 2708.

*Service R.F.G.*—Used for rifled small arms, except the Martini-Henry, Gatling guns, and pistols; for 7 pr. R.M.L. guns, and for the bursting charge of shrapnel shell when F.G. is exhausted. The size of grain is determined by its passing through a 12 and resting on a 20-mesh sieve. Density 1·6.

Service R.F.G$^2$ §§ 2487, 3065.

*Service R.F.G$^2$.*—Used for the Martini-Henry and Gatling cartridges. It is of the same size as the R.F.G. powder, but its density is greater, being 1·72; it is therefore a slower burning powder. It is to give a muzzle velocity, when fired from a Martini-Henry rifle, of between 1,290 and 1,340 feet per second, *i.e.*, a velocity of 1,315 ft. ± 25 ft.

Service F.G. § 2286.

*Service F.G.*—This powder is no longer made; it is used for S.B. small arms, for the charge of the 7-pr. R.M.L. gun, and for the bursting charge of shrapnel shells. The size of the grain is determined by its passing through a 16 and resting on a 36-mesh sieve.

Service Pistol.

*Service Pistol.*—Used for Colt's and Adams' pistols, and for the bursting charges of Shrapnel shell. The size of grain is determined by its passing through a 44 and resting on a 72-mesh sieve. As before pointed out, the quickness of its action renders it specially suitable for Shrapnel shell; and also it is suitable for the short barrels of pistols, where a slower burning powder would not be consumed. The siftings of the R.F.G$^2$. powder have been tried in the B.L. Adams' pistol cartridges, and found to answer well. The use of them for this purpose is now approved. They pass through an 18 mesh and rest on a 44 mesh sieve.

Adams' Pistol. §§ 3222.

The above are all the powders used in connection with projectiles.

Blank R.L.G. and L.G.

*Blank R.L.G. and L.G.*—For blank charges of all descriptions of rifled and S.B. ordnance, including the *reduced* charges for 9, 8, and 7-inch R.M.L. guns.

Blank R.F.G. and F.G.

*Blank R.F.G. and F.G.*—For blank small arm cartridges of every description (L.G. may also be used for blank S.A. cartridges). These powders can be used for blank charges of ordnance where there is a surplus store.

Shell L.G.

*Shell Powder L.G.*—For the bursting charges of all shells, rifled or S.B., except Shrapnel, and the 6, 9, 12, and 20-pr. segment shells.

Shell F.G. and R.F.G.

*Shell Powder F.G.*—For the bursting charges of 6, 9, 12, and 20-pr. segment shells. Shell R.F.G. to be used when shell F.G. is exhausted.

---

* The density of L.G. and early manufactured R.L.G. cannot be given; L.G. has been found to vary from 1·6 to 1·8.—Preliminary Report, Committee on Explosives, p. 10. Handbook of Manufacture of Gunpowder, p. 53.

† § 1967, Changes in War Stores, directed that rifled guns should be examined after every 25 rounds, when L.G. was used; but this was cancelled by § 2087.

See Extracts, Vol. X, p. 117. At 5° elevation, charge 3 lbs., the mean difference of range (10 rounds fired) of the 16-pr. was 13 yards with R.L.G. and 71 yards with L.G.

Failing shell powder the higher classes of powder must be used. The only reason for using the inferior powder is to prevent waste; no doubt the better the powder, the more effective would be the shell. The effect of firing a shell containing a bursting charge of powder is to set back the latter in a hard dense mass; unexploded shells have been recovered with the powder converted into a solid body, so hard as scarcely to be cut with a copper tool.

Mealed powder is largely used in the R.L. in the manufacture of ammunition. There are two kinds:—

**Mealed Powder.**

(1.) *Mealed Powder.*—Passes through an 120-mesh sieve. It is ordinary powder reduced to an impalpable dust by drumming it in a revolving barrel with gun-metal balls, or by rubbing it between hard wooden surfaces by hand; its use is due to its easy ignition and rapid rate of burning. It is used where great regularity of burning is not required, for instance, in quick match, portfire composition, friction tubes, &c.

(2.) *Pit Mealed Powder.*—So-called because it is made from gunpowder specially prepared, the charcoal having been charred in pits instead of in cylinders, as is the case with the service powders. It passes through an 120-mesh sieve. It is used for fuzes, because it has been found to burn with great regularity.*

**Liability of gunpowder to explode by a blow or by friction.**

Too much stress cannot be laid on the fact that gunpowder can be exploded either by a blow or by friction; this forms a clue to most of the precautions which should be taken in dealing with it. When they are carefully observed the risk of an explosion is but small; indeed, in a magazine where it is unnecessary to handle loose powder, an accident can hardly take place unless the rules are violated.

Some attempts have been made experimentally to determine the amount of force required to explode gunpowder. It is necessary to consider, (1) the hardness of the surfaces with which the powder is struck; (2) the intensity of the blow, which will depend on the size of the surface struck; (3) the thickness of the layer of the powder and the condition it is in, whether in grains, or as mealed powder.

Supposing the force of the blow to be constant, such as is given by a weight dropping a fixed height, then the harder the surfaces, the smaller the area struck, and the thinner the layer of powder, the more likely the powder will be to explode.

Experiments carried out in the Chemical Department of the Royal Arsenal showed that a 50lb. weight falling 36 feet on a surface of one square inch, the surfaces being brass, having mealed powder $\frac{1}{20}$th of an inch thick spread between them, exploded the gunpowder.

The same weight falling on the surface of $\frac{1}{4}$ of a square inch, exploded the powder with a drop of 10 feet.

This result coincided as nearly as possible with what might be expected; the surface being diminished to $\frac{1}{4}$ we should have

---

* No doubt the charcoal prepared in cylinders is much more uniform in quality, and would naturally be expected to answer best. The reason is possibly due to the fact that the pit charcoal is less thoroughly charred than the cylinder, and hence will burn more rapidly. The object aimed at in adding mealed powder, is to increase the rate of burning of fuze composition, and it is found as a rule that quick burning compositions burn more regularly than slow burning ones, thus a composition can be made to burn with more regularity at the rate of one inch in five seconds than at the rate of one inch in ten seconds. Possibly the weak burning composition does not give a sufficient rush of gas to keep the vents clear, and if this is the case, the varying pressures would alter the rate of burning.

anticipated an explosion with a fall of $\frac{1}{4}$ of that it previously required, or a fall of 9 feet, whereas it required 10 feet.*

The important fact to remember is, that the more the surface is diminished the more readily will the powder explode; thus, if in the above experiment the surface be lessened to the $\frac{1}{36}$th of a square inch, we should expect the weight to explode it with a drop of 1 foot.

In dealing with gunpowder, there is sure to be some powder dust about, and by striking this, explosion may take place, especially if the surfaces are hard, and so occasion a serious accident.

The effect of various surfaces is well shown in an experiment carried out in the Royal Arsenal in 1872:—

"A 25 lb. weight was allowed to fall 2 feet on small packages of " gunpowder consisting of about 5 grains of Government powder wrapt " in tinfoil, so as to exclude any possibility of a spark reaching it." The packages were placed between two metal plates, and 10 blows were struck, using different metals. When steel was used the powder exploded every time; when one brass and one steel plate were used it exploded 4 times out of 10 trials; when both plates were brass it exploded twice out of 10. When lead plates were used there was no explosion even with a fall of 40 feet.

**Artillerist's Manual, p. 52, 11th Edition.**

This shows the advantage of covering the floors with some soft material when moving powder, and of using soft metal for tools, such as copper, (though this alone will not afford complete safety, as if gunpowder is hit hard enough between copper surfaces, it will go off); it shows above all, the necessity for perfect cleanliness, as if loose powder is allowed to escape there will always be risk. If any grit or sand is allowed to collect in the passages, the danger becomes great, as grit (*i.e.*, silex) is harder than almost any metal, and the sharp points coming in contact with powder dust, would present the most favourable conditions for an explosion.

In a magazine such a state of affairs could only be caused by neglect of the regulations laid down, which are given in the Appendix, p. 311.

There have been no experiments to determine the amount of friction that is necessary to explode gunpowder, but we have abundant evidence that gunpowder will ignite by friction against the walls of rifled shells when fired, as will be seen further on at p. 31.

The property of igniting from a sudden concussion is made use of in Palliser projectiles; no fuze is employed, and the powder is found to explode on the impact of the projectile against iron plates.

Gunpowder possesses the property of standing climate well, when properly made. Saltpetre is not deliquescent to any extent when pure, the glaze and density of the powder aid to preserve it from damp; still in very damp climates, or in a damp magazine, it is necessary to keep powder in metal cases, the air being excluded by luting. (See Chap. VII., p. 74).

Gunpowder should not be allowed to remain in direct contact with metal, as if there is the least damp (especially if heat be also present) the saltpetre will attack and corrode the metal, thus spoiling the powder and damaging its envelope. This effect may be especially noticed in the earlier patterns of B.L. Snider ammunition, in which the brass case

---

* The arrangements for the falling weight were somewhat rough, there was some oscillation, which led to jamming between the guides, and the friction must have been considerable. The powder was placed between two brass plates, and work must have been expended in flattening the upper plate, thus cushioning the blow to some extent.

was not protected by tissue paper from the powder. On opening the cartridges, especially if they have been in a damp hot climate, minute perforations may frequently be seen on unrolling the case and holding it up to the light, and in more aggravated cases, the brass in contact with the powder is almost entirely eaten away. This applies however, as a rule, only to cartridges which have been manufactured some time. Considerable difficulty was experienced from the same cause with the earlier patterns of Hales' rockets, in which the interior of the iron or atlas-metal case was imperfectly protected from the action of the powder. Again, paper, if of any thickness, is found to cause powder to deteriorate, as it is given to absorb and retain damp. This causes part of the saltpetre to be absorbed into the paper, which becomes more or less impregnated with it, at the expense of the powder. The difficulty can be overcome by varnishing the paper. The paper cylinders contained in certain B.L. cartridges were found to deteriorate the powder until they were varnished.

**Researches of Noble and Abel on Explosives—Fired Gunpowder, pp. 136, *et seq.***

Before concluding this chapter, it may be interesting to quote the summary of the results arrived at after a long and exhaustive series of experiments by Captain Noble (late R.A.), and Professor Abel, on the pressure, &c., of fired gunpowder.

The results are given in the two cases of gunpowder fired in a space entirely confined, and of gunpowder fired in the bore of a gun.

In the first case the results are calculated on 1 gramme of powder occupying 1 cubic centimetre, and are given as follows:—

1.—The products of combustion are 57 per cent. by weight of matter which ultimately assumes the solid form, and 43 per cent. by weight of permanent gases.

2.—At the moment of explosion the fluid products of combustion occupy about 0·6 cubic centimetres, and the permanently gaseous ones about 0·4 cubic centimetres; so that both the fluid and gaseous matters are of approximately the same specific gravity.

3.—The permanent gases generated at 0° C and 760 m.m. pressure occupy 280 times the volume of the original powder.

4.—The chemical constituents of the solid products are: Potassium carbonate, sulphate, hyposulphite, monosulphide, sulpho-cyanate and nitrate; Ammonium sesqui-carbonate; Sulphur; Charcoal.

5.—The composition of the permanent gases is Carbonic anhydride and oxide; Nitrogen; Sulphydric acid; Marsh Gas; Hydrogen.

6.—The tension is about 6,400 atmospheres (about 42 tons per square inch), but varies with the mean density of the products of combustion.

7.—The temperature of explosion is about 2,200° C. (about 4,000° F.).

In the case of powder fired in the bore of a gun, it was found that:—

1.—The products of explosion, at all events as far as regards the proportions of the solid and gaseous products, are the same as in the case of powder fired in a close vessel.

2.—The work done on the projectile is effected by the elastic force due to the permanent gases.

3.—The reduction of temperature due to the expansion of the permanent gases is, in a great measure, compensated by the heat stored up in the liquid residue.*

* The solid products are in a liquid form for a sensible period of time after explosion. See pp. 93, 94, of Researches of Noble and Abel.

Also generally—

1.—Very small-grain powders, as F.G. or R.F.G., furnish decidedly smaller proportions of gaseous products than a large-grain powder (R.L.G.), while the latter furnishes somewhat smaller proportions than pebble powder, though the difference between the gaseous products of these two powders is inconsiderable.

2.—The variations in the composition of the products of explosion furnished in close chambers by one and the same powder, under different conditions as regards pressure, and by two powders of similar composition under the same conditions as regards pressure, are so considerable that no value whatever can be attached to any attempt to give a general chemical expression to the metamorphosis of a gunpowder of normal composition.

3.—The proportions in which the several constituents of solid powder residue are formed are quite as much affected by slight accidental variations in the conditions which attend the explosion of one and the same powder in different experiments as by decided differences in the composition, as well as in the size of grain of different powders.

4.—In all but very exceptional cases, the most important components of the solid residue are Potassium carbonate, sulphate, hyposulphite, and sulphide.

In accounts of experiments with gunpowder, the term "wave action" will frequently be met with. It is a term applied to the abnormally high pressures which are found to occur in a gun when very high charges are used. They appear to be local, and do not give increased velocity to the shot, though of course they tend to strain the gun at the points where they occur.

# CHAPTER II.—GUN-COTTON. INGREDIENTS USED IN LABORATORY COMPOSITIONS, THEIR PROPERTIES, AND METHOD OF MIXING THEM.

Gun-cotton.

All gun-cotton is now manufactured at Waltham Abbey, although a considerable amount has been obtained from the trade. The use of it for the various descriptions of torpedoes, and its handiness for demolitions of all kinds, especially hasty ones, have rendered it of late years a most important substance in warlike operations.* The composition will best be understood by giving a brief sketch of its manufacture. Fuller details will be found in "Short Notes on Gun-cotton," by Capt. Geary, R.A., Assistant Instructor, R.M. Academy.

The cotton is thoroughly cleansed from all fatty and foreign matter, and thus becomes almost pure cellulose ($C_6H_{10}O_5$). It is then steeped for a short time in the strongest nitric acid, to which three parts by weight of strong sulphuric acid have been added, the use of the latter being to take up the water which results from the action of the nitric acid on the cellulose, and which, if left unabsorbed, would weaken the strength of the nitric acid, and thus prevent the proper conversion of the cotton into true gun-cotton. The cotton, now converted into gun-cotton,† is very thoroughly washed to rid it of the free acid, which, if left in, even in minute quantities, would not only be fatal to the keeping qualities of the gun-cotton, but would make it dangerous to store, owing to its liability to decomposition and spontaneous combustion. After the washing process the gun-cotton is converted into a pulp, and in this state can be moulded and pressed, into any required form.

Gun-cotton when dry *may* ignite at 277° F., it *must* ignite at 400° F.; its mean igniting point may be taken at about 340° F. Gunpowder ignites at about 600° F. The low igniting point of gun-cotton is taken advantage of in using loosely twisted strands of it as additional priming for fuzes, when very small charges are used. (See p. 24) Gun-cotton varies considerably in its rate of burning, according to its mechanical condition and mode of firing If it be merely washed without being pulped, it retains its original colour and appearance, and can only be distinguished from ordinary cotton by the peculiar harsh feel of its fibres. In this state it can be twisted into strands, woven into fabrics, or treated in fact in any way in which ordinary cotton can be treated. For torpedo and mining operations it is in our service pulped, moulded and compressed by hydraulic power into discs or slabs, which may of course be of any size or shape required.

---

* Much information will be found in Bloxam's Chemistry, 1st Edition, p. 500; Abel's recent Investigations and Applications of Explosive Agents, 1871; Abel's Contributions to the History of Explosive Agents (from the Proceedings of the Royal Society, No. 150, 1874); Extracts, Vol. VIII, p. 59, Vol. IX, p. 37, Vol. X, pp. 28, 179; Report of Committee on Storage and Transport of Gun-cotton, 1872.

† The chemical equation for the production of gun-cotton by the action of nitric acid on cellulose is

$$C_6H_{10}O_5 + 3HNO_3 = \underbrace{C_6H_7(NO_2)_3O_5}_{\text{Gun cotton}} + 3H_2O.,$$

or 162 grains of cotton are transformed into 297 grains of gun-cotton. In practice the weight of gun-cotton produced is less than this.

When dry, gun-cotton, perfectly unconfined, is ignited by a flame, or by a heated body; it burns quietly and very rapidly* with a bright yellow flame; if, however, the cotton is confined in a strong case, even of wood, the action is very different, it explodes with great violence, and the strength of the explosion will depend upon the thickness of the case; to develope it fully a strong iron case is required.

Compressed gun-cotton may be detonated,† even when unconfined, by the action of various detonating bodies of which fulminate of mercury is found the most suitable. It is this property that renders gun-cotton so valuable for torpedoes, destroying stockades, bridges, &c.

When about 5 grs. of fulminate of mercury are enclosed in a tin tube, and ignited in contact with a disc or slab of gun-cotton, a most violent action is set up; to guard against any chance of failure, 20 grs. are used. Various means are used for igniting the fulminate; a description of the service detonators will be found at page 92.

It is essential that the gun-cotton should be in a compact form, such as that produced by the pulping and subsequent pressing process, as light flocculent gun-cotton cannot be detonated by the fulminate.

When the detonation is required to spread from one disc to others, it is not necessary to have them absolutely in contact, rows of discs from 30 to 40 feet in length, with intervals of from $\frac{1}{2}$ an inch to 1 inch, have been detonated by a disc at one end, set in action by a fulminate detonator.

The discs or slabs have one or more perforations in them to take the detonator or detonators.

The chief products of the explosion of gun-cotton are carbonic oxide, carbonic acid, water in the form of aqueous vapour, and nitrogen; the first is highly dangerous in confined spaces, such as mines, &c., as it is an active poison and very inflammable.

Gun-cotton cannot be injured by wet. It may be kept under water, and yet when dried again it will possess all its original qualities of easy ignition and susceptibility to detonation, so long as its mechanical cohesion be not destroyed. Not only this, but wet gun-cotton may itself be detonated by fulminate if a considerable quantity of the latter be used; or, as is the safest and most convenient plan in practice, by detonating a small quantity of dry gun-cotton in contact, or nearly so, with the wet mass, by means of an ordinary detonator. One or two half-pound dry discs of gun-cotton, when thus detonated, will communicate their action to a large mass of wet gun-cotton. For instance, a charge of gun-cotton may be suspended in water, entirely unprotected from the latter, and detonated by means of a dry primer protected in a waterproof case. Wet gun-cotton is actually more powerful in its effects when detonated than dry.

---

* The rapidity of its ignition may be shown by placing a small piece of flocculent gun-cotton on the top of, or even under, a few grains of powder. The cotton may be then ignited by a hot rod, or other means, without firing the powder; the latter is only thrown about.

† It is difficult to find a clear definition of the words *explosion* and *detonation*, but their general sense is sufficiently understood. The main difference seems to be in the rate of ignition; in fact, *burning* runs into *exploding*, and *exploding* into *detonating*. In the last action, so rapid is the decomposition, that no confinement is necessary to develope its power. If a small piece of loose gun-cotton be hit on an anvil by a hammer it will detonate, but the detonation is local and does not extend to contiguous portions. For instance, if one end of a loosely twisted strand of gun-cotton, such as is issued for priming for fuzes be thus struck, the detonation will not extend along the strand.

At the same time, it is impossible to ignite wet gun cotton by any flame. If a disc of wet gun-cotton be put in a fire it will gradually smoulder away as it dries, but no explosive effect will be produced.

These properties are most important. They allow the gun-cotton to be stored wet, in which condition it is absolutely safe from any danger of explosion by fire, and it can be redried at will for use in small quantities, or left wet for use in large charges, subject to the use of a dry primer as above mentioned.

When gun-cotton is detonated, its action is so very rapid* that no confinement is required, thus there ceases to be any necessity for using a strong case for torpedoes, or for tamping mines, and it can be used to cut down stockades, &c., by simply attaching the discs loosely to the obstacle.

The force of detonated gun-cotton is stated to be about four times that of exploded gunpowder, weight for weight.† Extracts, Vol. IX, p. 38.

From the experiments given in the footnote, it seems that ordinary stockades would be disposed of by using about 2 lbs. of gun-cotton per foot-run., the discs being in contact, and exploded by a detonator. As the discs must be touching, or nearly so, a handy plan would be to have the discs attached to a rod or rods, which could be fastened or laid against the stockade; or simply to string them on a piece of line, of which the extremities were fastened by nails or other means to the stockade.

Gun-cotton is issued to the service in the following shapes and sizes for land service:— Issue. § 2922.

I.—Discs and Primers for Submarine Mines.

| Diameter in inches. | Length in inches. | Weight in ounces. | Remarks. |
|---|---|---|---|
| $3\frac{1}{10}$ | 2 | { 8‡ / 9 } | With two perforations for detonators, in the case of the primers, which are supplied in the proportion of 25 discs per ton. |

II.—Discs and Primers for Blasting, &c.

| Diameter in inches. | In what sized hole to be used, inches. | Length of disc in inches. | Weight in ounces. | Service. | Remarks. |
|---|---|---|---|---|---|
| $\frac{7}{8}$ | 1 | 2 | 1 | Blasting, &c. | With one perforation in each for detonator. |
| $1\frac{1}{4}$ | $1\frac{1}{2}$ | $1\frac{1}{4}$ | 1 | ,, | |
| $1\frac{3}{4}$ | 2 | $1\frac{3}{8}$ | 2 | ,, | |

* The rate at which detonation is transmitted from mass to mass in a row of *wet* gun-cotton discs ranges from 18,000 to 20,000 feet per second; the rate is slower with *dry* gun cotton, being about 17,000 feet.

† The following experiments illustrate the force of detonated gun-cotton:—A stockade was formed of four wooden balks, each 12 inches square, let into the ground about 2′ 6″ and well strutted behind. A 1 lb. disc of gun-cotton was placed in the centre of each balk, the intervals being filled with ½ lb. discs (10), all the discs were touching. One of the extreme discs was fired by a detonator, when the whole exploded with great violence, cutting down the stockade level with the ground, and hurling the pieces all round.—Extracts, Vol. VIII, p. 60.

An 18-inch brickwall can be destroyed by employing slabs of compressed gun-cotton simply placed against it at the rate of 1¼ lb. per foot-run.—Extracts, Vol. IX, p. 37.

‡ These constitute the earlier supplies, but this particular size will not be again manufactured.

### III.—Slabs for Field Operations, &c.

| Dimensions in inches. | | | Weight lbs. | Service. | Remarks. |
|---|---|---|---|---|---|
| Length. | Breadth. | Thickness. | | | |
| $6\frac{1}{8}$ | $6\frac{1}{8}$ | $1\frac{3}{4}$ | $2\frac{1}{2}$ | Torpedo and submarine mining services | With four perforations in the case of primers. |
| $6\frac{1}{8}$ | $6\frac{1}{8}$ | $1\frac{3}{8}$ | 2 | Demolition of walls, stockades, &c. | With three perforations in each for detonators, or for lacing slabs together. |
| $6\frac{1}{8}$ | $6\frac{1}{8}$ | $1\frac{1}{8}$ | $1\frac{1}{2}$* | .. .. | With four perforations for the same purpose. |

Instructions for packing, stowing, and periodical inspection of damp gun-cotton, issued with A.C., Nov. 1876. §§ 2402, 2594, 2685, 2964.

The damp cotton is to be moved from one station to another in service transport boxes (28 lbs. or 50 lbs.), having the lids screwed down. If destined for a foreign station, the edges are to be luted with tape and marine glue. It is to be put in the store tanks, each of which holds about one ton, immediately on arriving at its destination. The contents of boxes having the same dates, &c., are to be kept as much as possible together in the tanks. The tanks, slightly inclined towards the plug hole, when packed with gun-cotton, are to be filled with fresh water till the cotton is all submerged. The plug (which is an ordinary G.S. plug) is then to be taken out and the water to be allowed to run off. The plug is to be replaced and the lid of the tank screwed down. Every three months the tanks are to be opened and the contents examined; a sample is taken of ten pieces from the top layer and weighed. If the weight shows that it contains less than 20 per cent. of water, the contents of the tank are to be re-wetted.†

Cases, wood, for gun-cotton. §§ 2082, 2682.

Gun-cotton is issued in wooden cases. The first pattern held 28 lbs. in the form of discs, or about 36 lbs. in the form of slabs. Of these no more will be manufactured. The present pattern is coated on the interior with crude paraffin, and holds 50 lbs. of gun-cotton discs, or about $52\frac{1}{2}$ lbs. of slabs.

Report of Committee on storage of gun cotton. Extracts, Vol. XI, p. 153, 1872.

It must be remembered that the safety of the gun-cotton depends on its being kept damp, if dry it would have to be treated as any other explosive. An experiment carried out in May, 1873, showed that a ton of wet gun-cotton did not explode when exposed to the action of fire in a strongly built magazine.

Abel's investigations on explosive agents, p. 22. Extracts, Vol. VI, p. 83.

From what has already been said as to the action of gun-cotton, it will be seen that it would be a most valuable agent for bursting shells, the qualities which render it unsuitable for the charges of guns are just those required for the bursting charges of shells. Experiments have been carried on from time to time with a view to utilize it in shells. Some 13″ mortar shells were fired safely with gun-cotton, but on trying it with 7″ and 8″ rifled shells, the gun-cotton exploded with

* Future supplies will be made with only three perforations.

† In cold countries precautions should be taken to prevent the water in the tanks freezing, as the cotton is liable to split and flake off in the act of freezing. Frozen gun-cotton can be detonated like dry gun-cotton. Curiously, pure frozen nitro-glycerine cannot be detonated, at least so states a large American manufacturer (Mr. G. M. Mowbray). He gives facts in support of this statement.

great violence in the gun; one gun was burst and the other rendered unserviceable; the friction against the walls of a rapidly rotating rifled shell being greater than against the sides of a spherical one, accounts for the explosion.

Experiments have been carried on to ascertain whether wet gun-cotton can be used in rifled shells in conjunction with a fulminate detonator.* See p. 242.

Drying gun-cotton. § 3056.

Wet gun-cotton may be dried by simple exposure to the air of a dry room till it ceases to lose weight, or by leaving it in the open air in dry weather exposed to the sun and wind. Even without sun it will dry in about five days. Of course the actual time taken to dry any given specimen will depend on climate, state of atmosphere, &c. If it be required to dry gun-cotton quickly, it is to be dried in a special apparatus constructed for the purpose, for which full details and instructions are given in the paragraph in Changes in War Stores, referred to in the margin. This apparatus will dry large slabs in from 18 to 24 hours. It is simply a portable hot air chest heated by steam.

### Ingredients used in Laboratory Compositions.†

The combustible compositions may be divided into two classes, those which detonate and those which do not; as an aid to memory it may be remarked that chlorate of potash will be found in the detonating class, and saltpetre in the compositions which do not detonate.

Sulphur.

*Sulphur* (S) burns at a low temperature (about 500° F.), and gives out great heat. This is useful in enabling the other ingredients of the various compositions, of which sulphur forms a part (*e.g.*, gunpowder), to ignite in the first place, and the heat given out by the burning sulphur increases the rapidity and power of action of the whole. Sulphur melts at 239° F. This substance is found in all the burning compositions, and, owing to its property of detonating with chlorate of potash or sulphide of antimony, in the greater part of the detonating compositions.

Sulphur has been used in the R.L. in two forms, viz., sublimed sulphur, known generally as "flowers of sulphur," and ground distilled sulphur. A full account of the preparation and properties of these two forms of the element will be found in the Handbook of Manufacture of Gunpowder, chapter III, and in Bloxam's Chemistry, pp. 120, *et seq.* It is sufficient here to say that the sublimed sulphur consists of spherical granules composed of insoluble, or electro-positive sulphur, enclosing soluble, or electro-negative sulphur.‡ The distilled sulphur, on the other hand, consists almost entirely of the soluble variety which crystallizes in the form of rhombic octahedra. The use of sublimed sulphur will be discontinued in future, on account of the deleterious influences found to be exerted on some of the compositions of which it formed a part by the free acids ($SO_2$ and $SO_3$) which, from the mode of its manufacture, it is liable to contain. These may be removed by washing, but it has been found advisable on the whole to make use of ground distilled sulphur only.

Saltpetre.

*Saltpetre*, or *Nitrate of Potash*, or *Nitre* ($KNO_3$), is used as a source of oxygen. One cubic inch of saltpetre contains about 207 grains weight

---

* An account of experiments with Picric powder in shells, and of its composition, will be found in Abel's Investigations of Explosive Agents, p. 25.

† For a table of the ingredients used in various compositions, see p. 319.

‡ The terms "soluble" or "insoluble" refer to the behaviour of sulphur in bisulphide of carbon ($C.S_2$).

of available oxygen, equivalent to that contained in about 3,000 cubic inches of air. As most of the laboratory compositions burn in a more or less confined space, some source of oxygen is indispensable. This need is supplied by saltpetre, which is found in all non-detonating compositions.

Saltpetre is supplied to the R.L. from Waltham Abbey, a full account of the process of refining will be found in the Handbook of the Manufacture and Proof of Gunpowder, chapter II. The presence of any salts containing chlorides is carefully provided against, as they would render saltpetre liable to deliquesce, a fault from which pure saltpetre is comparatively free.*

Ground saltpetre is used in the R.L. It is supplied in a powder fine enough to pass through an 80-mesh sieve.

Chlorate of potash.

*Chlorate of Potash* ($K.ClO_3$) contains, weight for weight, more available oxygen than saltpetre. As, however, it has the property of detonating on being rubbed or struck when mixed with sulphur or sulphide of antimony, it cannot with safety be used in the ordinary burning compositions. It would manifestly be highly dangerous to use it instead of saltpetre in gunpowder, for instance. This property, however, renders it most useful in detonating compositions, in all of which (except when fulminate of mercury alone is used) it is to be found. It is also used in the manufacture of coloured lights, and of stars for the 7-pr. star shell. The great heat developed by its powerful action, causes the metallic salt, on which the colour depends, to be thoroughly burnt. In mixing these compositions, great care is necessary to avoid accidents from detonation. If chlorate of potash and sugar be mixed together, and touched with a drop of sulphuric acid, the mixture burns violently. Fougasses and mechanical torpedoes have been constructed to be exploded by the fracture of a glass tube or globule containing sulphuric acid, and imbedded in a mixture of chlorate of potash and sugar, which, when inflamed by the action of the acid, in its turn ignites the charge. The action is somewhat slow, but is much improved by the addition of a proportion of ferrocyanide of potassium ($K_4Cy_6Fe$).

Sulphide of antimony.

*Sulphide of Antimony*, or *Crude Antimony* ($Sb_2S_3$), like sulphur, detonates with chlorate of potash, and gives a more violent explosion. It burns with a long flame or flash, which renders it useful in compositions intended to ignite other bodies at a little distance; thus it is used in caps, friction tubes, and in the detonating composition for Pettman's fuzes. The peculiarly long flash is probably due to the volatilization of the metal.

Fulminate of mercury.

*Fulminate of Mercury* ($C_2HgN_2O_2$), is a most dangerous substance to handle, and should never be dealt with by inexperienced men. It detonates readily and with great violence on being rubbed or struck. It will also explode when heated to 360° F., or on being touched with nitric or sulphuric acid. It is kept wet till required for use, in which condition it is harmless. It is prepared by the action of alcohol on a solution of mercury in nitric acid. The products of its explosion are metallic mercury, carbonic oxide, and nitrogen. It is used alone in detonators for gun cotton, and in company with other ingredients in cap composition.

Magnesium.

*Magnesium* (Mg) burns with a very brilliant white flame, and is used

* The presence of chlorides in a sample of saltpetre is easily recognised by the formation of a white precipitate of chloride of silver on adding a few drops of nitrate of silver to an aqueous solution of saltpetre.

in a powdered state in light compositions. It is prepared with paraffine to preserve it from oxidation.

Red Orpiment.

*Red Orpiment*, or *Realgar* ($As_2S_2$), is a sulphide of arsenic. When used in conjunction with saltpetre it burns with a brilliant white flame, producing arseniate and sulphate of potash. It is thus very useful in light compositions.

Turpentine, Spirit of.

*Turpentine* is used as a solvent of rosin in carcasses, forming a sort of cement which binds the mass together. It is an ingredient of the lacquer used for the interior of all rifled shells. It is highly inflammable.

Methylated Spirit.

*Methylated Spirit* is alcohol, or spirits of wine ($C_2H_6O$), of Sp. Gr. ·83, mixed with 10 per cent. of wood spirit or methylic alcohol ($CH_4O$), the latter being derived from the destructive distillation of wood. The only object of adding the methylic alcohol is to render the mixture nauseous and undrinkable.* The spirit is used in the R.L. for damping detonating compositions, so that they can be handled in the form of a paste. It evaporates without injuring the composition, of which it forms no part in their ultimate form. It is used also as a solvent for shellac.

Charcoal.

*Charcoal* is carbon containing more or less oxygen and hydrogen. The longer the charcoal is burnt in its manufacture, the more nearly will it approach pure carbon (C). It acts as fuel in gunpowder, &c., combining with the oxygen of the saltpetre, and, in the combination induced by ignition, forming a highly heated, expansive gas. It is used also in rocket composition.

Mealed Powder.

*Mealed Powder* causes composition to burn readily and quickly, and the rate of burning may to some extent be regulated by the quantity of mealed powder employed. See also p. 25.

Rosin.

*Rosin* or *Colophony* is the resin obtained from the turpentine or viscous exudation obtained by incising the bark of various species of pine trees. The rosin is held in solution in the oil or "spirits" of turpentine, and is freed from it by distillation. It is used in lacquer for the interior of shells, in carcasses, &c., and is run in a melted state into the interstices between the balls of rifled Shrapnel shell to bind them together.

Shellac.

*Shellac*, or more correctly *shell-lac*, as distinguished from *stick* or *seed lac*, is the exudation of certain tropical trees punctured by an insect (*coccus lacca* or *ficus*). It is a complex resin. It is largely used in the R.L., dissolved in methylated spirits of wine, for various varnishes, and also for damping various detonating compositions. The spirit evaporates, and the shellac, remaining behind, acts as a sort of cement.

## Method of Mixing Laboratory Compositions.

The leading points to be attended to in making up laboratory compositions are:—

1. Purity of the ingredients used.†
2. The proper proportion of each ingredient to be accurately weighed.
3. The thorough mixing, or incorporation of the ingredients.

In order to ensure the mixing being complete, it is necessary to have

---

* Methylated spirit was allowed to be free from the duty on spirits as being only available for manufacturing or scientific purposes. However, a process has been discovered which renders it palatable, so the exemption no longer exists.

† Samples of the ingredients are analyzed in the Chemical Department, Royal Laboratory.

the ingredients in a state of very fine division, this enables the different bodies to be brought into close contact with each other, and thus ensures their acting on one another when inflamed.

Such bodies as are not supplied in a state of fine division, are reduced to an impalpable powder by being placed in a revolving barrel along with a number of gun metal balls; after being "drummed" for a sufficient time, the ingredients are passed through a very fine sieve,* to ensure the powders being sufficiently fine.

There are two different methods of mixing commonly used in the R.L., one consists in placing the ingredients in a revolving barrel, with gun metal balls, and "drumming" them. The second consists in placing them in a revolving barrel, fitted up with wooden arms or fans, in the interior, so arranged that when the barrel revolves in one direction, the fans revolve in the other.

While the above methods are being used for what may be called the *burning compositions*, it is found necessary to adopt another method for the detonating compositions, which would explode under friction or pressure. It consists in brushing the ingredients through a fine sieve with a strong brush;† such bodies as sulphur and chlorate of potash may thus be mixed, but when fulminate of mercury is used, even this method is too rough, and the ingredients are mixed by a man using a fine badger's hair brush, all proper precautions being taken to guard against explosions.

When the ingredients are thoroughly incorporated in proper proportions with a view to form a large volume of gas on explosion, there should be little or no solid matter left, as is the case with gunpowder, but the laboratory burning compositions are, as a rule, required to burn for a considerable time, and the volume of gas evolved is of no importance, hence the proportions employed are chosen with a view to cause them to burn a certain definite time, and there being a residue of solid matter, or slag, is unimportant.

The most important point aimed at is regularity of burning. By increasing the density of a composition, its time of burning will be increased, as there will be more matter to be burned in a given volume, therefore, by subjecting the composition to a very heavy pressure,‡ the time of burning is prolonged.

A table of the various compositions used in the R.L. is given on p. 319.

Many of the compositions possess the property of burning under water; as they do not depend on air for the supply of oxygen, water will only extinguish them when it comes into sufficient contact with the burning matter, to reduce its temperature below the point of combustion. When the composition is enclosed in a case, having a vent or vents, and a sufficient amount of gas is generated to give a high pressure when issuing from the vent, the water cannot make its way to the composition, which will therefore continue to burn. This may be illustrated by plunging a wood time fuze, when well lit, into a bucket of water, it will continue to burn: the larger natures of carcasses, when

---

* In some cases an 80-mesh, in others an 120-mesh is used.

† This method seems to have been the old plan for most compositions; it is necessary to brush, not to sift, the ingredients through the sieve, as the last would separate the ingredients instead of mixing them, the lightest finding its way to the top. It is not easy to insure a thorough incorporation by this plan, it is also slow and cumbrous.

‡ The pressure is generally applied in the R.L. by means of an hydraulic press; in some cases, such as war rockets, the pressure reaches several tons per square inch.

well lit, will also burn, and so will the light balls; with the smaller natures, such as the 12-pr. carcass, the rush of gas from the vents is not always strong enough to prevent the entrance of water.*

For the precautions to be taken in laboratory operations, see p. 314.

---

## CHAPTER III. — FUZE HOLE GAUGES. GENERAL REMARKS ON TIME FUZES. CAUSES WHICH ALTER THEIR TIME OF BURNING. METHOD OF PACKING FUZES. CAUSES OF BLIND SHELL AND OF PREMATURE BURSTS.

---

Fuze-hole gauges.

BEFORE entering on the general question of fuzes, it is necessary to mention the various gauges or sizes of the fuze holes of shells which are found in the service. Each size of fuze hole requires a special fuze to fit it, and thus complicates the question of fuzes.

§ 1238.

It has therefore been determined to adopt one size of fuze hole for all rifled shells, and in time, as the ammunition for smooth bores becomes obsolete, all fuzes will be made of the general service gauge, the only exceptions which exist at present in rifled ammunition are the fuze holes for common and segment shells for B.L.F.S. guns. The table below shows the gauge of the fuze holes, and the shells and fuzes of each gauge.

FUZE Hole Gauges of all Shell and the Fuzes which fit each Gauge.

| Gauge. | Shell. | Fuze. |
|---|---|---|
| Large mortar .. | Large mortar, 13″, 10″, 8″ .. | Large mortar, time. |
| Common .. .. | S.B. common and diaphragm, Shrapnel | Common, diaphragm, and small mortar, time. Pettman land service percussion. |
| General service .. | S.B. naval shells and all rifled shells, except B.L. field service, segment and common | 5, 9, or 20 secs. B.L. or M.L. time. Royal Laboratory percussion, Pettman G.S. percussion, and Sensitive fuze. |
| Armstrong field service | B.L. field service, segment and common† | B.L. plain percussion, and for S.S. Armstrong E. time. |

N.B.—This table only shows the gauges of both shells and fuzes, not the fuze which is to be used with each nature of shell. Tables giving the fuze for every shell will be found at p. 58.

---

* Rockets will burn and move rapidly under water; but no effectual means have as yet been found to control the erratic course of submarine rockets.

† The 20-pr. B.L. common shell for S.S. has the G.S. gauge.

Adapters.

All shell having the obsolete Moorsom gauge can be converted to the G.S. gauge by using an adapter. This consists of a gun-metal bush which screws into the Moorsom gauge fuze hole, the interior being of the G.S. gauge. There are two distinct adapters, one for spherical shell, 1″·38 long; the other for rifled shell, 1″·75 long. There are three Marks of the adapter for rifled shell: Mark I. screws in nearly flush with the top of the fuze hole; II.* and III. leave a countersink of ″·2, which is necessary to receive the naval wad, see p. . Mark I. can be converted to II., but is available for L.S. without alteration. For the 20-pr. B.L. common shell a "flanged" socket converts the Moorsom into the F.S. gauge. For method of fixing adapters, see p. 146. The above are the only serviceable adapters, obsolete adapters, which do not convert the Moorsom to the G.S. gauge, are ordered to be broken up. The details of obsolete patterns will be found in §§ 693, 284, 771.

§ 1396.

§§ 1238, 1427.

§§ 1345, 2843.

§ 1838.

It is not necessary to give any account of fuzes manufactured for the service prior to 1855.†

General remarks on time fuzes.

In that year General Boxer introduced his valuable time fuzes which are still used in the service, and which were greatly superior to the fuzes they replaced, both as to accuracy and facility of preparation.

Wood‡ has been adopted for the body of the fuze, a hard durable wood with a grain suitable for turning is required; beech is found to answer well. The wood is seasoned, and is afterwards desiccated by artificial heat.

All Boxer fuzes are conical, this shape having a great advantage over the cylindrical form, as there is no risk of the fuze setting back into the shell on the shock of firing, at least not when the angle of the cone is sufficiently great; also, if the wood expands or contracts the fuze will only project or go in a little more, while with a cylinder the result would be either that the fuze would not enter the fuze hole when expanded, or would fall through into the shell when shrunk.

---

* Mark II. is simply Mark I. altered.

† Prior to this date some Boxer fuzes with a small cone were introduced, but were withdrawn as they were found to set back into the shell on firing. The shells suited to this form of fuze had their fuze holes subsequently enlarged, and shells so altered bear the letters L.C. (large cone).

‡ Time fuzes for naval shells were at one time made of metal.

The advantages claimed for metal over wood were chiefly safety, preservation of the composition, and causing the shell to act more violently, as the fuze hole is more securely closed.

Experience showed that the advantages as to safety and preservation of the composition were imaginary, while the closing of the fuze hole was found to be unimportant except in very small calibres of shells. As wood is much cheaper it is now used, and the metal fuzes 7½ and 20 seconds Moorsom gauge are now obsolete. Vide clause 143, A.C. 1869, for detailed instructions as to their disposal.

§ 1838.

The Boxer (wooden) time fuze for breech-loading rifled ordnance was tested before its introduction into the service as follows:—

1st trial. Four 40-pr. shells with fuzes fixed were placed round a 10lb. cartridge, two of them leaning upon it, and the other two standing fuze uppermost. The charge was exploded and none of the fuzes ignited.

This experiment was repeated five times.

2nd trial, with fresh fuzes.

3rd trial, with fresh fuzes.

4th trial, with fuzes of 3rd trial.

5th trial, with fuzes of 2nd and 4th trials.

None of the fuzes exploded, or were blown out of the shell, or loosened in any way.—Extracts, Vol. II, p. 62.

The same pitch of cone is used in all Boxer fuzes. The cone increases at the rate of about $\frac{1}{10}$th inch in diameter for each inch in length.* The different sizes are obtained by taking different sections of the cone

The fuze composition is contained in a channel which is not bored completely through the wood, as it is necessary to support the composition to prevent it from setting back on firing. This channel is placed centrally in the body of the fuse, if there are no powder channels, eccentrically when there are, so as to leave room for them. Powder channels are found in fuzes intended for a time of flight of 10 seconds or under. They are essential in fuzes for Shrapnel shells, where the bursting charge is not immediately surrounding the fuze, and consequently a strong flash is required, they also would be necessary in the case of common shell when the fuze is bored short, because the flash would be obstructed by the side of the fuze hole.

If only one powder channel were used it is obvious that there would not be room for the side holes to be bored in a fuze reading to tenths of inches of fuze composition, the diameter of each side hole being 0″·125. By the use of two powder channels we are therefore enabled to graduate the fuze just twice as finely as we could do with one powder channel.† The powder channels are, except in the common fuze, connected at the bottom by a groove filled with quick-match, to cause them both to act at the same time, giving a strong downward flash.

In all time fuzes the last hole in the row is bored through into the composition to ensure the action of the fuze when fixed in the shell without preparation.‡

In all time fuzes it is desirable that they should not be liable to be ignited before they are "uncapped," as otherwise fuzed shell would be endangered by sparks, or by a neighbouring explosion,§ this is attained in the case of fuzes for S.B. ordnance by covering the head with a cap of block tin, in most of the M.L. fuzes by a copper strip covering the

---

* The exact rate of increase is 1 in 9·375.

† Experimental fuzes have been tried with six and eight powder channels, and one with pistol powder covering the whole circumference of the fuze. In the case of the latter the powder is protected by varnished paper, and over the paper is a sort of skeleton covering of gun-metal with eight longitudinal slots, through which the figures corresponding to eight powder channels appear. The exterior of the metal covering is tapped to fit the G.S. gauge. This fuze and the one with eight powder channels contain 3″·3 of composition, burning at about half the rate of ordinary fuze composition. They are graduated to ¼ seconds, and read from 7¾ and 7½ seconds respectively to 30 seconds. The fuze with six powder channels also relates to ½ seconds and is graduated from 1, *i.e.*, ½ second to 30, *i.e.*, 15 seconds.

‡ Where there are powder channels, the bottom hole of each channel is bored through, as it is convenient to support the powder by passing quick-match through these holes, of course the fuze acts when it burns down to the first of those holes.

§ An experiment was made in the Royal Laboratory at Woolwich in March, 1865, to determine how far the M.L. fuzes, protected by a copper strip and tape, were secure in this respect; the result was satisfactory. The following are the details of the experiment: Four 7-inch B.L. common shells were placed upright and a 10lbs. cartridge exploded in the centre, and this repeated a second and third time with the same fuzes; none of the fuzes ignited. Four 7-inch B.L. common shell were placed round a 10lbs. cartridge, with the fuzes resting on the cartridge, and the cartridge exploded, this repeated a second and third time; one fuze out of 12 may be expected to light. An experiment made in the R.L., on September 9, 1874, showed that the Mark III. M.L. fuzes were as safe as those of Mark I. from the risk of being ignited by a neighbouring explosion.

priming;* while the B.L. fuzes are protected by papier mâché wads and copper discs covering the escape holes.

Stress is laid in drill on the importance of not uncapping the fuze till the shell is placed in the bore.

It will be remarked that the fuzes for rifled ordnance have their heads closed, otherwise they would probably be extinguished when the shell struck point foremost,† they would also burn much quicker on account of the increased pressure of the air when flying point foremost. See also p. 27.

In the M.L. guns the fuze is ignited by the flash of the charge, which ignites a quick match priming, this priming is contained in the head of the fuzes for S.B. ordnance, and in a grove at the side of the head in the M.L. fuzes. In B.L. fuzes the quick match is used to convey the flash from the detonator to the fuze composition.

It is an essential point in all time fuzes to drill a small hole in the top of the fuse composition, this roughens the surface and renders ignition more certain, if it were not done the hard polished surface of the fuze composition would often fail to be ignited by the priming. The hole is also of use to fix the zero point, or the point from which the burning of the fuze is reckoned, the length of the composition being measured from the bottom of the hole.‡

Gun-cotton priming. §§ 2285, 2409.

For R.M.L. field guns, gun-cotton priming has been introduced for use when firing at high angles with small charges, as the quick-match failed to ignite.

The gun-cotton is issued in tin cylinders containing 20 feet of loosely twisted gun-cotton and strands of silk for attaching it to the fuze, the cylinder holds enough for about 20 fuzes, it is closed by a band of tape.

The directions for use are enclosed in each cylinder and are as follows:—

Uncap the fuze as usual, open out the priming, and wind about 10 or 12 inches of the gun-cotton round it—bringing the ends of the priming between the strands of gun-cotton; tie the two ends of the latter together leaving about two inches loose, then fix the whole firmly by tying over it a piece of silk.

It is well to remember that when the gun-cotton is attached, the fuze is liable to be ignited by the least spark, therefore the shell should be placed in the gun as soon as possible after the gun cotton is attached.§

* An experiment carried out in the R.L., in 1870, showed that fuzes having the priming protected by a tape band were not so secure against an explosion near them as those having copper bands. Four fuzes were placed in shells, and the shells standing upright were arranged round a 10 lbs. cartridge, three at a distance of two feet, one within a foot of the cartridge; the cartridge was exploded and none of the fuzes ignited. The experiment was repeated with the same fuzes, and the nearest fuze ignited. On a third trial, with new fuzes, none ignited. Finally, the shells were laid with their heads resting on a cartridge; all the fuzes ignited on exploding the cartridge.

† It was hoped that this would get over the difficulty of the effect of the pressure which causes a varying rate of burning in rifled guns. This, however, has not been attained.

‡ The depth of this hole varies from 0″·7 in the 9-seconds R.M.L. fuze, Mark III., to 0″·1 in the 9-seconds R.B.L. fuze, Mark II.

§ It was found that even with extra quick-match priming the M.L. time fuze would not ignite with certainty when a 10-oz. charge was used in a 16-pr. gun; when gun-cotton was used, a 4-oz. charge was found to ignite the fuze. See Extracts, Vol IX, p. 208.

The time fuzes of the G.S. gauge have a paper lining to the fuze composition bore.

In hot climates the wood is liable to shrink, and by doing so may leave a space between the wood and the fuze composition, thus exposing the sides of the fuze composition to the action of the flame of the ignited fuze; should this take place, the fuze will have a large surface ignited at once, and will burn very rapidly, so causing premature explosion.

The paper lining prevents this, as it is not liable to shrink; the lining has, however, the disadvantage of injuring the keeping qualities of the fuze, because a porous paper absorbs moisture. The old fuzes of the common gauge keep far better,* it has already been pointed out, p. 11, that paper in contact with gunpowder is apt to injure it. It is hoped that this defect has been remedied in the last pattern of time fuze, see p. 36.

The ingredients of fuze composition are given in the table, p. 319. This composition, however, will vary slightly from the proportions given there, no doubt there is always some slight difference both in the purity and the mechanical condition of the ingredients, and so two mixings, though prepared as far as possible in the same way, will not give exactly the same time of burning. The rate of burning of each mixing is ascertained, if it is found too slow a little mealed powder is added, if too quick a little more saltpetre and sulphur, until it is brought to burn at exactly the required rate.

Fuze composition burns at the rate of 1 inch in 5 seconds. This slow and regular rate of burning is due not only to the proportion of the ingredients but to the fixed amount of pressure to which they are subjected.

Mealed powder is used for the 5-seconds fuzes; here again the composition varies, as generally a little saltpetre has to be added to bring down the rate of burning (about 1 oz. of saltpetre to 1 lb.) of mealed powder. Mealed powder burns at the rate of an inch in 2½ seconds. twice as fast as fuze composition; hence it will take only half as long for the fuze to burn from one side hole to the next, when mealed powder is used (the wood blocks for the 5 and 9 second fuzes are perfectly similar), and the mealed powder fuze can be bored to act at half the intervals of time that the fuze having fuze composition can be bored to act.

There is no difficulty in introducing compositions which burn slower, as this can easily be done by reducing the proportion of mealed powder.

The fuzes which were made for the 7-pr. R.M.L. gun illustrate this well, they were all made from the same size wood block and burned respectively 5, 10 and 15 seconds, the last named having the smallest proportion of mealed powder.† It has, however, been remarked that the same amount of regularity can hardly be attained when very slow burning compositions are used. The intervals of time also, taken in burning from one side hole to another, are longer the slower the fuze composition burns. This defect was met in some experimental fuzes

* Reports received from Mauritius, the West Indies, and other foreign stations, show that the M.L. and B.L. fuzes, which have been in store a few years, burn long, while the common fuzes, though of much older date, generally burn within the limits allowed.

† These fuzes are now obsolete for the 7-pr., any existing may be used up with S.B. common shell.

for the 8″ rifled howitzer by adding a third powder channel, and thus increasing the number of side holes.

In all fuzes, except the mortar and parachute fuzes, the numbers refer to the time of burning in half-second units, thus 2·5 on the 5-seconds fuze, means 1¼ seconds, 5 on the 9-seconds fuze, means 2½ seconds, 20 on the 20-seconds fuze, means 10 seconds. The 5-seconds fuze reads to ¼ seconds, the 9-seconds fuze to ½ seconds, and the 20-seconds fuze to seconds.

With Shrapnel shell fired from field guns, it is essential to have a fuze which acts at short intervals of time; thus, suppose a shell to be flying at the rate of 1,200 feet or 400 yards a second, in this case the space corresponding to ½ a second is 200 yards; it is evident that to develope the full powers of Shrapnel shell, we should have a fuze which can be bored to act at shorter intervals than ½ seconds. Hence the advantage of the 5-seconds fuze which can be made to act at intervals of ¼ seconds corresponding to 100 yards in flight. Even this is rather too long an interval for accurate practice with Shrapnel shell of the smaller calibres.

Only the even numbers are marked on the 20-seconds fuzes, so they can be bored to act at intervals of 1 second. Short intervals are not so essential in this case, as the fuze is used with garrison shell at long ranges, and therefore, as powder channels are not necessary, the side holes are arranged spirally to gain space for marking.

The mortar fuzes are marked with figures indicating inches of composition, the side holes are only marked by indentations (arranged spirally), the inch space having 5 side holes, each hole corresponds to 1 second in time of burning.

By adding a cypher to the figures on the mortar fuzes they will then read to the same unit as the other fuzes, thus 6 on the large mortar fuze will indicate 60 half seconds or 30 seconds, 3 on the small mortar fuze will indicate 30 half seconds or 15 seconds.

The marking in all the short range fuzes begins at 2, except in the 5-seconds, where it begins at 1, in the 20-seconds (or long range) fuzes the marking begins at 20, where the 9-seconds ends, the last hole which is bored through into the composition is not numbered in Marks I. and II. fuzes, but the marking of the last hole has been approved for the Mark III. M.L. fuzes.*

§ 2485.

In the large mortar fuze the marking begins at 2″, *i.e.*, 20 half seconds, and in the small mortar fuze at 1″, *i.e.*, 10 half seconds.

The mode of preparing and fixing fuzes is given on page 65.

It will be observed in the instructions for preparing fuzes that all time fuzes are to be hammered into the fuze holes so as to fix them securely, except the B.L. fuzes which have detonating composition in the head, these are to be screwed in as firmly as possible by hand.

---

* All fuzes having powder channels have a hole additional to the required length of fuze. This is unnumbered. The reason for the existence of this hole is as follows: *Both* bottom side holes must be bored through, or the quick-match support could not be introduced to them. In the case of the 9-seconds M.L. fuze which burns 10 seconds at rest, the bottom hole on one side is marked 20; if the hole marked 19 in the other row of holes was treated as the bottom hole and bored through, the fuze would obviously burn only 9½ seconds; hence the introduction of the additional hole which is bored at a length which, if marked, would be 21. It would be simplest to bore this additional hole directly opposite to the 20 hole, but there is not room for this. In speaking of the "bottom side hole," in future it is always to be understood that the hole whose marking corresponds to the full time of burning of the fuze is referred to.

**Proof of time fuzes.**

Time fuzes are proved as to their time of burning before passing them into the service, 15 fuzes out of every thousand made are selected; 10 of these are burnt at rest, the limits allowed for these new fuzes are very narrow. See p. 44.

The remaining 5 fuzes are fired at Shoeburyness, and the time of burning noted.

As has been before stated, fuzes are found to deteriorate in keeping,* especially those with paper linings, the tendency is always to burn long. To allow for this, the limits of burning are enlarged in the high direction for firemaster's proof at out stations (see p. 307), an excess of 10 per cent. being allowed on the nominal time of burning. The low limit, however, remains unaltered, as there is no necessity to increase it; moreover, it is more objectionable for a fuze to burn short than long, as prematures are apt to injure our own troops.

The limits of the common and diaphram Shrapnel fuzes are the same as those of the 9-seconds and 5-seconds fuzes.

The limits of the mortar fuzes are all given in the high direction; as a mortar shell carries a large charge and moves with a low velocity, a short fuze would be most objectionable.

Hand grenade fuzes also have only a high limit allowed, the danger of a short burning fuze in this case being obvious.

**Influence of the atmosphere on the time of burning of fuzes.**

Besides the effects of climate, there are other causes which alter the time of burning of fuzes, increased atmospheric pressure† causes fuzes to burn more rapidly, while diminishing the pressure causes them to burn more slowly. This point has been established both by the experience of officers in the service, and by careful experiments carried out to determine the rate of increase of the time of burning due to diminished pressure. The same class of results has been remarked at Dartmoor and in the highlands of Abyssinia.

The following rule is sufficiently correct for practical purposes, viz.: Each diminution of atmospheric pressure to the extent of one mercurial inch in the barometer, increases the time of burning by ·03, or what is nearly the same thing, by $\frac{1}{30}$. The barometer falls about one inch for an increase of 1,000 feet in elevation. Thus at 5,000 feet elevation, the time of burning of a large mortar fuze would be increased by $\frac{5}{30}$, and would therefore burn about 35 seconds.‡

The effect of varying pressures on the rate of burning, explains to some extent the important fact that fuzes burn at sensibly different rates when fired out of different guns; as a rule, they are found to burn quicker in large than in small guns, probably because the projectiles from the former keep up their velocity better. The following are the average times of burning of various fuzes in different guns, taken from the proof records in the R.L.

---

* See Extracts XII, 164, for an account of deterioration of time fuzes at Bermuda.

† Quarter-Master Mitchell, R.A., first brought this fact to notice, and it has been confirmed by experiments carried out by Dr. Frankland, F.R.S. Proceedings, Royal Society, Vol. XI, p. 137. See also Extracts, Vol. X, p. 297.

‡ The reason given for the retardation of the time of burning, due to diminished pressure, is briefly this. Each layer of the fuze composition must be raised to the temperature necessary for combustion by heat transmitted from the burning layer above it. when the pressure is diminished the incandescent gases can expand more

The 9-second M.L. fuze burns 10·4 seconds in 9-pr. R.M.L. gun.
,, ,, 9·3 ,, 64 ,,
,, ,, 9 ,, 12 in., 25 ton gun.*

Experiments have been made to ascertain the effect of rotation on fuzes. It was found that they burnt quicker when rotated than when at rest. For example: the mean time of burning of certain fuzes at rest was 10·1 seconds; when similar fuzes were rotated at a rate of 151 revolutions per second, the mean time of burning was 9·74 seconds. Again, fuzes whose mean time of burning at rest was 11·10 seconds, when rotated at the same rate as those above gave a mean time of burning of 10·75 seconds. It will be noticed that the difference in each case is almost identical.

It is to be noted that the so-called 9-second fuzes really burn 10 seconds at rest. The name was originally given as these fuzes were supposed to burn 9 seconds only in rifled guns, and though this is now known not to be generally correct, yet, to avoid confusion, the nomenclature has not been altered.

In manufacturing fuzes, it is found necessary to avoid using oil on the tools employed in boring the fuze composition channel, as the oil coming in contact with the composition increases the time of burning.

When time fuzes are used with shells fired from rifled guns, they are found to act on direct impact against a bank of earth, or some solid obstacle; the projectile striking point foremost, the fuze is probably driven in. This action is more certain with fuzes having powder channels, than with the long range fuzes.

Paint.

All the service wood fuzes are painted in black and drab,† except the 5-seconds, which is painted red and drab, in order that this quick burning fuze may readily be distinguished from the 9-seconds fuze of similar dimensions. The composition of the paint is given on p. 322; it is really more a lacquer than a paint, as it mainly consists of shellac dissolved in spirits. It aids to protect the fuze from moisture.

Marks.

§ 1006.

Besides the numbers of the side holes, each fuze will be found to be marked with a Roman numeral, indicating the pattern or mark, the number of thousand of manufacture, and the date of the month and year on which it was made:—thus II. 88, 2/68 would show that the fuze was one of the 88th thousand of Mark II., and that it was made in February, 1868.

The most important marks are the numeral of the pattern and the date; the No. of thousand is chiefly of use to the manufacturer.

§ 1999.

Fuzes for B.L. guns bear a label directing that "the safety pin is not to be removed before fixing the fuze in the shell." The proper time for removing the pin is just before placing the shell in the gun.

Packing fuzes.

§ 1810.

The old system of packing fuzes was in zinc cylinders, secured by a tape band, and the common and diaphragm Shrapnel fuzes will still be found thus packed in cylinders holding 25 or 50;‡ but future issues will be made in a similar way to that given below for the G.S. gauge time fuzes.

§ 1843.

In 1869 the system of packing fuzes, lights, and tubes, and all such combustible stores in cylinders containing small quantities was approved. This method has the advantage of not leaving any large

---

* The fuze is not now a service fuze for the 12-inch gun in the L.S., but is inserted to show how its rate of burning is altered.

† The special fuzes for the parachute lights are painted blue.

‡ Issues made for the Gold Coast (1873) were by fives in tin cylinders.

quantity exposed to the action of climate after the cylinders are opened.

In 1870 tin cylinders, with lids secured by a tin band soldered round, and thus hermetically sealed, were introduced, and after a comparative trial with zinc cylinders secured by a tape band was finally decided on in 1872.*

§ 1871.

§§ 2055, 2217.

Cylinders have been made to contain 20, 10, or 5 fuzes; they are almost always issued in the cylinder containing 5, this cylinder being used invariably in the field service is the most convenient for general issue, and has the advantage of only leaving a few fuzes exposed when the cylinder is opened. When new, these cylinders are tested in the Royal Laboratory by placing them in water. A cylinder properly soldered can be placed under water for some days with safety to its contents.

The cylinders of fuzes are issued to the O. S. Department in packing cases, the number of fuzes and size of the case depending on the demand. The cylinders containing 5 and 9 seconds M.L. wood time fuzes are placed in a wood case containing six cylinders; the R.L. percussion are similarly packed.

The cylinders have a label on the top showing the nature of the fuze, the number contained in the cylinder, the mark or numeral of pattern of the fuze, and the service for which the fuzes are intended; also a caution *not to open the cylinder until the fuzes are required for use or special inspection.* When special inspection is required a fixed percentage of fuzes should be examined (see p. 307), only a sufficient number of cylinders being opened to furnish the required number of fuzes. Instructions for use are given on a label on the side of the cylinder, as well as instructions for opening the cylinder.† In the case of B.L. fuzes a caution is added against *placing the fuzes in a magazine.* This precaution is necessary as they contain detonating composition in the head.

A.C. 1874, Cl. 159.

The cylinders are painted black, the top label is printed in red ink for the 5-seconds fuze; the top of the cylinder itself being also painted red, while the cylinders for the 9 and 20 seconds fuzes have the top of their cylinders painted black and the labels printed in black ink.

The chief defects to be guarded against in fuzes are their causing blind shells or permature explosions.

Causes of blind shell.

The first defect may fairly be attributed to some fault in the manufacture of the fuze or in its preparation,‡ such a mistake as firing a shell without a bursting charge will so rarely occur that it may be neglected. In the description of the M.L. fuzes it will be found that Mark I. fuze will not always act when small charges are used; thus a *blind shell* might be caused by improperly using this fuze in field service. The most frequent source of *blinds* seems to be due to boring the fuze too long. If this is done the shell on striking the earth may shake out the

---

* There is a difficulty in soldering a zinc cylinder so that it may be readily opened; the solder combines so closely with the zinc as almost to form an alloy. In opening, the zinc is apt to be broken.

Fuzes packed in cylinders secured by tape may still be issued to Garrison Artillery and to the Navy, A.C. 1874, C.S. 159. The lids of the cylinders are closed with beeswax and turpentine, and a strip of calico varnished on. The tape or calico is generally painted over.

† The tin strip securing the box is to be pulled sharply from left to right, the bottom of the cylinder resting against the body.

‡ Too short a bit may cause a blind fuze, or the bit not being properly fixed in the hook borer. It is possible to bore so as not to pierce the fuze composition. This is very likely to happen when boring the small mortar or the 20-seconds fuzes with the hook borer, unless care is taken to keep the fuze in the proper position.

fuze; the velocity of the shell being suddenly checked the fuze has naturally a tendency to fly forward; if the practice is over water the fuze may be extinguished by striking the water. It is to be remarked that such a grazing action is very different to direct impact, which, as before stated, will generally make the fuze act.

Practice reports received from out-stations show that *blinds* are sometimes due to fuzes which have deteriorated by keeping, being bored in accordance with the range tables.*

Suppose a fuze which should burn 10 seconds has increased its time of burning to 12 seconds, it is plain that allowance must be made by boring the fuze somewhat shorter than the length laid down. If at practice shells are found to be blind, the first remedy to try is to shorten the length of the fuze. The exact amount of allowance to be given can be readily determined by a "rule of three" sum; thus, suppose that we want to know where to bore a fuze which burns 12 seconds instead of 10, so as to correspond to 6 in the range table, by proportion we find it should be bored at 5. It may often take place that a number of fuzes which burn slowly but regularly may be found at an out-station, and these could be utilized on an emergency by correcting the range table so as to make it suit those fuzes as above indicated.

*Blinds* may occur when using Mark I. B.L. fuze (which has a kamptulicon disc at the bottom) with Shrapnel shell, as the disc sometimes interferes with the action of the primer. This might be remedied by removing the disc before inserting the fuze.

Independent of the fuze, *blinds* have been caused by the use of Mark I. primer in Shrapnel shell. See p. 133.

The chief causes of blind shells may be briefly summed up as follows:—

1. From the time fuze not igniting. This would be likely to take place if a Mark I., M.L. fuze was used with a F.S. gun.
2. From the fuze being bored too long, and being extinguished on graze.
3. From a Mark I. primer being used with Shrapnel shell. See p. .
4. From the primer being covered with any foreign substance, for instance, by rosin which may have worked its way from the interior of the shell.
5. Mark I. 9-seconds B.L. fuzes with a kamptulicon disc at the base, when used with Shrapnel shell.
6. From the hole not being bored through into the composition.†

In the above causes of blinds we have only been considering the time fuzes. When using percussion fuzes a premature or blind may of course be due to any of the causes before mentioned relating to shells or primers, but as no preparation is required with a percussion fuze, exeept the removal of the safety pin, blinds or prematures caused by this fuze could only be due to defective manufacture, or to defects inherent in the fuze, unless the fuze was used with a projectile for which it was not designed.

Causes of premature explosions.

Premature explosions are even more serious than blind shells, as Artillery frequently fire over their own troops; a shell with a heavy bursting charge, exploding in a large gun, is almost certain to disable

* See Extracts, Vol. XI, p. 82.

† A defect in the borer used may possibly cause a blind; in all cases care must be taken to bore till the bit is stopped by the shoulder.

the gun. Unfortunately, *prematures* may be due to many causes, but frequently all the blame is laid on the fuze. It will be convenient here to state all the most probable causes of prematures, both those which belong to the fuze, and to the projectile and bursting charge.

It is generally considered that prematures may occur from shells not being properly filled.*

Captain Hewlett, R.N., reports the premature bursting of four 7″ B.L. common shell which were half filled with 4 lbs. of powder; a shell fired with a bursting charge of 8 lbs. of powder did not burst prematurely.

Proceedings, O.S.C., 1861, p. 450.

Sir W. Armstrong stated that a vacancy, however small, should be avoided, as a tendency to premature explosion would inevitably be the result. The Ordnance Select Committee agreed with Sir William, and an order was issued not to fire shells with less than the authorised charges; the rule now existing is, that shell, except Shrapnel, are to be completely filled.

§ 954.

The great friction due to rotation and the setting back of the powder in an elongated rifled shell renders prematures much more likely in them than in S.B. shells.

Lacquer, necessity for.

Hence the necessity of lacquering the inside of the shell was soon found; at first black lacquer, consisting of equal parts of pitch and asphalte, was used in shells for B.L. guns; the R.L. early recommended the employment of red lacquer (see p. 322), however the black was used for some years on the score of simplicity and economy.

From time to time prematures occurred and were referred to various causes, until an experiment was carried out in the 7″ shunt gun, charge 20 lbs., with a number of shell having black lacquer. Five which were fired, filled and securely plugged, burst; the same happened to 5 fired with time fuzes; the Committee concluded that the fault lay in the lacquer, which was very rough. Red lacquer has since been used in shells, but many B.L. shell exist with black lacquer, which we see may possibly cause a premature.†

Extracts, Vol. I, p. 252; Vol. III, pp. 142, 241; Vol. VIII, p. 158.

It is curious to remark that the frequent prematures were attributed to the fuze, before the experiment showed that they also happened in plugged shells,

Recently attention has been called to the risk of premature explosions by accidents happening when firing common shells from the Woolwich guns. Thus, on the 30th May, 1872, 3 10-inch common shell burst when fired on board H.M.S. "Hercules" with a full charge of "P" powder, the steel tube of one of the guns was cracked, disabling the gun; wood fuzes were thought to be the cause, and their use was ordered to be discontinued with common shells for R.M.L. guns

Extracts, Vol. X, p. 260, 339.

---

* In one year's practice at a foreign station several prematures occurred. An officer was then told off specially to superintend the filling of the shells, and to see that they were properly *filled*. No prematures then occurred. The shells in question were 7-inch B.L. common.

† 50 64-pr. shell were fired without any lacquer; none burst. This experiment does not seem to have been carried farther. Probably much depends on the core being thoroughly removed in manufacture. Sand and iron combined would be just the conditions for an explosion.

Extracts, Vol. III, p. 331.

In 1870, 20 of the worst black lacquered 7-inch B.L. shell in store were fired without a premature. This was considered satisfactory, and the shell were considered serviceable. Extracts, VIII, 159.

Two 7-inch B.L. guns in Madras were rendered unserviceable by the bursting of the shell in the bore. Extracts, Vol. IX, 291. Other instances of prematures will be found in Extracts, Vol. VIII., 37, 158, 160.

of 7-inch and upwards, except when firing a 14lb. charge from the 7-inch gun.*

Experience at Shoeburyness confirmed the risk of prematures, and numerous experiments were carried out to trace their origin. They were found to occur in shells filled with powder and securely plugged, thus showing that the fuze was certainly not the only cause at work. Attention was then turned to the shell; it was thought possible that a spongy texture of the iron might give a passage to the flash, and a number of shell were tested by water pressure. 12 9-inch shells were selected, that stood a ton pressure on the square inch without leaking, and 24 were selected through which the water oozed under this pressure; 9 of the sound shell were fired, and gave two prematures; 12 of the leaky ones gave one premature; all the above shells were filled in the usual way, no bag being used. The remaining 12 leaky shells were fired, having their charges contained in serge bags; there were no prematures; this experiment was confirmed by selecting 20 10-inch shell and 20 9-inch shell, through which the water oozed under a pressure of 1,000 lbs. on the square inch; these shells had their bursting charges contained in serge bags; they were fired with battering charges, and there were no prematures.†

§ 2493.

It was concluded that the prematures arose from no fault in the shells, and the use of bags was recommended and adopted 21/3/73 for all R.M.L. shells of 7-inch calibre and upwards. The above experiments clearly prove that the friction of the powder against the sides of the shell is a frequent cause of prematures; the question still remained whether the time fuzes were not also a source of danger. To test this a number of 9-inch common shells were fired with M.L. time fuzes, the bursting charge being contained in a serge bag; 50 rounds were fired with Mark I. M.L. fuzes, and 50 with experimental fuzes, in which the powder channels are so arranged as to leave a greater thickness of wood outside them (see p. 36). Mark I. gave 4 prematures; the experimental fuzes gave no prematures in the gun, but one occurred about 200 yards off, when using a 5-seconds fuze (bored to act at 2 seconds).‡ In these fuzes the powder channels came much higher up than in the Mark III. fuze, and were not sufficiently protected by the wood of the fuze. See description of the different Marks, p. 36.

The order as to the discontinuance of time fuzes for the Woolwich guns when firing common shell is still in force, and it is well to remember that in firing Shrapnel from a heavy gun it is advisable to use Mark III. fuze, which will eventually supersede the other marks for L.S. See p. 37.

A possible cause of prematures in small shell is their being so full that the time fuze cannot be properly fixed; therefore, in filling these shell, care should be taken to leave space for the fuze. See p. 62.

It has already been pointed out that prematures may be caused by the wood shrinking away from the composition in fuzes where no paper lining is used.

Of course boring the fuze too short will cause a premature.

---

O.S.C., 1869, p. 272.

* There have been several instances of prematures. In 1869, two 9-inch shell exploded prematurely on board the "Royal Alfred" (the Superintendent R.L. points out that when shells burst in the gun the fault is not due to the fuze).

† See Extracts, Vol. X, pp. 53, 55, 134, 261, 262, 263, 336, 339, in support statement in the text.

‡ See Extracts, Vol. XI, p. 58.

To sum up we may divide the probable causes of prematures into two classes.

Prematures due to causes connected with the shell may arise from—

1. Bad lacquer, iron or grit in the shell, or in the R.M.L. shell of 7-inch and upwards from no bag being used.
2. From the shell not being filled.
3. From a weak or defective shell.

The above are probably the leading causes to which prematures are due; it is difficult to overrate the importance of getting rid of them. Our most powerful ships now carry only a few guns, and their armament might be disabled by one or two common shell bursting in the gun. In the land service, artillery must frequently fire over the heads of their own troops, and a few prematures might cause disastrous results.*

The third cause of premature is the one most likely to affect our F.S. Shrapnel shell, where, in order to get a large capacity, the strength is less than that of the common shell; the numbers fired at proof of necessity are small, nor is it possible to test each shell as a gun is tested before issue into the service; the great test is from the number of shell fired at practice; it is desirable that every premature should be reported to the Royal Laboratory, and if possible traced to its source.

Prematures due to the fuze may arise from—

1. A fuze improperly bored.
2. A fuze not home (as when too large a bursting charge is used).†
3. A fuze which is too high in gauge, so as to throw the side holes above the bush.‡
4. A fuze without a paper lining where the wood has shrunk away from the composition.
5. The powder channels coming high up, and not being sufficiently protected by the wood of the fuze when very heavy charges are fired; this defect has been remedied in Mark III. fuze.
6. If a fuze be low in gauge it *may* set back so as to cover the fire-holes, and so cause a premature.

---

* Not only by actually killing and wounding our own men, but more especially from the moral effect produced. Nothing demoralises troops more than being exposed to the fire of their own artillery.

† Nos. 1 and 2 are not strictly due to the fuze, but to the unskilfulness or want of care on the part of the persons using them.

‡ Fuzes of the common gauge are sometimes so enlarged by damp swelling the wood, as to be unserviceable from this cause. This never occurs with fuzes of the G.S. gauge.

# CHAPTER IV.—TIME FUZES.

LARGE MORTAR FUZE.—SMALL MORTAR FUZE.—DIAPHRAGM SHRAPNEL FUZE.—COMMON FUZE.—M. L. FUZES, 5, 9, AND 20 SECONDS.—B. L. FUZES, 5, 9, AND 20 SECONDS.—RULES AS TO LENGTH OF FUZE AT VARIOUS RANGES.—ARMSTRONG E. TIME FUZE.—HAND GRENADE FUZE.—PARACHUTE FUZES.

Large mortar fuze.

L. MORTAR.

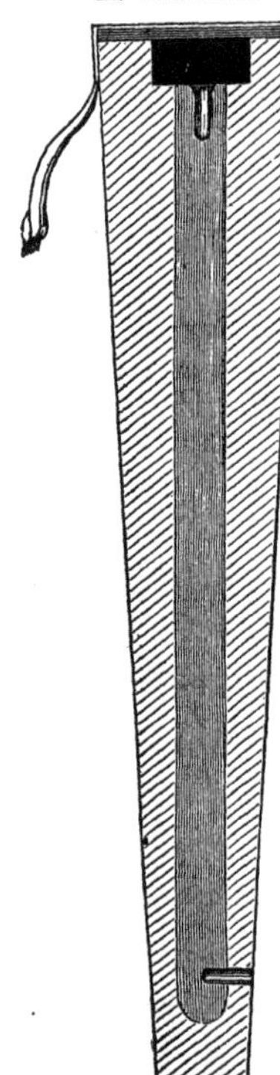

The *large mortar fuze* is used with 8″, 10″, and 13″ mortar shells, its construction is shown by the figure (see cut), the head is protected by a pure tin cap and disc of pasteboard which are removed by means of the tape when the shell is placed in the mortar. Ignition is secured by means of quick-match priming, and also by the hole in the top of the fuze composition. The composition is driven in a channel bored in the centre of the wooden cone; its length being 6 inches, the fuze burns 30 seconds. The figures on the fuze refer to the inches of composition, but, as before pointed out, by adding a cypher they will refer to the general half-second unit; only five divisions are marked to the inch, so the fuze reads to seconds, there are no holes bored at the marks on the fuze, but only indentations. The fuze is bored through at 6 inches (in future manufacture a pellet will be inserted as in the 20-seconds fuzes, p. 38), so that it will act without boring at this length: the first hole marked for boring is at 2″.

The first ring on this fuze serves to mark the depth it will enter the fuze hole of 13″ and 10″ mortar shell, below this at a distance of ·9″ another ring marks where the fuze is gripped by 8″ shell. The first hole will only act in the 10″ shell, as it falls against the metal in the 13″ and 8″ shells. For preparation, and implements required in preparation, of this fuze, see p. 67.

Action.

The fuze is ignited by the flash of the discharge and burns till the flame reaches a bored hole, through which it passes and explodes the shell.

Issue.

Packed in whole size metal-lined case holding 330.

S. MORTAR.

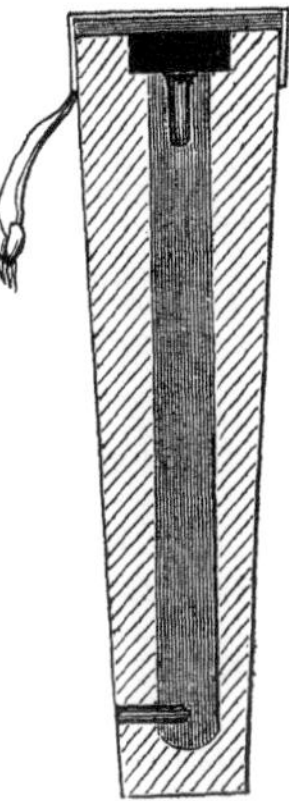

Small mortar fuze.

*Small Mortar Fuze*, used with 24-pr. and 12-pr. common shell fired from $5\frac{1}{2}''$ royal, and $4\frac{2}{5}''$ coehorn mortars respectively at long ranges (see cut). It is of the same gauge as the common fuze but longer, containing 3 inches of fuze composition burning 15 seconds; its marking and construction are similar to the large mortar fuze, therefore the intervals between the holes correspond to one second of time in burning, the first mark for boring is at 1 inch.

When used with the 12-pr. shell a piece of rag, paper, &c., must be wrapped round the fuze to make it fit.

Issue.

1,000 in metal-lined case, but in future will be issued in tin cylinders holding five fuzes.

Diaphragm Shrapnel fuze.

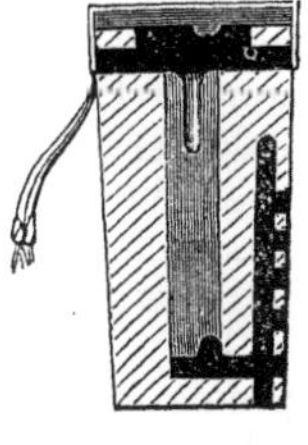

*Diaphragm Fuze*, used for diaphragm Shrapnel shell. The arrangement for protecting the head, and for the ignition is similar to that given above. Four holes are bored through the top of the fuze to secure the quick match. The channel for the fuze composition is bored eccentrically so as to give space for two powder channels, one of which is shown in the section (see cut). The main channel is driven with fuze composition, the powder channels are filled with pistol powder, and the powder in each channel is supported by a piece of quickmatch passing through the lowest hole, by means of which the flame explodes the shell if the fuze is not bored. A groove cut in the bottom of the fuze with a piece of quickmatch laid in it unites the two powder channels. Holes are bored through into the powder channels, filled with powder, and protected by clay being pressed upon them. They are covered externally by varnished paper.

The powder channels are useful, because (1) in case of the hole bored coming in contact with the metal of the shell they carry the flash to the charge; (2) by having the odd numbers marked on one channel and the even on the other they allow smaller divisions to be marked, reading to half-seconds; (3) because with Shrapnel shell a powerful flash is necessary to ensure the powder igniting, owing to the construction of the shell.

The length of fuze composition is one inch, the fuze burns 5 seconds, and it is marked on the side channels up to 10 half-seconds.

For preparation, and implements required in preparation, of this fuze, see p. 64.

Action.

The flame from the fuze composition ignites the powder in one of the channels at the point where the fuze has been bored, which explodes, giving a strong flash through the bottom holes, and both channels explode together as they are connected by the quickmatch at bottom.

Issue.

See page 28.

Common fuze.

*Common Fuze*, used with S.B. common shell * (see cut). The con-

---

§ 2097.

* The 5, 10, and 15 seconds fuzes, which were issued specially for the 7-pr. gun, may be used with S.B. common shells, as they are obsolete for the 7-pr., which has now the G.S. gauge. They resemble Mark II. M.L. fuzes, but are of the common gauge.

COMMON.

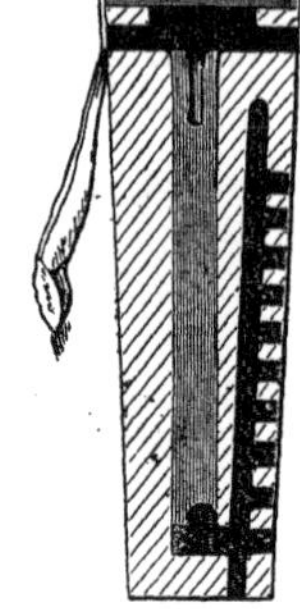

struction is the same as the diaphragm fuze, except that the powder channels are not connected at the bottom by quickmatch, but are stopped with shellac putty. It contains 2 inches of composition and burns 10 seconds; the preparation and action are identical with the diaphragm fuze. This fuze would be used instead of the small mortar fuze, when 12-pr. and 24-pr. shells are fired at short ranges out of the $4\frac{2}{5}$ and $5\frac{1}{2}$-inch mortars. The common fuze may also be used with 100-pr. diaphragm Shrapnel shell.

Issue.

See page 28.

M.L. fuzes.

It is important to have a clear idea of the difference between the various Marks of M.L. fuzes as they cannot be used indiscriminately. The difference between Marks I. and II. is in the amount of priming and in the manner of protecting it. The priming in Mark I. was found not to ignite with certainty when small charges were used, therefore Mark II. was introduced, having an increased quantity of priming wrapped round the groove in the neck of the fuze, this caused a projection, over which the copper strip used in Mark I. could not be placed,* so Mark II. has the priming covered with a tape band only, and is not so well protected against risk from fire, when placed in the shell, as Mark I.; it has, however, the advantage of acting with small charges, such as are used with field guns. It was obviously an inconvenience to have two fuzes, one for field service, and the other for garrison service, therefore Mark III. has been introduced, which will ultimately become the only pattern for land service.

§ 2485.
§ 2622.

Several important changes have been introduced in Mark III.

They are more certain of ignition, will be less likely to cause prematures, and stand climate better than the previous patterns, from which they differ as follows:—

1. The head projects a little farther from the shell,† has a larger groove containing the quickmatch priming; and differs from II. in having a copper band‡ protecting the priming.
2. The paper lining is reduced by one-half its thickness, and is coated with varnish. It is hoped that this may prevent the fuzes from deteriorating, as Marks I. and II. are liable to do in keeping.
3. The powder channels are brought nearer to the centre of the fuze, and are slightly reduced in length; thus protecting the powder from the chance of being ignited by the discharge of the gun. The side holes are not bored beyond the powder channel. The bottom side hole is numbered.

The changes in the 20-seconds fuze are similar to those in the short-range fuzes, excepting of course the powder channels.

Regulations for issue.

The following are the regulations for the issue of the various Marks of M.L. fuzes.

---

* The copper strip used in Mark I. fuzes is ·3″ broad, the one used with Mark III. fuzes is ·42″. Possibly this increased width would have enabled it to be used with the Mark II. fuzes.

† The length of 5-seconds and 9-seconds fuzes is increased by ·15″ at the top, that of the 20-seconds fuze by ·1″. The diameter of the head of the 20-seconds fuze is increased by ·021″. The increase of length facilities ignition and makes the fuze easier to uncap.

‡ The protecting band is of copper, coated with india-rubber solution, between two thicknesses of tape.

Fuzes of Mark I. (§§ 1417–1953) will be retained in the service, and supplied to the Navy for use with all rifled M.L. guns, other than boat and field guns; and also, so long as these fuzes are available in the district, for land service for the 64-pr. and 80-pr. R.M.L. guns only.

§ 2485.
§ 2622.

Fuzes of Mark II. (§§ 2064–2071) will, until the store is exhausted, be supplied for land service, and for naval service for boat and field guns.

Mark III. will be the only fuze manufactured in future for L.S.*

§§ 2485, 2622.

Marks II. and III. are to be supplied to the Navy when Mark I. is not available.

§ 2707.

*5-seconds M.L. Fuze.*—For shells with which this fuze is used, see table, p. 58. For construction and general arrangements see plate, p. 351.

5-seconds M.L. fuze.
§§ 1953, 2064, 2484.

They resemble those of the common fuze, but the head is closed by a gun-metal plug, round the pin of which quickmatch is looped and led through two fire holes to a groove.

This arrangement of the head obliges the fuze to be longer than the common fuze.

A paper lining is introduced to prevent the formation of a space between the wood and the composition in the event of the wood shrinking, which would cause the fuze to act prematurely.

The clay stopping in the side holes is dispensed with, and varnished paper alone covers them and the powder channels, which are united by quick match at the bottom of the fuze.

The fuze is driven with mealed powder, which, as it burns at twice the rate of fuze composition, allows the fuze to be graduated, so that the interval between two consecutive holes corresponds to quarter-seconds time of burning. The marking of the fuze commences at 1, and the side holes are numbered 1, 1·5, &c., thus enabling the fuze to be bored to quarter-seconds; the integral figure referring as in other fuzes to half-seconds.

It is necessary to be able to burst F.S. Shrapnel shell with great accuracy to develope their power, hence the advantage of this fuze for F.S. See p. 159.

*9-seconds M.L. Fuze.*—For shells with which this fuze is used, see table, p. 58. For construction and general arrangements, see plate, p. 351.

9-seconds M.L. fuze.
§§ 1236, 2064, 2622.

This fuze, though called a 9-seconds fuze, will when at rest burn 10 seconds. It contains 1″·8 fuze composition, above which is driven a pellet ·4″ long of mealed powder (equal in time of burning to ·2″ fuze composition). This is done to obviate the risk of cracking the composition, when boring for short ranges, a result liable to occur when there is only ·2″ of composition over the top side hole. The construction and size of this fuze are identical with those of the 5-seconds fuze, except that fuze composition is used instead of mealed powder, and consequently the side holes are marked to half-seconds only, as shown in the plate, p. 351.

---

* This also applies to S.S. for boat and field guns.

§§ 2485, 2622.

20-seconds M.L. fuze.

*20-seconds M.L. Fuze.*—For shells with which this fuze is used, see table, p. 58. For construction and general arrangements, see cut.

§§ 1417, 2071, 2485.

20-Sec. M.L. Mark I.

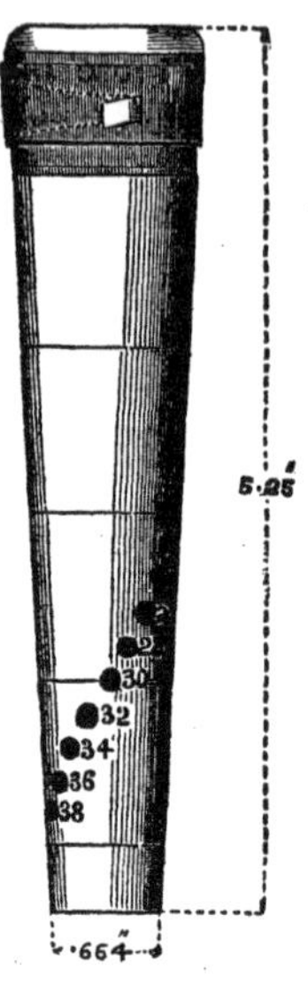


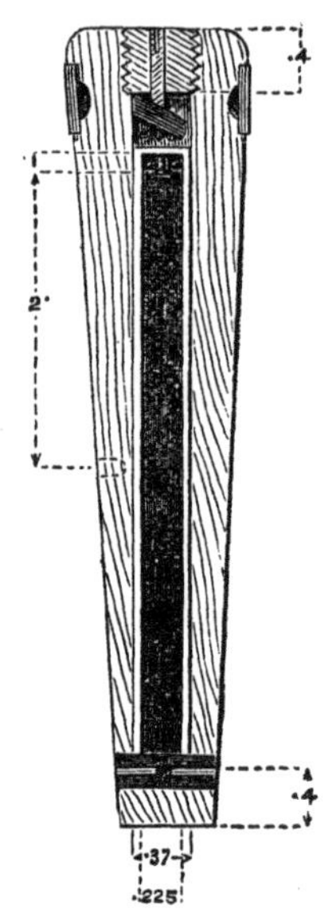


It has 4 inches of fuze composition, on the top of which is ·15″ (in Marks I. and II. ·1″) of mealed powder, through which the small hole to ensure ignition is bored.* Its general construction and action resemble those of a mortar fuze, as it has no powder channels; but the arrangements as to priming, paper lining, &c., are the same as in other M.L. fuzes. It has a pellet of mealed powder, pierced, as shown in the section, to carry the flash from the bottom of the fuze into the bursting charge of the shell.

The marking begins at 20, and reads to even half-seconds only, as it runs 20, 22, &c.

Action of the M.L. fuzes.

When the fuze composition becomes ignited, it burns out of the two fire holes provided for the purpose, in other respects the action resembles that of the diaphragm fuze (p. 35), except the 20-seconds, which acts like the large mortar fuze (p. 34).

B.L. fuzes.

The B.L. fuzes resemble the M.L. Marks I. and II., except in the construction of the head.

As there is no windage in B.L. guns, their fuzes have to be ignited by a detonating arrangement which is described below. There is only one Mark of the 5-seconds fuze, but of the 9 and 20-seconds fuzes there are two. Mark I. 9 and 20-seconds had no safety pin, but had a kamptulicon disc on the top and bottom, so as to lessen the chance of their exploding if accidentally struck. Mark II. has a safety pin as described below; † the two patterns are interchangeable, but Mark II. has a

---

* There is no necessity for a ·4″ length of mealed powder in this fuze, as the marking only commences at 2″, *i.e.*, 20 on the fuze.

† The safety pin is made of copper. Brass is employed in the R.L. B.L. plain and sensitive percussion fuzes for this purpose, and it is probably preferable, see p. 45.

decided advantage as to safety. As before pointed out, Mark I. 9-second fuze may cause a blind shell when used with Shrapnel (see p. 30). There is a difficulty in fixing these fuzes properly; they are directed to be screwed in by hand and not hammered (see p. 65). No doubt it would be dangerous to hammer in Mark I., and it is contrary to regulation to do so with either pattern.

5-seconds B.L. fuze. § 1984.

5-*seconds B.L. Fuze.*—For shells with which this fuze is used see table, p. 58). For construction and general arrangement, see cut. They are similar to those of the 5-seconds M.L. Marks I. or II., except as regards the arrangement of the head. As there is no windage, the fuze has to be ignited by a detonator. A cylinder, of an alloy* resembling gun-metal, screws into the head of the fuze; this cylinder contains a hammer, supported by a copper wire, below the hammer is a hollow in the cylinder containing a detonating composition, viz., chlorate of potash, 6 parts; fulminate of mercury, 4 parts; sulphide of antimony, 4 parts. A hole is bored through the cylinder for the passage of the flash. A copper safety pin passes through the head of the fuze between the hammer and the detonating composition, so that the fuze cannot be accidentally fired, even if the suspending wire be accidentally sheared. This pin is withdrawn by the braid just before placing the shell in the gun.

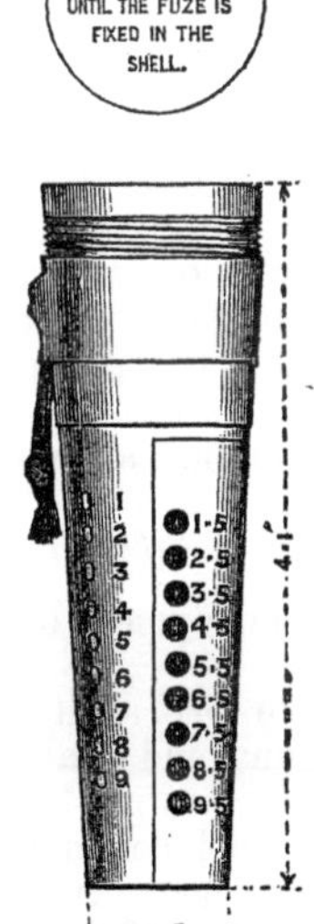


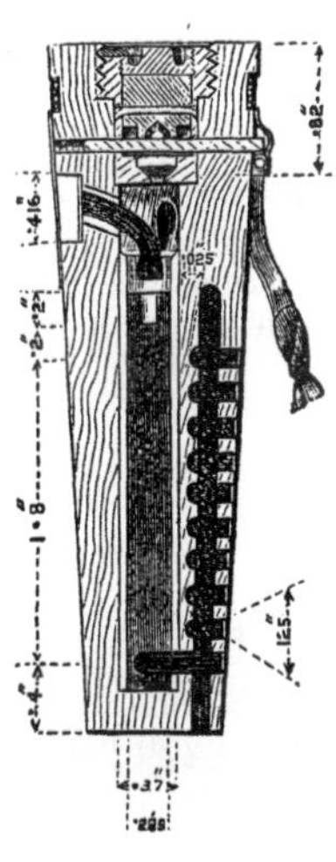


There are three escape holes, one of which is shown in the section, to allow of the escape of gas. These holes are protected by thin copper discs and papier mâché wads, which are forced out by the gas when the fuze is lighted; quickmatch leads up to these holes. The exterior of the head is woolded with copper wire, soldered over, to keep it from splitting when the detonator is being screwed in.

---

* For details of alloy, see p. 332.

The ignition is produced by the hammer setting back on the shock of discharge, and thereby shearing the copper wire and falling on the detonating composition; the latter is thus exploded and the fuze ignited.

9-seconds B.L. fuze, Mark II., § 1999.

*9-seconds B. L. Fuze.*—For shells with which this fuze is used, see table, p. 58. The construction is identical with that of the 9-seconds M.L. fuze, Marks I. or II., except the arrangements in the head, which are the same as those described above for the 5-seconds B.L. fuze.

20-seconds B.L. fuze, Mark II., § 1999.

*20-seconds B.L. Fuze.*—For shells with which this fuze is used, see table, p. 58. The construction is identical with that of the 20-seconds M.L. fuze, Marks I. or II., except the arrangements in the head, which are the same as those described above for the 5-seconds B.L. fuze.

9 and 20 seconds B.L. fuzes, Mark I.

For Mark I. of the 9 and 20-second B.L. fuzes, see §§ 873, 1417, and p. 38. If Mark I. 9-second fuze be used with Shrapnel shell, the kamptulicon disc should be first removed; otherwise, as previously pointed out, the shell may be blind.

N.B.—All the B.L. fuzes are rather longer than the corresponding M.L. fuzes, owing to the room taken up by the detonating arrangement.

*Short Rule for getting the Length of Fuze.*

Divide the number of hundreds of yards in the range by 2, and add 1 up to 1,000 yards, 2 up to 2,000 yards, and so on for length of fuze in tenths of inches; this will be found nearly correct with rifled ordnance. Example: to find length of fuze at 2,600 yards for 16-pr. gun, $\frac{26}{2}+3=16$. Shrapnel require to be bored a little shorter.

*Short Rule for getting Length of Fuze for Mortars.*

Add 17 to the number of hundreds of yards in the range for the length of fuze in tenths of inches, thus the fuze for 1,700 yards will be $3''\cdot 4$.

It must be remembered that all the lengths of fuze given by rules or by range tables must be taken as approximations only to be corrected by practice. As before pointed out (p. 27), the age of the fuze, the height of the barometer, and a variety of other causes affect the rate of burning. When blind shell occurs, a shorter fuze should be tried.

*Rules for use of the various Fuzes for Rifled Guns.**

The following are general rules for the employment of time fuzes for rifled ordnance:—

1st. The 5-seconds fuze is for use with Shrapnel shell for ranges up to about 1,700 yards with R.M.L. guns, and to about 1,600 yards with B.L. guns.

2nd. The 9-seconds fuze is used for Shrapnel, common, 7″ and 7-pr. double shell, and garrison segment shell. It is available for ranges up to about 2,800 yards with R.B.L. guns, and up to about 3,100 with R.M.L. guns.

3rd. The 20-seconds fuze is for garrison common shell and for garrison segment shell at long ranges, and for the 7-pr. double shell. It is not available for Shrapnel, because there are no powder channels to conduct the flame to the primer.

---

* This gives *general* rules only, for details, see table p. 58. It is to be remembered that time fuzes are not used with the Woolwich guns in the L.S.; and in the S.S., only with Shrapnel; and with common and double shell from the 7-inch gun with a 14lbs. charge.

In all cases the nature of the gun, whether B.L. or M.L., must be considered. On an emergency, B.L. fuzes might be used with R.M.L. guns.

## Armstrong "E." Time Fuze.

*Armstrong "E." Time Fuze.*—It is only used by the Navy with B.L. segment shells, F.S. It was withdrawn from the Royal Artillery in June, 1870. § 1907.

It represents a large class of time fuzes employed by continental nations with Shrapnel shell.*

The construction is complicated, and its cost about double that of the Boxer B.L. time fuzes, but it has several important advantages, specially as a fuze for Shrapnel. It can be set to very small intervals, a point of the greatest importance with Shrapnel shell, it can be altered again after setting, and it is open to inspection, so that the officer or the No. 1 of a gun can see that it is correct, instead of depending on those employed in preparing shells at the limber or in the shell room.

It is not necessary to give the various patterns which have been introduced, or to dwell on the various changes of manufacture, as only one nature of fuze is now issued to the service, and may be known by the word "cap" stamped upon the base of the fuze. Various marks may be found in combination with this word, as old fuzes are repaired by having cap composition substituted for the amorphous phosphorous composition which did not stand exposure to climate and deteriorated by keeping.† § 2178.

The fuze being complicated in its construction demands very careful manufacture, and in this respect compares unfavourably with the Boxer fuze, as there are more sources of failure.

Many defects existed in the early patterns,‡ and so brought the fuze into disrepute. The chief faults are:—

(1.) The fuze occasionally fired when carried in the limbers. This was due to the pellet containing the detonating composition being supported by lead feathers, which gave way under the jolting motion of the limbers. This has been remedied by using a cup-shaped support of thin brass, as shown below. (The cup was proposed by Col. Freeth, R.A.)

(2.) The fuze sometimes failed to ignite when the detonating arrangement fired. This was due to the hard surface of the fuze composition; the flash from the pellet failed to light it with certainty. This has been overcome by boring a small

---

* A full account of the fuzes used by European nations up to 1871 will be found in *Recherches théoriques et pratiques sur les Fusées, par H. Romberg.* The gradual development of the Armstrong class of fuzes can be traced there, starting from the Bormann and Breithaupt fuzes, to those used by the Prussians, Austrians, and Swiss. There is also an interesting appendix to the above work, entitled, *Fusées Prussiennes, modifications proposées, par H. Romberg.* Both works are published by H. Merzbach, Brussels.

† This composition consisted of amorphous phosphorus (with 10 per cent. calcined magnesia) 8 grs., chlorate of potash, 16 grs., ground glass, 6 grs. The deterioration was caused by the effect of moisture, which caused the phosphorous to oxidise at the expense of the chlorate of potash, thereby injuring or destroying, according to circumstances, the detonating character of the composition.

‡ Much information will be found in a book by Col. Wray, C.B., R.A., giving the "Changes of pattern and modifications in the Armstrong time and F.S. percussion fuzes since their introduction in 1860," printed at the D.G.O. Office, Woolwich, in 1869; also see §§ 90, 609, 1,294, 1,472, 2,178 2,496, Changes in War Stores.

hole in the ring of fuze composition close to the channel which conducts the flash.

(3.) The fuze sometimes became unserviceable from the phosphorous composition deteriorating. This has been overcome, as pointed out before, by using cap composition.*

No doubt a prejudice was created by the bad results obtained when using this fuze with segment shell, but these shell should be used with a percussion fuze; no time fuze will give good results with segment shell (see p. 148).

"E." time fuze, Mark III. §§ 1472, 1790, 1791, 2178, 2496.

The following is a short description of the fuze at present issued to the Navy:—

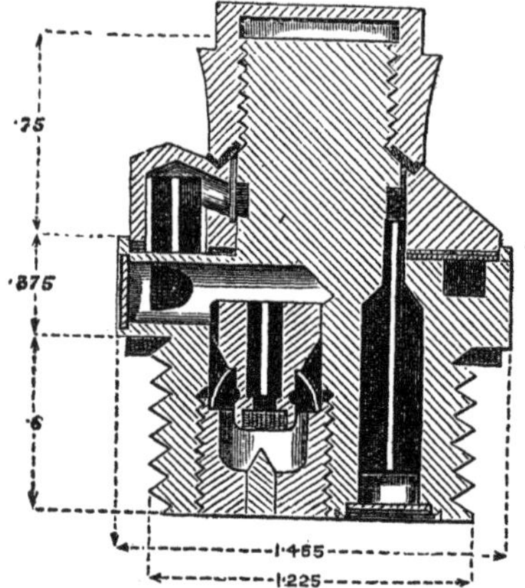


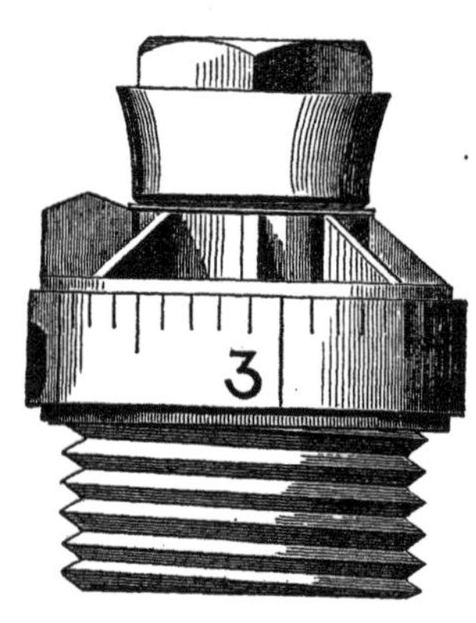


Both body and nut of the last pattern (E. III.) are made of gun-metal, and the graduations for length of fuze in inches and tenths are marked on the metal rim instead of on paper, as in former patterns. The pellet which is supported by a brass cup is filled with R.F.G. powder, secured by thin paper fastened on its base; the detonator in the head consists of cap composition (fulminate of mercury, chlorate of potash, and sulphide of antimony), instead of the amorphous phosphorous composition which deteriorated in damp climates. In those lately made a disc of brass, ·001″ thick, covers the detonating composition. The word "cap" is stamped on the base of the fuze.

§ 2178.

The channel by which the flash from the pellet reaches the ring of fuze composition is enlarged in this pattern, and a strand of quickmatch is placed in it; a little hole is bored in the ring of fuze composition to ensure its lighting. The fuze composition is pit mealed powder pressed into a ring or groove which runs round close to the exterior of the fuze body; this composition burns at the rate of 1 inch in two seconds, and, owing to a metal stop, can only burn in one direction, *i.e.*, from left to right.

A leather washer and movable gun-metal collar cover the ring of composition. At one part of the collar a channel, primed with mealed powder driven and pierced, and marked on the outside with an arrow, communicates with a groove round the neck of the fuze, which contains mealed powder; this groove is connected by a channel with the blowing

---

* An account of an extended trial of the old pattern of "E." time fuze will be found in Extracts, Vol. VIII., p. 183.

chamber, which is primed with mealed powder, driven and pierced; a small brass disc closes the chamber.

The movable collar is kept in its place by a nut which screws on to the neck. The body has a small hole in the side to fit a projection in the Armstrong key used in screwing in the fuze. See p. 65.

Preparation.

Stress must be laid on the importance of screwing the nut tightly home when the fuze is adjusted, otherwise the washer will not be tightly pressed down on the ring of fuze composition, and a premature may occur.

Action.

On firing the gun, the brass cup is crushed in, the pellet strikes the needle, which explodes the detonating composition, the ring of fuze composition is ignited by the flash and burns till it comes to the channel marked by the arrow head, leading to the groove in the neck primed with mealed powder, the flash is then instantaneously conveyed into the blowing chamber, and thence into the shell.

The changes recently introduced, particularly the cap composition, and the ensuring ignition by piercing a hole in the ring of fuze composition, have greatly improved this fuze.

Short rule for finding length of fuze.

Divide number of hundreds of yards in range by 6 for length in inches, thus for 1,200 yards, length of fuze = 2 inches.

Issue.

One in a waterproof bag placed in a cylindrical tin box wrapped in brown paper, 72 boxes in a deal case, placed on the sides or heads, the bottom of each tin box is marked "top" to prevent it being placed downwards. Since June, 1875, the tin boxes containing "E." time fuzes, primed with cap composition, are stamped "cap," and the same word is stencilled on the waterproof bag. Since July, 1875, in the case of repaired fuzes the date of repair will be labelled on the boxes containing them.

§ 2751.

§ 2761.

An "F" time and percussion fuze was introduced in 1867, but was not a success, as the percussion action often failed. It was ordered in February, 1869, to be regarded simply as a time fuze, and in June, 1869, the manufacture was discontinued, the "E" time fuze, Mark III., previously approved for India, being then brought in for general service.

§§ 1473, 1566, 1720, 1790.

## *Special Fuzes.*

Hand grenade fuze.

*Hand Grenade Fuze.*—Used with the 3-pr. or 6-pr. hand grenades. They are readily known from all other fuzes by their small diameter, absence of paint, and by the head being secured only by a paper cap. They contain 1·5 inches of fuze composition, which burns 7·5 seconds; the composition is unsupported at the base.*

Preparation.

No boring is required; the fuze must be firmly fixed in the grenade, uncapped, and lighted by a portfire.

They are now rarely demanded. When men are using them they should be cautioned not to retain the grenade too long in their hands.

Issue.

1,200 in a whole metal-lined case.

Fuzes for parachute lights.

These fuzes in their general construction, preparation, and action, resemble the common fuze. The figures, however, give the time of burning in seconds instead of half-seconds; a paper label is pasted above the row of figures with the words "time of burning," and the figures have "″" above them indicating seconds. The fuzes are painted blue. Each nature of parachute light has a special fuze.

---

* As the grenades are thrown by hand, support at the base is unnecessary.

10-inch parachute fuze.

*10-inch Parachute Fuze.*—Has three inches of ordinary fuze composition and burns 15 seconds. The first side hole is numbered 6, indicating 6 seconds; the second row of side holes give the half-seconds, beginning at 6·5; the last hole is numbered 15. The fuze is about four inches long.

8-inch parachute fuze.

*8-inch Parachute Fuze.*—Has 2 inches of slow burning composition, which burns 13 seconds. The marking is peculiar, the first side hole is numbered 7, indicating 7 seconds, the top side hole of the second row is numbered 7·75, the next hole in the first row is numbered 8·5, and so on, the fuse reading to intervals of $\frac{3}{4}$ second, the last hole is marked 13. The fuze is about three inches long.*

5½-inch parachute fuze.

*5½-inch Parachute Fuze.*—Has 1·5 inches of slow burning composition, which burns 10 seconds. The first side hole is numbered 4, indicating 4 seconds; the succeeding holes are numbered 4·75, &c.; the last side hole is numbered 10. This fuze is about $2\frac{1}{2}$ inches long.†

Fuze for life-saving rocket.

*Fuze for life-saving rocket.* See p. 282.

The following are the limits of time of burning allowed in manufacture. These must not be confounded with the limits employed in testing fuzes at out-stations given in Instructions for Proof, p. 307.

| | Time of burning. |
|---|---|
| Large mortar fuze .. .. .. | 30 to 32 seconds. |
| Small " " .. .. .. | 15 to 16 " |
| Common fuze .. .. .. .. | 9·7 to 10·3 " |
| Diaphragm Shrapnel fuze .. .. | 4·8 to 5·2 " |
| 5-seconds M.L. and B.L. fuze .. | 4·8 to 5·2 " |
| 9 " " " .. | 9·7 to 10·3 " |
| 20 " " " .. | 19·4 to 20·6 " |
| Hand-grenade fuze .. .. .. | 7·5 to 8·0 " |
| Parachute light, 10-inch fuze .. | 14·4 to 15·6 " |
| " " 8 " " .. | 12·5 to 13·5 " |
| " " 5½ " " .. | 9·7 to 10·3 " |

The reason why no minus limits are allowed for the mortar fuzes is because the shells are fired with a low velocity and contain large charges; hence a fuze burning at all short is to be avoided, as pieces of the shell might fly back to the battery; a minus limit would evidently be most objectional in a hand-grenade fuze, so none is allowed.

---

* The 10 and 8-inch parachute fuzes may on any emergency be used with common shells for S.B. guns.

† This fuze would not be used with S.B. common shell as its diameter is too small.

Manby's fuze.

*Manby's Fuze.*—Is obsolete, as the Manby's shot has been superseded by the life-saving rocket. Their use was to give light so as to allow the course of the shot to be seen at night when firing over a wreck. They burned 12½ seconds.

# CHAPTER V.—PERCUSSION FUZES.

GENERAL REMARKS.—PETTMAN L.S. AND G.S.—B.L. PLAIN, R.L. FUZES, AND SENSITIVE FUZE.

General remarks.

Wood, though suitable for time fuzes, cannot be used for percussion fuzes, where great nicety of fit is necessary, and where even a trifling alteration of form would prevent the fuze from acting.*

The metal employed must not be liable to oxidise readily; iron was at first tried and proved to be quite unsuitable. The so-called gun-metal, consisting of copper and tin, to which a little zinc has been added† (see p. 332), answers well for the body.‡ Copper wire has been found suitable for suspending any part of the fuze which has to "set back" on the shock of the discharge, while brass wire, being less soft and more elastic, is suitable for safety pins.§

Lead answers well for checking rebounds, owing to its softness and lack of elasticity, but these qualities render it unfit to be the support to the internal parts of a fuze, at least when it has to sustain the shock caused by the jolting motion of a limber; it answers however as "feathers" or projections destined to be sheared on the shock of discharge.

When percussion fuzes are employed in rifled shells which strike point first, a very simple construction answers, but when shell may strike in any accidental position, the problem is more difficult, as may be exemplified by the construction of the obsolete Moorsom fuze, which had no less than 3 hammers and 5 patches of detonating composition. The difficulty has been successfully met in the Pettman fuzes.

Percussion fuzes are employed for two distinct purposes; they are

* Freeburn's concussion fuze was made of wood. The principle is simple and ingenious. The fuze contains a central channel, the lower portion of which is driven with fuze composition, the upper with mealed powder. Three holes closed with three conical plugs with the large part of the cone towards the interior of the fuze lead into the portion of the fuze driven with mealed powder. On firing, the conical plugs are supported by the mealed powder, this burns away rapidly, and when the shell strikes the object the plugs being unsupported fall into the fuze, the flame from the fuze composition passes through the holes and explodes the shell. See Extracts, Vol. VIII, pp. 85, 153, 348.

† Adding a little zinc or lead renders the metal easier to turn.

‡ It is essential that the bodies should be strong, as otherwise the fuze will fail with even moderately heavy charges. Hence the introduction of Mark II. R.L. percussion fuze. See p. 55.

§ Wire is not suitable when a "pellet" is desired to set back on the shock of discharge, and to move forward again on striking an obstacle, as the broken ends of the wire are liable to form burrs and check the action.

used in shells intended exclusively to act against solid obstacles, such as earthworks, brick, or moderately thin stone walls, or wooden ships; and they are also used in shells employed against troops in the field, as well as in shells directed against buildings and material generally.

For the first purpose it is desirable that the fuze should only act on direct impact, and a very instantaneous action is not required. It will be seen (p. 48) that the Pettman G.S. percussion fuze is specially designed to act only on direct impact.

For the second purpose, when used against troops in the open, it is necessary that the fuze should act on graze, and that it should act almost instantaneously, as otherwise the shell has time to rise to a considerable height before bursting, and thus its effect against troops is diminished. Such an action is secured in the R.L. and B.L. plain percussion fuzes and in the Sensitive fuze. This nature of fuze is absolutely necessary to develope the effect of segment shell when used against troops in the field, and good results have been obtained from Shrapnel when using such a fuze at moderate ranges.

Safety pins are used in the fuzes which act on graze. They serve to protect the "feathers" in the fuze by taking the shock caused by the jolting motion of the limbers. The "feathers" only come into use when the safety pin is withdrawn, guarding against risk while the shell is rammed home.

It is important to employ a detonating composition which keeps well; the earlier percussion fuzes frequently failed in this respect. Experience has proved that cap composition (p. 320), properly pressed and varnished, resists climate well. The most trying climate is one where great changes of temperature are to be met with in combination with hot damp seasons; parts of India are very trying in this way. The changes of temperature are likely to open the joints of the fuze by alternate expansion and contraction, while the warm damp air is sure to penetrate it, if it finds the least opening. Even in such climates the ordinary percussion caps are found to keep well; this is due probably more to the mechanical condition of the composition, than to any speciality in the composition itself. A very heavy pressure is applied to the composition when placed in the cap, thus forming a hard polished surface not liable to absorb moisture: in addition the composition is protected by varnish.

In all fuzes where pressure can be applied to the detonating composition this system is carried out, but it cannot be done in all cases, and where it is impossible, the greatest care must be taken to close the fuze hermetically. This is illustrated in the construction of the Pettman G.S. fuze.

All percussion fuzes in the service at present (except the B.L. plain, which fits inside the shell), are tapped with a screw thread to fit into the conical fuze hole of the shells.* A fuze having its screwed portion conical can be screwed home much more rapidly than when it is of a cylindrical form, as the fuze will enter some distance into the conical fuze hole before the screw bites. There is no necessity for a shoulder, as a conical screw can not be screwed too far home.

Percussion fuzes are useful in the absence of range finders when firing trial shell, It is difficult when time fuzes are used to judge exactly whether the shells are bursting under or beyond the object, but when

---

* A number of obsolete fuzes, such as the Moorsom, 7½ and 20-seconds metal fuzes, pillar fuzes, and Pettman S.S. fuze, may readily be distinguished by their being tapped to fit the cylindrical Moorsom gauge.

the shells are burst on the ground it is easy to determine on which side of the object they are bursting.

Percussion fuzes are useful when firing at artillery where a local action is desirable; they are less effective than time fuzes against an open formation of troops.

A great advantage of this class of fuze is that they require no preparation, beyond withdrawing the safety pin in those fuzes which act on graze. The safety pin of the sensitive fuze is not withdrawn, but the fuze is uncapped.

For the Navy, a proportion of the shell are issued filled, and fuzed with the Pettman G.S. fuze. Shell thus fuzed are very safe, and not likely to be injured by an accidental explosion in their neighbourhood. See p. 132.

The following brief account of the percussion fuzes in the service will be sufficient, in connection with the drawings, to explain their action:—

*Pettman Percussion Fuze, L.S.*

Pettman percussion fuze, L.S. § 414.

*Pettman Percussion Fuze*, L.S., used with common shell of common gauge. The fuze hole must be tapped throughout to receive this fuze; this is indicated by a cross cut on the plug.*

The body of the fuze is made of gun-metal; the ball, cone plug, and steady plug, of a harder alloy to strengthen them; the remaining parts are gun-metal, except the lead cup.

The section shows the construction.

A. Body.
B. Top Plug.
C. Steady Plug.
D. Detonating Ball.
E. Cone Plug.
F. Lead Cup.
G. Bottom Plug.

A strand of quickmatch closes the fire hole in the bottom plug.

No preparation is wanting with this fuze, which explodes the shell on its striking a hard body, such as a wooden ship, walls, &c. It is not intended to act on graze.

The ball is the only part of the construction which is not quite explained by the sketch; it is roughened by vertical grooves and has a horizontal groove as well; it is coated with a detonating composition composed of chlorate of potash, 6 oz.; sulphide of antimony, 6 oz.; sulphur sublimed, $\frac{1}{2}$ oz.; mealed powder, $\frac{1}{2}$ oz.; made into paste with methylated spirit and shellac. Varnished gut is tied over it, and a cover of varnished silk over that; these covers both keep off damp and prevent premature explosion.

Action.

On firing, the shock of discharge crushes up the lead cup, the ball, cone plug, and steady plug setting back; the sketch shows how the lead cup dovetails on to the cone plug and bottom plug, preventing rebound; the steady plug prevents the ball touching the sides as it sets back, and the irregular motion of the shell in the air causes it to disengage from the ball. On the shell striking the object, the ball, now unsupported, is dashed violently against the side of the body, explodes the detonating

* The fuze is fixed in the shell by the key, fuze, and plug, G.S., see pp. 66, 69.

composition and fires the shell, the flash passing through holes in the cone and bottom plugs. The hole in the latter has a thin paper disc over it.

Issue.

These fuzes are packed by fives in tin cylinders.

This fuze was introduced for garrison service; it is not suitable for field guns as it does not act with certainty on graze.

Proof.

5 per cent. of the different parts are most carefully gauged before being put together; samples of the metals are selected and submitted to the Chemist, War Department, for analysis. After the fuzes are finished 15 per 1,000 are selected for examination and proof. The whole of these are carefully gauged and examined.

10 are dropped in a 32-pr. common shell 20 feet on an iron block, in any accidental position. These should not explode. Three of the dropped fuzes are opened and the balls taken out and a weight of 7½ ounces allowed to fall upon them from a height of 22 inches. These should fire.*

The 5 remaining fuzes are fired at Shoeburyness from a 32-pr. gun, common shell, 3 with 10-lb. charges at about 3° elevation to ricochet on water. The object of this proof is rather to test the fuzes under a high charge than to determine whether or not they explode on striking water. As a general rule they do not explode, but they are not specially constructed to stand such a test.

The other 2 are fired with a 4-lb. charge from the same gun, at an oak butt, 200 yards. These should explode.

*Pettman Percussion Fuze, G.S., Mark I.*

Pettman percussion fuze, G.S., Mark I. § 1235.

This fuze is arranged so as to act equally well from a S.B., from a B.L., or R.M.L. gun.

For shells with which this fuze is used, see table, p. 58.

It is specially designed so as to act on impact, not on graze: it will not explode on a shell passing through a wave, but will explode on the shell striking a wooden ship.

Its construction is shown in the plate. Page 352.

The body and top plug are made of gun-metal, the cone plug, detonating ball, and steady plug are also made of gun-metal, but of a harder alloy to prevent them from altering their shape; the plain ball is made of brass, and the suspending wire is made of copper.

Body. Figs. 1, 21.

The body is conical, tapped throughout with a screw, to screw into the G.S. gauge fuze hole, it is about ·2″ thick, a strong case being essential to resist the shock of a heavy charge. It is slightly hollowed out in the centre to allow sufficient play to the detonating ball, and is also hollowed out at the base to allow the lead cup to dovetail into the recess when it is crushed up. There are two slots cut in the top of the body to allow the fuze to be screwed into the shell, it is tapped at the top to receive the top plug. There is a hole in the centre of the base which serves to allow the cone plug to set back.

Top plug. Figs. 2, 3.

The top plug is a small disc having two holes in the upper part to enable it to be screwed into the fuze, and a cup-shaped recess in its lower part, into which the plain ball fits; it is tapped with a screw thread to fit the body.

Plain ball. Fig. 4.

The plain ball is a small solid ball, turned from brass wire.

Steady plug. Figs. 5, 6.

The steady plug is a disc, recessed at the top and roughed to receive

* This proof is quite distinct from that used by fire masters at out-stations, see p. 306.

a ring of detonating composition, and having a cup in the centre to receive the plain ball; three fire holes are pierced through it to allow the flash to pass down. The bottom of the central hole is enlarged to receive the projection of the detonating ball. A detonating composition (see p. 320) is pressed into the recess and is covered over by a thin copper washer.

Figs. 16, 17.

Detonating ball.
Figs. 7, 8, 11.

The detonating ball is roughed by a number of vertical grooves and has a deep horizontal groove near the centre, these grooves serve to retain the composition with which the ball is coated (see p. 320), and also render ignition certain when the ball strikes against the body. At the top of the ball is a cylindrical projection which fits into the steady plug, and at the bottom is a smaller rounded off projection which fits the cone plug. Over the composition two thin copper hemispheres are placed and united by a piece of shellaced paper, the ball is further protected by covers of varnished gut and silk.

Figs. 9, 10.

The object of the copper hemispheres and of the copper washer over the steady plug is to bring the sensitiveness of the composition within such limits that the shock of grazing will not explode it, while the shock of direct impact against a solid body will make it act.

Cone plug.
Figs. 12, 13.

The cone plug is so called because it has a conical form in the Pettman L.S. fuze. It is pierced by three fire holes, the central one being slightly enlarged to support the detonating ball, the bottom of the plug contains a chamber which is filled with mealed powder, driven and pierced like a tube, it is recessed near the top of the cylindrical part to allow the lead cup to dovetail on to it, pierced near the base for the suspending wire, and closed at the base by a small cardboard disc.

Figs. 18, 19.

Lead cup.
Figs. 14, 15.

The lead cup is a hollow cylinder, having a flange on the head to fit the recess on the cone plug.

Securing or cementing Pettman G.S. fuze.
§ 1719.
Fig. 20.

When the fuze has been put together in a perfectly dry state, the top plug, having its edges coated with a waterproof cement,* is screwed in and allowed to stand until the cement sets. The slot holes in the head of the fuze are then carefully filled with cement, the hole in the base is closed by a cardboard disc, and finally the top and bottom of the fuze are coated with cement. These cementing operations secure the fuze from damp, their importance will be seen further on. Fuzes thus secured have been found to resist water when placed in it for some hours.

A label is attached to the top of the fuze, giving the Mark, number of thousand and date of manufacture; if the fuze has been secured or re-primed, the letters † S. or R. and date of the operation will be found on the label. P. on the label shows that the fuze has been packed in tin cylinders.

Pettman G.S. fuze, Mark II.

The Mark I. fuzes as above described were found to deteriorate to a very serious extent in store, especially in hot moist climates. Numerous complaints were received from foreign stations, and for a time it was hoped that when secured from damp, re-primed, and kept in hermetically sealed cylinders that the evil would be obviated. This result, however, was not attained, and even under these circumstances they deteriorated

* Venetian red, Stockholm tar, shellac, and methylated spirits.

† Pettman G.S. Fuzes, Mark I., which have not the letters S. or R., and are of manufacture previous to 434th thousand, are not to be considered serviceable; and are to be returned to Woolwich for exchange unless in shell on board ship.

§ 1634.
§ 2359.

Few, if any, of these are now likely to be met with, as the order above came out in October, 1872.

to a very large extent.* At the same time the Pettman L.S. fuzes, which were packed either singly in waterproof bags or simply in brown paper, and kept loose in a deal box, seldom failed to satisfy the proof.† This made it apparent that there was some defect inherent in the fuze itself. It turned out, as the most probable cause, that the sublimed sulphur ‡ used in the detonating composition was liable to contain free acid. The composition being on the gun-metal ball, and covered by the copper hemispheres, was injuriously acted upon in consequence partly by chemical and partly by the setting up of galvanic action. Hence Mark II. Pettman G.S. fuze has been introduced.

§ 3200.

It differs from Mark I. as follows: The detonating ball with the composition on it is covered first with gut, then come two thicknesses of thin silk, then the copper hemispheres, then another layer of gut, and lastly three thicknesses of silk. Each layer of silk and gut is varnished. This entails slight modifications in the ball, so as to allow of the extra thickness of protection, and in the copper hemispheres. The copper washer on the steady plug is also lacquered.

Mark I. is easily converted into Mark II., and will probably be entirely used up for this purpose.

Proof.

5 per cent. of parts gauged, samples of metal submitted for chemical analysis, &c.

15 per 1,000 of these fuzes are selected for proof, 10 of these are placed in an iron block and dropped a height of 20 feet on to iron, they should not fire on falling, but may do so when the fuze rebounds and falls a second time, as the first fall puts the fuze in action. This test proves that the interior parts of the fuze are properly arranged, as if the balls were out of position the fuze would fire on the first fall. 3 fuzes are fired from a 7-inch R.M.L. gun, charge 22 lbs., to ricochet on water; these should not fire.§ 2 are fired in a 40-pr. B.L. gun with a small charge, 4 lbs., against an oak butt, or against sand bags 200 yards off; these should fire.

The very ingenious arrangement of this fuze is necessary to meet the difficulty of getting a fuze to act with rifled guns of both natures and also smooth bores. For rifled guns only, a much simpler arrangement answers.

Action.

(1.) Suppose a shell fired out of a M.L. gun, the steady plug, ball, and cone plug set back on shock of firing, the suspending wire is broken, the lead cup prevents rebound, and the stem of the cone plug protrudes through the base of the fuze, the detonating ball being released from its pivots by the slight wobble of the shell; on striking, the action will be the same as in the L.S. fuze, the flash finding exit, through the holes in the cone plug, to the priming, and thence to the powder in the shell.

(2.) When fired from a B.L. gun the steady plug may not disengage, owing to the steadiness of the flight of these shells, and in this case the detonating ball will not act; the plain ball is released by the steady

* On the average over 30 per cent. failed under firemaster's proof at out-stations.

† In the course of nearly four years, as firemaster at Bermuda, I never knew a L.S. fuze to fail at proof, while on the other hand the G.S., which had been comparatively a very short time in store, even if left in the hermetically sealed cylinders, frequently deteriorated so much that the detonating composition might be scraped off the ball with a knife, and hammered on iron without detonating; and this when no signs of damp were visible.

‡ Ground distilled sulphur will be used in future.

§ This proof is quite distinct from that used by firemasters at out-stations. See p. 306.

plug setting back, and is caused by the centrifugal force to spin round the circumference of the body over the ring of detonating composition. On shock of striking the object, the ring is dashed against the plain ball and detonates, exploding the shell through the fire holes.

These fuzes are packed by five in tin cylinders.

Issue. § 2359.

This fuze being specially designed for firing over water is issued to sea fronts of fortifications and to the Royal Navy.

It is to be remarked that it is impossible to protect the composition on the detonating ball by pressure and varnish in the same way as can be done in a cap. It was early found that the damp affected these fuzes and the first remedy was to protect them by unscrewing the top plug, luting it with red lead moistened with shellac varnish, screwing it in again, and coating the top with the luting; this was to be done at out-stations as well as at Woolwich.*

§§ 1612, 1634.

§ 1743.

Experience has shown the Pettman G.S. fuze to be very safe from prematures.

Both the Pettman fuzes are designed to act on direct impact, but the L.S. may occasionally act on graze (see proof, pp. 48, 50), the G.S. on the other hand requires direct impact. The proof shows with what nicety the sensitiveness of this fuze has been brought within the desired limits, as it explodes when fired with a reduced charge from a B.L. gun against wood, and does not explode on striking water when fired with a battering charge from a 7-inch R.M.L. gun.

We now come to the fuzes which act on graze. The smaller the charge the more difficult it is to ensure the action of the fuze. The fuzes about to be described act very well in the ordinary field guns, such as the 9-pr. and 16-pr. R.M.L. and in the Armstrong B.L. field guns, the difficulty occurs in such a gun as the 7-pr. R.M.L. when using a 6-oz. or 8-oz. charge † where the velocity is too slow to ensure sufficient shock being given on discharge to shear off the feathers. Quickness of action is essential with these fuzes, as the effect of a burst on graze is lost if the shell has time to rise to any height before bursting.

A short range is the most trying to the fuze, as when the trajectory is flat the velocity of the shell is but little checked by grazing; these fuzes have however been found to act well at 400 yards and even at shorter ranges. The softer the ground the greater the chance of failure, but our present experience is that these fuzes may be depended on over all

---

* At some stations the fuzes were found to be carelessly put together, so that the fuze was in a most dangerous condition, the detonating ball being loose inside. All fuzes so repaired at out-stations were re-called. Should any still exist they may be known by the date of repair being marked on them *in manuscript*. They are to be looked upon as dangerous, and if in shell with an unloading hole, the bursting charge is to be drowned. If the shell has no unloading hole it is to be thrown into deep water.

A.C. 1870, Cl. 16.

A.C. 1869, Cl. 143.

† A letter from the Superintendent of Experiments, Shoeburyness, states that 55 common shell have been fired from the 7-pr. R.M.L. steel gun, 6 oz. charge F.G powder, using R.L. percussion fuze, Mark I.; 45 acted correctly, 4 on second graze, and 6 were blind. Mark II. acts worse, 6 out of 20 tried at proof were blind.

natures of ground.* It is also satisfactory to find that prematures are of very rare occurrence.

The fuze introduced in 1862 closely resembles the present B.L. plain percussion fuze in its arrangements, but the latter has two important advantages, viz., cap composition, which, as before pointed out, resists damp, and a safety pin.

*B.L. Plain Percussion Fuze.*†

**B.L. plain percussion fuze. §§ 1983, 2292, 2620.**

Used with field service B.L. common and segment shell. This is a modification of the Armstrong C. percussion, the improvement consisting in using cap composition pressed and varnished, as in gun caps, which experience has shown to stand damp climates well.

**Construction.**

The body and top are made of gun-metal, and the body has a rim projecting at the top which ensures the fuze being placed in the correct position in the shell. In the centre of the top, on the inside, is fixed a steel needle, point down; the top is pierced with four holes to allow of the action of the Armstrong E. time fuze, which is still used in the Navy in conjunction with the B.L. plain fuze. A washer of thin sheet brass closes these holes (it is blown in by the action of the time fuze). The body is pierced with two holes for the safety pin.

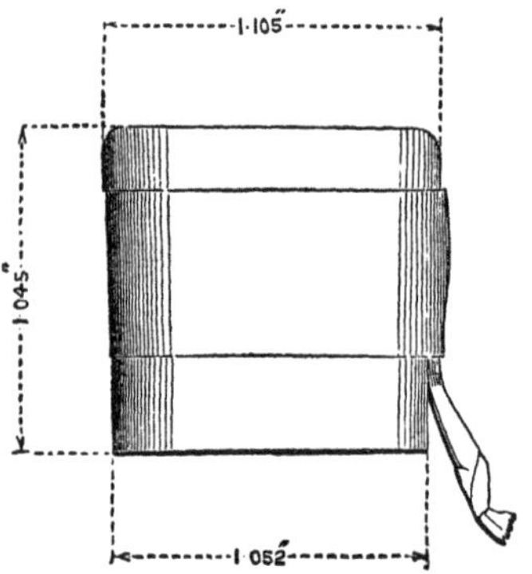


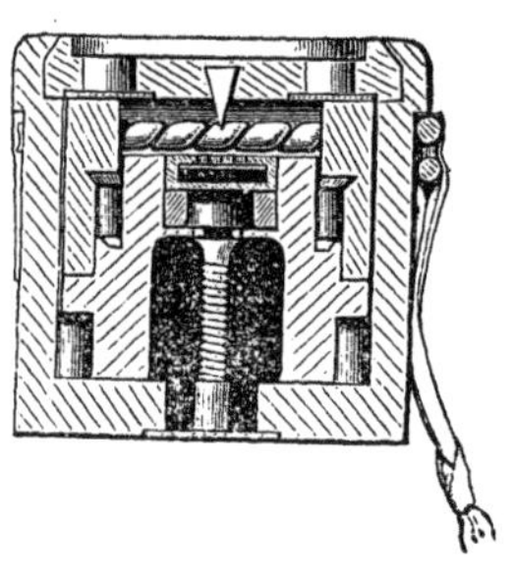

N.B.—The cut represents an earlier pattern converted. The present pattern has the bottom screwed into the body.

---

* See Extracts, Vol. X, p. 351.

A letter in the Royal Laboratory, $\frac{16,534}{\frac{15}{5}}$, gives an account of a trial of the R.L. percussion fuze, Mark I., from the 9-pr. R.M.L. gun. It states that the fuze acted well on all natures of ground and water at ranges from 825 to 2,250 yards; 138 out of 141 burst on first graze. On boggy ground 102 burst first graze out of 108, ranges 600 to 1,500 yards. On sands 137 out of 143 burst on first graze, range 800 to 1,800 yards. This practice was carried on using the hollow headed rammer (now obsolete) and to this cause three prematures which occurred during the practice were attributed; this rammer used to grip the fuze, and so cause it to come loose in ramming home.

† This fuze was known as the "C. Cap" percussion fuze, but the nomenclature was altered to "B.L. plain" by § 2620.

By § 2029 a number of obsolete fuzes were called in to be converted to the B.L. plain percussion fuze. They are easily known, as none of them have a safety pin with braid attached to it. When converted "Cap" is stamped on them. By § 2761 the date of repair is to be labelled on the boxes.

A "C." percussion screw fuze, which fitted the Armstrong F.S. gauge, is also obsolete.

For details of the various patterns of these fuzes, from the introduction in 1860 of Armstrong's original "Metal Concussion Fuze," see Colonel Wray's pamphlet, referred to in foot-note p. 41.

The guard is made of gun-metal, and is pierced with two holes for the safety pin; it fits inside next the top, and is recessed inside to receive the head of the pellet; there is a slight undercut at the top of the recess into which the pellet expands when the guard sets back. **Guard.**

The pellet is cast of equal parts of lead and tin; it is hollowed out and receives in its top the copper cap, which is primed with cap composition (fulminate of mercury, sulphide of antimony, and chlorate of potash), pressed* and varnished in the same way as in gun caps. The composition is further protected by a very thin disc of brass†; this has been found necessary to prevent premature explosions; a disc of paper coated with shellac is stuck on the top of the pellet covering the cap. **Pellet.**

**Cap.** The cap is pierced with three small holes, arranged in a triangular form round the centre; these holes allow the flash to pass down to the lower part of the pellet, which is filled with a pellet of pressed powder pierced like a tube and roughened so as to ensure ignition.

On the exterior of the pellet are four feathers or flanges, and below the pellet a disc of paper is placed to prevent its adhering to the bottom of the fuze.

The bottom consists of a gun-metal disc which screws into the base of the fuze; in the centre is a small hole which contains pressed powder driven and pierced as usual. This hole is closed on the exterior by a thin brass disc. **Bottom.**

The safety pin is made of twisted brass wire, and has a piece of braid attached to it to enable it to be withdrawn readily; a little beeswax is applied to seal the hole, and the braid is secured by a paper strip shellaced round the fuze. **Safety pin.**

Since 12/71 these fuzes have been painted with a black varnish similar to that used with friction tubes in order to exclude damp as much as possible. § 2169.

Remove the safety pin and drop the fuze into the shell, rim to the front; replace the plug in the shell, except for naval service, when the E. time fuze is used. **Preparation.**

When the safety pin is removed, the guard is supported by the feathers of the pellet. On the shock of discharge the guard sets back, shearing off the feathers, and its shoulder is jammed into the undercut in the interior surface of the guard. When the shell strikes an opposing object, or grazes, the pellet and guard fly forward as one mass, the cap comes violently against the needle and is thereby fired. The flash is taken up by the powder pellet, the brass disc at the bottom is blown out, and the bursting charge exploded. **Action.**

This fuze may be depended on to act on graze even on wet boggy ground or on water, and will act at 400 yards where case ceases to be effective. Before the addition of the safety pin the feathers of the pellet were found apt to give way under the jolting to which the fuze was exposed in limber boxes; but now the fuze is quite safe, as the pin takes the part of the lead feathers, which are only called into play when the pin is withdrawn.

In tin cylindrical box, holding two fuzes, each fuze in waterproof bag, 80 tin boxes, or 160 fuzes in a deal packing case. These fuzes will be issued in tin cylinders (5 in each) when the present store of cylinders is used up. **Issue.** § 2217.

---

* The pressure is 600 lbs. on the area of the cap.

† ·001″ thick.

§ 1966. For field service 20 fuzes are carried on their sides in a tin box fitted to receive them, and carried in the centre limber box.

*Fuze, Percussion, R.L., Mark I.*

§§ 2191, 2620. Used with R.M.L. field service 7-pr. and 9-pr. common and Shrapnel shell. This fuze exactly resembles the B.L. percussion fuze in its internal arrangements. The body is of gun-metal; both body and top are cast in one piece, the bottom is screwed in, a square hole in the head fits the G.S. key by which it is screwed into the shell. This fuze fits the G.S. gauge.*

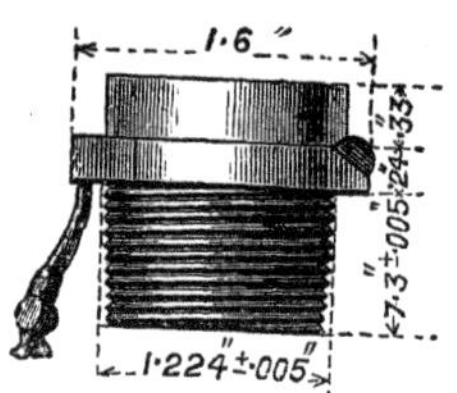


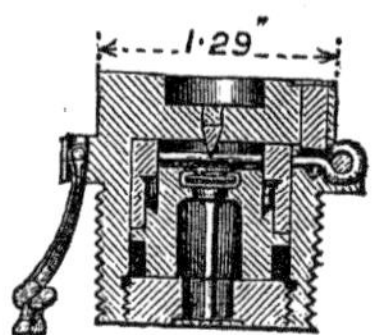


The safety pin is a plain brass wire. Through the loop at one end passes a brass wire ring which fits round the fuze in a recess cut for the purpose in the upper surface of the shoulder. This ring has a piece of braid attached to it, and the latter is kept in its place by a slip of paper pasted over it. *The pin is not to be withdrawn until the shell is placed in the gun.*

When the safety pin is withdrawn, the hole through which it passed, if left open, would probably admit of the passage of the flash from the discharge of the gun into the interior of the fuze, and so a premature burst would take place. To guard against this a small lead pellet slides freely in a recess cut in the head and closed by a thin brass disc, soldered on flush with the top of the fuze. When the shell is rammed home the pellet sets back and so closes the safety pin hole.

Preparation. See p. 66.

Action. As described above for the B.L. plain fuze.

This fuze of course acts on graze in the same way as the B.L. percussion.†

These fuzes resist damp well; out of 48 which had been wetted in a centre limber box only two failed, one from the cap being damp, the other from the guard working stifly.

---

* A leather collar used to be issued on the fuze under the shoulder, but s now discontinued as it is an advantage to screw the fuze home as far as possible.

† The action of this fuze from the 7-pr. is uncertain, a considerable number of blinds have occurred.—Extracts, Vol. IX, p. 26.

It acts very well from the 9-pr. R.M.L.—Extracts, Vol. IX, pp. 57, 119.

The lowest charge that will cause it to act with certainty from the 9-pr. appears to be 10 oz.—Extracts, Vol. IX, p. 120.

Trials were made as to whether the fuze would act without removing the safety pin, several blinds occurred when the safety pin was left in.—Extracts, Vol. IX, p. 297.

Five fuzes in a tin cylinder, each fuze wrapped in brown paper, and six cylinders packed in a deal case.

Issue. §§ 2201, 2328

For field service 16 fuzes are carried on their sides in a tin box for the 9-pr., and 12 fuzes for the 7-pr.

FUZE, PERCUSSION, R.L., MARK II.*

For shells with which this fuze is used, see table p. 58.

§ 2621.

It can be used with common, segment, or Shrapnel shells, and will act either on graze or impact.

It is tapped with a screw thread to fit the G.S. gauge of fuze hole.

Its construction is given in the plate, page 353.

This fuze has been introduced because the R.L. fuze, Mark I., was found to fail when fired with heavy charges, and differs from it in the following particulars:—

1st. The pellet and guard, *vide* plate, are smaller in diameter, to admit of greater thickness in the side of the fuze, and of a deeper screw thread at the bottom, thereby giving the base greater power to support the weight of the pellet and guard on the shock of discharge.†

2nd. The pellet has no powder pressed into it, and there is an increased quantity of detonating composition in the cap at the top of the pellet. The fuze is thus made quicker and more certain in its action.

3rd. The safety pin (of double twisted wire) passes through the head of the fuze, and is kept in its place by the two ends being opened out slightly, so as to bind themselves in a conical cup, as shown in the drawing. A thin disc of brass is then fitted in over the ends, and soldered over to keep the fuze water-tight. The head of the safety pin is fitted with a loop of string, by which it is withdrawn. There is no brass ring or recess round the head of the fuze.

Preparation.

See page 66.

Action.

The action is the same as in the B.L. and Mark I. fuzes, except that there is no powder to ignite in the pellet, the flash from the cap going direct to the bursting charge.

Issue.

As for Mark I. fuze.

Fuzes of Mark I. are only suitable for use with 7-pr. and 9-pr. guns; but for these natures they are as efficient as the Mark II. fuzes. All issues for these two natures of guns will therefore be taken from the existing store of Mark I. fuzes until it is used up.

The R.L. percussion fuze, Mark II., has acted well from the 9-pr., 16-pr., 40-pr., 64-pr., and 80-pr. R.M.L. guns, and also from the 40-pr.

---

* This fuze will replace Mark I. for the 7 and 9-prs. when the present stock of the latter is used up.

§ 2621.

D. N. O., 18/4/74, reports that R.L. fuze, Marks I. and II., answer equally when fired in 7-pr. common and Shrapnel shell, 8-oz. charge. With the double shell, 4-oz. charge, 13 Mark I. out of 14 were blind, and of 14 Mark II. all were blind. This shows that the 4-oz. charge does not set the fuze in action. Extracts, XII., p. 159.

† The thickness of the brass disc, which closes the firehole in the bottom, has been reduced in those made since November, 1873. The safety pin hole used to be covered with beeswax, and the bottom of the fuze painted, but in recent manufacture these operations are discontinued.

B.L. gun; if it was desired to use segment shell against troops in the field, this fuze would be invaluable.*

Proof of B.L. plain and R.L. percussion fuzes, Marks I. and II.

15 out of each 1,000 are selected, the "feathers" are cut off 10 of them, and the fuze dropped head down about four inches in an iron block; these should fire. The remaining 5 are fired, the B.L. plain from 9 or 12-pr. B.L. gun, the R.L. from various R.M.L. guns up to the 80-pr., to see that they act on graze.†

*Sensitive Fuze, Mark I.*

Sensitive fuze. § 3221.

A sensitive fuze has been approved for use with 7-pr. R.M.L. guns, and 8″ and 6″·3 R.M.L. howitzers. In general principle it resembles the Prussian percussion fuze.

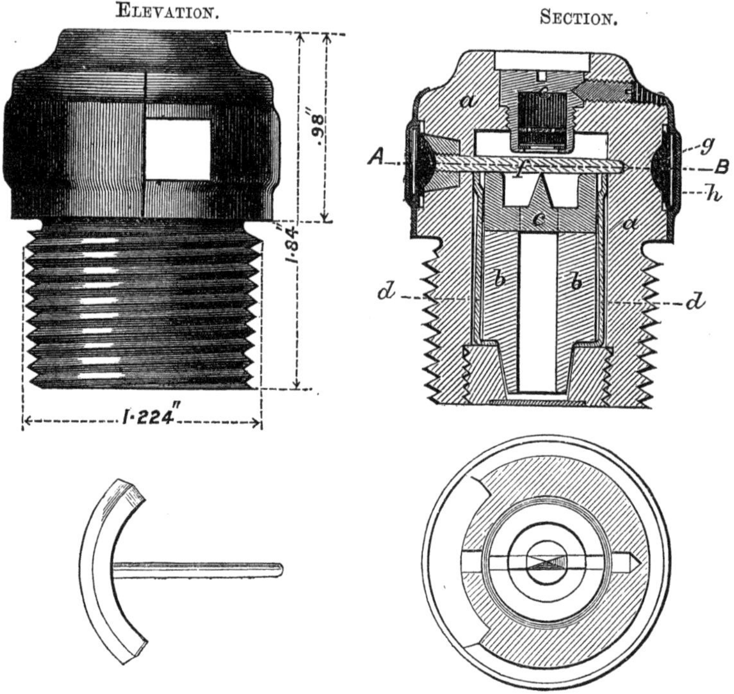


SAFETY PIN. SECTION ON A B WITH PIN REMOVED.

* See Extracts, Vol. XI, p. 67.

The Superintendent R.L. states the trials of the R.L. percussion fuze, Mark II., have been very satisfactory, that there appears to be no liability to premature explosions, that the fuze acts with at least as much certainty as any other percussion fuze yet tried under similar conditions, and that the rapidity of action is better than any yet tried.—Extracts, Vol. XI, p. 134.

Experiments have been carried out to ascertain whether this fuze would act when used in connection with the lowest charge fired from the 8-inch R.M.L. howitzer, it failed to act when fired with 5 lb. or 6 lb. charges at an elevation of 10°, but acted with charges of 7lbs. and over. The initial velocity is probably too slow to cause the "guard" to set back when the small charges are used.

A trial at Shoeburyness (January, 1874) showed that the R.L. Mark II. fuze acted with the 7-inch B.L. gun; 25 were fired, 23 burst first graze, and two burst second graze. It is obviously necessary to take out the safety pin before placing the shell in the bore when using the R.L. fuzes with B.L. guns.

The lowest charges which make the R.L. fuze Mark II. act in the 40-pr. and 64-pr. R.M.L. guns are 2¾ lbs. and 3¼ lbs. respectively.

† For firemaster's proof, see p. 306.

It is designed so as to be equally efficient with very low charges and with the highest charges used with the above pieces.

Construction.

The fuze consists of (*a*) body; (*b*) hammer or pellet; (*c*) steel needle; (*d*) thimble; (*e*) detonating cap; (*f*) safety pin; (*g*) outside primer of quickmatch, and (*h*) band. The body, hammer and cap are made of gun-metal, and so is the bottom plug which has a fire-hole closed with a thin brass disc spun over.

*The hammer* tapers slightly from top to bottom to allow it to move freely forward on impact of the shell. A thin steel plate, the centre portion of which forms the needle, is fitted as shown in the cut, into a slot in the top of the hammer, and a hole bored through the centre of the latter allows the flash from the detonator to pass on both sides of the steel plate and down into the shell.

*The thimble* is a thin brass cylinder, flanged at the bottom, and encloses the hammer. It allows the latter to move freely forward, and, going with it, prevents the hammer from being impeded by dirt, &c., which might otherwise enter through the safety pin hole on graze or impact on earth, &c.

*The detonating cap* is screwed into the head of the fuze and secured by a small side "stop screw." It contains about 7½ grains of pressed mealed powder covered with a perforated copper disc. Below this is pressed 3½ grains of cap composition, which is covered with varnished fine white paper and a thin brass disc; the latter has a hole ·1″ in diameter in the centre to allow the needle to pass through to fire the composition.

The use of the meal powder is to ensure the production of a sufficient quantity of flame to communicate with the bursting charge.

*The safety pin* is of brass wire (No. 15 B.W.G.) screwed to a heavy head of gun-metal, as shown in the cut. It is secured by a strand of six-thread quickmatch, fastened by silk thread and coated with meal powder priming. The whole is covered with a tape and copper band, like that used with wooden time fuzes for R.M.L.O. The safety pin fits easily through one wall of the body and into a recess in the opposite wall.

Preparation.

The fuze is "uncapped" by removing the tape and copper band. This is not to be done till the shell has been placed in the bore.

Action.

The flash of discharge burns up the quickmatch.* The safety pin, now free to move, is whirled out by the centrifugal force due to the rotation of the shell. On impact or graze the hammer and thimble fly forward and the needle point enters the cap. The flash passes down the pellet, blows out the thin brass disc which closes the base of the bottom plug, and so fires the shell.

Issue.

Five fuzes in a tin cylinder. Six cylinders in a deal box.

---

* Gun-cotton loosely twisted will very possibly be shortly substituted for quickmatch.

## Service Fuzes, Rifle and Smooth Bore.

| Name of Fuze. | Shells used with L.S. | Shells used with S.S. | Other Shells that the fuze can be used with on an emergency. |
|---|---|---|---|
| | **R.M.L. Ordnance.** | | |
| Time, wood, M.L. 5 seconds | Shrapnel up to 80-pr. inclusive; common 9 and 16-pr.; and 7-pr. star | Shrapnel, 9 and 7 pr. | All M.L. shell of G.S. gauge, when time of flight does not exceed 5 seconds. |
| Ditto ditto 9 seconds | Shrapnel up to 80-pr. inclusive (7-pr. excepted); common up to 40-pr. inclusive; and 7-pr. double | Shrapnel, 64-pr. and upwards; common 7, 9, and 64-pr., 7-inch and double 7-inch | Ditto ditto, time not exceeding 9 seconds. |
| Ditto ditto 20 seconds | Double, 7-pr. | Common, 64-pr. and 7-inch; and double 7-pr. | Ditto ditto, except Shrapnel, time not exceeding 20 seconds. |
| Percussion, Pettman, G.S. | Common 64-pr. and upwards; and double 7-inch | Common, 64-pr. and upwards; and double 7-inch | |
| Ditto R.L. | Common up to 80-pr. inclusive; Shrapnel up to 40-pr. inclusive | Common 7 and 9-pr. | |
| Ditto Sensitive | | | |
| | **R.B.L. Ordnance.** | | |
| Time, wood, B.L., 5 seconds | Shrapnel 9 and 12-pr. | ... | All shell of G.S. gauge, when time of flight does not exceed 5 seconds. |
| Ditto ditto 9 seconds | Common and segment 7-inch and 40-pr.; Shrapnel 9 and 12-pr. | Common, 7-inch 40 and 20-pr.; segment 7-inch and 40-pr. | Ditto ditto, time not exceeding 9 seconds. |
| Ditto ditto 20 seconds | Common and segment 7-inch and 40-pr. | Ditto ditto ditto | Ditto ditto, except Shrapnel, time not exceeding 20 seconds. |
| Percussion, Pettman, G.S. | Common 7-inch and 40-pr. | Ditto ditto ditto | |
| Ditto R.L. | Common and segment 7-inch and 40-pr.; Shrapnel 9 and 12-pr. | | |
| Ditto B.L., plain | Common and segment up to 20-pr. inclusive | Segment, 20, 12, and 9-pr. | |
| Time, E. | ... | Ditto ditto | |
| | **S.B. Ordnance.** | | |
| Time, wood, diaphragm, Shrapnel | All diaphragm, Shrapnel | Diaphragm, Shrapnel, 24 and 32-prs., and 68-pr. or 8-inch | S.B. common shell, when time of flight does not exceed 5 seconds. |
| Ditto common | All common* | Common, 24-pr. | |
| Ditto M.L. 9 sec. | Naval | Naval | |
| Ditto M.L. 20 sec. | Ditto | Ditto | 24 and 12-pr. common shells, when fired from $5\frac{1}{2}$-inch and $4\frac{2}{5}$-inch mortars, and time of flight exceeds 16 seconds with the $4\frac{2}{5}$-inch, and 10 seconds with $5\frac{1}{2}$-inch. |
| Ditto Mortar, large | Mortar 13, 10, and 8-inch | | |
| Ditto ditto, small | Common 24 and 12-pr. when fired from $5\frac{1}{2}$ and $4\frac{2}{5}$-inch mortars, at times of flight over 5 seconds with $5\frac{1}{2}$, and 10 seconds with $4\frac{2}{5}$-inch | ... | All other natures of S.B. common shell. |
| Percussion, Pettman, L.S. | All common shell having a × on the fuze-hole plug | Common, 24-pr. | |
| Ditto ditto G.S. | Naval | Naval | S.B. common shell. |

* Including 24 and 12-pr. common shells, when fired from $5\frac{1}{2}$-inch and $4\frac{2}{5}$-inch mortars, when time of flight does not exceed 10 seconds.

N.B.—In substituting any time fuze for another, care must be taken that the top side hole does not appear above the fuze hole, as the flash of discharge might enter.

Mark I. R.L. Percussion fuze to be used up with 7-pr. and 9-pr. guns.

# CHAPTER VI.—SHELL AND FUZE IMPLEMENTS.

## DISTRIBUTION OF, AND INSTRUCTION IN, THE USE OF SHELL AND FUZE IMPLEMENTS.—PAPIER MÂCHÉ WADS.

Shell and fuze implements are issued, as their name implies, for preparing shell and fuzes. There were formerly five sets for garrison service, four sets for field service S.B., and one set for rifled guns of position. By the last regulations, "Errata and Addenda to Army Equipment, 1876," issued July, 1877, and Cl. 138 A.C. 1877, "sets" of implements are abolished, and implements are to be demanded separately. The detail will appear in the following pages.*

### Implements for Rectifying Fuze-Holes of Common (S.B.) and Diaphragm Shrapnel Shell.

One set per station; two sets if large station.

| | |
|---|---|
| Blocks, wood, 10-inch, 8-inch, 32-pr. (of each) .. | 1 |
| Gauges, fuze-hole, common .. .. .. .. | 1 |
| Holder, shell, spherical .. .. .. .. .. | 1 |
| Key, iron, square .. .. .. .. .. | 1 |
| Levers, common .. .. .. .. .. .. | 1 |
| Rectifier, or rimer, common, fuze hole .. .. | 1 |
| Screws, coach, 4 in. by ½ in. .. .. .. .. | 4 |
| Tap, screw, Pettman, common .. .. .. .. | 1 |

### Implements for Rectifying Fuze Holes of Mortar Shell.

(Issue as above.)

| | |
|---|---|
| Blocks, wood, 13-inch .. .. .. .. .. | 1 |
| Gauge, fuze-hole, mortar .. .. .. .. | 1 |
| Holder, shell, spherical .. .. .. .. .. | 1 |
| Levers, mortar .. .. .. .. .. .. | 1 |
| Rectifier, or rimer, mortar, fuze-hole .. .. .. | 1 |
| Screws, coach, 4 in. by ½ in. .. .. .. .. | 4 |

* Instructions for use of implements, shell and fuze, are issued, printed on calico. There are three sets or sheets, viz.: For S.B. Ordnance, for Rifled Ordnance, Garrison and Naval Service; and for Rifled Ordnance, Field Service only. These instructions are given in the text of this chapter, almost *verbatim*.

§§ 2934, 2967, 2968.

All the above implements will be retained only where shells still require to be rectified, and will cease to be recognised as part of the equipment for future issue.

## Rectifying Fuze-Holes.

Instructions.

*S.B. Shell, Common.*—Unscrew the metal plug from the fuze-hole by means of the square key and lever [if the plug should be so firmly fixed in the shell that it cannot be unscrewed by the key, a few smart blows with the hammer will loosen it], insert the fuze-hole gauge—if the larger end of the cone is not flush with the exterior of the fuze-hole, place the shell in the holder, and fix it by screwing up the movable jaw—the smaller natures of shell can be supported by hand in the proper position until the jaws have a firm hold; but for the 32-pr. and heavier natures a block of wood is necessary for the shell to rest upon; the jaws should grip the shell about half-way between the fuze-hole and the bottom—insert the rimer into the fuze-hole, and turn it *gently* round with the lever until the fuze-hole is of the proper dimensions, *great care being taken not to make it too large*, then perfect the thread by means of the screw-tap. The shell-holder when in use must be screwed to a bench or table.

*S.B. Shell, Mortar.*—Remove the cork from the fuze-hole and insert the fuze-hole gauge. If the larger end of the cone is not flush with the exterior of the fuze-hole, place the shell, resting on a block of wood, in the holder, and proceed as above until the fuze-hole is of the proper dimensions, *great care being taken not to make it too large.*

## Implements, S. B. Garrison, for Fixing Wood Bottoms.

One set per shell-filling room, where spherical shells are prepared.

| | |
|---|---|
| Block, wood .. .. .. .. .. .. | 1 |
| Hammer, riveting .. .. .. .. .. | 1 |
| Pricker, removing wax .. .. .. .. .. | 1 |
| Punch, iron, riveting bottoms .. .. .. .. | 1 |
| Rectifier, common, for rivet-hole .. .. .. | 1 |
| Spanner, box .. .. .. .. .. .. | 1 |

## Fixing Wood Bottoms.

Instructions.

*S.B. Shell, Common.*—Remove the bottom from the iron bar by unscrewing nut at the end of the bar with the box spanner; place the shell on the wood block; remove the beeswax from the rivet-hole with the pricker; place a rivet in the rivet-hole of the wooden bottom, with the point projecting beyond the concave surface; place it on the shell, moving it about until the rivet drops into the rivet-hole; place the punch on the head of the rivet and give it a few smart blows with the hammer.

If the rivet-hole in the wood bottom is rough and jagged pass the rectifier through it, turning it round so as to bring the hole to its proper form.

The common shell, which are issued loose, have the fuze-hole secured by a metal screw-plug, and are prepared with rivet-holes for fixing the wood bottom. The wood bottoms are packed by twenties on iron rods, and secured by an iron nut.

IMPLEMENTS, S.B. GARRISON, FOR FILLING S.B. COMMON, DIAPHRAGM, SHRAPNEL, AND MORTAR SHELL. Implements.

1 set per shell-filling room.

| | |
|---|---|
| Drifts, wood, common .. .. .. .. .. .. | 1 |
| „ „ diaphragm, Shrapnel, large .. .. .. | 1 |
| „ „ „ „ small .. .. .. | 1 |
| Drivers, screw „ „ large .. .. .. .. | 1 |
| „ „ „ „ small .. .. .. .. | 1 |
| Funnels, leather, copper spouts, common, large .. .. .. | 1 |
| „ „ „ „ diaphragm, Shrapnel, large .. | 1 |
| „ „ „ „ „ „ small .. | 1 |
| Key, iron, fuze and plug, G.S. .. .. .. .. .. | 1 |
| Mallets, common and diaphragm .. .. .. .. .. | 1 |
| „ mortar .. .. .. .. .. .. .. | 1 |
| Instructions, printed sheets .. .. .. .. .. | 2 |

The following implements are supplied for filling common, Palliser, segment, and Shrapnel shell, for rifled ordnance of garrison calibres:—

| | | |
|---|---|---|
| Mallets, tent .. . .. | 2 per shell-filling room. | |
| Funnels, copper, shell, large .. | 1 per shell-filling room, 7-inch R.M.L. guns and upwards. | § 2493. |
| Drivers, screw, diaphragm, Shrapnel, large .. .. | 1 per shell-filling room. | |
| Extractors, fuze, small .. .. | 1 „ „ | |
| Funnels, common, large.. .. | 2 „ „ | |
| Holders, Palliser shell (see p. 64) | 1 per shell-filling room for each nature of Palliser shell. | |
| Instruction sheets .. .. | 2 per shell-filling room for R.M.L. or R.B.L. as required. | § 2967. § 1308. |
| Key, iron, fuze and plug, G.S. .. | 2 per shell-filling room. | |
| Wrench, removing plug .. | 1 per shell room for Palliser shell only. | § 1490. |
| Rods, brass, for filling shell .. | 1 per shell-filling room, 7-inch R.M.L. guns and upwards. | § 2493. |
| Block, wood, or shell holder .. | 1 per shell room, for Palliser shell only, to be made locally. | |

FILLING AND SECURING SHELLS.

*Shells, Spherical, S.B., Common.*—Remove the plug from the fuze-hole by means of the key; insert the funnel and pour in the bursting charge; carefully wipe every portion of powder from the fuze-hole, and drive in a papier mâché wad with the drift as far as the shoulder on the drift will allow;* then screw in the fuze-hole plug, or fix fuze as required.

*Shells, Spherical, S.B., Diaphragm, Shrapnel, Boxer.*—Remove the plug from the loading-hole by means of a screwdriver; hold the shell in a position with the loading-hole uppermost; insert the funnel and pour in the bursting charge; turn the shell from side to side to facilitate the filling; carefully wipe every portion of powder from the loading-hole, and drive in a papier mâché wad with the drift as far as the shoulder on the drift will allow, and screw in the plug; unscrew the fuze-hole plug, to which is attached a wood plug covered with serge (to prevent the bursting powder from passing into the socket in sufficient quantity to

* Applies to shells not wanted for immediate service.

cause inconvenience in fixing the fuze); and in order to ensure the small hole communicating with the powder-chamber being clear, shake a few grains of powder from the powder-chamber into the socket; then replace the fuze-hole plug.

*Shells, Spherical, S.B., Mortar.*—Mortar shells for L.S. are only filled when required for firing. Remove the cork from the fuze-hole, except in the case of the 10 and 13-inch, when it may be driven in; insert the funnel and pour in the bursting-charge, and insert the fuze as described under head of "preparing fuzes."

*Shells, Rifled, M.L., Common, under 7-inch.*—Remove the plug from the fuze-hole, insert the leather funnel and pour in the bursting-charge; the shell should be tapped with a mallet or a piece of wood to ensure its being completely filled, just leaving room for the fuze, if it is to be fuzed with a time fuze; this can be ascertained by inserting a piece of wood the same size as the fuze; after filling the shell, carefully wipe every portion of powder from the fuze-hole, then fix the fuze or plug as may be required.

In shells for siege, field, or boat service, not required for immediate service, or in the case of shells not using bags and not required for immediate service, insert the wad, papier mâché, G.S., with the side on which the shalloon is cemented downwards, *i.e.*, next the powder; drive it in with the "drift wood, G.S.," as far as the shoulder on the drift will allow, and then screw in the plug.

*Shells, Rifled, Shrapnel, Boxer.*—Remove the plug from the fuze-hole, and after seeing that the fuze-hole is clear of any dirt, &c., insert the leather funnel and pour in the bursting charge. This must be done gradually, for if the whole of the powder is put in at once the tube will probably become choked. Shake the shell from side to side on its base, until the whole of the bursting charge has passed down the tube, taking care that none of the powder is left at the bottom of the socket. Drop in the metal primer, and, by means of the large Shrapnel screwdriver, screw it tightly into the tube, and then screw in the fuze or plug as may be required.

*Shells, Rifled B.L., Common and Segment.*—The common and segment shells, 40-pr. and upwards, and the 20-pr. sea service common shell, are to be filled as directed for R.M.L. common shell under 7-inch. In the common shell, 20-pr. land service, 12-pr., and 9-pr., which are fitted with a metal socket to carry the B.L. percussion fuze, insert the leather funnel, and carefully pour in the bursting charge through a small hole at the bottom of the socket until the shell is thoroughly filled, then (care being taken that no powder remains in the socket), place the papier mâché wad, recessed part uppermost, in the hole at the bottom of the socket, and drive it in flush; any flat-ended piece of wood or stick larger in diameter than the wad can be used for this purpose, then fix the fuze or screw in the fuze-hole plug as may be required.

The object of this wad being used is to prevent the powder from working up into the socket after the shell has been filled.

The 20-pr., 12-pr., 9-pr. and 6-pr. segment shells have iron bursters, a wood plug covered with serge being placed on top to secure them whilst travelling: this plug is to be removed before inserting* the percussion fuze.

* The bursters are issued wrapped up in brown paper, this covering should not be removed, but only the top end should be torn off, to allow ignition. If the shell have not a small lead disc permanently secured at the bottom of the bursting chamber, one should be inserted before putting in the burster. Shell which have these discs permanently secured in them are marked with the letter D. on the coat. This does not apply to shell for S.S., in which service the E. time fuze is used.

*Shells, Rifled, M.L., Common, 7-inch and upwards.*—Remove the plug from the fuze-hole, place the filling-rod in the bag, and fold the latter round the rod, insert it through the fuze-hole, taking care not to force the end of the rod through the bottom of the bag; carefully push in the bag until the neck only is in the fuze-hole, a portion being kept outside, as the whole bag must not be allowed to slip into the shell during the operation of filling: then withdraw the rod and insert the funnel in the neck of the bag, pressing the funnel well down into the fuze-hole; pass the filling-rod down through the funnel and *gradually* pour in two or three pounds of powder; take out the funnel and rod, lift up the bag and jerk it, so as to "set" the powder well down to the bottom and to open the bag. Then re-insert the funnel and rod as before, and continue the filling.

The filling-rod should be moved up and down to facilitate the passage of the powder through the funnel; the powder in the shell being tamped on at the same time. The use of a large mallet against the side of the shell (or any piece of wood that would answer the same purpose), will materially assist in getting as much powder as possible into the shell.

When the shell is quite full,* withdraw the funnel and filling-rod and tie the neck of the bag with two half-hitches of twine close to the top of the fuze-hole. Cut off the superfluous choke and push the neck of the bag well down, and to one side of the fuze-hole; then screw in the fuze or plug as required.

When the shell is to be fuzed with a time fuze it must not be quite filled with powder, but sufficient space must be left for the insertion of the fuze when the choke of the bag is pushed in.

No preparation of the bag by pricking or otherwise is necessary when using either percussion or wood time fuzes, except in the case of the 20-seconds fuzes, which require that the bags should be pricked. A sharply pointed piece of wood should be used.

*Shells, Rifled, M.L., Palliser.*—The shell is to be placed upon its point, which may be inserted in a block of wood hollowed for the purpose, or in any convenient place to steady it. No special pattern of block is necessary; it can be provided on the spot and the recess cut by any carpenter.

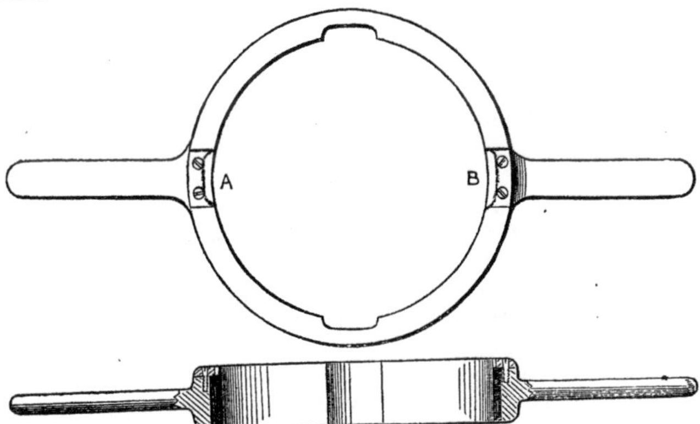


* It is very important, particularly with the larger natures of shell, that they should be completely filled. Great care must be taken to keep both the shell and mallets free from grit, to allow no loose powder to be spilled, and it is well to keep the floor damp.

HOLDER, SHELL, PALLISER.

| | 7″ § 1483. | 8″ § 1483. | 9″ § 1483. | 10″ § 1682. | 11″ § 2412. | 12″ § 1696. | 12″·5 § 2990. |
|---|---|---|---|---|---|---|---|
| Diameter at A. B. .. | 6″·97 | 7″·97 | 8″·97 | 9″·97 | 10″·97 | 11″·97 | 12″·47 |
| No. of grooves .. .. | 3 | 4 | 6 | 7 | 9 | 9 | 9 |

To remove the plug the "Holder, shell, Palliser" is to be slipped over the base until it rests on the bottom studs, and then held firmly by one man while another applies the "Wrench, removing, plug" to the plug and unscrews it. The bag is then inserted, and the shell filled as directed for the larger natures of R.M.L. common shells. Before replacing the plug, any grains of powder or grit adhering to either screw must be thoroughly removed. The plug is to be screwed home as hard as possible, and the key-hole filled with red-lead putty.

When empty Palliser shells are kept at the guns, where the plug is liable to become set fast by corrosion from the action of salt water, the base plug should be unscrewed once at least every six months, and the screw coated with a mixture of equal parts of cocoa-nut oil and finely powdered chalk.

*Fixing Plugs and Fuzes.*—When plugs or metal fuzes are screwed into shells they should be lubricated with a mixture of equal parts of cocoa-nut oil and finely powdered chalk.

*Distinguishing Marks.*—All filled shells must be marked with the words "Filled," and date, and also "Bag,"* if a bag is used.

IMPLEMENTS FOR PREPARING FUZES.

1 set per gun, 64-pr. or 80-pr. R.M.L. and 40-pr. or 7-inch R.B.L.

1 set for every 2 pieces of S.B. ordnance.

The set contains the following implements, which are to be demanded separately:—

| | |
|---|---|
| Bits, hook borer .. .. .. .. .. .. | 6 |
| Borer, hook, with handle .. .. .. .. | 1 |
| Cylinder, wood, common .. .. .. .. | 1 |
| Instructions, printed sheets, to every 2 pieces .. | 1 |
| Key, iron, fuze and plug, G.S. .. .. .. .. | 1 |

PREPARING FUZES.

*Fuzes, Time, Wood, Boxer.*—The Boxer wood time fuzes are prepared for any desired time of flight by boring through the side hole corresponding to the required time into the composition.

When using the hook borer, place the fuze in the hook in the proper position for boring the required hole,† enter the bit into the side hole,

* These words, &c., to be stencilled in red on the head of the shell, except in the case of the Shrapnel shell, which have red heads; on these the stencilling is to be black.

† Head of fuze should be towards the body, the handle of the borer towards the right hand.

When using the hook borer with such a fuze as the small mortar or 20-seconds, great care must be taken to prevent the fuze turning to one side; if this is not attended to the bit may pass to one side of the composition, specially when boring near the small end of the cone.

screwing up until the bit has entered as far as the borer will allow taking care not to press upon the fuze so as to prevent its bedding fairly in the hook.

Unscrew, and when the bit is quite clear remove the fuze from the hook. The length of the bit is so regulated that, when placed in the handle, it will enter sufficiently far into the composition when screwed down to the shoulder. If the bit should become unserviceable the handle must be detached from the shank and the tightening-screw unscrewed, the square hole in the hook being made for that purpose. Care must be taken when substituting another bit that it is properly placed in the handle, *and that the tightening screw firmly presses upon it, for if any space be left between the handle and the head of the bit the end will not enter a sufficient depth into the composition.* The borer should be occasionally examined and cleaned. The operation of preparing the fuze and fixing it in the shell takes, on an average, about 15 seconds; with a little practice these operations may be performed in a shorter time.

When using the gimlet-borer, hold the fuze in the hollow of the hand, enter the borer into the side-hole, pressing it in perpendicular to the axis of the fuze; when it has reached the bottom of the hole, use it as a gimlet to complete the communication with the composition, boring up to the handle; then pull the borer straight out.

*Fuzes, Time, Metal, Armstrong, E.*—The Armstrong metal time fuzes are prepared for the desired time of flight after they have been screwed into the shell.

Loosen the nut with the key, so as to free the collar. Hold the shell in the right hand and move the collar with the fore-finger and thumb of the left hand, until the arrow points to the required length of fuze; then screw up the nut with the right hand, steadying the collar at the same time with the left hand. Finally, tighten the nut with the key, or by inserting it in the socket on the axletree-box, giving the shell a slight turn towards the trail. Should the fuze be taken from the shell, the collar must be set at the "blank" before it is replaced in the box.

### *Fixing Fuzes.*

*Fuzes, Time, Wood, Boxer, Common, and Diaphragm*—Remove the fuze-hole plug and place the fuze in the fuze-hole of the shell, and give the head of the fuze two or three smart taps with the mallet, or against the gun carriage if more convenient. Before the fuze is placed in the socket of the diaphragm shell, care must be taken to remove any superfluous quantity of powder which may have passed into the socket through the communication hole. This is important, because the shell will burst prematurely if the powder in the socket prevent the fuze being securely fixed.

The fuze should not be uncapped until the shell is placed in the muzzle of the gun; this much reduces the chance of an accident and secures the priming from injury.

*Fuzes, Time, Wood, Boxer, Breech-Loading.*—The fuzes for B.L. shells should be screwed into the fuze-hole by hand; when they cannot be screwed any further they are properly secured. *In fixing the B.L. Boxer time fuzes, neither a mallet nor any other instrument is to be used.* The safety pin is not to be withdrawn until just before entering the shell into the breech.

*Fuzes, Time, Wood, Boxer, Muzzle-Loading.*—The fuzes for muzzle-loading shells are fixed in the fuze-hole by screwing the fuze round by hand until it is held firmly in the fuze-hole, or in the land service by giving the head of the fuze two or three smart taps with a mallet, or suitable piece of wood, or (with the smaller natures of shell) by striking them against the gun carriage, if more convenient. This operation should be performed fairly, and not so as to split or injure the top of the fuze. The fuze must not be uncapped until the shell is placed in the muzzle of the gun. The fuzes for rifled M.L. guns are "uncapped" by taking hold of the small end of the copper band, which is left exposed, and unwinding from left to right smartly, so as to thoroughly detach the band from the head of the fuse and to leave the priming fully exposed.

*Fuzes, Percussion, B.L., Plain.*—These fuzes are used in the B.L. 20-pr., 12-pr., 9-pr., and 6-pr. segment shells, and in the 20-pr. land service 12-pr. and 9-pr. B.L. common shells; they require no preparation except the removal of the safety pin, which is to be taken out just before dropping the fuze into the shell; care must be taken to insert it properly; the part with the rim round it is the top of the fuze; it should be firmly held in its place by the fuze-hole plug, or the E. time fuze, whichever may be used, being screwed tightly down on to it.

*Fuzes, Percussion, Pettman, G.S. and L.S.*—These fuzes require no preparation; they are simply screwed firmly into the fuze-hole by means of the "Key, iron, fuze, and plug, G.S." In the case of certain shells* filled for the Navy, the wad with loop, see § 2370, will be placed over the G.S. fuze.

*Fuzes, Percussion, R.L., and Sensitive*—The R.L. fuzes require no preparation, except the removal of the safety pin; they are screwed firmly into the fuze-hole by means of the "Key, iron, plug, G.S."

The safety pin is *not to be* removed until the shell is placed in the muzzle of the gun. In the case of B.L. shell, it must be removed just before loading. In the case of the sensitive fuze the safety pin is not removed, but the fuze is uncapped when the shell is in the bore.

*Wad, Papier Mâché, in Fuze-Holes.*—When fixing fuzes in shells having a wad in the fuze-hole, or in the bottom of the socket of 20-pr. land service, 12-pr. and 9-pr. B.L. common shells, it is not necessary to remove the wad, as the explosion of the fuze is sufficient to force it into the shell, if using percussion fuzes; and if using wood time fuzes, the wad should be driven into the shell in the operation of fixing the fuze.

## Extracting Wood Fuzes.

*Fuzes for Rifled Ordnance.*—Apply the fuze-extractor to the head of the fuze and unscrew; if the adapter which is in the fuze-hole of some B.L. shells should also be unscrewed, do not remove the fuze from it by striking it on the end, as a blow in that direction may weaken or break the wire that suspends the hammer in the breech-loading fuze.

*Fuze for Common and Diaphragm Shrapnel.*—Clear out the cup of the fuze with the projecting piece of metal on the handle of the fuze-extractor; take a firm hold of the head of the fuze between the jaws of

---

* 20-pr. B.L., common; 40-pr. and 7-inch B.L., common and segment shell; and common shell, for 64-pr. R.M.L. and upwards.

the fuze-extractor and turn from left to right. The small knob between the jaws fits into the cup of the fuze and prevents the top from collapsing or giving way.

An "extractor, fuze, large" is supplied for use with 13, 10, and 8-inch mortar shells on demand.

## Implements for preparing and fixing Large Mortar fuzes.

One set to every 2 Mortars.

A set contains the following implements, which are to be demanded separately:—

| | |
|---|---|
| Bits, mortar .. .. .. .. | 6 |
| Brace ,, .. .. .. .. | 1 |
| Cylinder, wood, mortar .. .. .. | 1 |
| Mallets, mortar .. .. .. .. | 1 |
| Instructions, printed sheets .. .. | 1 |

*Fuzes, Time, Wood, Boxer, Mortar.*—Hold the fuze firmly in the left hand; insert the point of the bit into the required hole; place the head of the brace against the body and turn with the right hand until the stop comes in contact with the wood; reverse the motion until the bit is clear of the fuze.

The wooden bottom of the fuze must on no account be cut off, as it supports the composition and prevents its being disarranged by the shock at the discharge.

The shell is filled in the ordinary way, and the fuze, previously bored to the length required, inserted into the fuze-hole and set home with the mallet.

Shell and fuze implements carried by Horse Artillery and Field Batteries. Clause 138, A.C. 1877.

The batteries R.H.A. and field batteries R.A. are provided with the shell and fuze implements in the proportions shown in the following table:—

| Implements. | R.M.L. | | | | | | R.B.L. | | | | | | | | Remarks. |
|---|---|---|---|---|---|---|---|---|---|---|---|---|---|---|---|
| | 16-pr. | | 9-pr. | | 7-pr. | | 40-pr. | | 20-pr. | | 12-pr. | | 9-pr. | | |
| | Gun carriage and limber. | Ammunition wagon. | Gun carriage and limber. | Ammunition wagon. | Gun carriage and limber. | Ammunition wagon. | Gun carriage and limber. | Ammunition wagon. | Gun carriage and limber. | Ammunition wagon. | Gun carriage and limber. | Ammunition wagon. | Gun carriage and limber. | Ammunition wagon. | |
| Bits, hook borer ... ... | 6 | 6 | 6 | 6 | 6 | 6 | 6 | 6 | — | — | 6 | 6 | 6 | 6 | For implements used with 40-pr. B.L. garrison service, see p. |
| Cylinders, wood, common ... | 1 | 1 | 1 | 1 | 1 | 1 | 1 | 1 | — | — | 1 | 1 | 1 | 1 | |
| Drifts, wood, G.S. ... ... | — | 1 | — | 1 | — | 1 | — | 1 | — | — | — | — | — | — | |
| Drivers, screw, diaphragm Shrapnel, large | — | 1 | — | 1 | — | 1 | — | — | — | — | 1 | 1 | 1 | 1 | |
| Extractor, fuze, small ... | 1 | — | 1 | — | — | 1 | 1 | — | — | — | 1 | — | 1 | — | |
| Funnels, leather, with copper spouts — common, large | — | — | — | — | — | — | 1 | — | — | — | — | — | — | — | |
| Funnels, leather, with copper spouts — ,, small | — | 1 | — | 1 | — | 1 | — | — | 1 | — | — | 1 | — | 1 | |
| Handles, hook borer ... ... | 1 | 1 | 1 | 1 | 1 | 1 | 1 | 1 | — | — | 1 | 1 | 1 | 1 | |
| Hooks, hook borer ... ... | 1 | 1 | 1 | 1 | 1 | 1 | 1 | 1 | — | — | 1 | 1 | 1 | 1 | |
| Instructions, printed, sheets, rifled ordnance | 1 | 1 | 1 | 1 | 1 | 1 | 1 | 1 | 1 | 1 | 1 | 1 | 1 | 1 | |
| Key, plug, G.S. ... ... | 1 | 1 | 1 | 1 | 1 | 1 | 1 | 1 | 1 | 1 | 1 | 1 | 1 | 1 | |

* In Appendix to Cl. 138 A.C., 1877, will be found a similar table for Field and Mountain Service, S.B. guns. As no S.B. batteries now exist it is not considered necessary to quote it at length here.

Hook Borer.*
Mark II.

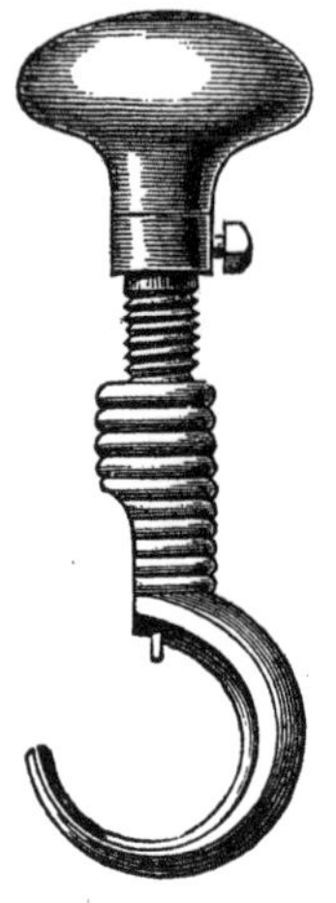

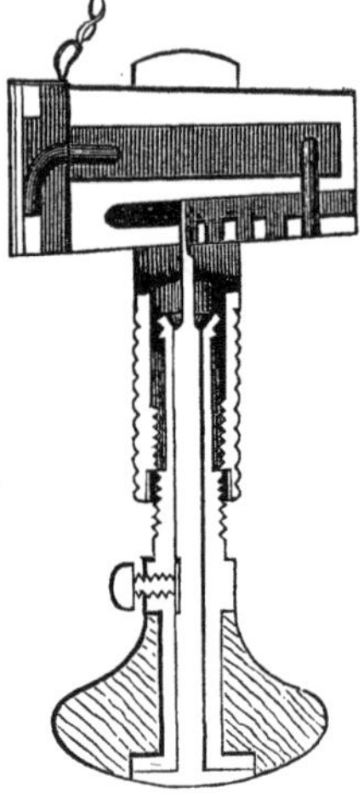

Cylinder, Wood, Common with Bits.

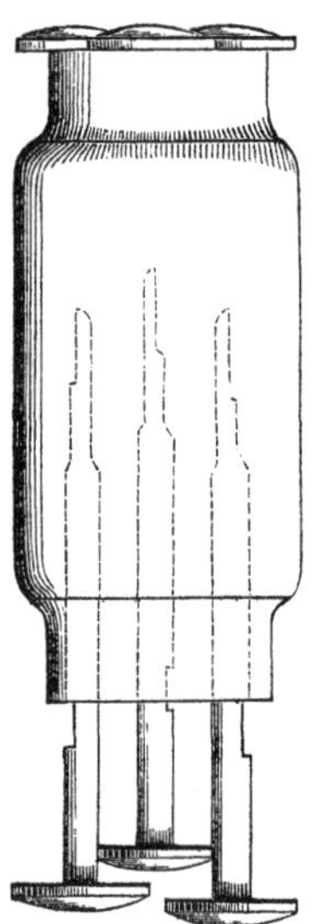

Bit for Hook Borer.

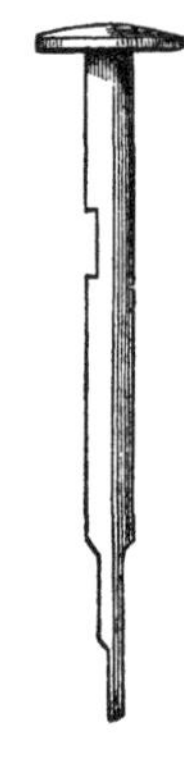

Key, iron, plug and fuze, G.S., § 1308.
Cl.91, A.C./73.

*Key, iron, plug, and fuze, G.S.*—One arm, *b*, intended for pillar fuzes, is used for the G.S. plug and adapters, the other arms, *a* and *c*, are

§ 3030.

§§ 1928, 2123.

* The hook-borer, Mark III., has the same bit as previous patterns. The handle and hook differ from Mark II. in having the screw thread continuous, and thus no longer allowing the bit to be pushed in or withdrawn without screwing or unscrewing. Mark I. had a continuous screw thread. Mark III. differs from both Marks I. and II. in having the interior of the hook somewhat V-shaped at the part where the fuze rests instead of being conical, and it has a piece of metal added on each side, so as to give an increased longitudinal bearing in the V for the fuze. This pattern is to be issued to R.H.A. and Field Artillery. The previous patterns are to be used up for other services. A gimlet-borer was issued to R.H.A. and Field Batteries, but is superseded by the above.

used for the common gauge plug, and Pettman's L.S. and G.S. fuzes. See cut.

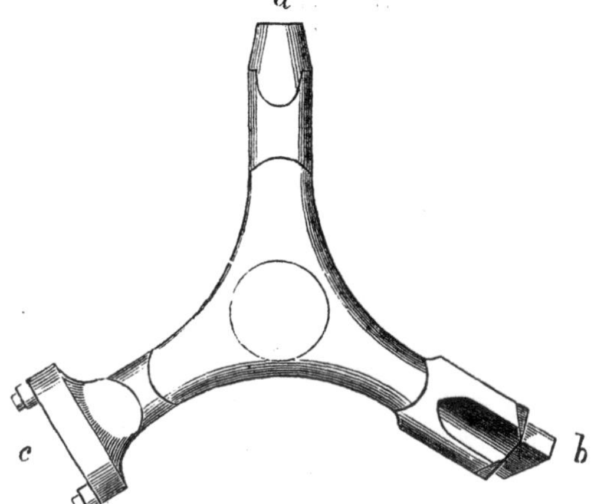


*Key, plug, general service*, is simply for screwing or unscrewing G.S. plugs and R.L. and sensitive percussion fuzes.

Key, plug, G.S., § 1992.

There is also a *key, Armstrong, Universal*, for use with the Armstrong E. time fuze, but this is only now issued for S.S.

Key, Armstrong, Universal, § 1683.

A *key, common plug*, was issued in the sets for S.B. guns.

Key, common plug.

A *G.S.drift* is issued to insert the wad in the common shell for rifled guns.

Drift, wood, G.S., § 2045.

IMPLEMENTS, FUZE AND SHELL, SERVICE AND PROPORTION OF.

For 80 and 64-pr. Rifled Converted Guns—

| | | |
|---|---|---|
| Implements, fuze and shell .. | Bits, hook borer .. .. .. .. .. .. | 6 per gun |
| | Cylinders, wood, common .. .. .. .. .. | 1 ,, |
| | Handles, hook borer .. .. .. .. .. | 1 ,, |
| | Hooks ,, ,, .. .. .. .. .. | 1 ,, |
| | Instructions, printed, sheets, rifled ordnance .. .. | 1 to 2 guns |
| | Keys, iron, fuze and plug, G.S. .. .. .. .. | 1 per gun |

For 7-inch and 40-pr. R.B.L. Guns—

| | | |
|---|---|---|
| Implements, fuze and shell .. | Bits, hook borer .. .. .. .. .. .. | 6 per gun |
| | Cylinders, wood, common .. .. .. .. .. | 1 ,, |
| | Handles, hook borer .. .. .. .. .. | 1 ,, |
| | Hooks ,, ,, .. .. .. .. .. | 1 ,, |
| | Instructions, printed, sheets, rifled ordnance .. .. | 1 to 2 guns |
| | Keys, iron, fuze and plug, G.S. .. .. .. .. | 1 per gun |

For Cast-iron Guns, Howitzers, and Carronades—

| | | |
|---|---|---|
| Implements, fuze and shell .. | Bits, hook borer .. .. .. | 6 per 2 pieces of ordnance. |
| | *Blocks, wood { 10-inch .. .. | 1 per station, 2 if large station. |
| | 8 ,, .. .. | ,, ,, |
| | 32-pr. .. .. | ,, ,, |
| | Cylinders, wood, common .. | 1 per 2 pieces of ordnance. |
| | *Gauges, fuze-hole ,, .. | 1 per station, 2 if large station. |
| | Handles, hook borer .. .. | 1 per 2 pieces of ordnance. |
| | *Holders, shell, spherical .. .. | 1 per station, 2 if large station. |
| | Hooks, hook borer .. .. | 1 per 2 pieces of ordnance. |
| | Instructions, printed, sheets, S.B. ordnance | ,, ,, |
| | Keys, iron, fuze and plug, G.S. .. | ,, ,, |
| | * ,, square .. .. | 1 per station, 2 if large station. |
| | *Levers, common .. .. .. | ,, ,, |
| | *Rectifiers, or rimers, common, fuze-hole | ,, ,, |
| | *Screws, coach, 4-inch by ½-inch .. .. | 4 per shell holder. |
| | *Taps, screw, Pettman's .. .. | 1 per station, 2 if large station. |

The implements marked * will be retained only where shells still require to be rectified, and will cease to be recognized as part of the equipment for future issue.

For 13, 10, and 8-inch Cast-iron Mortars—

| |  |  |
|---|---|---|
| Implements, fuze and shell .. | Bits, mortar .. .. .. | 6 per 2 mortars. |
| | *Blocks, wood, 13-inch .. .. | 1 per station, 2 if large station. |
| | Braces, mortar .. .. .. | 1 per 2 mortars. |
| | Cylinders, wood, mortar .. .. | ,, ,, |
| | Extractors, fuze, large .. .. | 1 per station. |
| | *Gauges, fuze-hole, mortar .. | 1 per station, 2 if large station. |
| | *Holders, shell, spherical .. .. | ,, ,, |
| | Instructions, printed, sheets, S.B. ordnance | 1 per 2 mortars. |
| | *Levers, mortar .. .. .. | 1 per station, 2 if large station. |
| | Mallets ,, .. .. .. | 1 per 2 mortars. |
| | *Rectifiers or rimers, mortar, fuze | 1 per station, 2 if large station. |
| | *Screws, coach, 4-inch by ½-inch .. | 4 per shell holder. |

For 5½ and 4⅖-inch Bronze Mortars—

| |  |  |
|---|---|---|
| Implements, fuze and shell .. | Bits, hook borer .. .. .. | 6 per 2 mortars. |
| | Cylinders, wood, common .. | 1 ,, ,, |
| | *Gauges, fuze-hole.. .. .. | 1 per station, 2 if large station. |
| | Handles, hook borer .. .. | 1 per 2 mortars. |
| | *Holders, shell, spherical .. .. | 1 per station, 2 if large station. |
| | Hooks, hook borer .. .. | 1 per 2 mortars. |
| | Instructions, printed, sheets, S.B. ordnance | ,, ,, |
| | Keys, iron, fuze and plug, G.S. .. | ,, ,, |
| | * ,, square.. .. .. | 1 per station, 2 if large station. |
| | *Levers, common .. .. .. | ,, ,, |
| | *Rectifiers or rimers, common, fuze-hole | ,, ,, |
| | *Screws, coach, 4-inch by ½-inch .. | 4 per shell holder. |

The implements marked * will be retained only where shells still require to be rectified, and will cease to be recognized as part of the equipment for future issue.

For use in Shell Filling Room—

| | Article | Allowance | Remarks |
|---|---|---|---|
| Implements, fuze and shell. | Blocks, wood, riveting .. .. | 1 per shell filling room | For S.B. ordnance, except mortars. |
| | Drifts, wood, common .. .. | ,, ,, | ,, ,, |
| | Drifts, wood, dia. Shrapnel, large .. | ,, ,, | ,, ,, |
| | Drifts, wood, dia. Shrapnel, small.. | ,, ,, | ,, ,, |
| | Drivers, screw, diaphragm, Shrapnel, large .. | ,, ,, | For R.B.L., 80 and 64-pr. R.M.L., and S.B. ordnance above 18-pr., except mortars. |
| | Drivers, screw, diaphragm, Shrapnel, small.. | ,, ,, | For S.B. ordnance 18-pr. and under, except mortars. |
| | Extractors, fuze, small .. .. | ,, ,, | For R.B.L. and 80 and 64-pr. R.M.L. only. |
| | Funnels, leather, with copper spout, common, large .. | 2 ,, ,, | For all ordnance except heavy R.M.L. guns, but only 1 allowed where there is S.B. only. |
| | Funnels, leather, with copper spout, dia. Shrap., large.. | 1 ,, ,, | For S.B. ordnance above 18-pr., except mortars. |
| | Funnels, leather, with copper spout, dia. Shrap., small.. | ,, ,, | For S.B. ordnance 18-pr. and under, except mortars. |
| | Hammers, riveting .. .. | ,, ,, | For S.B. ordnance, except mortars. |
| | Holders, shell, Palliser .. .. | 1 per shell filling room for each nature of Palliser shell | For 7-inch R.M.L. guns and over. |

| | | | |
|---|---|---|---|
| Implements, fuze and shell. | Instructions, printed sheets — Rifled ordnance | 2 per shell filling room | For R.B.L., and R.M.L. ordnance only. |
| | Instructions, printed sheets — S.B. ordnance.. | ” ” | For S.B. ordnance only. |
| | Keys, iron, fuze and plug, G.S.. | ” ” | For all ordnance except cast-iron mortars, but only 1 allowed where there is none but S.B. |
| | Mallets — common and diaphragm | 1 ” ” | For S.B. ordnance, except cast-iron mortars. |
| | Mallets — Shrapnel | | |
| | Mallets — mortar .. .. | ” ” | For cast-iron mortars. |
| | Prickers, removing wax.. .. | ” ” | For S.B. ordnance, except mortars. |
| | Punches, iron, riveting bottoms | ” ” | ” ” |
| | Rectifiers or rimers, common rivet | ” ” | ” ” |
| | Spanners or wrench box .. | ” ” | ” ” |
| | Wrenches, removing plug .. | ” ” | For 7-inch R.M.L. guns and over. |
| Mallets, tent .. .. .. | | 2 ” ” | For R.B.L. and R.M.L. guns only. |
| Blocks, wood, or shell holder .. | | 1 ” ” | For Palliser shell only. |
| Funnels, copper, shell, large.. .. | | ” ” | For 7-inch R.M.L. guns and over. |
| Rods, brass, filling shell .. .. | | ” ” | ” ” |

## Papier Mâché Wads.*

The following papier mâché wads are used with various shells:— *Wads, papier mâché.*

*Wad, Papier Mâché, Fuze-hole, Common.*—It is a disc of millboard, ·25-inch thick and ·95-inch across. They are soaked in melted bees-wax. They are used to place in the fuze-holes of filled common S.B. shells, when the latter are filled and not required for immediate service. § 853.

*Wad, Papier Mâché, Fuze-hole, Diaphragm Shrapnel Shell.*—These wads are of two sizes; the larger for use with diaphragm Shrapnel of 24-prs. and upwards; the smaller wads for the lower natures. They differ from the wads described above only in size, being made to fit the loading holes of the diaphragm Shrapnel shell. § 853.

*Wad, Papier Mâché, Fuze-hole, B.L. Sockets,* is used in conjunction with the sockets of 9-pr., 12-pr. and 20-pr. B.L. common shell,† and is used when the shell is carried loaded. They are placed in the hole in the bottom of the socket, recessed part upwards. § 1708.

*Wad, Papier Mâché, Fuze-hole, Naval, Plain,* is a plain millboard wad, saturated with beeswax, for insertion in the fuze-hole of all empty rifled shell (of garrison calibres) when issued for naval service. Its use is to protect the plug and prevent corrosion. The top is painted blue. § 1346.

---

* As there is a considerable vagueness about the use of the terms "papier mâché," "millboard," "pasteboard," I here introduce the definitions of them as understood in the trade, for which I am indebted to the courtesy of Messrs. Wiggins, Teape, and Co., wholesale stationers.

*Papier mâché* is formed by *glueing* together sheets of strong brown paper, the upper surface being sometimes coated with a surface of hard enamel. There is also a papier mâché made by running pulp into moulds, and subsequently coating it.

*Millboard* is produced of any required thickness by successive dips into the vat of pulp; in other words millboards are really solid single sheets of paper.

*Cardboard* is formed by pasting together as many sheets of fine paper as will give the required thickness; so that both "middles" and "facings" are of the same quality.

*Pasteboard* is made like cardboard, but the "middles" are of an inferior quality of paper, and the "facings" only of a fine quality.

† L.S. only for 20-pr., as 20-pr. S.S, has G.S. gauge.

§§ 2370, 2413.

*Wad, Papier Mâché, Fuze-hole, with Loop.* is a plain millboard wad, saturated in beeswax, and fitted with a small loop of raw hide on the top. It is placed over the G.S. Pettman fuzes when these are carried in filled shell for S.S., and is coated over with Venetian red cement.* The use of the loop is to pull off the wad when it is desired to extract the fuze.

§§ 2075, 2527, 2627.

*Wad, Papier Mâché, Fuze-hole, G.S.*—The *G.S. wad*, Mark II., serves

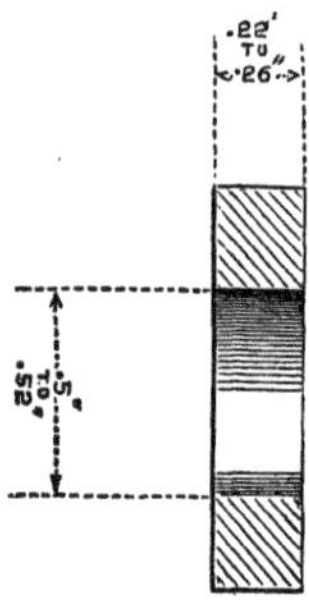


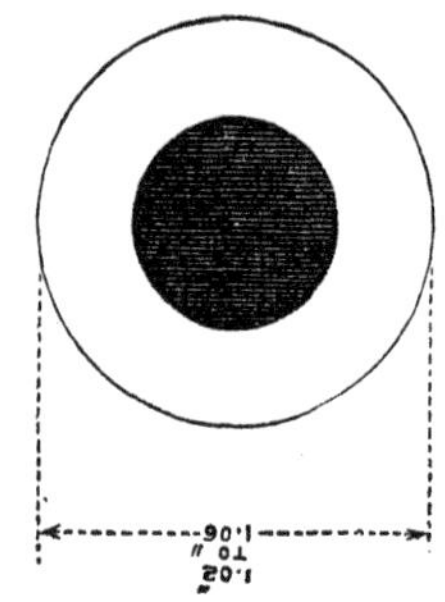


to prevent the powder from working up in the fuze-hole of common shell. It is made of papier mâché, and has a hole in the centre covered by shalloon cemented to one side. This wad is forced in by fixing the time fuze, and does not require removing when the percussion fuze is used. The side covered with shalloon is placed downwards in the shell.

Use.
§§ 3160.

It is used in all common and double shell for rifled guns which do not have their bursting charge contained in a serge bag, and which are not to be fired immediately after filling. In other words, it is to be used with all common shell up to the 80-pr. inclusive, and with the 7-pr. double shell, when these projectiles are carried or stored filled.

Copper scrapers.
§§ 2174, 2823.

*Copper scrapers* are used in removing powder from filled shell. There are three sizes. They consist of round copper rods, having both ends flattened down. One end is turned up nearly at right angles to the body of the rod, the other end has a slight bend in the opposite direction.

The largest size is 42 inches long, and is issued for use with 10-inch R.M.L. shells and upwards.

The next size is 32·25 inches long, and is issued for use with 80-pr. R.M.L. to 9″ R.M.L. with 7″ R.B.L., and with 13″ S.B.

The smallest size is 19·9 inches long, and is issued for use with 7-pr. R.M.L to 64-pr. R.M.L. with all R.B.L. calibres (7″ excepted), and with all S.B. calibres, excepting 13″.

Powder horns.
§ 1125.

One *miner's powder horn* is issued per battery for garrison guns. It contains $1\frac{1}{2}$ lbs. of F.G. powder, and is used for priming guns if required.

Pockets, buff.
§ 1994.

A *buff leather pocket* or pouch to hold 5 Boxer fuzes is issued for F.S. It is fitted with a strap to fasten round the waist. It also carries a Boxer.

---

§ 2370.

* See table of paints, &c., p. 322.

Directions for fixing :—

1. The edge of the wad is to be well covered all round with cement.
2. The wad is to be driven well into the recess over the fuze.
3. The cement is then to be spread over the whole surface of the wad and round the edge, putting it also on the rivet and under the loop.

N.B.—The cement must be well stirred before it is used, and if it has become too thick a little spirits of wine must be added

There are two patterns of this wad differing only in the mode of attaching the raw hide loop. In Mark I. it is riveted on with a copper rivet; in Mark II. it is sewn on with waxed thread. The change was made in consequence of the loops becoming deteriorated from the oxidation of the copper.

# CHAPTER VII.—MONOGRAMS OF STATIONS.—CASES AND BARRELS USED FOR FILLED CARTRIDGES AND GUNPOWDER.—HINTS ON COOPERING.

---

*Monograms of Stations.*

To enable the place where cartridges, shells, &c., are filled to be traced, each station where there is a Laboratory, &c., has a monogram. §§ 1633, 2294.

HOME STATIONS.

| Station | Monogram | Station | Monogram |
|---|---|---|---|
| ALDERNEY | A. | GOSPORT | G. |
| CHATHAM | C. | HARWICH | H. |
| CHESTER | H(R)C. | GUERNSEY | GG. |
| CORK | Я(K)C. | JERSEY | J. |
| DEVONPORT | D. | PEMBROKE | P. |
| DOVER | V(R)D. | SHEERNESS | S. |
| DUBLIN | Я(N)D. | TYNEMOUTH | T. |
| EDINBURGH | E. | UPNOR | U. |
| FORT GEORGE | FG. | WOOLWICH | W. |

FOREIGN STATIONS.

| Station | Monogram | Station | Monogram |
|---|---|---|---|
| BARBADOES | B(DS). | HALIFAX | H. |
| BERMUDA | B. | HONG KONG | H.K. |
| CAPE TOWN | C(T). | MALTA | M. |
| CEYLON | C(N). | MAURITIUS | ЯMS. |
| GIBRALTAR | GIB. | | |

*Cases and Barrels.*

As the same case may serve for cartridges for rifled and S.B. guns, it will be convenient to enumerate them all here.*

1. Metal-lined cases.
2. Pentagon cases.
3. Brass rectangular cases, plain.
4. " " " corrugated.
5. Gun ammunition barrels.
6. Boxes.
7. Zinc cylinders.
8. Powder barrels.

---

* Dell's case is obsolete; for disposal, see C. 75, A.C., 1869.

For capacity and weights of various cases, see tables, pages 172, 220, 238, 252, 325.

Luting.

A luting of equal parts of tallow and beeswax is used on the lid of the metal lined and metal cases to exclude the air as much as possible, and should always be carefully applied before stowing away the cases in magazines.

1. Metal-lined case.

*_Metal-lined cases_ are of three sizes,—whole, half, and quarter; they are rectangular cases of deal, the corners of oak, and the cleats of ash, lined with tinned copper. Their dimensions are:—whole 17″×20½″. Half 13½″×16½″. Quarter 10½″×14″.

A square lid opens on hinges on top of the case; it is screwed down by two gun-metal bolts by means of a gun-metal key; this lid covers a circular opening which is closed by a bung of tinned copper; the bung is luted into its place when the case is full.

The whole size will take all S.B. cartridges, and rifle cartridges up to the 9″ inclusive, except "P" powder; the two smaller sizes are generally used by the Navy for small combustible stores, and blank S.A. cartridges.

Metal-lined cases are used in magazines which are not very dry, sailing vessels, and siege trains.

The half and quarter cases are used to contain small arm ammunition. For the numbers of various small arm cartridges contained in them, see table p. 80.

§ 2290.

When filled at Woolwich a paper label is pasted across the edge of the lid immediately under the ring, having the packer's name and date of issue. On the case is stencilled the tare and gross weight, the lid is marked with the nature and number of cartridges, the station intended for, and date of their manufacture. Similar information would be given with the other cases.

2. Pentagon case. § 2483.

_Pentagon case_ of 2 sizes, whole and half. The lid hinges on a curved bolt; there are slots in the projecting rim of the lid and corresponding projections on the neck of the case; the lid will only open when the slots and projections are in a corresponding position. The dimensions of the whole size case are, 19·3″ × 15·5 × 11″.

Suppose the lid to be closed. To open it, first with the spanner unscrew the screw which presses on the curved bolt, then place the curved projection on the lever into the eye of the curved bolt, the other projection bearing against the lid, and turn from left to right; the lid of the case will then be opened. To close the lid, you turn from right to left.

There is a second socket furnished for the bolt in case the other should get broken. The body of the case is made of sheet brass, the top and fittings of cast brass.

The whole size takes all S.B. cartridges, and rifled up to 8″ inclusive ("P" powder excepted), and also the reduced charges of 15 lbs. for the 9″ R.M.L. gun,

The half case is produced by taking a section of the pentagon along a line bisecting the long side of the head and perpendicular to it. It has four sides, and is used by the Navy for convenience in stowage, generally used for small stores.

The shape enables the pentagon case to pack well in a ship's magazine. These cases were introduced for naval use.

3. Brass rectangular case, plain. § 975.

_Brass Rectangular Case_ is made of sheet brass, cast brass top and fittings. It opens on the same principal as the pentagon. The head

* Bags, calico, metal-lined, or pentagon cases (Mark I.), L.S., § 2431. When powder, not made up in cartridges or bags, is stored in these cases the calico bag will be used.

working on the curved bolt is a ring in this case. It has two holes which take the projections on the lever.

The case will take all S.B. cartridges and rifle cartridges up to the 9″ inclusive, "P" powder excepted. It was specially made for the 100-pr. and 150-pr. No more will be made. The dimensions are, $22'' \times 18\frac{1}{4}'' \times 11''$.

**4. Brass rectangular corrugated case.**

*Brass Rectangular Case, Corrugated.*—Made of corrugated sheet brass, with cast gun-metal top and fittings. The corrugations strengthen the case. These cases are used for S.S. only.

Powder Case, Corrugated Metal, Rectangular. "A." Mark II. *

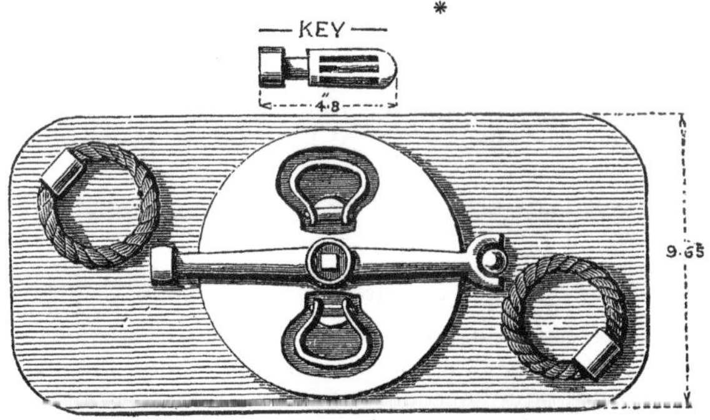


Top view.

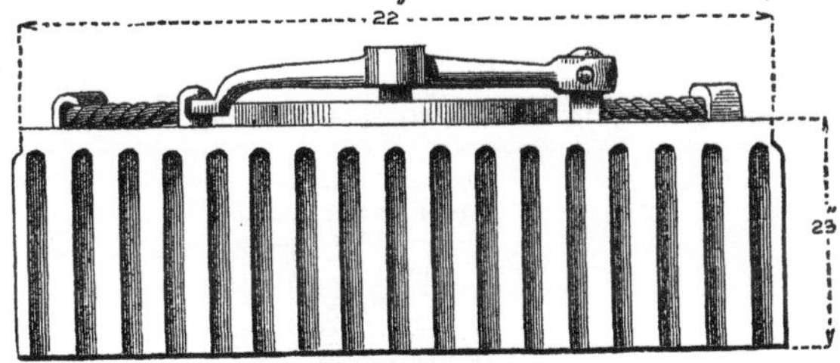


Side view.
$\frac{1}{8}$th full size.

* Mark II. Key made of gun-metal, with a cross handle bevelled off at one end for the purpose of extracting the ring handles of the lid of the case, if they should become fixed in the recess.

§ 2538.

§ 1864.

On 30/11/72 copper wire beckets, covered with leather, were approved for the D. case, and tarred rope provisionally for the other natures. (F case has the covered beckets.) The set screw in the cross bar, which fastens the lid, was found to be too weak. Stronger ones are substituted in cases made since 1869, and in old ones when repaired.

D, E, and F cases have a wood lining to the top to prevent the cartridges being cut by the lower ends of some of the top fittings.

It is opened by unscrewing the central screw bolt and pulling the bar to one side, thus allowing the lid to be lifted off.

§ 1909. There are six sizes, known as A., B., C., D., E., and F.

§ 1402. "A."* takes 7″, 8″, and 9″ R.M.L. cartridges. Dimensions 22″ × 23″ × 10″.

§ 1700. "B." takes 10″ R.M.L. cartridges; either two battering, or three full charges of "P" powder. 26″ × 20″ × 11″.

§ 1771. "C." takes 12″ 25-ton R.M.L. cartridges; either two battering, or three full charges of "P" powder. 22″ × 24″ × 13″.

§ 2208. "D." will take either two battering, or three full charges of "P" powder for the 12″ 35-ton gun. 24″ × 32″ × 13″.

§ 2776. "E." takes two battering, or three full charges of "P" powder for the 11″ R.M.L. gun. $20\frac{1}{2}$″ × 12″ × $28\frac{1}{2}$″.

§ 3061. "F." takes two battering, or two full charges of "$P^2$" powder for the 12″·5 R.M.L. gun; or two battering or three full charges of "P" powder for the 12″ R.M.L. gun of 35 tons. $24\frac{1}{2}$″ × $34\frac{1}{2}$″ × $13\frac{1}{2}$″.†

Before issue they are tested by a hydraulic pressure of 10 lbs. to the square inch. They will take all S.B. cartridges, and all the smaller cartridges for rifle guns.

5. Gun ammunition barrel. § 774. *Gun Ammunition Barrel.*—Of two sizes, whole and half; full bound; four copper hoops; staves of oak, or teak for tropical climates. The lid is generally teak, and has a circular opening, into which a wooden lid fits, working on a hinge and secured by a screw bolt. A gun-metal key is used to screw or unscrew the bolt. The whole size will contain all S.B. cartridges.

These barrels are the same size as powder barrels, but are not intended to contain loose powder. They are used in dry magazines to contain cartridges. No luting is used to close the lid.

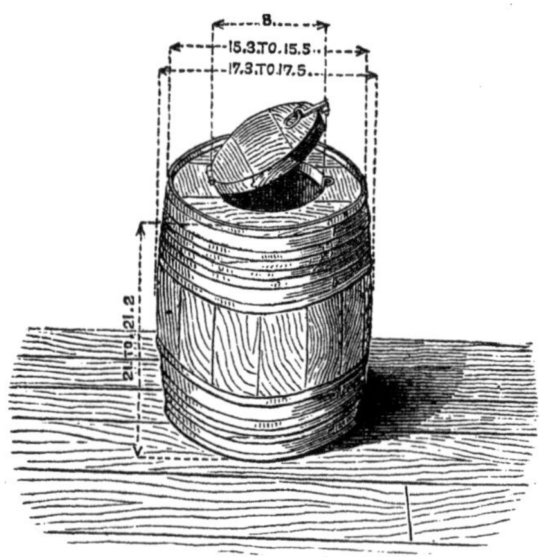


6. Ammunition box. § 1369. *Ammunition Boxes* are used for the issue of cartridges to field

* Mark I. "A." corrugated case opens in the same way as the plain rectangular case, the neck and lid projects beyond the sides, entailing the necessity of placing battens between the cases, thus causing loss in the magazine. On return to Woolwich they are altered to Mark II. before re-issue.

N.B.—The B., C., D., E., and S. cases have their openings at one end instead of the centre.

† For details as to exact number of cartridges contained in the cases, see table, p. 220. The "F" is not yet (December, 1877) sealed.

batteries at home. They are marked with the nature and number of cartridges.

7. Zinc cylinders. §§ 2171, 2172, 2173, 2185, 2270, 2760.

*Zinc Cylinders* (see table p. 79 for dimensions), are used in the L.S. to contain the cartridges of the Woolwich guns. Each cylinder holds one cartridge; the 7-inch, Mark II., will hold two 14 lb. charges, or one battering charge. They not only act as cases in the magazine, but also serve to bring the cartridge up to the gun. Their construction will be understood from the sketch. The 12″·5, the 12″, or 11″ are not to be stacked more than three deep, the 10″ and 9″ four deep, the 7″ five deep.

§§ 2437, 2469, 2295, 3187.

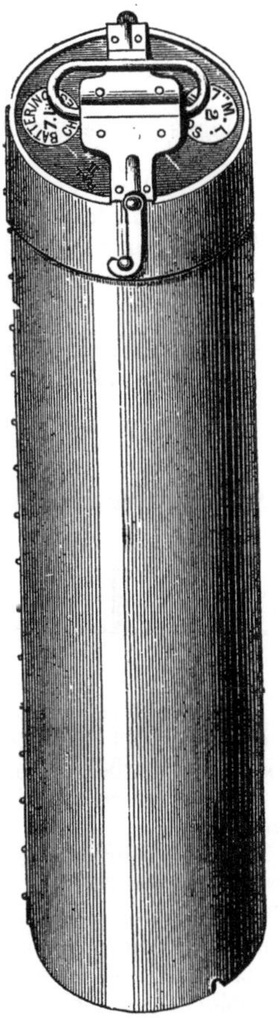

None have yet been sealed for the 8″ gun, which was introduced for S.S.

The 9, 10, 11, and 12-inch 25-ton guns have each two cylinders, one for the battering, and one for the full charge; the cylinder for the battering charge of the 12″ gun of 25 tons serves for the full charge of the 12″ gun of 35 tons. The latter has a cylinder for its battering charge. The 12″·5 gun has a cylinder, at present, only for the battering charge.

§ 3042.

Cylinders of recent manufacture will contain either pebble or R.L.G. cartridges.

§ 2891.

A hook fastening has replaced the slot which was not sufficiently strong, and to avoid a difficulty found in opening these latter in the ordinary manner, the latest pattern, Mark II., has the interior of the lid fitted for a felt ring saturated with beeswax,* which is to be inserted when the cylinders are packed with cartridges.

§ 2921.

Metal straps are to be issued to stations having Mark I. cylinders, with bayonet fastenings. By tightening up the straps by means of screws, one strap to the lid and another to the bottom of the cylinder, a good hold is obtained, and the lid can be twisted round.

§ 2173.

Scale 2 ins.—1 foot.

Zinc cylinders are painted stone colour, inside and out, like brass powder cases.

§ 2437.

The hooks, handles and rivets are made of galvanized iron.

The wire loops with string which attached the lids to the cases are discontinued, as the loops made it difficult to insert or extract the cartridge.†

§ 2891.

*Bearer, Wood, Cartridge Cylinder.*—This is an ash stave 3′ 6″ long, 1″ thick, $2\frac{1}{4}$″ broad in the centre, and tapered off to each end to $1\frac{1}{2}$.″

Bearer, wood, for cylinders. §§ 2120, 2944.

*Mark I.* was intended to carry 12″ cartridge cylinders only, and had

---

§ 3186.

* If the felt strip is not supplied, tow or wool, saturated with beeswax, is to be used. Mark I. 9″ and 7″ cylinders are to be lutened with ordinary luten (half tallow and half beeswax).

† When zinc cylinders are used in cartridge filling rooms, wadmiltilts should be demanded.—Army Equipment, p. 77.

one small grove on one side in the centre to receive the handle of the cylinder.

*Mark II.* is for use with all natures of zinc cartridge cylinders, and differs from Mark I. only in having two additional grooves on the side. It will take one large, two medium, or three of the smaller cylinders.

Cage Metal. §§ 2218, 2799.

*Cages, Metal, Lifting Cylinders.*—These are of two sizes. They are made of gun-metal, and are used for hoisting zinc cylinders containing cartridges up the powder lifts of magazines. The smaller size, stamped "A." is for cylinders for 12″ of 25 tons and under; it weighs about 14½ lbs. The larger size is for cylinders for 12″ of 35 tons, and 12″·5 guns, is stamped "B.," and weighs about 17 lbs. 14 oz.

§§ 2475.

Mark II. of "A." cage, has the top band made broader and carried inwards to prevent jamming when being hoisted up the lift of a magazine.

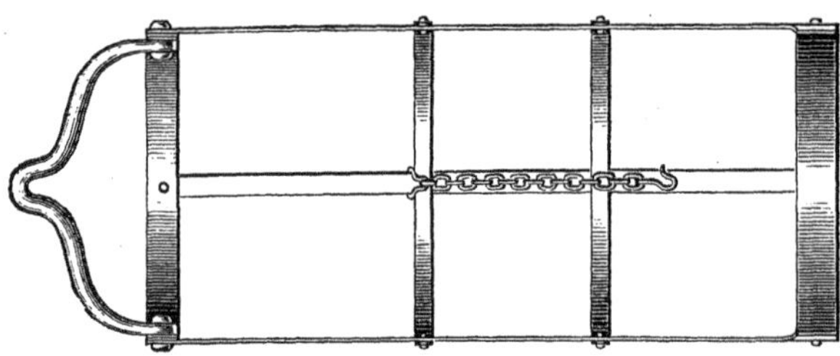

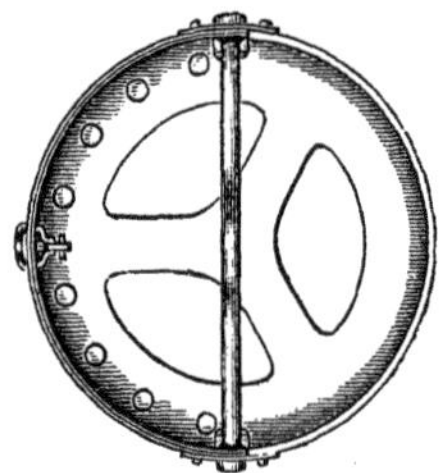

§§ 2293, 2548.

A 2″ white rope about 30 feet in length with a hook at one end and a counterweight zinc ball 1⅜ lb. weight fixed above the hook, is approved for use with the above.

Powder barrels. §§ 1884, 2553.

*Powder Barrels* used to be full bound with twelve ash and four copper hoops.

Machine made barrels are now issued with six ash hoops, they were for some time issued with four, but six are considered to be preferable. The ash hoops protect the barrel by keeping the bilge off the ground, and also keep the copper hoops from slipping. They are of three sizes, whole, half and quarter, and are used to contain loose powder, and occasionally to hold gun ammunition; they hold respectively 100 lbs., 50 lbs., and 25 lbs. of powder.*

Bags for cylinders.

When powder is sent by rail it is put in a flannel bag and placed in a half or quarter barrel. The barrel is covered by a canvas bag and placed in an iron case or cylinder. These iron cases are only made of two sizes, half and quarter; any other combustible stores are sent in the same manner. B.L. small arm cartridges may be sent without being placed in iron cases, as they are very difficult to ignite in any way, and will not explode in mass under any circumstances.

Budge barrels.

*Budge Barrels.*—A quarter powder barrel with only one head, the other being replaced by a leather bag, the mouth of which is closed by a leather thong. Used for holding loose powder for mortars.

Small arm cartridge barrels.

*Small Arm Cartridge Barrels* are of three sizes, half, quarter, and eighth; they have no copper hoops, and are used for conveyance and storage of S.A. ammunition. The half size is used for blank cartridges,

---

* The whole barrel holds 125 lbs. of "P²" or "P" powder owing to its greater density.

## Details of Patterns of Zinc Cylinders.

| Nature of Cartridge, and Numeral of Case. | Date of Approval. | § Changes in War Stores. | Length. | Diameter. | Marks on Cylinder. | Remarks. |
|---|---|---|---|---|---|---|
| | | | Inches. | Inches. | | |
| 12″·5 I. 1 Battering, 130 lbs. P². .. .. .. | 1/1/77 | 3042 | 29·5 | 12·75 | 12″·5, 130 lbs., R.↑L. I. | Two handles. |
| 12″ 35 ton I. 1 Battering, 110 lbs. P... .. .. | 9/5/76 | 2921 | 27·1 | 11·9 | 12″, 110 lb. P. ,, I. | Two handles. |
| 12″ 25 ton I. 1 Battering, 85 lbs. P., or 67 lbs. R.L.G. | 8/71 | 2172 2760 | 22·1 | 12·0 | 12″, 85 lbs. P., or 67 lbs. R.L.G., R.↑L. I. | The 12″ cylinders used with the 12″ gun of 25 tons were marked "Battering 12′ charge," and "Full charge, 12″" respectively, till the introduction of the 12″ gun of 35 tons, when it was altered, as the same cylinder serves for the battering charge of the 25 ton, and the full charge of the 35-ton gun. See § 2760. |
| 12″ 25 ton I. 1 Full, 55 lbs. P., or 50 lbs. R.L.G. | 12/71 | 2185 | 15·1 | 12·0 | 12″, 55 lbs. P., or 50 lbs. ,, ,, I. | |
| 1″ I. 1 Battering, 85 lbs. P., or 70 lbs. R.L.G. | 4/72 | 2270 | 26·5 | 11·0 | Battering charge, 11″ M.L., R.↑L. I. | |
| 1″ I. 1 Full, 60 lbs. P., or 50 lbs. R.L.G. | 4/72 | 2270 | 19·5 | 11·0 | Full ,, 11″ ,, ,, I. | |
| 10″ II.* 1 Battering, 70 lbs. P., or 60 lbs. R.L.G. | 23/12/71 | 2185 | 26·0 | 9·8 | Battering ,, 10″ M.L. II. | |
| 10″ II.* 1 Full, 44 lbs. P., or 40 lbs. R.L.G. | 23/12/71 | 2185 | 18·1 | 9·8 | Full ,, 10″ ,, II. | |
| 9″ to contain 1 43 lbs. R.L.G., cartridge I. .. | Provisionally 23/5/66 | 1269 | 26·196 | 8·65 | | |
| 9″ ,, 1 30 lbs. R.L.G. ,, I. .. | Provisionally 23/5/66 | 1269 | 17·196 | 8·65 | 9″ M.L. 30 lbs. I., W.↑D. | |
| 9″ II. 1 Battering, 50 lbs. P., or 43 lbs. R.L.G. | 22/9/71 | 2171 | 23·0 | 8·75 | Battering charge, 9″ M.L., R.↑L. II. | "A few 9″ Mark I. cylinders of large diameter were issued and marked *Battering charge*. The strap for these want a lengthened screw."–Extracts XII. p. 151. |
| 9″ II. 1 Full, 30 lbs. R.L.G. .. .. | 22/9/71 | 2171 | 17·0 | 8·5 | 9″ M.L. 30 lbs., R.↑L. II. | |
| 9″ III. 1 Full, 30 lbs. R.L.G... .. .. .. | 8/7/72 | 2295 | 17·0 | 8·75 | 9″ ,, ,, ,, III. | |
| 7″ to contain 1 22 lbs. R.L.G., cartridge I. .. | Provisionally 23/5/66 | 1269 | 19·696 | 6·75 | | |
| 7′ ,, 2 14 lbs. R.L.G., cartridges I. .. | Provisionally 23/5/66 | 1269 | 27·196 | 6·75 | 7″ M.L. 2 14 lbs. I., W.↑D. | |
| 7″ II. 1 Battering, 30 lbs. P., or 22 lbs. R.L.G., or 2 Full, 14 lbs. R.L.G. .. .. | 22/9/71 | 2171 | 27·0 | 6·65 | Battering charge, 7″ M.L., also 7″ M.L. 2 14 lbs., R.↑L. II. | Vide §2437, Changes in War Stores, for directions as to painting, packing, and securing. |

The cylinders approved 6/67 and 9/67 for 13″ cartridges are now obsolete.

*10″ I. { 1 60 lb. cartridge, R.L.G. / 1 40 lb. ,, ,, } several of these have been issued, vide § 2185.

## Packages of Cartridges, S.A., Breech Loading.

| Package. Name. | Dimensions, External, for Stowage. | Average Tare lb. oz. | Ball, B.L. Snider Arm, Charge 70 grs. Bullet 480 grs. Mark IX. Rounds. | Gross. lb. oz. | Ball. B.L. Martini-Henry Arm, Charge 85 grs. Bullet 480 grs. Mark III. Rounds. | Gross. lb. oz. | Ball, B.L. Martini-Henry Cavalry Carbine, Charge 70 grs. Bullet 410 grs. Mark I. Rounds. | Gross. lb. oz. | Blank B.L. 2½ drs. Mark IV. Rounds. | Gross. lb. oz. | Blank (Converted) B.L. 2½ drs. Mark IV. Rounds. | Gross. lb. oz. | Ball, Pistol, R. Revolver, Adams B.L., Charge 13 grs. Bullet 225 grs. Mark I. Rounds. | Gross. lb. oz. | Ball, B.L. Gatling, ·45 Bore, Charge 85 grs. Bullet 480 grs. Mark I. Rounds. | Gross. lb. oz. | Ball, B.L. Gatling, ·65 Bore, Charge 270grs. Bullet 1422grs. Mark I. Rounds. | Gross. lb. oz. |
|---|---|---|---|---|---|---|---|---|---|---|---|---|---|---|---|---|---|---|
| Bundle ... | — — | — | 10 | 1 0$\frac{10}{16}$ | 10 | 1 1$\frac{14}{16}$ | 10 | 1 0 | 10 | 0 4 | 10 | 0 5$\frac{6}{16}$ | 12 | 0 8$\frac{3}{16}$ | 10 | 1 2¼ | 10 | 3 7$\frac{3}{16}$ |
| Barrel, Cartridge, Half ... | L. 16$\frac{10}{16}$ D. 13$\frac{4}{16}$ | 12 2 | — | — | — | — | — | — | 2000 | 62 2 | 2000 | 79 5 | — | — | — | — | — | — |
| Barrel, Cartridge, Quarter ... | L. 14¼ D. 11$\frac{6}{16}$ | 7 12 | 700 | 81 4 | 700 | 86 0 | 700 | 77 12 | 1300 | 40 4 | 1300 | 51 7 | — | — | 750 | 93 5 | 230 | 87 1 |
| Box, Wood, Ammunition, Small Arm, with Tin Lining: Adams Revolver ... II. | 11¾ by 8 by 4¼ | 4 7 | — | — | — | — | — | — | — | — | — | — | 600 | 30 1 | — | — | — | — |
| Bullock ... I. | 20½ ,, 10¼ ,, 9⅜ | 16 8 | 780 | 98 0 | 840 | 110 5 | — | — | 1320 | 49 8 | 1320 | 60 13 | — | — | — | — | — | — |
| do. India Pattern ... | 19¼ ,, 9¼ ,, 10 | 14 4 | — | — | 600 | 81 4 | — | — | — | — | — | — | — | — | — | — | — | — |
| Camel ... I. | 27¼ ,, 14 ,, 13⅜ | 33 1 | 2400 | 284 9 | 2600 | 323 9 | — | — | 4130 | 136 5 | 4130 | 171 13 | — | — | — | — | — | — |
| do. India Pattern ... | 25 ,, 10½ ,, 18 | 34 8 | — | — | 1800 | 235 8 | — | — | — | — | — | — | — | — | — | — | — | — |
| Special ... | 27¼ ,, 14 ,, 12⅛ | 31 7 | 2000 | 239 4 | — | — | — | — | — | — | — | — | — | — | — | — | — | — |
| Service ... I. | 16¼ ,, 7 ,, 8½ | 9 0 | 420 | 52 8 | 480 | 62 10 | — | — | 740 | 27 12 | 740 | 33 14 | 1680 | 80 10 | 500 | 66 0 | 160 | 64 3 |
| do. ... II., IV., and V. / Special ... VII. | 22 ,, 7 ,, 8½ | 12 4 | 560 | 70 4 | 600 | 79 4 | 630 | 75 4 | 960 | 36 4 | 960 | 44 8 | — | — | 680 | 89 13 | 200 | 81 4 |
| without Tin Lining, Service... I. | 16¼ ,, 7 ,, 8½ | 7 10 | 420 | 51 2 | 480 | 61 4 | — | — | 740 | 26 6 | 740 | 32 8 | 1680 | 79 4 | 500 | 64 10 | 160 | 62 13 |
| without Tin Lining, Service... VI. | 22 ,, 7 ,, 8½ | 9 7 | 560 | 67 7 | 600 | 76 7 | 630 | 72 7 | 960 | 33 7 | 960 | 41 11 | — | — | 680 | 87 0 | 200 | 78 7 |
| Case, Wood, Powder, Metal-Lined, Half... | 14¼ ,, 13½ ,, 16¾ | 30 0 | 1440 | 181 0 | 1540 | 202 0 | — | — | 2400 | 90 0 | 2400 | 110 10 | 5364 | 258 12 | 1680 | 227 0 | 540 | 215 6 |
| Case, Wood, Powder, Metal-Lined, Quarter ... | 11 ,, 10¼ ,, 13¾ | 18 0 | 560 | 76 0 | 660 | 91 11 | — | — | 1020 | 43 8 | 1020 | 52 5 | 2220 | 112 11 | 680 | 95 9 | 220 | 93 12 |

Ball, Pistol, R. Revolver, Colt B.L., Mark I.—Charge 13 grs. pistol powder. Bullet 135 grains. Weight of bundle of 18 = 6$\frac{10}{16}$ oz. Weight of metal-lined quarter case with 2250 rounds, 69 lb. 12 oz.

Ball, B.L. Buck Shot, Snider Arm, Mark II.—Charge 54 grs. R.F.G., Buck Shot, 13 at 220 per lb. Weight of bundle of 10 = 15 oz. Weight of quarter barrel with 760 rounds, 79 lb. Weight of S.A.A. box, tin-lined, Mark I. with 420 rounds, 48 lb. 6 oz. Weight of quarter case with 560 rounds, 70 lb. 8 oz.

Ball, B.L. Martini-Henry Arm, Mark IV. Weight of bundle of 10 = 1 lb. 0$\frac{3}{16}$ oz. Weight of S.A.A. box, tin-lined, Mark IV. or V., with 630 rounds = 76 lb.

Ball, B.L. Martini-Henry Arm, solid case. Weight of bundle of 10 = 1 lb. 3¾ oz. Weight of S.A.A. box, tin-lined, Mark IV. or V., with 580 rounds = 82 lb. 15 oz.

PROOF, Blank B.L., for Snider Arm, issued in quarter barrels containing 850 each. Charge 5 drams of compressed powder. Weight of bundle of 10 = 7$\frac{5}{16}$ oz. Gross weight 46 lb. 9 oz.

Bullets and empty Cartridge Cases for proof of Martini-Henry Arm, are issued in Packing Cases containing 500 each.

| | | Tare lb. oz. | Gross lb. oz. |
|---|---|---|---|
| Cases, Cartridge, | Blank, B.L., strengthened for proof ... | 7 8 | 21 2 |
| | Ball, Service ... | 7 6 | 21 4 |
| Bullets ... | Service. Weight 480 grains ... | 3 0 | 37 8 |
| | 1st proof ,, 714 ,, ... | 3 10 | 55 2 |
| | 2nd ,, ,, 714 ,, (papered) ... | 3 10 | 55 10 |

Bullets and empty Cartridge Cases for proof of Snider Arm, are issued in Packing Cases containing 500 each.

| | | Tare lb. oz. | Gross lb. oz. |
|---|---|---|---|
| Cases, Cartridge, Ball, Service ... | | 8 15 | 20 14 |
| Bullets, | Service. Weight 480 grains ... | 5 8 | 49 8 |
| | 1st proof ,, 724 ,, ... | 5 14 | 57 10 |
| | 2nd ,, ,, 724 ,, (papered) ... | 5 14 | 58 2 |

Boxes, Wood, Ammunition, Small Arm, Service.

- Mark I. tin lined.—If iron nailed are for "Land and Home Service only"; if with brass screws and copper nails, may be issued for S.S.
- ,, II. tin lined,—Brass screws and copper nails, zinc fittings for straps for "Home Service."
- ,, III. tin lined,—Tinned iron screws, nails, fittings, and bands for "Land and Home Service."
- ,, IV. tin lined,— ,, ,, ,, ,,
- ,, V. tin lined,—Brass screws, copper nails and bands, but no fittings for straps for "Land and Sea Service."
- ,, VII. tin lined,—Brass screws, tinned iron bands, no fittings for straps, "Special, for India only."
- ,, I. not lined,—For "Home Service only."
- ,, VI. not lined,—Brass screws, no bands or fittings for straps for "Home Service only."

All service S.A.A. boxes issued for "Foreign Service," must be made of Mahogany or Teak; the boxes if made of Teak may weigh a few ounces heavier than shown in Table
Repaired boxes are re-issued with original numeral, except Mark III. which becomes Mark IV.
No new boxes of Mark I., II., III., or IV. will be manufactured in future.

**Note.**—Old face type figures are not regular Service Packages.

§ 3197. A Mark III. box has been introduced for Adams' Revolver Ammunition to hold 240 rounds for L.S. Weight, empty, 2 lbs. 6 oz.; filled, 12 lbs. 9 oz.

the weight would be inconveniently great with ball cartridge), the quarter size for ball, and the eighth for small supplies. For the number of cartridges contained in these barrels, see table p. 80.

For ball cartridge they are being superseded by boxes, but are still in use for home service.

**Box, wood, S.A. ammunition.** §§ 2175, 1991, 1877.

*Box, wood, S.A. Ammunition*, with sliding lid, made of teak* or mahogany, lined with tin for foreign countries, and sometimes of deal unlined for home service. The box weighs about 70 lbs. when packed with Snider cartridges, and 78 lbs. with Martini-Henry cartridges. The inner lid of tin is soldered to the tin lining.† A handle is attached to one corner by which the plate can be torn off. For the number of cartridges carried in these boxes, and the various patterns of boxes, see table, p. 80.

Mark IV.

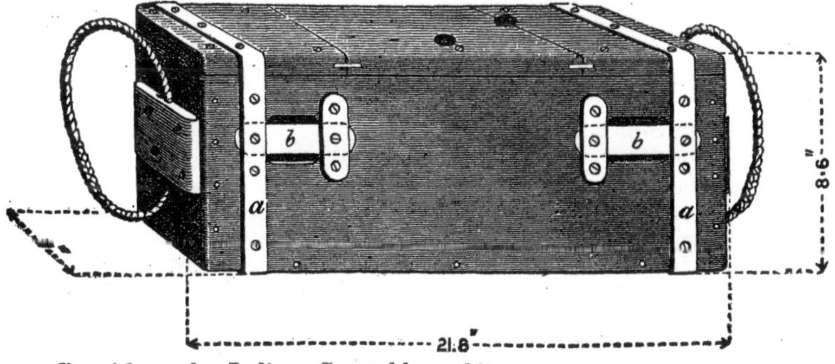


**Camel box.**

*Camel boxes* for India. See table p. 80.

**Bullock box.**

*Bullock boxes* for india. See table p. 80.

§§ 1990, 3197.

*Box, wood, S.A. Ammunition*, Marks II. and III., for Adams' pistol, lined with tin. See table, p. 80.

**Covers, paper, cartridge.** §§ 729, 1874, 2730, 3029, 3117.

*Covers, Paper, Cartridge*, are issued to cover cannon cartridges which are not contained in metal or metal-lined cases, or which are intended for field service. They are simply brown-paper bags on which are marked the contents of the cartridge which they cover. They are not issued to the Royal Navy for field or boat service.

§ 3116.

They are made of sixteen sizes, suitable for all S.B. and R.B.L. cartridges. Covers marked for two or more charges are, when used for any particular charge, to have the markings for the other charges crossed out. These covers have also been made for all the various charges used with R.M.L. guns up to the 80-pr. inclusive.‡ They are made of 90 lb. brown paper (*i.e.*, weighing 90 lbs. to the ream) for all sizes, except those for the 7-pr., which is of 45 lb. paper. They fit tightly over the cartridge, the bottom being notched, turned in, and pasted over a circular disc of paper, so as take up as little room as possible in packing.

In packing cartridges covered with the above covers in a barrel or box, the interior of the latter is also lined with brown paper.

In field batteries the covers to some extent preserve the cartridge from wearing out, they are also stated to preserve the cartridge from damp. Moreover, they prevent powder dust working through the cartridges into the limber boxes.

---

§§ 1616, 1866.

* Deal has been used for the sides and bottoms.

† The early patterns of S.A. ammunition box are smaller. Mark III. differs from IV. only in minor details.

‡ Also for the 8″ and 6″·3 R.M.L. howitzers.

Waterproof bags may be specially demanded; they are made of two thicknesses of fine white paper cemented together by india-rubber dissolved in naphtha.

§ 204
§ 2439.
§§ 3185.

Waterproof covers for cartridges, 7″ R.M.L. guns and upwards, are provided for the Navy; they are bags of fine cambric, waterproofed by vulcanized india-rubber, attached on one side (not made in the R.L.).

§§ 954, 1581, 1686.

As shells are now carried filled, bursters do not require to be made up. As before stated, Shrapnel are filled by measure or weight to ensure their having the full charge, as less would fail to open the shell. Shells issued to field batteries would be issued empty, except in first equipments. Bursters made up in paper bags enclosed in calico are, however, still retained for S.B. field batteries; this applies to 6-pr., 9-pr., 12-pr., 18-pr., and 24-pr. S.B. diaphragm Shrapnel shells.

Leather cases.

*Leather Cases for Cartridges* are ten sizes,* viz.:—

No. 1, for 13-inch L.S. mortar, 10-inch S.B. gun, howitzer, or mortar.

No. 2, for 7-inch R.B.L., and 68-pr., 56-pr., or 42-pr. S.B. guns.

No. 3, for 8-inch S.B. gun.

No. 4, for 80-pr. and 64-pr. M.L., or 40-pr. R.B.L. L.S., and 32-pr., 63 cwt. to 56 cwt. S.B.

No. 5, for 32-pr., 50 cwt. to 39 cwt., 24-pr., 50 cwt. or 48 cwt., and 8-inch howitzer, 40-pr. and 20-pr. B.L.

No. 6, for 32-pr., 32 cwt. or 25 cwt., 24-pr., 20 cwt. guns, 8-inch, $5\frac{1}{2}$-inch, and $4\frac{2}{5}$-inch mortars, and all carronades, and all smaller natures.

No. 7, for 13-inch S.S. mortar, or 100-pr. S.B. gun.

A. for 7″ R.M.L. guns, S.S.
B. „ 8″ „ „ „
C. „ 9″ „ „ „

The case is a leather cylinder with lid, which slides up the handle, and is used to bring cartridges up to the gun from the magazine; it is not uncovered until the sponge is out of the bore.

Cases of Clarkson's material. §§ 1798, 2076, 2645, 2816, 3039.

*Cases of Clarkson's Material*† are of eight sizes, viz.:—

A. for 7″ R.M.L., S.S.
B. „ 8″ „ „
C. „ 9″ „ „
D. „ 10″ „ „
E. „ 11″ „ „
F. „ 12″ of 25 tons
G. „ 12″ „ 35 „
H. „ 12″·5

(E. to H.) Has side handles so as to be carried in an horizontal position.

Leather hides. Wadmilltilts. Hair cloths.

*Leather Hides* are used to cover the floors of magazines. Wadmilltilts are also used for the same purpose, in order to diminish accidents from grit or sand. Cloths, hair, are provided in equipment.

N.B.—It may be mentioned here that no one is allowed to enter a

* Cases are not required in the L.S. for R.M.L. 7″ guns and upwards, as the cartridges are brought to the gun in zinc cylinders.

† They are made of strips of cork cemented together between two layers of canvas, and strengthened on the outside by cork bands. The whole is covered with split leather, specially prepared. The cement used is india-rubber solution.

These cases are made in the R. C. D.

magazine without putting on leather slippers, which are kept for the purpose.

When men are employed in magazines they should be made, when other circumstances permit, to change their clothes on entering them, to avoid risk from their having matches, &c., in their possession.

*Hints on Coopering.*

The following *hints on coopering* are introduced, as likely to be useful Powder barrels consist of three parts, viz.:—

1. Staves.
2. Heads.
3. Hoops.

**1. Staves.**

The most protuberant part of the barrel is known as the "bilge," and the centre of the bilge is distinguished as the "pitch."

Between the bilge and the end of the barrel is the "quarter."

The extreme end is known as the "chime."

To distinguish one end of the barrel from the other, that which is opened (when required) is known as the "top end," the other as the "back end." The top end may be known by having the staves bevelled off close to the chime to facilitate heading.

There are thus also the "top" and "back bilge," the "top" and "back quarter," the "top" and "back chime."

**2. Heads.**

The heads are known as the "top heads" and "back heads" respectively. When a head is in three parts, the "dowels" having been broken or pulled asunder, the two outside pieces are known as the "cants" or "outsides"; the other part is known as the "middle" piece.

**3. Hoops.**

Barrels used to be either "full bound" or "quarter bound," according to the number of hoops. All powder barrels, either full or quarter bound, have four copper hoops, the remainder ash. These hoops are situated about the chime and round the bilge of the barrel, and are known as the "copper chime" and "bilge hoops."

On the "full bound" barrel there were also six ash hoops at each end, situated one below the copper bilge hoop; four at the quarters and one above the copper chime hoops.

§ 2553.

Powder barrels are now made with four copper and six ash hoops, each copper bilge hoop has an ash hoop on each side of it, the copper chime hoops have ash hoops outside them. The wood hoops protect the barrel, which would otherwise rest on the bilge, they also keep the copper hoops from slipping.

**To unhead a barrel. First method**

A barrel can be unheaded in two ways. The first and more common method is to place the barrel with the top end uppermost, and then to remove the top chime hoops and loosen the top quarter hoops. The left hand is then pressed upon the middle piece of the head, which is struck gently with the adze or mallet close to the chime on the side nearest the cooper, until it is started out of the groove and falls into the barrel.

**Second method, or "boxing."**

The second method is called "boxing out" the head, and is adopted when the groove is deeper than usual, or when from other causes, such as the barrel being incorrectly made and having too sharp a curve, the head cannot be readily removed by the first method. The hoops are loosened and removed as before, and the left hand placed upon the head, and a few smart blows are struck with the mallet round the pitch of the barrel, by which means the staves are, as it were, sprung back, and the head being thus released falls through.

**Heading a barrel-head whole.**

To head a barrel.—The head, if whole, is placed with its bevelled edge (on the side away from the cooper) in the groove, the left hand is then placed upon it, and the head slightly struck, as much as possible in the direction *away* from the workman, with the adze or mallet; in this manner it is driven into the groove all round. The chime hoops are then replaced. If on heading a barrel the head should accidentally be driven a little below the groove, it can generally be jarred back into its place by laying the barrel on its side and tapping the top end of the staves.*

**Head in two pieces.**

When the head is in two pieces, the dowels (if still adhering) must be cut away. The larger piece is then placed with the whole of the left front in the groove to the left hand side, away from the workman. The small piece is then placed alongside the larger, its further edge also entering the groove; the left hand is then placed over the junction, and by means of a few gentle blows, given with care and at the spots where they may seem to be most required, the head is driven into the groove.

**Head in three pieces.**

If the head is in three pieces, the dowels must be cut off and the pieces matched according to the lettering on the head. One of the "cants" or "outsides" is then placed as the larger piece in the last case, and supported by the left thumb, which is brought over the side. The middle piece is then placed against it, its further edge in the groove, and its straight edge pressing hard against the side of the "cant." The other cant is then placed in the groove; and proceed as when the head is in two pieces.

**Flagging.**

Sometimes when the barrel is headed the head will be found to be a little out of round or injured at the edge, thus leaving an opening between the head and staves. It then becomes necessary to use the "flagging tool." One of its teeth is pressed against the inside of the stave where the opening appears, and the other tooth outside the stave to the right. By pressing against the handle, and using it as a lever, the opening is widened, and a little "Dutch rush" or "flag" (if not procurable, paper or rag will serve) is placed inside the gap; the flagging tool is then removed, and the stave being released springs back into its place, pinching in the rush against the head.

To avoid using a knife (*which should never be allowed to enter a magazine*) the rush should be placed as much as possible flush with the top of the head of the barrel.

**To alter the size of a hoop.**

If the ash hoops are too large they may be reduced in diameter to the required size by placing a small three-sided prism or wedge of wood, called a "Dutchman," between the shoulders or notches of the hoop. If the hoop is too small, it may be enlarged by cutting away part of the shoulders.

**Taking to pieces or "shaking" barrel for stowage.**

Before taking a barrel to pieces for stowage (called "shaking" a barrel) the staves must be numbered round the inside with a piece of chalk or a pointed tool.† The hoops are then removed and laid aside. The ash hoops (if the barrel is to be sent away) are seldom packed with it; the copper hoops are doubled up. The head is divided into two or three pieces by pulling open the joints without breaking the dowels.

The staves are then packed round the copper hoops and the "ends," and the pack secured with twine or with some of the wooden hoops.

---

* It is forbidden to use nails in re-heading a barrel; sometimes copper nails have been used, but these are objectionable even when the barrel is empty, as in unheading they are apt to get into the barrel and so find their way to the powder when the barrel is refilled.

† This need only be done with hand-made barrels; with machine-made barrels it is not necessary.

**To put a barrel together.**

To put the barrel together again the pack is untied and the copper hoops unbent. One of the copper chime hoops is then taken in the left hand and held at about the height of the barrel from the ground, the cooper kneeling on his right knee. The staves, as numbered, and with their top ends uppermost, are then arranged round the inside of the hoop, their lower ends resting upon the ground, the first few staves as they are arranged being supported by the outside of the left leg and left foot. In this manner the barrel may be built up, when the upper bilge hoop is slipped on. The barrel is then turned round and the other bilge hoop slipped on. The head is then put together and the back head is placed into the barrel (working chiefly from the inside); the back chime hoop is then placed on. The barrel is then headed up, the top chime hoop being previously removed to admit of this being done; the chime hoop is then put on again.

**To remove a stave without "shaking" the barrel down.**

All the hoops, except the bottom chime hoop and the top bilge hoop, must be removed; remove the required stave and replace it by another, and then replace the hoops.

**Heading and unheading vats.**

The heads of vats (chiefly used for the conveyance of clothing, harness, &c.) are secured by means of two hoops nailed round the inside of the chime of the vat, the head being between them.

To unhead a vat thus secured the "outside lining hoop" (as the hoop above the head is called) must be removed, and this is done by "prising" out the nails with a chisel or lever of any convenient sort, commencing at the "lap" of the hoop.

To head the vat.—The head is laid upon the inside lining hoop, and the outside lining hoop is nailed over it.

---

# CHAPTER VIII.—FRICTION TUBES, COPPER AND QUILL. — LANYARDS. — COMMON QUILL AND PAPER TUBES.—ELECTRIC TUBES, FUZES, AND DETONATORS. — PORTFIRES AND LIGHTS.—QUICK AND SLOW MATCH.—BICKFORD'S FUZES.

---

**Friction tubes**

Friction tubes are one of the most important stores manufactured in the R.L.; their action when new is very certain, their keeping qualities are however not equally satisfactory. Experience has shown that tubes about 10 years' old should be regarded with suspicion.*

Both copper and quill friction tubes are manufactured, the latter exclusively for the Navy, as copper is found dangerous where men work with bare feet, and also the copper tube rebounding from the beams of the deck is apt to cut men's faces, probably this inconvenience will be found troublesome in casemated batteries. The quill tube, requiring a upport, is not suitable for L.S. guns, but on an emergency any armourer could fit up a support for the loop.

A copper tube has been introduced for special naval service, which is not liable to fly about, if the ordinary copper tubes are found inconvenient in casemates this special tube might be demanded.

The use of the various tubes is given below, but it is as well to point out when firing the $4\frac{2}{5}$ or $5\frac{1}{2}$-inch mortars with the short copper tube difficulties have been experienced from the tube expanding and sticking in the vent. The thickness of metal in the mortar being less than three inches, the tube is unsupported at the bottom and hence would expand; probably the special 7-pr. tube would be the best to issue for these pieces.

Full directions for the proof of tubes by firemasters will be found on page 303.

In reporting upon the condition of tubes, all the marks on labels, the condition of the cylinders, and the metal of which the cylinder is made, whether tin or zinc, should be stated.

Tubes may fail in different ways† and firemasters should state how the failure occurred.

---

* See Extracts, Vol. VIII, pp. 195, 196, and Vol. VI, p. 21, where the result of the examination of tubes at various stations will be found.

§ 1420. § 1871. The caution given in the text refers especially to friction tubes manufactured before May, 1867. Since that date they have been lacquered internally, and will probably possess better keeping qualities. Since January, 1870, they have moreover been issued packed in small quantities in tin cylinders hermetically closed.

† Occasionally bad tubes may be restored by drying them for about fifty-six hours in a room heated to a temperature of 95° Fahr., this plan might be tried on an emergency, but little dependence can be placed on it.

A tube may fail—

(1.) From the friction bar breaking.
(2.) From the detonating composition failing to ignite when the friction bar is drawn.
(3.) From the mealed powder failing when ignited, to fire the puff

As a rule the detonating composition is found to keep well.

*FRICTION tubes of copper, about ·2″ diameter, are used for firing guns in the L.S.

Tubes, copper friction.

There are three sizes, viz., the short friction tube about 3″ in length for guns in general, a special tube about 2″ long for the 7-pr., and the long friction tube about 5″ long for 7″ R.M.L.† guns and upwards in the L.S.

§ 2049.
§ 2443.
Army Equipt 1876.

There is a special 5-inch tube with a copper wire attached to keep it from flying, issued for all the Woolwich guns in the Navy, when waterproof cartridges are used with guns on main decks or in turrets, a small lanyard is hooked on to the wire and hitched on to the gun carriage.

Special 5″ tube.‡

For sketches of all these tubes, see plate, p. 354.

A copper friction tube was provisionally approved for firing 9-pr. Hale's rockets (§ 1513). It was made obsolete by § 2783. It resembled the 7-pr. tube, but had a thick envelope of paper to fit the vent of the rocket.

The friction tube consists of a copper tube, (since May, 1867, lacquered inside to prevent the powder and metal from acting on each other), driven with mealed powder and pierced with a central hole. The top is stopped with shellac putty, and the bottom closed by a disc of varnished paper. A hole is bored through, near the top of the tube, and over this hole is secured, by woolding with fine copper wire and subsequent soldering, a short cylinder or "nib-piece," containing a copper friction bar, roughened and slightly turned up at one end. The roughened portion of the bar has detonating composition of chlorate of potash, sulphur, and sulphide of antimony (see table, p. 320), smeared on both sides.§ The nib-piece is pinched down so as to press on the friction bar, the projecting part of which has a slightly turned up eye, into which the hook of the lanyard fits. The exterior of the tube is varnished black.

On pulling the lanyard (which should be stretched and then sharply pulled) the friction bar is drawn out, igniting the composition and firing the tube.

Action.

The hole pierced up the centre of the tube composition is important, it gives a passage for the flash, and causes the tube to act instan-

---

* Proportion issued, one per round and 10 per cent. spare.—Army Equipment, 1876.

† The short friction tube might be used with 7″ R.M.L. guns and upwards when the long are not available, but an occasional missfire may take place, especially when using P powder.

‡ The designation of this tube was altered from "Tube, friction, copper, service, long, for waterproof cartridges," to "Tube, friction, copper, long, with wire loop." § 2784.

§ Previous to 3/5/75 the priming composition was placed on both sides of the friction bar in the form of small dry discs. It was found to fail when, as occasionally happens, a friction bar broke. The composition is now smeared on in a damp state, the damping being given by shellac varnish, and the defect noted above is obviated. § 2746.

taneously; without it the mealed powder would burn like a squib and fail to ignite the cartridge.*

Issue.
§§ 1810, 1871, 2001, 2055, 2217.

Copper friction tubes are now issued in tin cylinders, hermetically sealed by a tin strip soldered around the junction of the lid and cylinder, containing 25; formerly zinc cylinders holding 25, and secured by a varnished band of tape, were used, but no more of these will be made, as tin has been found superior to the zinc cylinder. The tin cylinder has a loop formed by a corrugated tin strip, soldered around the inside, so as to form a rack for each tube.† Four cylinders of tubes are placed in a wooden case.

On each cylinder will be found several labels, one on top showing the nature of its contents, and one on the side giving directions that it is "*not to be placed in the magazine on any pretence whatever*," and also directions as to method of opening, and that the cylinder is "*not to be opened until required for use or for special inspection.*"

§ 2405.

For garrison service, a tin tube box, 4″ × 4″ × 3″, with strap is used for serving the gun; it contains 100 tubes. A Mark II. box is issued which takes the long tube.

§ 1622.

For field service a leather tube pocket with strap is used.

Quill friction tubes.
§§ 1148, 1613, 2192.

*Quill Friction Tubes* of two sizes, about 2¾″ and 4″ long, are used by the Navy. The general principle of construction is nearly similar to that of the copper friction, but differs slightly in several details; a little mealed powder is added to the detonating composition, which is put on one side only of the friction bar; the bar passes through the tube. To support the tube when the lanyard is pulled, a leather loop is attached which fastens on to a crutch or pin screwed into the gun near the vent.

Tube, friction, quill, short. Use.

*Quill Friction Tube, Short*, is used for Naval S.B. and rifled guns, under 10″ calibre, except when firing reduced charges of 8-inch R.M.L. guns and over, and using waterproof cartridges; it is also used for the signal rocket gun and with life-buoy portfires.

---

* Foreign nations employ friction tubes constructed on a similar plan, but containing large grained powder, the interstices between the grains act the same part as the hole in the mealed powder, giving free passage to the flash. The central hole should be as small as possible, consistent with freedom from choking. Suppose $l$ to denote the length of the tube, and $r$ the radius of the small hole. Then obviously the strength of the tube (*i.e.*, strength of flash) will vary directly as the length multiplied by the circumference of the circle, whose radius is $r$, and inversely as the length multiplied by the area of that circle. Thus—

$$\text{Strength of tube} \propto \frac{l \times 2\pi r}{l \times \pi r^2} = \frac{2}{r}.$$

In other words, the strength of the tube varies inversely as $r$.

The diameter of the hole in all friction tubes is No. 18 B.W.G., *i.e.* ·049.

Some tubes made on this principle were tried in the R.L., using Curtis and Harvey's No. 6 powder, they did well, but as our own tube answers equally well no extended trial was made. Possibly the grained powder would resist damp better than the mealed.

† The loops were introduced to prevent the tubes exploding when the cylinder fell. A cylinder exploded in 1870, prior to the introduction of the present mode of packing, on falling a height of about 20 ft. on asphalte.

SHORT QUILL FRICTION TUBE, MARK II.*

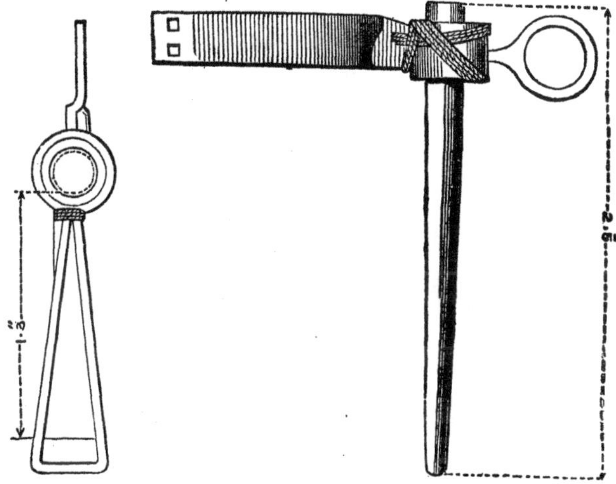


*Quill Friction Tube, Long*, formed by cementing two quills together, is used for 10″ R.M.L. guns and upwards; for firing reduced charges in R.M.L. guns of 8″ calibre, and over, and also for the 24-pr. Hales' war rocket for sea service.

**Tube, friction, quill, long. §§ 1613, 1856.**

The last patterns of quill friction tubes, viz., Mark IV. for the short tube, and Mark II. for the long tube, differ from previous patterns in having their heads closed with shellac putty instead of beeswax, and have no parchment cap on the top. The beeswax was found to run in hot climates, and spoil the tubes. The substitution of shellac putty rendered the parchment cap unnecessary. The friction composition is improved by the addition of ½ oz. ground glass, and there is ½ oz. less of mealed powder. It is now put on damped with spirit instead of dry as formerly.

**§ 2785.**

Quill frictions have the tubes varnished black. The head and friction bar are shellac varnished.

Quill friction tubes are packed on their sides by 25, in tin cylinders. On each cylinder will be found labels, one on top giving nature of contents, and one on the side directing that it is "*not to be placed in the* "*magazine on any pretence whatever*"; also directions for opening the cylinder and when it should be opened, with instructions for fitting the landyard and using the tube.

**Issue.**

Two boxes are made to contain cylinders of friction tubes when issued for naval service. There is a large and a small size of box.

**§ 3059.**

A vent-server, with lanyard, L.S., Mark I., has been introduced. It is used in lieu of the thumb to serve the vents of 64-pr. R.M.L. guns and upwards. It consists of a conical piece of brass with a gun-metal

**Vent-servers, Mark I. § 2856.**

---

* Mark II. differs from I. only in having a leather wad which keeps the loop open fastened by a single wire. **§ 1148.**

Mark III. differs from II. in being a ¼-inch longer to ensure more certainty in its action. **§ 2192.**

head, the part which enters the vent being covered with a thick conical piece of leather.

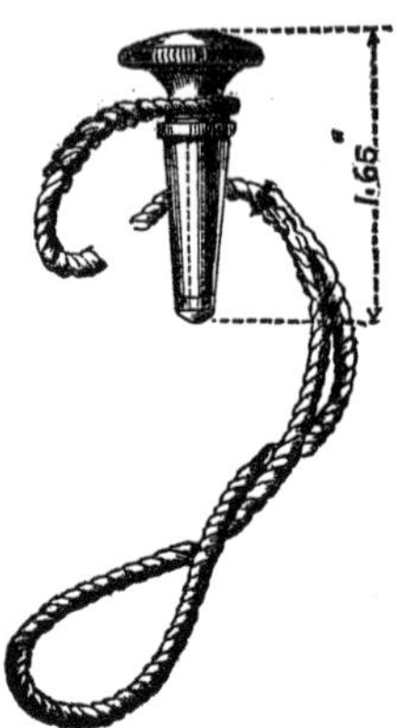


The lanyard, which attaches the vent-server to the gun, is of whipcord, and is to be fitted locally. The loop at the free end is about 2″ long, and is intended to be slipped over the hind sight. This governs its length.

**Mark II. §§ 3207.**

Mark II. differs from the above in being slightly longer, in being forged out of Bessemer steel, and having the lanyard fastened through two holes in the head instead of round the neck. It is for L.S. and S.S.

In L.S. it is to be used with all R.M.L. guns and howitzers from 25-pr. gun inclusive. Also with all S.B. garrison ordnance.

In S.S. it is to be used with all guns.

Vent-servers will be issued complete with lanyards for S.S., and for siege guns and batteries of position; but for garrison service the whipcord will be demanded separately and the lanyards fitted locally.

The length of whipcord required for each lanyard, so as to admit of a loop of about 2″ in length being spliced at the end, is as follows:—

| | Inches. |
|---|---|
| 12″·5 gun of 38 tons .. .. .. | 45 |
| 12″ " 35 " .. .. .. | |
| 12″ " 25 " .. .. .. | 35 |
| 11″ " .. .. .. | |
| 10″ " .. .. .. | |
| 9″ " .. .. .. | 25 |
| 8″ " .. .. .. | |
| 80-pr. converted guns .. .. .. | |
| 64-pr. wrought-iron guns .. .. .. | |
| 7″ guns .. .. .. .. .. | 20 |
| 64-pr. converted guns .. .. .. | 17 |
| 40-pr. guns .. .. .. .. .. | |
| 25-pr. guns .. .. .. .. .. | |
| 9-pr. and 7-pr. guns S.S. .. .. | |
| R.M.L. howitzers .. .. .. .. | |
| S.B. ordnance .. .. .. .. | 27 |

**Lanyards. §§ 683, 1257, 1532, 2083, 2697, 2722.**

Friction tubes are fired by means of lanyards, of which there are five descriptions. They differ chiefly in length. Those for the Navy are made of white instead of tarred line, and have a loop to allow of "half cocking," and a wooden toggle at the end.

The following is the present nomenclature, &c., of lanyards:—

| Lanyards, friction tube | | | Length. |
|---|---|---|---|
| field and siege | | | 5′ 4″ |
| garrison | plain (*a*) | | 12′ 0″ |
| garrison | with loop (*b*)* | | 12′ 0″ |
| Naval | long | turret vessels | 18′ 0″ |
| Naval | long | 64-pr. and upwards | 9′ 6″ |
| Naval | long | 64-pr. converted (*c*) | 8′ 6″ |
| Naval | short | | 8′ 4″ |
| Naval | Rocket machine with block and pulley. | | |
| Line with loop for garrison landyards (*d*) | | | 2′ 4″ |

**Mark II. § 3134.** Mark II. lanyards differ from the above, only having the hook made of one uniform pattern as shown in the cut.

This hook has been specially designed for use on board ship, to prevent the friction bar of the tube, after firing, from being thrown about the decks; the bars remaining on the hook, and not being removed until after a number of rounds.

**Common quill tube.** *Common Quill Tube*, about 3″ long, can hardly be considered a service tube; it is used when the gun is fired by a portfire. It would be easily manufactured on an emergency. The point of the quill is cut off and a head is formed by slitting the top into 7 prongs and passing a piece of worsted alternately over and under each prong, so as to form a small cup about ½-inch in diameter. The tube is driven with mealed powder, damped with methylated spirits, and a hole pierced by a wire, the cup filled with a priming of mealed powder, gum and water, made into paste, and sprinkled on the top with dry mealed powder. A paper cap is twisted on to the head of the tube, and the wire passed up again after capping to ensure the hole not being choked up. The tubes should be thoroughly dried. They require to be uncapped before firing.

**Match, or Fynmore's tubes.** *Match, or Fynmore's Tubes*, have been used as primers for B.L. guns. They are the same as common quill tubes, having in addition 8 strands of worsted covered with a paste made of mealed powder, gum arabic, and methylated spirits fastened to the cup.

**Paper tube.** *Paper Tube*, about 2½″ long. A strip of paper rolled into a cylinder of ·2″ diameter; on to the top of the tube another piece of paper is rolled spirally, so as to make a cup. The tube is driven with mealed powder, damped with methylated spirits, and pierced, and the cup filled with priming of mealed powder and water. The tube is capped by a piece of paper soaked in saltpetre and water, and tied on by silk. After capping, a wire is passed up to ensure the hole not being choked. The tubes should be thoroughly dried. The cap need not be removed before

(*a*) For every smooth-bore garrison gun and for all rifled garrison guns under 7-inch.
(*b*) For all rifled guns, 7-inch and upwards.
(*c*) For the 64-pr. M.L. gun converted, 71 cwt. See § 2609.
(*d*) For fitting to plain garrison lanyards, when required for 7-inch guns and upwards (§ 2697). Loop to be 18-inch from the hook (§ 2821).
* The loop is slipped over the centre hind sight to prevent the hook from flying back when the gun is fired.

firing. This tube is not a service one, but might be made on an emergency to be used with portfires.

Coated with black varnish, shellac dissolved in methylated spirits would answer.

**Dummy steel friction tube.**

*Dummy Steel Friction Tube*, Mark I., issued for drill, consists of a steel pin, which fits the vent, having a prong in the head, into which a V-shaped spring fits, representing the friction bar. The lanyard is spliced on to the spring, which can be drawn through the prong.

**§ 3033.**

Mark II. dummy tube has the head formed of two spring clips, which represent the eye of the friction bar of the service tube, so that ordinary service lanyard can be used with it. The strength of the clips is so adjusted that it requires a pull about equal to that necessary to fire a service tube to draw the hook of the lanyard between them.

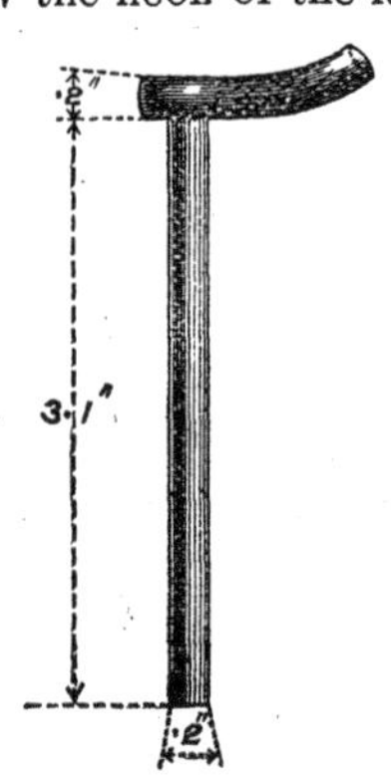


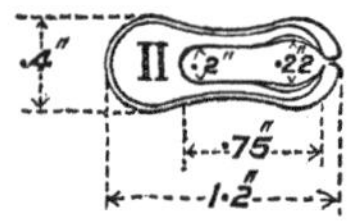


**Apparatus for proof of tubes. §§ 2031, 2101, 2296, 3023.**

Apparatus for the proof of tubes are issued to inspectors of warlike stores, and directions are issued with the stands. See Appendix, p. 303.

## Electric Fuzes, Tubes, and Detonators.

Of these stores there are at present in the service several varieties, but they may all be placed in two classes, viz., "high tension" and "low tension."*

**Comparative advantages of "high" and "low" tension fuzes, &c.**

Speaking generally we may say that "high tension" fuzes are ignited by the passage of electricity through an inflammable semi-conducting material used as priming, and interposed between the terminals; while those of "low tension" are ignited by the incandescence of a thin metallic wire bridging over the interval between the terminals. In both

* The use of the terms "high tension" and "low tension" as applied to electricity, though demurred to by many electricians, will be adhered to in this work as being the recognized service terms; and although perhaps theoretically wrong, they serve to convey the general ideas implied in their use. Valuable information as to the rise and progress of the ignition of explosives by electricity will be found in Major Majendie's Treatise on Ammunition, and in Professor Abel's "Recent Applications of Electricity to the Explosion of Gunpowder," from which Major Majendie largely quotes.

cases, then, there is a "bridge" between the ends of the conducting wires, *i.e.*, the terminals, and in one case this bridge is composed of a chemical mixture of substances, while in the other it is a thin, and therefore highly-resisting, metallic wire.

It is to be clearly understood that the electrical arrangements are independent of the fact whether the store treated of is a fuze, tube, or detonator; for the sake of brevity in these preliminary remarks we shall speak of this class of stores simply as electric fuzes.

"High tension" fuzes have the great advantage of being available without the use of voltaic batteries, which more or less entail the use of fluids, &c. They can be fired by any kind of electricity, that is to say, by frictional electricity, by magnetic exploders, by an induction coil in conjunction with a weak voltaic battery, or by a *powerful* voltaic battery. They are, however, of varying resistance, and not only does the resistance of different fuzes vary considerably, but the same fuze is liable to vary its resistance on being stored for any length of time, or exposed to climatic influences. The testing of them *in situ* in a system of mines is therefore difficult, and may be dangerous.* They require high insulation.

The "low tension" fuzes have the great advantage of being independent of climatic influences, and they are not at all likely to deteriorate in store. Their resistance moreover will remain constant. A comparatively "low tension" current (the actual strength required depending on the material, length, and sectional area of the thin wire bridge between the terminals) is sufficient to explode them. There is no liability to inductive action on other neighbouring mines, and, what is most important in the case of a system of submarine mines, they can at any time be readily tested *in situ*, without any fear of exploding the fuze. They are easily improvised on emergency, and a high degree of insulation, though very desirable, is not so essential with these fuzes as with those of "high tension." They require, however, for their ignition some form of voltaic battery, and are open to the serious objection that, if it be desired to explode a series of mines simultaneously, one or two fuzes may be accidentally more sensitive than the others in circuit. The effect of this would be that the mines connected with the more sensitive fuzes (unless furnished with a disjunctor) would be fired, the circuit would thus be broken, and the remaining mines rendered, for the time being, useless.†

Speaking generally, we may say that for mines, &c., which are required to be ready for use for an indefinite period of time before firing the "low tension" fuzes, on account of the facilities which they afford for testing, are the best; while for immediate operations, such as blasting, hasty demolitions, firing guns at proof, &c., the high tension fuzes are handiest on account of the portability of a frictional machine, or a magnetic exploder; bearing in mind always the increased necessity for high insulation when using these forms of electricity, and the possible deterioration of these fuzes from the influence of climate or time.

**Electrical resistance.**

The term "resistance" has been frequently made use of above. To understand this term, let us suppose a stream of water to flow through a pipe and act on a piston moving in that pipe. We will suppose the pipe to be filled throughout its cross section by the water, and will

* Special provision is made in No. 6 Detonator for testing *in situ*, and will be hereafter described.

† In the service "low tension" fuzes are individually carefully tested, so as to have the same resistance within very narrow limits.

neglect the friction of the water against the sides of the pipe. The weight of the piston, and its friction against the sides of the pipe, will offer a certain *resistance* to the onward flow of the water. If the piston be light and move easily in the pipe, there will be little or *low* resistance to the flow of the water; if it be heavy or move with great friction, it will present a comparatively great or *high* resistance.

It is easy to imagine the piston so heavy that a moderate pressure of water would not move it at all, and thus the current would cease to flow, but it might be moved if the pressure of water were increased. Substituting the electric current flowing along a wire for the water in the pipe, and a good or bad conductor for the light and heavy pistons respectively, we have a very fair analogy. A feeble electric current, for instance, experiences so much resistance in attempting to traverse a "high tension" fuze that it is practically stopped altogether in its flow, while a strong current has power to traverse the resisting medium. In the case of a "low tension" fuze the resistance is comparativly small and so is overcome by a comparatively feeble current. The analogy might be pursued to meet the case of an actual spark passing, but this is perhaps rather foreign to the subject in hand.

High and low tension fuzes, &c.

Taking, then, the service terms "*high*" and "*low*" tension, we see that they really signify comparatively large and small resistance. A sheet of glass, for instance, is easily perforated by an electric spark of "high tension" *i.e.*, of power to overcome great resistance; while the thinnest film of glass is an impassable barrier to an ordinary current of "low tension" electricity, such as would be given by an ordinary telegraph battery.*

Electrical measurements

This electrical resistance has to be measured in terms of some definite unit. The unit in use is the "Ohm,"† and is usually denoted by the Greek letter "ω." Thus a resistance of 50 ohms would be written $R=50^{\omega}$. Taking as an "absolute unit" a force capable of giving in one second a velocity of one centimètre per second to a gramme mass‡, one ohm is equal to absolute unit $+ 10^{9}$. It is about equal to the resistance of 48·5 mètres of pure copper wire, one millimètre in diameter, at 0° C.

We may now proceed to the description of the various electrical fuzes, tubes and detonators in the service. The latest patterns will be described, and the differences between them and earlier patterns noted.§

No. 1, Fuze, electric Abel's Mark III.

*Fuze, Electric, No.* 1, *High Tension, Abel, L.S., Mark III.*—This fuze has a body of well-seasoned beechwood 1″·22 long. The head is 0″·7 in diameter and has a hemispherical top. A hole about 0″·18 diameter is bored down the centre to receive a "primer." Two eyes are bored laterally through the head on each side of the centre to receive two triangular copper tubes for conducting the current to the wire of the

---

* The actual *quantity* of electricity given by a very powerful frictional machine is vastly inferior to that given by a very weak voltaic cell.

† So called after the distinguished physicist of that name.

‡ The absolute unit is equal to $\frac{\text{foot-pound}}{13825g}$, where $g$ is the accelerating force of gravity in centimètres.

§ In all the *last* patterns the colour of the paint denotes the service for which the fuze, &c., is intended.

*Red* denotes presence of fulminate of mercury.
*Yellow* denotes that the fuze, &c., is for naval service.
*Blue* denotes that the fuze, &c., is for submarine mines.
*White* denotes that the fuze, &c., is for L.S. and "low tension."
*Black* denotes that the fuze, &c., is for L.S. and "high tension."
This, however, does not apply to the *bodies* of electric tubes which are *black*.

primer, and small grooves are cut over the top connecting the centre hole with one end of each lateral eye at opposite sides. The fuze is hollow and contains a charge of F.G. powder, secured by a millboard disc at the bottom, coated externally with gutta percha cement, and is covered to a length of 0″·63 from the bottom with a thin brass cap. The electric primer consists of two fine copper wires, insulated from each other, in a column of gutta percha. The ends of these wires are bared at the top, and are ultimately bent round into the two small grooves in the head and passed into the two eyes, where they are jammed by two small triangular copper tubes forced into each eye, and filed down flush with the exterior of the head. The bottom ends of the fine wires are also bared, and project very slightly from the column of gutta percha. They are $\frac{1}{16}$″ apart. A paper cylinder, closed at bottom by a piece of gut, contains at its bottom the priming composition of chlorate of potash, and sub-sulphide and sub-phosphide of copper.* This cylinder fits on to the gutta percha cylinder, so that the projecting ends of the fine wires are imbedded in the composition, and is varnished over outside. The complete primer, when placed in the body of the fuze, has a little shellac putty over the top of it. The same substance is pressed into the small grooves over the fine wires. The action is as follows: The conducting wires from the battery (or exploder) are placed in the eyes; on the circuit being completed, the current passes into each of the fine insulated wires, and ignition is caused by the actual passage of a spark between the two terminals, or by the semi-conducting priming composition becoming so heated by the resistance experienced in it to the passage of the current that it ignites. This of course ignites the powder, and so the charge in which the fuze is placed.

The fuze is varnished black; its resistance is from 1,500 to 2,000 ohms.

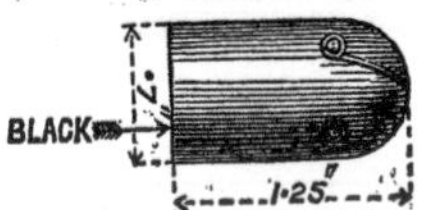


The Mark II. fuze differs from the above in having no brass cap, and in minor details. It is represented in the cut above. §§ 1512, 2484.

The Mark I. fuze was considerably larger than Mark II., and had the body containing the powder of smaller diameter than the head. § 1201.

These fuzes may be fired by Wheatstone or Browning's magnetic exploders (the latter is the most powerful) by dynamo-electric, or by frictional electric machines, or by a powerful voltaic battery.†

*Fuze, Electric, Submarine, Mark I. (for instructional purposes only).*— No. 2, Fuze, electric, submarine, Mark I. §§ 2344, 2484

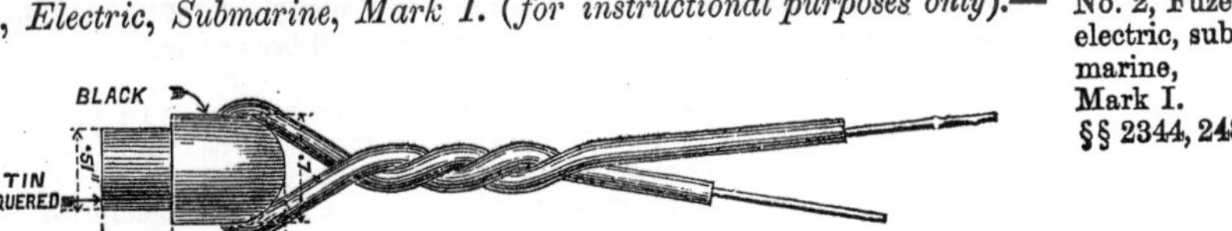


This fuze consists of a wooden body, the lower part of which is covered with a lacquered tin cap. The electrical arrangements generally

---

* Prepared by heating copper in vapour of sulphur and phosphorus respectively.

† About 20 cells of Walker's battery will fire them. They can be tested by 24 cells of a water battery (no acid being used), or three or four Daniell's cells, with an astatic galvanometer in circuit.

resemble those of No. 1, but the fine wires, instead of being imbedded in gutta percha, are insulated in a "pillar" made of 5 parts sulphur and 1 of lime cast round them. The priming is a mixture of fulminate of mercury and graphite. The body contains chlorate of potash, sulphide of antimony and sulphur, The body of the fuze is varnished black, the tin being lacquered. Its resistance is from 3,000 to 15,000 ohms.

§ 2787. This fuze may be fired by 20 cells of Walker's battery and tested by one cell. It is superseded, in the naval service only, by No. 9, Detonator.

No. 3, Fuze, electric, platinum wire, Mark I. §§ 2344, 2484.

*Fuze, Electric, Platinum Wire, Mark I.* (*for instructional purposes only*).

This fuze is a low tension one, and contains in a wooden body a thin platinum wire bridge between the terminals, surrounded by gun cotton. Below the gun-cotton is powder. The base is closed by a gutta percha wad.

The fuze is painted white, and stamped "P." Its resistance is ·2 ohm rising to ·8 ohm at firing point.* Probably no more will be made.

This fuze may be fired with one or two cells of Walker's battery, and tested with one cell, the galvanometer being in circuit.

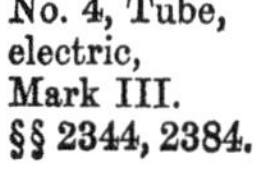

No. 4, Tube, electric, Mark III. §§ 2344, 2384.

*Tube, Electric, Abel, Mark III.*—This tube is now chiefly used for the

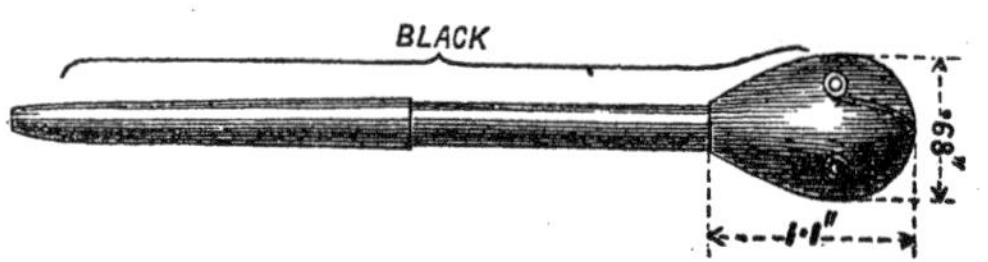


---

* The resistance of the wire bridge increases with the temperature.

*Instructions for the Use of Walker's Battery.*—1. The cells are to be filled with diluted sulphuric acid up to the red line on the plates, and a small quantity (about 4 oz.) of mercury is to be placed in each cell.

2. The acid should be in the proportion of 1 sulphuric acid to 8 water, by measure, and should be kept in a bottle when not required for use. The battery cannot be used until the mixture of acid and water has become cool; the same mixture will remain efficient for some months.

3. In testing fuzes or detonators, platinum wire, with the detector galvanometer; this instrument must be placed so as to be included in the circuit. The movement of the index needle shows that the fuze, &c., is correct. When the battery is used without the galvanometer, the platinum wire will become red hot, and the fuze or detonator will be fired.

4. Either one or both cells of the battery may be used. In the firing test, the wires should not be longer than necessary; one at least should be insulated with gutta percha, or other material, in order that the circuit might not be accidently completed. After the operations are over the cells are to be emptied, and the plates dipped in water before being replaced in the box.

5. Three distinct wires are required when the detector is used, one to proceed from one pole of the battery to one of the binding screws of the detector, another from the second binding screw of the detector to one of the wires of the fuze, and the third from the other pole of the battery to the other wire of the fuze.

6. When detonators are being tested it is advisable, on account of the greater intensity of the bursting charge, that the detonator should be covered up by a box, and be at a safe distance from the operator.

7. When the firing key is used it must be made a part of the circuit, and of course an additional wire will be necessary.

proof of guns in the Royal Arsenal. The electric arrangements, priming, &c., are the same as in the No. 1 fuze, Mark II. The head is, however, of a different shape (see cut), and the body is a long quill tube of the ordinary pattern. It is painted black all over. Mark II. of this tube was 1″·35 shorter than the preceding; the alteration to Mark III. was made to ensure the ignition of the cartridge in the heavier natures of guns. The existing store of this pattern is to be used up for L.S. Mark I. of this tube had a small space between the priming and the top of the tube. The object of doing away with this space in Mark II. was to render the ignition of the tube more complete.

§ 1237.

§ 2384.

§ 1201.

No. 5, Detonator, Abel, electric, Mark II.

*Detonator*, *Electric*, *No.* 5, *High Tension*, *Abel*, *L.S.*, *Mark II.*—The electrical arrangements in this detonator are the same as in No. 1, Mark III. To the head is fitted a tube of tin containing about 20 grains of fulminate of mercury, with a small plug of loose gun cotton on the top; the tube is 2″·29 long. This tube is closed at the bottom by a plug of solder, and is cemented to the head with gutta percha cement.* It has the head varnished black and the tube red.

Mark I. differs only in minor details from Mark II. It is painted red all over.

§§ 2344, 2484.

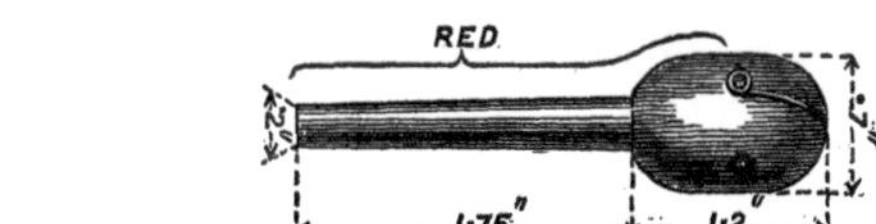


This detonator is superseded, in the naval service only, by No. 9.

§ 2787.

No. 6, Detonator, electric

*Detonator*, *Electric*, *No.* 6, *High Tension*, *Submarine*, *Mark III.*—This detonator has an ebonite† head with hemispherical top. It has the eyes bored and grooves cut in the head, as in those previously described. The lower part of the head has a brass socket fitted over it, and cemented with gutta-percha cement, to which is attached a tin tube. The electric primer consists of two fine platinum wires, No. 28, B.W.G., insulated in a sulphur and lime pillar, like that used in No. 2. The ends of the wires are flush at the bottom of the pillar, and have a streak of black lead between them, whose electrical resistance is about 2,000 ohms.‡ The priming composition is of fulminate of mercury and graphite, and is placed over the graphite streak and kept in position by a thin cardboard disc. The lower part of the socket and the tube are filled with fulminate of mercury, and the bottom of the tube is closed by cold cement of red lead and gold size.

The ebonite head is left black, the socket is varnished blue and the tube red. The resistance is from 3,000 to 15,000 ohms.

Mark II. of this detonator differs from the above in minor details only. Its head is varnished blue and the socket and tube red.

§ 2966.

---

* This consists of pitch and gutta-percha, 2 parts each.
Venetian red and resin, 1 ,, ,,
Beeswax ½ ,, ,,

Venetian red is a ferric oxide ($Fe_2O_3$), and is known also as colcothar, or jeweller's rouge. It is obtained by the calcination of the green sulphate of iron ($FeSO_4$).

† When a sheet of caoutchouc or india-rubber is allowed to remain for some time in fuzed sulphur at 250° Fahr., it absorbs 12 or 15 per cent. of that element, without suffering any material alteration; but if it be heated for a short time to 300° Fahr. it becomes *vulcanized*, and when still further heated is converted into the black horny substance called *vulcanite*, or "ebonite."—Bloxam's Chemistry, Second Edition, page 476.

‡ This is to enable the fuze to be tested with safety when placed in a charge.

§ 2344. Mark I. differs from the preceding in having a wooden head and fine copper instead of platinum wires. It is 0″·35 longer. It is painted like Mark II.

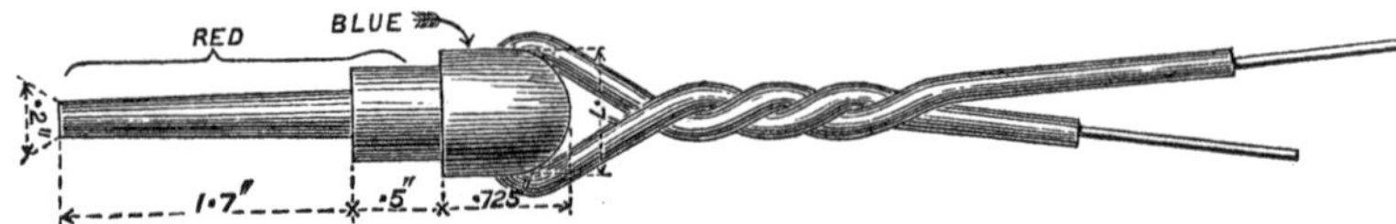


§ 2787. This detonator is superseded, in the naval service only, by No. 9, detonator.

No. 7, Detonator, electric, platinum wire.
§§ 2344, 2484.

*Detonator, Electric, Platinum Wire, Mark I.*—This detonator is simply No. 3 head with a tube containing fulminate of mercury. The wood head is painted white and stamped with the letter "P." The tin socket and tube are painted red. The resistance is of course the same as that of No. 3.

It may be fired with one or two cells of Walker's battery, and tested with one, the galvanometer being in circuit.

No. 8, Detonator, for Bickford's fuze, Mark II.
§ 2788.

*Detonator for Bickford's Fuze, Mark II.*—This detonator, though not fired by electricity, is classed with the electric fuzes, &c., for the sake of convenience. It consists of a tin tube containing fulminate of mercury; to the tube is attached a thin brass tube of diameter sufficiently large to allow of the easy introduction of a piece of Bickford's fuze into it. On the top of the fulminate of mercury is a small plug of wood through which passes a strand of quickmatch into the fulminate of mercury. The other end of this strand extends about a couple of inches up the brass tube, and is intended to convey the flash from the Bickford fuze to the fulminate of mercury. The exterior is painted red all over.

This detonator is intended for hasty demolitions. A proportion will probably be issued to cavalry pioneers,* who will also carry a proportion of gun-cotton slabs.

§§ 2344, 2484. Mark I. of this detonator differed from the above in having the brass tube rather longer and less in diameter than Mark II.

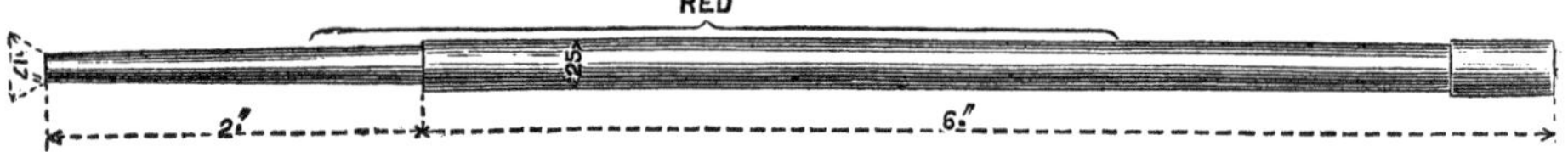


No. 9, Detonator, electric, L.T., S.S., Mark II.

*Detonator, Electric, No. 9, Low Tension, Naval Service, Mark II.*—The head is of ebonite, with an hemispherical top. The lower part of the head receives a brass socket attached to a tin tube. The conducting wires are of three strands of copper (No. 21, B.W.G). Two short pieces of copper wire, enlarged and recessed at one end, are fixed on the lower ends of the conducting wires; the strands are

---

* Six detonators, with Bickford fuze ready fitted, are fitted on a long slip of wood, which is packed in a tin cylinder.

soldered into the recess. The wire below the junction is then nipped so as to form a flange on either side, to prevent the wire from turning when fitted into the ebonite head. The two solid poles thus treated are dipped into hot gutta percha cement, and forced down into the holes prepared for their reception in the ebonite head. When cool the conducting wires are twisted together above the head, and whipped near the head with black thread. The two poles above-mentioned, project about ·1″ beyond the bottom of the head, and are ·25″ apart at the ends. They are connected by a bridge of fine platinum-silver wire (·21 grs. to the yard); the ends of the bridge being carefully soldered into fine incisions made on the flat ends of each pole. Round the bridge is a priming composition, prepared by mixing intimately 2 parts by weight of gun-cotton dust* and 3 parts of mealed powder. The detonator has a brass socket and tin tube attached to the head. The tube contains fulminate of mercury with a plug of gun-cotton at the top, immediately under the priming composition.

The head and shoulders of the detonators are varnished yellow, the tube red. The resistance of this detonator is 1·65 (±·15) ohms.

Mark I. § 2787.

Mark I. has the conductor wires solid instead of being stranded. They pass into the ebonite head through a short thick plug of sulphur and lime. This pattern differs from Mark II. in other minor details. It has a black head, and red varnished shoulder and tube.

Tube, electric, No. 10, Mark I. § 2296.

*Tube, Electric, No.* 10, *Low Tension, Naval Service, Mark II.*—This tube consists of two insulated wire terminals, with a platinum-silver wire bridge, an ebonite head, and an ebonite tube.

The two insulated wires fit into a small ebonite plug, which is screwed and cemented into the ebonite head. The portions which pass through this plug project ·1″ at the base, are ·25″ apart, and are "*feather-edged*," or "flanged" (see cut), to prevent their turning in the plug. The platinum-silver wire bridge is formed as in No. 9, Mark II. The insulated wires are twisted together above the head of the tube, and whipped a short distance with black thread. Each wire terminates in a spiral ·6″ long and about ·08″ internal diameter. The spirals are coated with tin, and covered with oiled silk secured by shellac varnish. The tube is pierced and driven in the ordinary way. It is of increased diameter at the upper end and is fitted into the head as shown in the cut. The bottom of the tube is closed with a varnished disc of demy paper. The priming is the same as that for No. 9.

The tube is black all over. The resistance is, of course, the same as that of No. 9, viz., 1·65 ohms (±·15).

Shortly after the introduction of this tube complaints were made by the Navy that the fragments of ebonite, which flew about on the firing of the guns, were inconvenient and dangerous to the gun's crew. A Mark II. tube was then made of rolled paper instead of ebonite, the small top plug still being of ebonite. This pattern was never sealed, and only a few were experimentally issued. In all probability a tube consisting of a quill head and a xylonite † bottom will be introduced. The quill head being of small diameter so as to fit into the vent, the wire bridge will be diagonal, so as to retain the same length of bridge, viz., ·25″. The whole tube will go into the vent, a stop on the insulating wires preventing it from being pushed in too far.

---

* Prepared by rasping or grating a piece of compressed gun-cotton, and sifting through a 120-mesh sieve. § 2866.

† Xylonite is a patent substance mainly consisting of compressed soluble gun-cotton. It is of course inflammable.

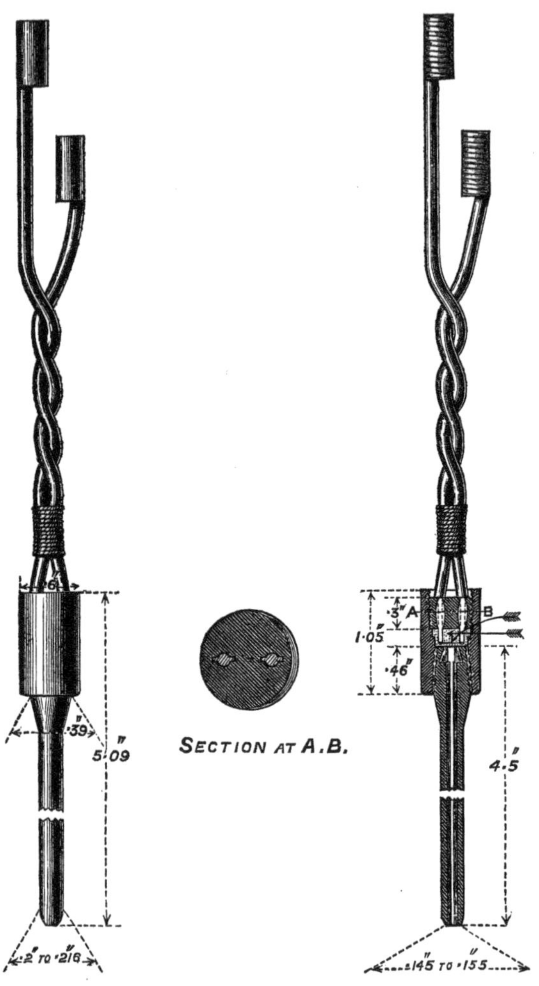


**No. 11, Tube, electric, L.T, Naval Drill, Mark II.**

*Tube, Electric, No.* 11, *Low Tension, Naval, for Drill Purposes, Mark II.*—The body of this tube is of gun metal. Into the top is screwed an ebonite plug, which has a deeply milled shoulder, and can be removed at pleasure. The milled shoulder is to give purchase to the fingers in screwing or unscrewing. The conducting wires and electrical arrangements generally are the same as in No. 10. The hollow head of the tube below the ebonite plug is filled with priming of the same composition as that for No. 10.* In the head are two side holes covered with discs of thin varnished paper, to allow of the escape of the flash when the tube is fired. Into the base of the head is screwed a solid gun-metal pin 3″·7 long, and of a diameter to fit easily into the vent of a gun. From the construction of this tube it is evident that the body can be used over and over again, only needing repriming, and, if

* This priming is used in glass bottles containing 2 oz. each to ships where drill tubes are used. § 2866.

necessary, a new wire bridge. The tube has its head lacquered, and the pin is varnished black.

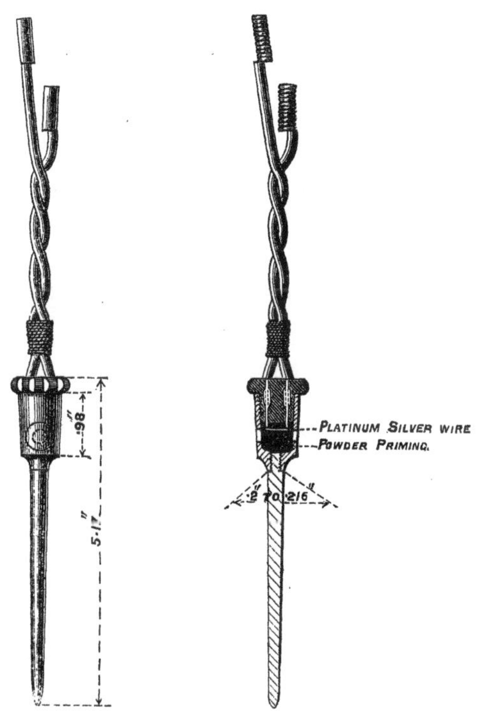


Mark I.
§ 2832.

Mark I. differs only in having the head varnished black.

No. 12, Detonator, electric.

*Detonator, Electric, No.* 12, *Low Tension, Submarine, Mark I.*—This detonator has the same ebonite head as No. 9, and generally resembles it in the electrical arrangements. The bridge is of iridio-platinum wire (10 Ir. to 90 Pt.), weighing 1·55 grains per yard. A small piece of gun-cotton yarn is tied round the bridge. The tube and lower part of the socket are filled with fulminate of mercury. The head is varnished white, the shoulder blue, and the tube red. The resistance is 0·32 ohms.

No. 13, Detonator, electric.

*Detonator, Electric, No.* 13, *Low Tension, Land Service, Mark I.*—This detonator resembles the preceding, but its bridge of iridio-platinum wire is finer, the wire weighing only ·45 grains per yard. The head and shoulder are varnished white, the tube red. The resistance is 1·05 ohms.

No. 14, Fuze, electric.

*Fuze, Electric, No.* 14, *Low Tension, Land Service, Mark I.*—This fuze has a wooden body and brass cap like No. 1, but differs from it in the construction of the electric primer. The fine wires are of the same size, but the gutta-percha cylinder is rather shorter. The bottom ends of the wires are bared, and project about ·3″. They are passed diagonally through a small piece of cork, so that the poles are ·25″ apart. The bridge is of the same wire as that used in No. 13. It is attached by twisting round the ends of the poles, and fixing by solder. A small piece of gun-cotton yarn is tied on to the bridge. The remainder of the fuze is like No. 1, except that two short insulated conducting wires are permanently jammed in the eyes by small copper wedges, and twisted together above the head of the fuze.

The fuze is varnished white. The resistance is the same as that of No. 13, viz., 1·05 ohms.

Annexed is a table showing the principal points connected with the fuzes, &c., above described.

Issue.

All electric fuzes, &c., are packed in tin cylinders, hermetically closed as usual, containing 25. On the top is a label descriptive of the number and nature of the contents.

Boxwood rectifier. § 2975.

*Rectifier; Gun-Cotton Primers, for Electric Detonators, Mark I.*—This is a simple piece of boxwood with a rounded handle, and its lower part, which enters the perforation made in the primer to receive the detonator, is the same in form and dimensions as the tube of the detonator. It is used to "rectify" the hole in the primer before inserting the detonator, and is in all cases to be used. One rectifier will be packed in each tin cylinder containing electric detonators, Nos. 5, 6, 9, 12, or 13, and each cylinder contains the following label:—

"CAUTION."

"ELECTRIC DETONATORS must *on no account* be *forced* into the gun-"cotton primers by screwing or twisting.

"This cylinder of Detonators contains a 'Rectifier.'

"Before inserting a Detonator into a Primer, force the Rectifier into "the hole for the Detonator, up to the full extent to which the Deto-"nator should enter, and then withdraw it by twisting."

CAUTIONS.

Cautions.

Nos. 13 and 14 electric detonators are only to be tested with the test cell specially made for the purpose, unless it is known for certain that a current not exceeding one-twentieth of a Weber * is used.

It is especially important to remember that detonators contain a large proportion of fulminate of mercury, and that it is *most dangerous* to treat them roughly, or to entrust them to unskilled persons. They should be handled with care, and only by persons well conversant with their properties. In testing detonators they should always be either removed to some distance, or strongly confined, in case of one being accidentally fired.† The consequences of one being fired by carelessness or rough treatment when in a gun-cotton primer are obvious.

Common portfire.

*The Common Portfire* ‡ consists of a cylinder 16″ long, and rather more than half-inch in diameter, made of stout paper pasted, rolled up,§

* As previously stated, an *ohm* applies to resistance, and is equal to an absolute unit (centimètre, gramme, second), $\times 10^9$. A *volt* applies to electro-motive force, and is equal to absolute unit $\times 10^8$.

A *Weber*, or, as it is sometimes called, a "*Farad*" $= \frac{\text{volts}}{\text{ohms}}$, and is a measure of quantity. A Daniell cell gives about one volt.

† A few days ago (July, 1877) an accident happened to an officer while testing an electric detonator, which he held in his hand. The loss of one eye was the immediate result, and further injuries were sustained. The present writer narrowly escaped a similar fate by the accidental ignition by a spark of some fulminate of mercury inclosed in a stout copper capsule, some feet away from him.

‡ Issued in the proportion of four per 100 rounds.

§ The case is rolled on a steel former, but any smooth cylindrical stick would do

## Electric Tubes, Fuzes, and Detonators.

| Number. | Date of Approval. | § in Changes. | High or Low Tension. | Priming. | Resistance in ohms. | Paint. | Service. | Remarks. |
|---|---|---|---|---|---|---|---|---|
| No. 1, Fuze, Mark I. ... | 27/1/66 | 1201 | High | Chlorate of potash, subsulphide and subphosphide of copper | 1500 to 2000 | Black ... ... | L.S. | |
| ,, ,, ,, II. ... | 27/12/67 | 1512 | ,, | ,, ,, ,, | ,, | ,, ... ... | ,, | Much shorter and body same diameter as head. |
| ,, ,, ,, III. ... | — | — | ,, | ,, ,, ,, | ,, | ,, ... ... | ,, | Has brass socket over body, and minor differences. |
| No. 2, Fuze, Mark I. ... | 13/6/72 | 2344 | ,, | Fulminate of mercury and graphite | 3000 to 15000 | Head black, tin part lacquered | Instructional only (Submarine) | Superseded in S.S. by No. 9. |
| No. 3, Fuze, Mark I. ... | 13/6/72 | 2344 | Low | Gun cotton, loose ... ... ... | ·2 (·8 at firing point) | White, stamped P... | ,, | |
| No. 4, Tube, Mark I. | 26/3/66 }<br>17/4/66 } | 1201 | High | Chlorate of potash, subsulphide and subphosphide of copper | 1500 to 2000 | Black ... ... | L.S. and S.S. | |
| ,, ,, ,, II. ... | 23/5/66 | 1237 | ,, | ,, ,, ,, | ,, | ,, ... ... | ,, | Top of tube close to priming. |
| ,, ,, ,, III. ... | 19/10/72 | 2384 | ,, | ,, ,, ,, | ,, | ,, ... ... | ,, | Tube 1″·35 longer. |
| No. 5, Detonator, Mark I.... | 2/12/71 | 2344 | ,, | ,, ,, ,, | ,, | Red ... ... ... | ,, | Superseded in S.S. by No. 9. |
| ,, ,, ,, II.... | — | — | ,, | ,, ,, ,, | ,, | Head black, tube red | L.S. | Differs in colour and minor details. |
| No. 6, Detonator, Mark I.... | 13/6/72 | 2344 | ,, | Fulminate of mercury and graphite | 3000 to 15000 | Head blue, body and tube red | Submarine | Superseded in S.S. by No. 9. |
| ,, ,, ,, II.... | 21/7/76 | 2966 | ,, | ,, ,, ,, | ,, | ,, ,, | L.S. | Ebonite head instead of wood. |
| ,, ,, ,, III.... | — | — | ,, | ,, ,, ,, | ,, | Head black, shoulder blue, tube red | ,, | Differs in colour and minor details. |
| No. 7, Detonator, Mark I.... | 2/12/71 | 2344 | Low | Gun-cotton, loose ... ... ... | ·2 (·8 at firing point) | White head, stamped P, shoulder and tube red | L.S. and S.S. | |
| No. 8, Detonator, Mark I.... | 13/6/72 | 2344 | — | — — | — | Red ... ... ... | L.S. | Not electrical, for use with Bickford's fuze. |
| ,, ,, ,, II.... | 5/3/75 | 2788 | — | — — | — | — | ,, | Brass tube shorter and wider. |
| No. 9, Detonator, Mark I.... | 30/6/75 | 2787 | Low | Gun-cotton dust and mealed powder | 1·65 | Head black, shoulder and tube red | S.S. | |
| ,, ,, ,, II.... | — | — | ,, | ,, ,, ,, | ,, | Head and shoulder yellow, tube red | ,, | No sulphur plug in head. Differs in colour and material. |
| No. 10, Tube, Mark I. ... | 24/2/76 | 2926 | ,, | ,, ,, ,, | ,, | Black all over ... | ,, | |
| ,, ,, ,, II. ... | — | — | ,, | | ,, | | ,, | |
| No. 11, Tube, Drill, Mark I. | 13/9/75 | 2832 | ,, | ,, ,, ,, | ,, | Black all over ... | ,, | |
| ,, ,, ,, ,, II. | — | — | ,, | ,, ,, ,, | ,, | Head yellow, pin black | ,, | Differs in colour. |
| No. 12, Detonator, Mark I. | — | — | ,, | Gun-cotton yarn ... ... ... | 0·32 | Head white, shoulder blue, tube red | Submarine | |
| No. 13, Detonator, Mark I. | — | — | ,, | ,, ,, ... ... ... | 1·05 | Head and shoulder white, tube red | L.S. | |
| No. 14, Fuze, Mark I. ... | — | — | ,, | ,, ,, ... ... ... | ,, | White ... ... | ,, | |

and when dry turned in at one end to form a bottom. The empty case or cylinder is supported in a mould and driven with portfire composition, consisting of saltpetre 6 lbs., sulphur 2 lbs., mealed powder 1 lb. 4 oz.

The top has a small hole bored in the composition, and is primed with mealed powder to make it light easily; the exposed end is secured by a paper cap, tied on with twine. They burn from 12 to 15 minutes, and are generally lighted by slow match.

Issue.

In bundles of 12, packed in deal boxes.

Life-buoy portfire. §§ 2654, 2741.

*Life-buoy Portfire* is used in the Navy in connection with a somewhat complicated life-buoy, which may be seen hanging over the stern of most men-of-war. The life-buoy is released by pulling a sort of bell-handle on deck; in its fall it fires a quill friction tube attached to the portfire; the latter is fired and gives a light for some twenty minutes, thereby enabling a man overboard to direct his course at night to the life-buoy, and a boat's crew to steer to his rescue. The portfire consists of short lengths of portfire joined in a sort of "key pattern" by papier mâché elbows, and arranged horizontally on a gun-metal plate. The composition used is much the same as for ordinary portfires, the first 3½ inches burning quicker than the remainder. There have been four patterns of this portfire, they differ in minor details only. Water will not extinguish the portfire.

Slow or blue portfire. § 2624.

*Slow or Blue Portfire* could be easily made on an emergency, and has the advantage of not showing flame at night or dropping sparks. Made of blue, porous, unsized paper (any porous paper would do, blotting paper, &c.), saturated with a boiling solution of 3 oz. saltpetre dissolved in one or two quarts of water; the more porous the paper the more water is required. When dried the paper is rolled into a solid cylinder, about 16″ long, and a little over half-an-inch in diameter. Burns two to three hours. It is removed from the list of service stores.

Coast Guard light. § 1724.

*Coast Guard Light, Mark I.*, has superseded the coast guard portfire; it burns about five minutes. The spike at the end is to enable the light to be struck in the ground, as a man holding it might be fired at by smugglers.

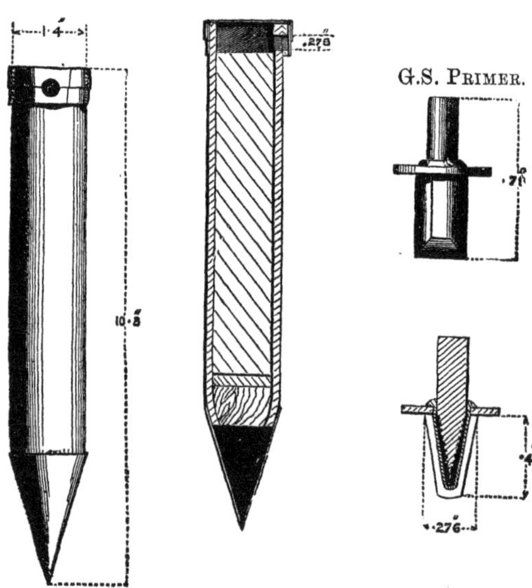

The composition is, saltpetre 7 lbs., sulphur 1 lb. 12 oz., red orpiment 8 oz. The cap of this light need not be removed before lighting. The top of the composition is primed with mealed powder and the flash blows off the cap.

In tin cylinders containing 4. — Issue.

The light is ignited by placing a G.S. primer in the hole in the head marked with a black dot, the wedge-shaped paper-covered part of the primer is inserted and the pin projects; a sharp blow struck with this pin on a hard substance ignites the light. — G.S. primer. § 1725.

The primer (see cut) is made on a similar plan to the head of the copper friction tube, the pin is roughed and coated with the friction tube composition, and the blow driving it through the wedge-shaped copper case explodes it. The case is open at one end and protected by varnished paper. Primers are issued in tin boxes holding 5, 10, or 15. — § 1727.

*Light, Long, G.S.*, burns about five minutes, ignited by G.S. primer as described with coast guard light, it has about $6\frac{1}{2}''$ of the same composition. A hollow wooden handle with a screw at the end holds eight primers, the handle fits into the projecting socket of the light and is fastened by a wooden pin which is tied on to the handle. The old long light was lit by a cap. — Light, long, G.S. § 1721. § 1726.

They are used for signalling and illuminating both in L.S. and S.S.

In tin cylinders containing 4. — Issue.

*Light, Signal, Magnesium*, burns one minute with a very brilliant white flame; its method of ignition, &c., is the same as described with the coast guard and long light, it has about $2\frac{1}{4}$ inches of composition, containing saltpetre 14 lbs., sulphur $3\frac{1}{2}$ lbs., red orpiment 1 lb., magnesium containing 25 per cent. of paraffine, $\frac{1}{2}$ lb. The handle is the same as that for the long light. They are used for signalling or illuminating, — Light, signal, magnesium. § 1723.

In tin cylinders containing 4. — Issue.

Patterns have been sealed to govern supplies for the Board of Trade.

Light for illuminating Wrecks. (Mark I). Stand for do. (Mark I). § 2488.

The light is about 28½ inches in length and 2·65 inches in diameter.

It consists of a cylindrical case of 1 X tin sheet in 6 lengths of 4½ inches each, fitted together and connected by small bands of tin sheet, half-an-inch in width, soldered over each joint.

The case is filled with the following composition, viz.:—

| | | | |
|---|---|---|---|
| Saltpetre, ground | .. | .. | 7 lb. |
| Sulphur, sublimed | .. | .. | 1¾ lb. |
| Orpiment, red .. | .. | .. | 0½ lb. |

One end is fitted with a piece of wood, with a loop of iron wire attached to it for suspending the light; the other end is primed with mealed powder, and covered with a kit plaster.

§ 2598.

The stand is a simple tripod, consisting of three wooden legs, about six feet in length, connected at the top by a piece of iron wire, having a small hook attached to it, on which the light is suspended; there are three iron rods which are hooked to and connect two of the legs, forming an incline for the light to rest on, so as to hang in a sloping direction—not vertically downwards.

The light, if hung as described, clears itself of dross when burning, and is kept further clear by the case separating each joint, as the heat of the burning composition sucessively melts the soldering of the bands.

The time of burning is about 30 minutes.

This light must not be roughly handled or thrown about, as it is liable to be broken across at the junction of the segments.

Care must be taken in removing the cap before lighting.

The case must be grasped firmly at the capped end, whilst the cap is torn off by means of the string loop; if there is any difficulty in removing the cap it must be eased off round the edge by inserting the blade of a knife.

Mark II. § 2723.

Mark. II. of this light differs from the above in various details. It is 1″ longer, and is built up in 10 segments instead of 6. These segments are of sheet iron (20 W.G.), and are each about 3″·25 long. A sheet-iron diaphragm, having a hole 1″·375 diameter; in the centre is fixed in each segment at 0″·5 from the top end, which is slightly increased in diameter, so as to form a cap to fit over, and make a junction with, the lower end of the segment to which it is attached. Each segment is filled, up to two-thirds of its length, with the following composition:—

| | | | |
|---|---|---|---|
| Saltpetre, ground | .. | .. | 7 lbs. |
| Sulphur, ground | .. | .. | 1¾ lb. |
| Orpiment, red .. | .. | .. | 9 oz. |

and the remaining third with a composition containing 1 oz. less orpiment. The greater quantity of orpiment makes a quicker-burning composition, which is necessary as the light is more confined as it burns up the case. The segments are secured with soft solder. The diaphragms prevent the composition from burning too rapidly up the sides of the case, and the segments fall off as the solder melts.

§ 2839.

A small hole is made in the adjoining faces of composition in the segments to ensure continuous burning of the whole.

Quick match.

*Quick Match.*—Is made of cotton wick boiled with a solution of mealed powder and gum.* For proportions, see table, p. 320.

Unenclosed it burns at the rate of about 1 yard in 13 seconds; when

* A portion of the powder is kept dry and dusted over the quick match.

enclosed in a tube of any kind it burns much more rapidly, being as instantaneous as a train of gunpowder, the pressure causing the gas to rush forward and fire the mass explosively. Quick match is made up in paper or calico tubes when this rapid action is required, and when so made up is termed a "leader." See p. 278.

The proportions of powder, &c., will vary with the number of threads in the wick, those given above are for six-thread wick. Quick match is demanded by weight, 1 lb. of six-thread match would be about 360 feet long. Quick match is largely used for priming fuzes, &c.

Issue.

Either in long packing or in metal lined cases, and should be demanded by weight.

Slow match.

*Slow Match.*—Is made of pure hemp slightly twisted and boiled in a ley of water and wood ashes in the proportion of water 50 gallons, wood ashes one bushel; this serves for 100 lbs. of yarn. It burns at the rate of one yard in eight hours, it is used for lighting portfires, &c. Slow match may be equally well made by boiling in a solution of 8 oz. saltpetre to one gallon of water.

Issue.

Loose, in skeins or parts of skeins, placed in a case with other stores. When large quantities are demanded it is issued in bales or casks. It should be demanded by weight; about four yards go to 1 lb.

Bickford's patent fuzes.

*Bickford's Patent Fuzes*, burn at the rate of about one yard in 70 seconds, they are made of flax with a column of fine gunpowder in the middle. There are about nine kinds, which vary in the amount of protection given to the flax according as it is wanted to burn in the air, dry or damp ground, or water. A portfire is the readiest way of igniting this fuze.*

The following is a brief description, mainly taken from W.O. Circular 406, of the fuzes most generally used in the service.

No. 2, Patent safety fuze.

*No.* 2, *Patent Safety Fuze.*—The flax is protected by a coating of tar, and the fuze "is adapted for all blasting in dry ground, being made "with an ample quantity of the best materials. When kept at a "moderate temperature and carefully used, its certain operation is "warranted."

No. 2*, Patent safety or dry soil fuze.

*No.* 2*, *Patent Safety Fuze,* "is the same fuze as the preceding, and "is only adapted to the same kind of blasting, but is specially varnished "for any given climate. That varnish which is suitable for a cold "country becomes soft and sticky if exposed to much heat, while that "which is suitable for a hot country becomes hard and brittle if exposed "to great cold. This inconvenience is remedied as much as possible "by the special preparation of the varnish to suit any given tempera-"ture, and which should be specified in the order."

No. 5, Patent taped sump or wet soil fuze.

*No.* 5, *Patent Taped Sump Fuze*, is covered with varnished tape, twisted round the fuze over the tar. It "is adapted for use in wet "ground, and is specially protected, so as to operate efficiently even "when the tamping is saturated with water. In such cases the charge

---

* The gutta-percha covered fuzes must have the gutta-percha removed and the powder laid bare where the fuze is placed in contact with the charge, the same must be done at the end which is to be lit by the portfire. Old fuzes should be tested, as the fuze sometimes deteriorates. Instances have occurred when the fuze has burnt very much too rapidly.

It seems at first sight strange that the action of Bickford's fuze should be so different from the action of quick match inclosed in a leader, its slow action is probably due to the small amount of powder and to the space allowed for the expansion of the gas in the flax coating. Accidents may occur from people supposing the fuze to have become extinguished, and, going up to look, thus exposing themselves to the explosion.

" of gunpowder should be placed in a cartridge, the end of the fuze " should be inserted into the centre of the charge, and the junction of " the fuze with the cartridge should be properly protected with a " waterproof varnish. If employed in this manner, its certain operation " is warranted."

No. 7, Patent gutta-percha or water fuze.

*No. 7, Patent Gutta-Percha Fuze*, is covered with gutta-percha, and " is adapted to subaqueous blasting, where it is not liable to much " motion from waves or currents, nor subjected to much pressure. It " has answered its intended purpose after it has been under water for " 24 hours, with a pressure of 40 lbs. to the inch; this is equivalent to " the weight of water at the depth of 88 feet."

No. 9, Patent taped gutta-percha or water fuze.

*No. 9, Patent Taped Gutta-Percha Fuze.*—" This is the same fuze as " No. 7, and is adapted to the same use and duty; but, having an " exterior coating of tape and varnish, which delays the oxidation of " the gutta-percha, it retains its efficiency for a much longer time. It " is therefore well adapted for service in distant countries."

Issue.

In tin cases of three sizes, containing 8, 24, and 50 fathoms.

§§ 2438, 2217.

Bickford's fuze when required for issue to the Navy will be packed in in tin cylinders containing two lengths of four fathoms each. This order does not affect the use of the case for boat magazines (§ 1701).

Ord's mining hose.
§§ 1102, 2619.

*Ord's Mining Hose*, has quick match protected by paper and calico hose cemented together with solution of india-rubber and coated by plaited cotton yarn. It is used for firing mines, and acts as a water-proof leader, having instantaneous action.

Issue.

In lengths as required, packed in a zinc or tin cylinder, or any suitable case.

Hose, calico.
§ 3211.

*Hose, calico, for gunpowder* is made of strips of strong calico sewn together, so as to make a tube or hose of about $\frac{3}{4}$ inch diameter when filled.

Issue.

Empty, in lengths of 100 feet.

# CHAPTER IX. — PROJECTILES AND CARTRIDGES FOR SMOOTH-BORE GUNS.*

SHOT, SHELL, AND MISCELLANEOUS PROJECTILES. — GAUGES. — WOOD BOTTOMS.—CARTRIDGES.

## SHOT.

Service shot.

1. Shot, solid, spherical, common.
2. „ case.
3. „ grape.
4. „ sand.

1. Solid shot.

1. *Solid Shot* are made of all calibres except the 10″, from 3-pr. to 100-pr. inclusive.†

They are attached (see p. 119) to wood bottoms for bronze guns and guns of position, but smooth-bore guns having disappeared from the field batteries and batteries of position, shot with wood bottoms are almost obsolete.

§ 2375.

Shot are used against masonry, wooden shipping, and masses of men. Hot shot may be considered obsolete as the furnaces are withdrawn.

Shot are fired from guns and carronades and the 12-pr. S.S. bronze howitzer, they are not fired from shell guns 8″ and 10″.‡

Issue.

Loose for garrison service. Loose, prepared for bottoms for Indian F.S. Riveted and boxed for F.S.

Shot bearers for use when painting shot are made in the Royal Laboratory.

Steel and chilled iron shot were issued for 100-pr. and 68-pr. guns for use against iron-clad ships, but as smooth-bores are powerless

---

* The calibres of the S.B. guns are—100-pr. 9″, 68-pr. 8″·12, 56-pr. 7″·65 42-pr. 6″·97, 32-pr. about 6″·35, 24-pr. about 5″·82, 18-pr. 5″·29, 12-pr. 4″·62, 9-pr. 4″·2, 6-pr. 3″·67, 3-pr. 2″·913. The windage generally lies between ·1″ and ·2″.

Paint.

All service projectiles are painted black; such as are issued with wood bottoms, together with case shot, grape shot, and filled mortar shell are packed in boxes, otherwise they are issued loose.

† The 100-pr. is a built-up gun of wrought iron. There are only a few in the service, and they are used for naval drill purposes only. They may be considered as practically obsolete.

‡ When fired under 3° of elevation, grummet wads are used, consisting of rope bent in a circle, and held in position by two cross pieces of small rope. The cross pieces go outside next the rammer head, otherwise the wads may be withdrawn by the rammer. Junk wads are made up of old junk beaten into a solid cylinder and woolded over. They are now only used in connection with tampeons.

against the iron-clad of the present day, these shots are become obsolete and are to be used as common shot. Steel shot have an S stamped on them, and are painted white. Chilled shot are painted black with a white belt.

Shot, case.
§ 1192.

2. *Case Shot* are made of all calibres; they consist of sand shot made up in cylinders, and packed in wood shavings or saw dust.

They may be divided into three classes, viz.:—

1. Iron case, with iron ends and iron handle on top.

This class includes case for 32-pr. gun, 8″ gun and howitzer, 68-pr. gun, 10″ gun or howitzer, and 100-pr.

2. Tin case, with iron bottom and rope handle on bottom.

This class includes case for all iron ordnance (except 5½″ howitzer) above 12-pr., not included in Class 1, namely, 18-pr. gun and carronade, 24-pr. gun and carronade, 32-pr. carronade, 42-pr. gun and carronade, 56-pr. gun, 68-pr. carronade.

3. Tin case, with wooden bottom.

This class includes case for all bronze ordnance, for all iron ordnance below the 12-pr. inclusive, and for the 5½″ howitzer (both iron and bronze).

The object of the wood bottom is to guard against injuring the bore of the bronze ordnance. and, for the sake of uniformity, the same case is used for the iron ordnance of corresponding calibre in the instances mentioned above.

Carronade case is not issued. Gun or howitzer case is used instead. Case for bronze howitzers has H stencilled in white, an inch and a-half long, to distinguish it.

In loading, the rule is always to put the handle away from the charge, except when it is made of rope, when the reverse is the case. The wood bottom goes next the charge.

Case shot are filled with sand shot, which vary in weight from 1 lb. to 8 oz. with heavy guns down to 32-pr., and vary from 8 oz. to 2 oz. with the smaller guns.

Use.

Case shot are fired from all natures of guns, howitzers, and carronades, against troops in masses, for flanking ditches, &c., and against boats and rigging of ships. They are effective up to 350 yards. Case which have wood bottoms follow the rule laid down as to shape given, in page .

Issue.

Case are issued in rough deal boxes with elm ends; the number in a box varies with the calibre of the shot.

3. Shot, grape.
§ 1689.

3. *Grape Shot*, of all calibres, from 6-pr. to 10″ inclusive, is being superseded by case. The pattern at present in use is "Caffin's" pattern.

The sand shot are held in position by four iron circular plates pierced with holes to grip the shot, an iron spindle passes through the plates, and a nut which screws on to the head of the spindle binds the plates and shot together.

Carronade grape is to be broken up when returned to Woolwich, it was made up in tin cylinders and painted red. At out-stations it is retained for local issue if serviceable.

The 10″ gun has special grape made up in cylinders like case, but has larger balls; it is known by having G. stencilled on it in white, an inch and a-half long.

The sand shot vary in weight from 4 lbs. to 13⅛ oz.; it will be seen that they are heavier than the sand shot used for case, and consequently there are fewer of them.

Use. Grape was used for the same purpose as case; it is effective at rather longer ranges than case, and would probably be more destructive to boats and rigging. It may be used up to about 600 yards.

Issue. Made up in boxes. the number in a box varies with the calibre of the shot.

4. Shot, sand. § 1582. 4. *Sand Shot* are seldom used as a projectile by themselves; they were formerly sometimes fired out of a mortar as pound shot, in which case a wooden bottom was necessary, but this mode of employing them was done away with by § 1582, and neither sand shot, nor the special hemispherical wood bottoms, now form part of equipment. They are cast-iron balls varying in weight from 4 lbs. to $1\frac{1}{2}$ oz.; their chief use is in the manufacture of case and grape and also Shrapnel for the Woolwich rifled guns.

Paint. Black if used for grape, otherwise unpainted.

Issue. When issued for use as pound shot they were packed in boxes containing 100 or 50.

Hollow shot. *Hollow Shot* may still be met with; no more will be made. They are to be used at practice, and are known by their weight.*

## Shells.

§§ 1819, 1880, 2204. All shell filled in any arsenal or dockyard have the monogram of the station, the word "filled," and the date of filling stencilled on them; the boxes also would be marked "filled" in red. Shells for field batteries are carried filled in the limbers and wagons.

Classes.

1. Common.
2. Naval.
3. Mortar.
4. Hand Grenade.
5. Diaphragm Shrapnel.

1. Common shell. 1. *Common Shell*, gauge, common. † Fuzes, common time and Pettman's L.S. percussion. They are fired from guns, howitzers, and carronades, and $5\frac{1}{2}$ and $4\frac{2}{5}''$ mortars, and are made of all calibres from 12-pr. to 10″ inclusive, except 100-pr., which has naval shell: they are about $\frac{1}{6}$th of their diameter in thickness, and weigh empty about $\frac{2}{3}$ the weight of solid shot of the same calibre; the 10″ is $\frac{1}{7}$th in thickness; they have wood bottoms in accordance with rule on page 120, and have a fuze-hole of the common gauge tapped throughout, so as to take Pettman's L.S. fuze, the fuze-hole is closed with a gun-metal plug marked with a + to show that the thread is tapped throughout; the plug has a shoulder fitting into a recess, a leather collar fits under the shoulder to make the joint tight, the fuze-hole is countersunk, which enables them to be used as shot without bottoms.‡ Mark II. common plug differs from the previous pattern in having no shoulder, and may be known from the G.S. plug by having a + marked on it. § 1603.

---

* The only sizes likely to be met with are the 10″ and 8″, and in the latter case only can any mistake arise, owing to their being externally like the 68-pr. solid shot. The latter projectiles were formerly painted red to prevent mistakes arising.

† Proportion of fuzes. One common fuze to two common shells and 10 per cent. spare; and one Pettman's L.S. percussion fuze to every two shells.—Army Equipment, 1876.

‡ Plug, common, with lanyard attached, is issued for drill purposes. § 773.

The 12-pr. shell has a gun-metal socket fitted into the fuze-hole, extending some way into the interior of the shell. Without this the shell failed to burst, as the powder was not sufficiently confined, the size of the fuze-hole being large compared to the size of the shell.

Blowing charges.

All common shell are completely filled with powder (L.G. shell), see page 61. For practice in places where the full charge might be dangerous, blowing charges are used, 3-oz. up to the 24-pr. inclusive, and 4 oz. for the higher natures; they are put in loose, except when the small mortar fuze is used with the 12-pr. and 24-pr. shells when fired from $4\frac{2}{5}''$ and $5\frac{1}{2}''$ mortars, in which case the charge is placed in a small red shalloon bag, vide p. 113.

Full directions for filling shell and securing them are given on p. 61. It may be pointed out that a shell with its plug and wad in it is secure against being ignited by an explosion near it;* the wad alone has been found to protect the powder when the plug has been shaken out. It is also to be remarked that the wad does not require removal before inserting the fuze, as the fuze drives it in on being set home; except when the Pettman L.S. fuze is used, when the wad should be driven in with the handle of the key. Directions for boring and fixing fuzes are given on pages 64, 65.

§§ 1698, 1841.

12-pr. and 24-pr. common shell are used with the $4\frac{2}{5}''$ and $5\frac{1}{2}''$ mortars; they are to be of small gauge, and it is directed that at out-stations shells for this purpose are to be tried by testing them with the mortars themselves; the common fuze would be used with these shells when fired from mortars, except at long ranges, when the small mortar fuze would be used.†

The 10″ being a weak gun causes some exceptional arrangement with its shell, thus the gun may not be double shotted; so with 10″ naval shell a special bottom is used, having two rivet holes, besides the ordinary hole for L.S., in order that it may be attached by either the naval or L.S. method; also its common shell is thinner, as the gun has a light charge; it has no Shrapnel, as the weight would strain the gun too much. 10″ shell which weigh over 85 lbs. are to be condemned, and may be converted to mortar shell.

Use.

Common shell are used against men in masses, houses, buildings, shipping, and material generally; they may be used either by bursting them in flight, when they act both by the velocity with which the shell is moving and by the force of the bursting charge; but they scatter too much in this way, and are not so effective against men as Shrapnel. They may also be used by bursting the shell when at rest, when they act as a mine; they are most destructive against wooden shipping: they would also be available against men in hollows or sheltered by buildings where Shrapnel would be powerless.

§ 1610.

Recovered shell, that have been fired at practice, may be used again, filled with sand only; they should have a red ring round the fuze-hole to show that no powder is to be used (recovered mortar shell may be used with blowing charges), and also the yellow line for practice.

§ 1778.

Recovered 32-pr. naval shell are not to be used because of the projecting bush, which would necessitate the employment of a wood bottom.

---

* They might, however, explode if struck directly by a projectile, or by a piece of a burst shell.

† The proportion of fuzes issued for $5\frac{1}{2}''$ shells is one common fuze to three small mortar fuzes for 4 shells; for $4\frac{2}{5}''$ nine common fuzes, and one small mortar fuze for 10 shells; 10 per cent. to be spare in all cases.—Army Equipment, 1876.

Empty, loose, prepared for bottoms for garrison service, and for India, garrison, and field service.* **Issue.**

Filled, riveted, and boxed for field or boat service. When for boat service, issued fuzed with Pettman's L.S. percussion.

*Naval Shell.*—Calibres, 100-pr., 10″, 8″, or 68-pr., 32-pr. Gauge, G.S. Fuzes, 9 or 20 seconds M.L. and Pettman's G.S. percussion. Naval shell differ from common in having the G.S. gauge (an adapter converts them from the old Moorsom gauge), and in having their bottoms attached by two rivets. **Naval shell.**

The fuze-hole is closed by the G.S. plug, which has no shoulder. The object of having the bottoms hollowed out is to have the iron of the projectiles in contact, otherwise they are likely to split when double shotted. The 100-pr. has a top attached in place of a bottom. Space in stowage is saved by this plan, and the fuze is well protected.

Naval shells are generally issued filled, riveted to bottom or top, and fuzed with G.S. percussion fuze; those with bottoms are boxed. For L.S. would be issued empty, prepared for bottoms.† **Issue.**

The adapter for naval spherical shells is made of gun-metal, tapped externally to fit the Moorsom gauge, and internally with the G.S. gauge.

*Mortar Shell.*—Calibres, 8″, 10″, and 13″. Gauge, large mortar. Fuze, large mortar.‡ Fired from 8″, 10″, and 13″ mortars. The fuze-hole is not regularly tapped, but is roughed. The gauge is much larger than the common gauge. The fuze-hole of the 8″ is a little smaller than the others, as otherwise the fuze would touch the bottom of the shell before it was fixed in the fuze-hole. Of course this makes the fuze protrude farther. The 10″ and 13″ shell have lugs; hooks fit into the lugs to enable the shell to be carried. The 13″ have the hooks hung by chains from a beam, and are called beam hooks. Hand hooks are used with the 10″. In future manufacture lewis holes would be used. The holes incline inwards, and the iron plugs at the end of the chains bite into them, when the chain to which the plugs are attached is tight, and can be removed when the chain slackens. The advantage of lewis holes is that there is nothing projecting which is liable to be broken off in piling or transit. The 8″ shell weighs $46\frac{1}{16}$ lbs.; the 10″, $87\frac{2}{16}$ lbs.; and the 13″, $195\frac{3}{16}$ lbs. **Mortar shell.** **Hooks, beam and hand.**

As these shells require no bottoms they have no rivet holes.

For L.S. they are issued loose, the fuze-hole closed with a beeswaxed cork. This cork may be driven in in the 10″ and 13″, but must be pulled out of the 8″, as it might otherwise hinder the fuze from being driven home; a corkscrew is provided for this purpose. For sea service they are issued filled and boxed. The fuze hole is closed with a cork and kit plaster. Kit plaster is stout canvas prepared with pitch, tallow, beeswax, and rosin. See table, p. 321. **Issue.** **Kit plaster.**

To prepare shell, see p. 62.

Mortar shells are completely filled for service, and sometimes for practice but for practice, where full charges would be dangerous, blowing charges are used, which are made up in red shalloon bags, narrow at the mouth, and with a brass ring to prevent the bag **Blowing charges.**

---

* For 10″ and 8″ guns and howitzers, and for 68-pr. guns and carronades, two wood shell boxes are provided to carry the shell up to the gun.—Army Equipment, 1876, p. 70.

† Naval shell may be issued for L.S.—Army Equipment, 1876, p. 72. They, of course, require fuzes of G.S. gauge.

‡ Proportion of fuzes issued is one per shell and 10 per cent. spare.—Army Equipment, 1876, p. 74.

falling into the shell. They are jammed into the fuze-hole by the fuze.

4 oz. used for 8″, 10″, and 13″.

3 oz. used for $5\frac{1}{2}$″ and $4\frac{2}{5}$″.

Use.

Mortar shells are used for high angle fire, and employed for the bombardment of towns, forts, entrenched positions, &c. They may be employed against shipping, but are too inaccurate to give good results on a small object. The $5\frac{1}{2}$″ and $4\frac{2}{5}$″ mortars are used against troops under cover. For this purpose the fuze should be bored rather short to ensure the shells bursting before penetrating the earth. On the other hand, the larger shells used against material should have their fuzes bored long.

Hand grenades.

*Hand Grenades* are of two sizes, 6-pr. and 3-pr., taking bursting charges of about 5 oz. and 3 oz. respectively. Small blowing charges are sometimes issued for drill purposes. They resemble common shells, but the walls of the shell are not so thick, being about $\frac{1}{7}$th of the diameter. The fuze-hole is much smaller, and is not roughened. They are generally issued empty, loose, for L.S., and filled and fuzed for S.S., the fuze being covered with a kit plaster. These are boxed.

Issue.

Use.

They are used chiefly for the defence of places against assault, being thrown among the storming parties in the ditch. They are useful in the defence of houses; they have been fired out of mortars instead of pound shot.* They can be thrown by hand about 20 or 30 yards.

Diaphragm Shrapnel shell.

*Diaphragm Shrapnel Shell.*—Gauge, common. Fuze, diaphragm.† For all calibres except the 10″ and 3-pr.‡ Fired from guns, howitzers, and carronades.

The shell is a thin cast-iron shell (see cut), weakened by four

12-pr. Boxer Diaphragm Shrapnel Shell.

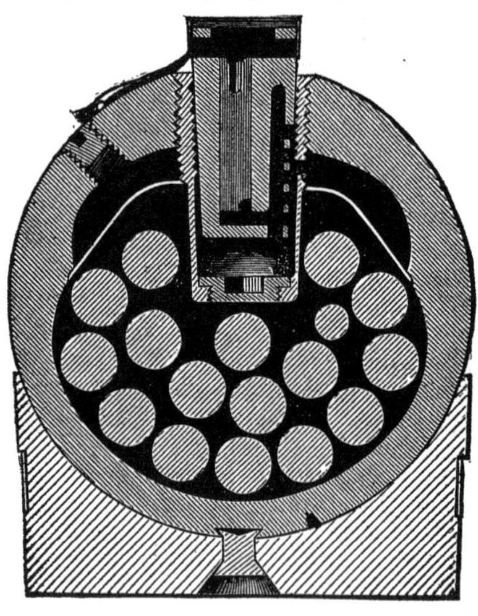

* The hemispherical wood bottoms, which were necessary in this employment of hand grenades, now form no part of the equiqment.

† Proportion issued is one per shell and 10 per cent. spare.—Army Equipment, 1876, p. 70.

‡ Shrapnel made prior to 1858 are not serviceable; they may be known by their projecting socket. The details of construction were not matured until 1858. Many of earlier manufacture were made by contract and were liable to break up.

grooves down the sides to make it open out easily and uniformly, thickened at the junction of diaphragm and shell, as otherwise it would split into two pieces instead of four or five, as desired; thickened at the fuze-hole to support the socket, and thickened at the base in all natures above 12-pr., to withstand the shock of discharge.

A wrought-iron cup or diaphragm divides the shell into two unequal parts, the smaller forming the powder chamber, and the larger being filled with lead and antimony bullets,* (lead six parts, antimony one part) packed in coal dust. The antimony hardens the lead and prevents the bullets losing their form by being pressed together.

The diaphragm has a hole in the centre, through which a gun-metal socket passes, which serves to contain the fuze. This socket is not countersunk, as is the case with common shell, but is flush with the surface of the shell. (These shell would not be used as hollow shot, so countersinking is not necessary.) Through this socket the bullets are introduced and the bottom of the socket is then screwed in. The socket communicates with the powder chamber by a fire-hole. The gun-metal plug which screws into it has a wooden plug covered with serge attached which prevents powder working in and filling up the space for the fuze. The powder chamber is filled with pistol R.F.G. or F.G. powder through the loading hole. The loading hole varies in size, being smaller for the lower natures of shell up to 18-pr. inclusive.

§ 2286, and A.C. 7|74.

The main advantage gained by separating the powder is to avoid premature explosions.

For instructions for preparing shell and fuzes, see pages 61, 64, 65.

Use.

As a small charge of powder is used merely to open the shell, the effect depends wholly on the velocity with which the shell is moving. The shell should be burst at a distance of 50 to 20 yards in front of the object, and from 15 to 10 feet above the plane. It is most destructive when used against columns, but may be used against troops in line. As the quantity of powder which the chamber holds is only just sufficient to open the shell, it is necessary to measure or weigh the charge to ensure the shells having the full amount of powder.

Issue.

Empty, loose, prepared for bottoms for India.

Empty, riveted, and boxed generally.

Filled for field, naval, and boat service, and boxed.

N.B.—Shrapnel are always issued with their balls. Mistakes seem to have arisen from Shrapnel being demanded "filled" under the idea that this term referred to balls, whereas it refers to the powder.

Improved Shrapnel.

*Improved Shrapnel* have long been discontinued, but may still possibly be met with at out stations. They may be recognized by their projecting socket and the large hole in the side of the shell through which the bullets were introduced. They scattered the bullets too much, owing to the position of the powder. They are now obsolete.

---

* Musket balls mixed with pistol, to fill up the intervals, are used with the larger natures of Shrapnel, viz., to 18-pr. inclusive, and carbine balls with the smaller.

As lead is half as heavy again as iron it is much better suited for bullets; sand shot lose their velocity quicker.

## MISCELLANEOUS PROJECTILES.†

Carcasses.
Ground light balls.
Parachute light balls.
Smoke balls.

Carcasses.

*Carcasses.*—Of all calibres from 12-pr. inclusive upwards, except the 100-pr. Fired from all natures of guns, howitzers, carronades, and mortars. They are shells with three vents (see cut) rather thicker than

12-pr. Carcass.

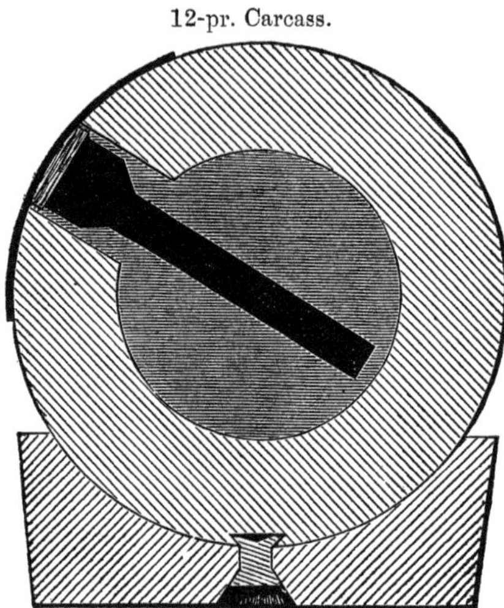

common shell (about $\frac{1}{5}$ diameter) to compensate for the weakness caused by the vents. They are a little heavier than common shell of the same calibre. They are filled with composition consisting of—

| | | |
|---|---|---|
| Saltpetre .. .. .. | 6 lbs. | 4 oz. |
| Sulphur .. .. .. .. | 2 „ | 8 „ |
| Rosin .. .. .. .. | 1 „ | 14 „ |
| Sulphide of Antimony .. .. | 0 „ | 10 „ |
| Turpentine .. .. .. | 0 „ | 10 „ |
| Tallow .. .. .. .. | 0 „ | 10 „ |

This composition is put in hot, and three holes made in it in prolongation of the vents. These holes are driven with fuze composition, and matched with quick match to insure ignition. The vents are plugged with brown paper and further secured by kit plasters.

Preparation.

Before firing, the plasters and plugs must be removed and the priming exposed. They burn with a violent flame and are difficult to extinguish. Water does not put them out. Earth is the best thing to check their action.

Carcasses have been known to burst.

Use.

To fire buildings, shipping, &c. Carcasses fired from 13″ S.S. mortar and 10″ gun are to be fired with charges not exceeding 16 lbs. and 8 lbs., respectively, to avoid straining the pieces with such heavy projectiles.

13″ carcasses burn 12 minutes and the others a shorter time down to the 12-pr. which burns three minutes.

† Are not now included in equipment, and would only be issued on special demand, showing for what purpose they are required. These stores are specially apt to deteriorate, hence the rule. Martin's shell was made obsolete by Cl. 18, A.C. 1869. It was introduced to contain molten iron, and was intended to destroy wooden ships.

Carcasses are generally issued filled in boxes marked in red as given above for shells. **Issue.**

*Ground Light Balls.*—Calibres, 10″, 8″, 5½″, 4⅖″, fired from mortars only. **Ground light balls.**

4⅖-in. Ground Ball.

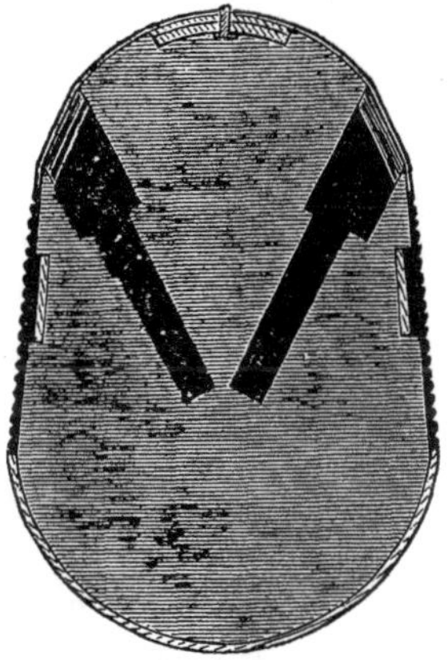

They have a wrought-iron skeleton frame (see cut), partially covered with canvas, filled with composition, consisting of saltpetre, sulphur, rosin, and linseed oil, which is put in hot, and holes made in it, driven with fuze composition and matched as given above for carcasses. **Construction.**

The body is woolded over with twine. The 8″ and 10″ have five vents in the top; the others have four. The vents are secured with plugs and kit plaster, which have to be removed before firing. The 10″ and 8″ have lugs to facilitate loading.

They are used at night to discover working parties, &c., of the enemy, and might, failing carcasses, be used in their place. **Use.**

As they are required to remain where they fall they are only suited to high angle fire.

The composition is not a very good one, but is hard to extinguish, water having little effect on it. A few shovelfulls of earth will hide its light.

Sometimes shells have been placed in light balls to deter men from putting then out, therefore *light balls of foreign or doubtful origin should be examined and burned with caution.*

Ground light balls weigh from ½ to ⅗ the weight of common shell of same calibre.

Time of burning varies from 9 to 16 minutes.

They are fired with very reduced charges, varying from 2 lbs. to 1 oz.

| — | 10″ | | 8″ | | 5½″ | | 4⅖″ | |
|---|---|---|---|---|---|---|---|---|
| | lbs. | oz. | lbs. | oz. | lbs. | oz. | lbs. | oz. |
| 200 yards .. .. .. | 0 | 10 | 0 | 4 | 0 | 2 | 0 | 1 |
| 300 ,, .. .. .. | 0 | 11½ | 0 | 6 | 0 | 3 | 0 | 2 |
| 400 ,, .. .. .. | 0 | 13½ | 0 | 8½ | 0 | 4½ | 0 | 3 |
| 500 ,, .. .. .. | 1 | 0 | 0 | 11 | 0 | 5½ | 0 | 4½ |
| 600 ,, .. .. .. | 1 | 2½ | 0 | 14½ | 0 | 8½ | 0 | 6½ |
| 700 ,, .. .. .. | 1 | 6 | 1 | 3 | 0 | 12 | 0 | 9½ |
| 800 ,, .. .. .. | 1 | 12 | 1 | 10 | 1 | 0 | — | |
| 850 ,, .. .. .. | 2 | 0 | 1 | 14 | — | | — | |

Issue.

Ground light balls are issued filled and finished and packed in boxes, No. per box varying with nature.

Parachute light ball.

*Parachute Light Ball.*—Calibres, 10″, 8″, 5½″, fired from mortars. Consists of two outer and two inner tinned iron hemispheres, the two outer are lightly rivited together, the two upper hemispheres are connected by a chain, the inner upper hemisphere has a depression at the top to admit the bursting charge and fuze. A quickmatch leader conducts the flash from the bursting charge to the fuze composition in the lower inner hemisphere; the inner upper hemisphere contains the parachute tightly folded up; to ensure its opening, a cord is passed between its folds and through a hole in the top of parachute, and is fastened to the upper inner hemisphere, so that when the hemisphere is blown away, the cord is pulled through and the parachute expanded.

Ball, Light, Parachute.

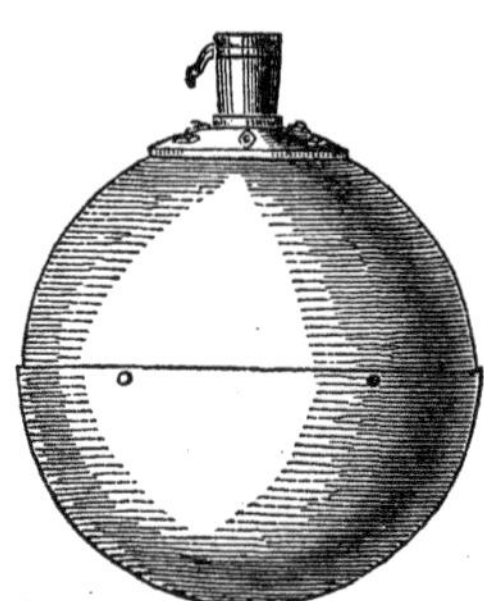

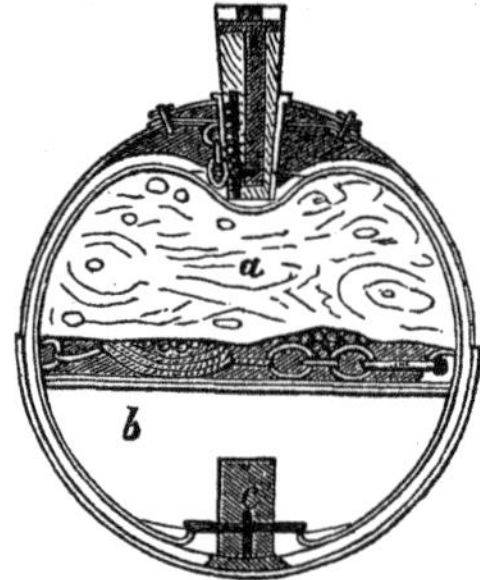


*a.* Parachute.
*b.* Composition.

The lower inner hemisphere contains a composition of saltpetre, 7 lbs.; sulphur, 1 lb. 2 oz.; red orpiment, 11 oz. A hole is bored and driven with fuze composition and matched as usual; this hemisphere is connected with the parachute by cords and chains.

Parachute open after firing.

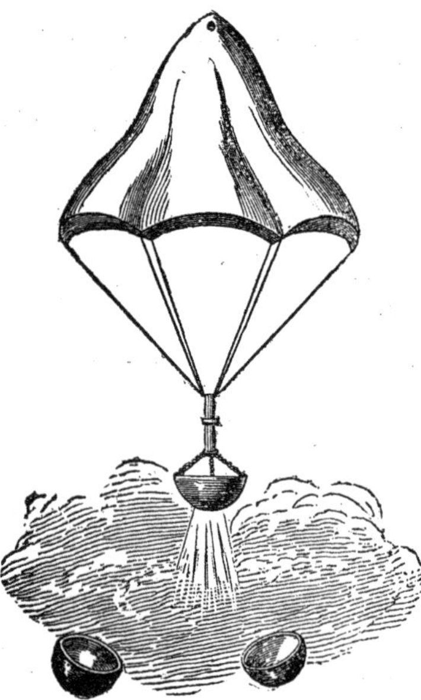

The bursting charge* is issued in the parachute, the fuze is bored to the required length and well hammered in, the parachute placed in the mortar and fired. **Preparation.**

The fuze ignites the bursting charge, the outer hemispheres are blown away, and the inner upper hemisphere, which is chained to the outer one is blown away with it, the parachute is opened by the cord and expands, the composition in the lower hemisphere being ignited by the quick match leader which ignites the fuze composition. **Action.**

The composition burns about 3 minutes in the 10″, 1 minute 40 seconds in the 8″, 1 minute in the 5½″. The 10″ weighs about 30 lbs.; the 8″ 15 lbs.; and the 5½″ 6 lbs.

The fuze should be so regulated as to open the parachute just as it begins to descend. No fuze tables have been laid down, but about half the length of fuze should be given which is given to a mortar with similar charge.†

Extreme charge for 10″, 2½ lbs.; range, 1,400 yards; 8″, 1 lb.; 5½″, 8 oz.

To throw light on the enemy's working parties, &c., at night; it has the advantage of being out of reach, so cannot be extinguished. Careful allowance for wind must be made. **Use.**

In wooden boxes, one in each. **Issue.**

*Smoke Balls.*—Calibres, 13″, 10″, 8″, 5½″, and 4⅖″, may be fired out of mortars with very light charges. **Smoke balls.**

A paper shell filled with L.G. powder, saltpetre, coal dust, pitch and tallow, the vent driven with fuze composition, and matched, and covered with kit plaster, a layer of sulphur, and coal dust is sprinkled in three times during filling; in burning, this clears the vent. **Construction.**

These balls appear to be useless as projectiles, they are intended (1) to put in enemy's mines, (2) to conceal operations from the enemy, (3) for signals in the Arctic regions; they burn from one to eight minutes.

It is very doubtful whether smoke balls have ever been fired.

Filled, finished, and boxed, number per box varying with nature. **Issue.**

*Gauges*, high and low, see table, p. 323, are issued for every kind of S.B. gun, and for stations of inspection; they are simply iron rings with handles, the high gauge should pass over the shot, the low should not. **Gauges. §§ 1314, 1698.**

All S.B. projectiles are below their nominal diameter, while the gun is a little above it, thus an 8″ shot or shell has a mean diameter of 7·85″.

*Windage* is necessary not only to allow the shot to load easily, but also to allow for the increased size of the shot caused by rust, &c. With shells windage is useful by allowing the flash of discharge to ignite the fuze. **Windage.**

When shot are repainted it is necessary to scrape off the old paint, which can be done with knives or a piece of an old sword or cutlass.

## Wood Bottoms.

Used with shot fired from bronze guns to save the guns, with shot

---

* The bursting charges are for the 10″ 3 drams.
„ „ „ 8″ 1½ „
„ „ „ 5½″ 1 „

† The fuze for a 10-inch mortar with a charge of 1lb. is 1·9 inch, the time of flight is about 10 seconds. So firing a parachute light ball with a 1lb. charge we should bore the fuze about 5 seconds; the first number marked on the 10-inch parachute fuze is 6 seconds, so the fuze has to be bored to 6.

carried with iron guns of position to steady them in the limbers, also used with all shells except mortar shells, and the 100-pr. naval shell which has a top, and with carcasses except when fired from mortars. They are necessary with shells to keep the fuze, and with carcasses to keep the vents, in the proper position. They are shaped so as to fit the pieces of ordnance for which they are intended; thus, they are conical as a general rule, for all the shell guns and howitzers which have gomer chambers, and for all unchambered ordnance of corresponding calibre (the 68, 32, 24, and 12-pr. gun shell would have conical bottoms though the guns have not gomer chambers); hemispherical for the $5\frac{1}{2}''$ and $4\frac{2}{5}''$ howitzers which have cylindrical chambers, and cylindrical for all other ordnance. They are stamped with the calibre or weight of the projectile with which they are to be used. Carronades have cylindrical chambers, but if shells were required to be used with them any kind of gun or howitzer shell would be used without regard to the shape of its bottom.

Common and Shrapnel bottoms.

They are made of well-seasoned elm or alder, or of teak for tropical climates. The grain runs plankways, except for Shrapnel up to 24-pr., where the grain runs endways, the bottom is carried higher up on the shell, and is secured from splitting by a tin strap, this ensures the bottom breaking up, and so there is less risk when firing over troops in front, also the shooting is said to be improved by the bottom quitting the shell readily.

It is hardly necessary to mention such an unimportant exception to the rule as the $4\frac{2}{5}''$ and $5\frac{1}{2}''$ Shrapnel shells, which have the same hemispherical plank bottoms as the common shells for the $4\frac{2}{5}''$ and $5\frac{1}{2}''$ howitzers, but the $4\frac{2}{5}''$ has the small Shrapnel rivet hole.

Rivets.

Bottoms for land and boat service are fastened by a single expanding gun-metal rivet; the rivets are of two sizes, small for Shrapnel up to the 18-pr., and large for all other shells, and also for shot; these rivets being hollowed out at the base expand into undercut holes in the shell or shot.

Naval bottoms.

For naval service, to allow of the practice of double shooting, it is found advisable to have the metal of the shells in contact, otherwise they are liable to break up. The single rivet cannot therefore be used, a piece being cut out of the centre of the wood bottom. The bottom is fastened to these shells by means of two inclined rivets, which are simply cylindrical pins of copper not hollowed out and having no heads.* Here

10″ bottom.

the weakness of the 10″ gun leads to its having a peculiar form of bottom; it would be dangerous to double shot the gun, so the bottom is not hollowed out, but it might be inconvenient to have a different plan of riveting for the naval service, so the bottom is prepared to take naval rivets, and to avoid having two kinds of bottoms for this gun it is also prepared for the L.S. central rivet.

To fasten on bottoms, see p. 60.

Issue.

When not issued fastened to the shells they are strung by 20 on an iron rod for land, and on a wooden rod for sea service; 10 per cent. over the number of shells are allowed.

Mortar bottoms. § 1582.

Mortar bottoms, 13″, 10″ and 8″, are hemispherical in shape, they are a special store required on firing pound shot or bouquets of small shell from mortars of above calibre. They now form no part of equipment.

Tops.

Tops were introduced for naval service; they save space in stowage, protect the fuze and facilitate loading, but owing to the introduction of

---

* Naval rivets are of two sizes. Medium for 100-pr. tops and 32-pr. bottoms; short for 10″ and 8″ bottoms.

rifled guns they have only been applied to the 100-pr., they are fastened to the shell by four rivets of the same description as are used for naval bottoms.

## General Remarks on Cartridges.

This appears to be a fit place for making a few remarks, which will apply to all cartridges in the service. Details which apply only to individual cartridges, or to certain classes of cartridges, will be found under the various ordnance for which they are used.

For safety, rapidity, and convenience of loading, the charge of powder which is placed in a piece of ordnance is enclosed in an envelope called a cartridge.*

A good cartridge for service should possess the following qualifications:—

**Requisite properties of material used.**

1. The material should be strong enough to bear reasonable knocking about when filled, and to stand the wear and tear of travelling.
2. It should be so close in texture as not readily to admit of the powder, if slightly dusty, working its way through, and yet be permeable to the flash from the tube, &c., intended to fire it, even if not well "pricked."
3. Lastly, and this is of the greatest importance, the material should consume entirely in the gun when fired; or, if this end cannot be obtained, it should leave no smouldering fragments, or sparks, in the bore.†

**Materials used.**

These conditions are very well fulfilled by the materials in use in the service, *i.e.*, serge and silk cloth. All cartridges in our service are made of these materials, with the sole exception of the 4-oz. charge for the 7-pr. gun, which is contained in a red shalloon bag. The terms *flannel* and *serge* are used indifferently in the service nomenclature.‡

Experience has shown that serge is hardly strong enough for the heavy charges now used, cases having occurred where the cartridges have broken when being packed.

---

* Mortars are sometimes loaded with loose powder, though cartridges are issued for them. The reason of this is that they are fired at a fixed elevation, so that the range depends on the amount of powder used, and hence the charge will vary from round to round if firing at an object in motion, and also slightly when firing at a fixed object till the right range is obtained.

† If much residue is left in the gun the vent is apt to be choked. If sparks remain, and the gun is re-loaded almost immediately, a serious accident will probably occur, Several accidents have thus occurred, especially when using blank serge cartridges, generally due to carelessness in sponging or serving the vent. See in text below for the rules as to the use of silk cloth cartridges.

‡ Flannel is the word used in the Priced Vocabulary. Strictly speaking, the following are the definitions of *flannel*, *serge*, and *shalloon*. I am indebted for them to the courtesy of Messrs. Shaw and Sons, Halifax.

*Flannel* is the term used to denote a thin fabric, woven plain, and made entirely of "short" wools, both warp and weft. The wool is first carded on a carding machine, and afterwards spun to the requisite fineness on mules.

*Serge* is woven twilled. The warp is made from "long" wool, which is first combed on a combing machine, and then spun into a worsted thread on a worsted spinning frame. The weft is made from "short" wool, carded on a carder, and afterwards spun on mules. The warp thread is named "worsted," and the weft thread named "woollen" by way of distinction.

*Shalloon* is made entirely of "long" wool, and is woven twilled. It is entirely a worsted fabric, prepared like the warp of a serge.

N.B. The warp threads are those running the length of the fabric, the weft threads those running across.

In all cases where serge is employed as a material for cartridges, care is taken to ensure its being of good quality and made entirely of pure wool. Samples are tested by the Chemist, W.D. Any admixture of cotton would be most objectionable, as increasing the chance of smouldering.

Silk cloth. §§ 1780, 1822, 1829, 1868, 2047, 2116, 2633.

*Silk Cloth* or, as it was at first called, *amiantine*, is made of the refuse silk from the outside of the cocoons. It is stronger* and of closer texture than serge, and is not so liable to "hold fire" or smoulder. It was originally introduced on the score of safety for blank charges, as in firing these there is not so much heat and pressure as when shotted guns are fired, and therefore less chance of the cartridge being entirely consumed. Its use is now much extended, as will be seen when we come to treat of cartridges for R.M.L. guns.†

Cartridges for which serge is used.

§ 1780, 1868.

§ 1829.

§ 2838.

All S.B. and B.L. cartridges are made of serge,‡ with the exception of blank charges for S.B. guns,§ which, for the sake of safety are, under certain conditions, contained in silk cloth. These silk-cloth cartridges would be used whenever a gun has to be loaded again after firing a blank charge, unless a long interval of time should intervene, as in the case of morning, mid-day, and evening guns. The rule laid down is, that silk cloth cartridges are to be used, (1), saluting, where the number of guns is less than the number of rounds fired; (2), where garrison guns are allowed to be fired at reviews by special order; (3), for exercise in dismissing recruits. These silk-cloth cartridges are at present made up for all guns, except the 100-pr. and 3-pr. Cartridges for all R.M.L. guns have been made of serge up to and including the battering charge for the 12-inch gun of 35 tons. In future, however, all new cartridges for R.M.L. guns will be of silk cloth.‖ In order, however, to utilize the very large existing store of serge cartridges, the following rules are given. All filled and empty R.M.L. cartridges made of serge now in store, with the exception of empty cartridges of 85 lbs. and upwards, will remain unaltered, and will be issued or used up, as a rule, before any store of silk cloth is made use of. The empty cartridges

---

* Some trials made showed the comparative strength of silk cloth and serge to be as follows:—

Serge, with the warp, 270; with the weft, 250.
Silk cloth, with the warp, 400; with the weft, 300.

† The silk cloth used in the manufacture of cartridges will, in future, probably be divided into three classes, viz:—

1. For charges up to 14lbs. Width of silk, 47″. To stand a strain of 150lbs. weft, and 200lbs. warp.
2. For charges from 14lbs. to 85lbs. Width of silk, 54″. To stand a strain of 180lbs. weft; and 230lbs. warp.
3. For charges above 85lbs. Width of silk, 54″. To stand a strain of 420 lbs. weft, and 520lb. warps.

‡ Two natures of serge are used for cartridges.

No. 1, width 54″ for all battering charges, and the 12″, 11″, and 10″ cartridges.

No. of strands per inch { warp, 42. weft, 41.

Weight per yard, 1lb.

No. 2, width 54″ for 9″ and under, except battering charges.

No. of strands per inch { warp, 42. weft, 38.

Weight per yard, 14oz.

§ In the case of a B.L. gun, from the nature of the construction there is little fear of sparks being left even with a blank charge. Silk cloth cartridges, however, have been made, but not sealed for blank charges for B.L. guns. Patterns of silk cloth cartridges for blank charges with S.B. howitzers, up to the 32-pr. inclusive, have been also made but not sealed.

‖ 4-oz. charge for 7-pr. excepted.

for 85 lbs. and upwards are to be exchanged and returned for conversion, and all future issues of such cartridges will be of silk cloth.

The store of empty cartridges to be held in reserve for R.M.L. field and siege artillery will be all of silk cloth.

**Conversion of serge cartridges. §§ 2177, 2838.**

Surplus stores of the undermentioned serge cartridges will be converted to meet demands as follows:—

| Nature | Charge | | Convert to |
|---|---|---|---|
| 32-pr. | 10 lbs. | convert to | 7-inch M.L. 10 lbs.<br>80-pr. M.L. 10 lbs.<br>64-pr. M.L. 10 lbs. |
| 32-pr. | 8 lbs. | ,, ,, | 64-pr. M.L. 8 lbs. |
| 32-pr. | 7½ lbs. | | |
| 32-pr. | 7 lbs. | | |
| 32-pr. | 6 lbs. | ,, ,, | 64-pr. M.L. 6 lbs. |
| 24-pr. | 8 lbs. | | |
| 68-pr. | 16 lbs. | ,, ,, | 16-pr. M.L. 3 lbs. |
| 10-inch | 12 lbs. | ,, ,, | |
| 8-inch | 10 lbs. | ,, ,, | 9-pr. M.L. 1¾ lbs. |
| 8-inch | 8 lbs. | ,, ,, | |

The 32-pr. and 24-pr. cartridges merely require to be turned and re-marked, the others require to be ripped, cut to proper shape, and re-made.

It is to be understood that *only* cartridges which have never been filled, and which are only marked on one side, will be utilized in this way.

In the conversion or utilization of such cartridges for 7-inch, 80-pr., 64-pr., &c., which do not require to be cut up and re-made, the hoops will continue to be of worsted, but care must be taken that they are filled with the proper description of powder for the service required; that the old marking is crossed out by diagonal lines of printer's ink (*not* obliterated by a daub or patch of paint), and that the charge and the nature of gun for which each is intended is properly marked on them before issue. When utilizing filled cartridges which are the correct size and contain the proper charge, it is not necessary to empty and turn them if they can be marked properly without doing so.

**Filling cartridges.**

All cartridges are filled by actual weighing instead of by measure as formerly, the Caffin's machines used for this purpose being now obsolete. A copper funnel, formerly called the "royal naval funnel," is used in filling them.

**Choking and hooping.**

All cartridges are choked and hooped. The choking closes the mouth of the cartridge; hooping is necessary to preserve its shape. Details as to the mode of choking and hooping various cartridges will be given in proper order.

It is as well to remember, with regard to choking and hooping, that:—

All S.B. serge cartridges are choked and hooped with worsted, except that for the 100-pr., which is hooped with worsted braid.

All S.B. silk cloth cartridges are choked and hooped with silk twist, except the 5-lb. cartridge* for the 32-pr., which has silk braid hoops.

All B.L. cartridges are choked with twine and hooped with blue worsted braid, except blank cartridges, which are choked with worsted, as twine is liable to carry fire.

All R.M.L. serge cartridges (except some converted ones) are choked with worsted and hooped with blue worsted braid.

All R.M.L. silk cloth cartridges are choked and hooped with silk, the hoops being braid for some, silk twist for others.

* Last pattern.

**Marking of cartridges.**

§ 1998.

All cartridges are marked in black, with the nature of the gun, &c.,* for which they are intended, and with the weight of their contents, and also with the mongram of the station at which they are filled. Cartridges filled by the R.A. have no station monogram marked upon them. Cartridges made up with L.G. powder for B.L. or M.L. rifled guns are marked with the letters L.G. in red letters one inch long. Similarly when S.B. cartridges are made up with R.L.G., they have R.L.G. marked in red upon them. Up to the full or service charge for the 12-inch R.M.L. gun of 35 tons, and down to the battering charge for the 7-inch R.M.L. gun, cartridges are marked both for P. and R.L.G.,† as shown in the woodcut, the actual contents being left unerased, the alternative charge erased by lines run through it.

§ 2103.

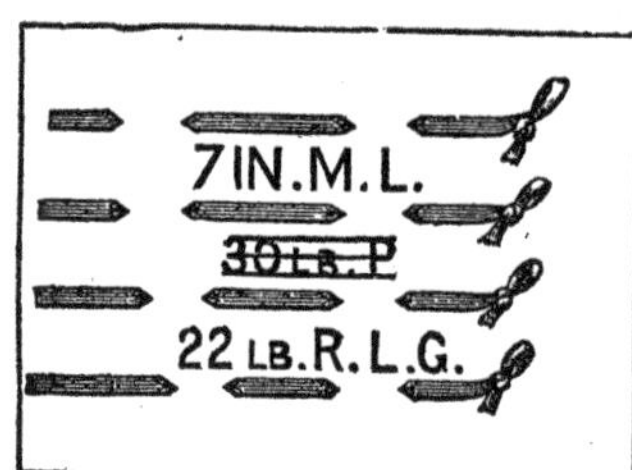


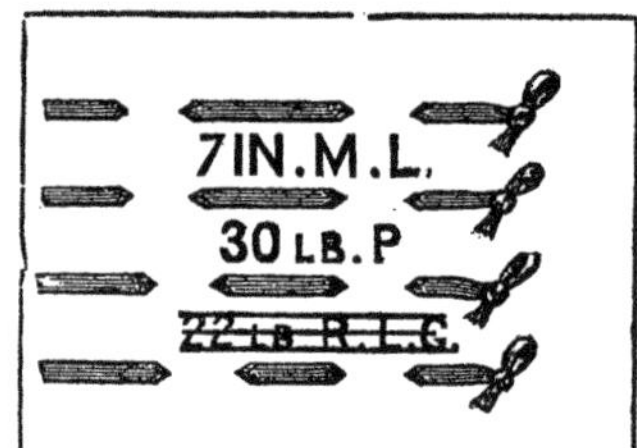


All marking on cartridges is done with printer's ink. Paint would be most objectionable, as it renders the cartridge liable to hold fire.

The diameter of the cartridge is made considerably smaller than the diameter of the gun to ensure easy loading; the interval between the cartridge and the gun also facilitates the passage of the flash, causing more rapid ignition. On the other hand, as the pressure varies inversely as the space occupied by the powder, the cartridge must not be unduly lengthened, as the force of the powder would be lost. A long cartridge will evidently strain the gun less than a short one, other things being equal.‡

---

* The 10lbs. cartridge for the 65 and 60 cwt. 8-inch guns, and the 10lbs. cartridge for the 58 and 56 cwt. 32-pr. gun, are marked D (distant), to show that they are not to be used for lighter guns of these calibres.

Cartridges for mortars are issued of a size to hold the maximum charge; thus a S.S. mortar, 13″, has a 20lbs. cartridge, and also a 16lbs. for carcasses.

† Some cartridges approved on the 11/1/71 were marked for P powder only. See table, p .

‡ Minute of D. of A., 1/10/74. Pressure reduced by lengthening cartridge and reducing diameter. Trial in 8-inch gun, firing a proof cylinder of 180lbs. Charge, 35 lbs. P.

| Cartridge. | | Initial Velocity. | Pressure in Tons. | | |
|---|---|---|---|---|---|
| Length. | Diameter. | In foot-seconds. | A. | B. | C. |
| 19″ | 7″·9 | 1,408 | 13·8 | 14·9 | 14·3 |
| 26″ | 7″·0 | 1,385 | 12·6 | 13·1 | 12·9 |

The following table requires no explanation, but shows that *within certain limits* the work obtained per lb. of powder is greater with charges comparatively low with reference to the weight of the projectile, than with charges where the proportion of powder to weight of the projectile is higher. This appears to have an exception in the case of 12″·5 gun, but it is to be noted that the proportion shown in the last column of the table is higher in the case of this gun.

The position of the vent affects the rate of ignition,* but this does not directly affect the cartridge.

| Nature of Gun. | Initial Velocity. | Proportion of weight of charge to that of projectile. | Work done in foot-tons per lb. of powder. | Length of bore ÷ length of cartridge. |
|---|---|---|---|---|
| 12″·5 | 1,425 | ·163 | 86·64 | 6·53 |
| 12″ (35 tons) | 1,300 | ·157 | 74·53 | 5·80 |
| 12″ (25 tons) | 1,300 | ·141 | 82·70 | 6·59 |
| 11″ | 1,315 | ·160 | 75·47 | 5·68 |
| 10″ | 1,364 | ·175 | 73·71 | 5·70 |
| 9″ | 1,420 | ·200 | 69·92 | 5·43 |
| 8″ | 1,413 | ·194 | 71·20 | 5·48 |
| 7″ (6½ tons) | 1,525 | ·260 | 61·83 | 4·82 |

"Air spacing" with regard to cartridges has lately received much attention, especially in the case of the experimental guns of 80 and 38 tons. To exemplify the meaning of this term, let us suppose that we have a cylinder capable of containing 100 cubic inches of powder tightly packed, lying horizontally. If, now, we wish to introduce, say, 80 cubic inches of powder into this cylinder we can either put them loosely in, so as to fill the cylinder, but leave considerable spaces between the grains, or we can pack them tightly into a smaller cylinder, and insert this into the large cylinder, in which case there will be space between the two cylinders, or we can put a tube or cone down the middle of the cylinder and pack the powder tightly round it. For a given number of cubic inches read a given number of pounds of powder, and we have the three modes of air-spacing that have been tried. The method of having an internal cone made of various materials has been very successful, and this method allows the cartridge to be ignited well in the centre, arrangements being made for the flash of the tube to communicate directly or indirectly with the charge through perforation in the cone. The best material for the latter is found to be zinc. In the case of some cartridges tried with the 80-ton gun the zinc cone was 48″ long and about 1 foot wide at its base. At a distance of 33″ from the base was a diaphragm, on which was cemented a mill-board disc, containing a small shalloon bag of L.G. powder. The bag was surrounded by small bundles of quick match, and hard by, in the sides of the cone, were perforations to allow of the flash of the L.G. powder and quick match, when ignited by the flame proceeding from an axial vent and travelling up the cone, inflaming the charge.

The effect of "air spacing" is to allow of the use of a heavier charge, and thereby to obtain higher muzzle velocity without increasing the strain on the gun. The following may be taken as a good example with the 38-ton gun :—

| Charge. | | | Cubic inches to the lb. | Muzzle velocity for an 800lb. shot | Mean pressure in tons per square inches of chamber. |
|---|---|---|---|---|---|
| Weight. | Length. | Volume. | | | |
| 130 lbs. | 28 in. | 3,217 c.i. | 24·6 | 1,449 | 24·3 |
| 130 ,, | 33·25 ,, | 3,900 ,, | 30·0 | 1,391 | 19·1 |
| 160 ,, | 41·5 ,, | 4,917 ,, | 30·7 | 1,477 | 19·8 |
| 180 ,, | 45·37 ,, | 5,400 ,, | 30·0 | 1,541 | 22·4 |

The pressure throughout was very moderate with the elongated (*i.e.*, air-spaced) cartridges, and, comparing the results obtained with the service charge of 130 lbs., rammed home in the usual manner, there is a gain in velocity of 90 feet, with a reduction in mean pressure of about two tons, by increasing thecharge to 180 lbs. and reducing the grammetic density, Extracts. XIV, p. 80. Gas checks were used with the projectiles, making their total weight 810 lbs., and the powder was $P^2$ from the same batch.

In the experimental 12-pr. gun an expanding cartridge has been tried. The gun is chambered, and the cartridge is filled in the ordinary way so as to fit the bore, but has a fold longitudinally, so that, if rammed into the chamber the latter will be loosely filled by the cartridge expanding, owing to the powder filling the hitherto empty fold. The cartridge is confined by a cover of white fine paper, and over this is a brown paper cover to preserve it from injury in travelling, &c. On loading the brown paper is torn off, and the cartridge confined by the fine paper only inserted into the bore. On entering the chamber the action of the rammer breaks the fine paper, and the cartridge expands.

* Ballistic Experiments, by Capt. W. H. Noble, R.A., p. 85.

The velocity is greatest when the vent is ·4 of the length of the cartridge from the rear, and least when the vent is at the rear of the cartridge. It has been urged in favour of the rear vent that the cartridge is more likely to be completely expelled.

**Issue.**

Cartridges are issued filled to the Navy * and empty to the artillery, except for field service; the filled cartridges for field service are issued in gun ammunition boxes. Empty cartridges are made up in bales pressed together by hydraulic pressure, covered with oiled canvas, and outside with stout Hessian stitched together. It may be remarked that the cartridges are not nearly so liable to be attacked by moths as long as the bales are unopened.†

The bales are marked with the nature of cartridge, weight of charge, number of cartridges, and date.

For the number of cartridges of various natures issued in bales, see table, p. 321.

### *Smooth Bore Cartridges.*

**S.B. cartridges.**

Cartridges are made conical for ordnance with gomer chambers; cylindrical for all other ordnance. The cylindrical cartridge is made in one piece, the conical in two; the edges are made to overlap and are sewed together, which avoids the risk of having three thicknesses of serge under the vent.

They are hooped and choked with worsted; the choke is cut off to 1″ for all guns except the 100-pr., which is limited to 3″.

**Choking.**

Choking consists in drawing together the mouth of the cartridge into several pleats with a brass needle threaded with three strands of worsted, three turns being taken round the pleats; the choke thus formed being further secured by passing the needle five times through it, alternately above and below the turns of worsted, thereby stitching down the worsted at four points equidistant from one another.

**Hooping.**

Hooping is necessary to preserve the shape of the cartridge; it is done as follows: In making the last stitch in choking, the needle is turned downwards and carried through the powder and out at the seam, dividing the space between the shoulder and bottom of the cartridge according to the number of hoops required; the worsted is then carried tightly round the cartridge forming a hoop, which is stitched to the cartridge at two or three points in the same way as the turns of worsted at the choke were secured; one, two, or three hoops as required are thus made.

For charges for various guns and packages, see table, pp. 325, &c.

The following are approximate rules for the ratio of weight of charge to weight of projectile ‡—

Shot guns, service charge from $\frac{1}{3}$ to $\frac{1}{4}$ weight of projectile.
Shell guns, $\frac{1}{6}$th to $\frac{1}{12}$th.
Carronades, $\frac{1}{12}$th.

The 5-lbs. charge for the 8″ gun is too short for the chamber, and is brought up to length by a wad consisting of coal dust contained in a blue serge bag.

Serge bags are used for containing 10 or 15 lbs. of loose powder or less for filling shell.

*Cartridges, Silk Cloth.* See p. 122 and 302.

**Drill cartridges.**

*Drill Cartridges* are hollow blocks of wood, covered with raw hide

---

* The Navy carry a certain proportion of empty cartridges, except for boat guns.

† It has, however, been asserted that the oiled canvas acts rather as an incentive to the moth. Many experiments have been made to preserve the cartridges by impregnating them with carbolic acid, corrosive sublimate, &c., but without satisfactory results.

‡ Reduced charges used when firing at angles of depression; half charge from 15° to 30°; quarter from 30° to 50°, in order not to dismount the pieces.

and fitted at one end with an imitation choke and rope handle; they are marked like the service cartridge, which they are intended to represent.

*Gauges, Filled Cartridge, Brass Ring*, are made of fourteen different sizes. The gauge is a gun-metal ring with a handle, on which is stamped the nature and size. They are used in the examination and making up of filled cartridges, over which they should pass freely.

**Gauges, filled cartridge. § 1695.**

Loose, in numbers, as demanded.

**Issue.**

| | Diameter, Inches. |
|---|---|
| 100-pr. gun .. .. .. .. .. .. | 8·25 |
| 10-inch gun and howitzer, and 68-pr. gun .. .. | 7·76 |
| 8 " and 56-pr. gun .. .. .. .. | 7·2 |
| 8 " howitzer and 42-pr. gun .. .. .. | 6·67 |
| 32-pr. gun .. .. .. .. .. .. | 6·09 |
| 24 " .. .. .. .. .. .. | 5·52 |
| 5½-inch and 24-pr. howitzer .. .. .. | 5·1 |
| 18-pr. gun .. .. .. .. .. .. | 5·02 |
| 32 " howitzer and 12-pr. gun .. .. .. | 4·39 |
| 12 " " .. .. .. .. .. .. | 4·2 |
| 9 " gun .. .. .. .. .. .. | 4·02 |
| 6 " " .. .. .. .. .. .. | 3·5 |
| 3 " " .. .. .. .. .. .. | 2·78 |
| 4⅖-inch howitzer .. .. .. .. .. | 2·2 |

# CHAPTER X.—INTRODUCTION TO AMMUNITION FOR RIFLED ORDNANCE.

THE great advantages resulting from the use of elongated projectiles fired out of rifled guns have for some time been duly recognized, and accordingly steps have been taken by all civilized nations to ascertain experimentally the system of rifling best adapted for the various requirements of field, garrison, and naval guns.

The two systems of rifling best known among European nations are*:—

1. A breech-loading system, using a soft-coated projectile slightly larger than the bore of the gun, which is forced by the explosion of the charge to take the rifling, which usually consists of a large number of shallow grooves cut spirally in the gun: for example, the Armstrong and the Prussian systems.
2. A muzzle-loading system, using projectiles with studs or ribs, shaped to the same general form as the spiral grooves in the gun, but with sufficient play to allow of facility and celerity of loading: for example, the British R.M.L. field guns and Woolwich guns, which are modifications of what has been called the French system.

Before proceeding further, let us answer the question, "What is the use of rifling a gun?" This question is best replied to by another one, viz., "What do we chiefly require in a projectile when fired from a gun?" To this we answer, "Range, and accuracy combined, with as flat a trajectory as possible." Now, by rifling a gun, we secure these advantages to a great extent; to how great an extent depends partly on the gun, partly on the projectile.

How, then, are these advantages secured by rifling a gun? It is not by increasing the muzzle velocity, because, *cæteris paribus*, a shot from a smooth bore gun will probably have at least as high a muzzle velocity as a shot of equal weight from a rifled gun of the same calibre. The reason is this: by rifling a gun we are enabled to use an elongated projectile, as we possess the power of giving it "spin" on its longer axis by means of the twist of the rifling. Did we not spin the projectile, it would, on leaving the gun, have a tendency to immediately turn over and rotate on its shorter axis, and thus its flight would be most inaccurate and its range most limited. We must, therefore, give it sufficient spin to ensure its travelling approximately *point first* throughout the entire range. By this means the resistance of the air to a projectile of given weight is greatly reduced, and the muzzle velocity is not nearly so rapidly reduced as in the case of a spherical projectile of equal weight fired under

* For certain pieces of ordnance, notably the 6"·3 M.L. howitzer and the 12-pr. M.L. field gun, it is highly probable that the rifling will be polygroove (*i.e.*, consist of a number of narrow, shallow grooves), and that the necessary rotation will be given to the projectile by means of a gas-check, the projectile itself having neither studs nor lead coating.

similar circumstance, and the difference is much greater if we take a spherical projectile of the same calibre.

For instance, the muzzle velocity of a 400lb. projectile, fired from the service 10″ R.M.L. gun is 1,364 feet per second, the charge of powder being 70 lbs. Now, at 2,000 yards range, the projectile will have a remaining velocity of 1,120 feet, while, at that range, a spherical solid shot of the same *weight* would have a velocity of 854 feet, and a spherical solid shot of the same *calibre* would have a velocity of 742 feet only. This presumes all the three projectiles to start with the same muzzle velocity, which would not in practice be the case; but the instance will serve to show how the velocity is kept up in the case of the elongated projectile.

Now this power of keeping up their velocity possessed by elongated projectiles, obviously means increased range and increased power of doing work at any given range, in the penetration of iron plates for instance. It also means comparative flatness of trajectory, and thereby the chances of hitting an object are increased, provided the line of fire be correct.

As to accuracy;—inaccuracy in a S.B. gun is chiefly due to the bounding of the shot more or less in the bore, and to the fact that, in consequence of this, the shot leaves the gun rotating on an accidental axis. It is true that the use of wood bottoms somewhat obviates this evil; but the projectiles are generally more or less "eccentric," and hence arises inaccuracy.*

In the case of a projectile fired from a rifled gun, rotation on a fixed axis, coincident with, or nearly coincident with, the axis of the bore of the gun,† equalizes the effects of irregularities of mass or surface, and prevents revolution upon an accidental axis, and thus accuracy of fire can be obtained. It is true that this rotation gives rise to a certain deflection or deviation; but this is constant and can be allowed for.

Rifling a gun does not *necessarily* entail the use of elongated projectiles; but the collateral advantages attained by their use are so important that they are universally used. Some of the advantages so attained have been mentioned; but we will briefly recapitulate them here, and mention a few others not noted above.

1. We have a diminished surface for the resistance of the air to act upon, and thus obtain greater range and greater power at a given range.
2. The trajectory is flatter; and thus we have a greater probability of hitting an object.
3. The head may be of any required form or weight; as in the case of Palliser and Shrapnel shell.
4. By varying the length, different kinds of projectiles for the same gun can be brought to the same weight; and thus complications in range tables, &c., are avoided.
5. On the other hand, if desirable, a specially heavy (or light) projectile may be fired; as in the case of the 7″ and 7-pr. double shell.
6. The capacity of the projectile for powder or bullets is increased.
7. A great saving of powder is effected, *e.g.*, the charge is from $\frac{1}{8}$ to $\frac{1}{5}$ the weight of the projectile instead of $\frac{1}{3}$, as in S.B. guns.

* Space forbids any attempt at an elaborate discussion on rifling here, the broad facts only are given. Those who wish to go more deeply into the subject will find ample scope in the many well-known books and treatises bearing on this subject.

† This is known as "centring" the projectile.

Besides all the above there are two advantages which are common to all rifled projectiles, whether elongated or not, viz.: Percussion fuzes may be of simple construction, as it is only necessary to provide for their action in one direction; and, in the case of Shrapnel, the bursting charge can be kept behind the bullets.

It has been found necessary to protect the bursting charges of powder against the risk of premature explosions arising from the rotation of the shell, either by lacquering the interior of the shell or by placing the powder in a separate case or bag. Red lacquer is now used, but some shell may be found with black lacquer.*

Lacquering has not been found a sufficient protection to the powder in R.M.L. common and Palliser shells of 7″ and upwards.† Serge bags are now used, see p. 203.

The bursting charges of common, garrison segment, and Palliser shell and Palliser shot (when used as a shell) follow the same rule as those for spherical shell, *i.e.*, the shell are completely filled with powder. Shrapnel bursters are to be weighed or measured to ensure their having enough powder. Segment bursters for field service shells are loosely filled with shell F.G. powder.

§ 1580.

§ 2089.

When rifled common shells are fired at practice as blind shells, the gun-metal plug is to be removed and returned to store, a wooden plug, supplied on demand, being substituted for it. The shell is to be *filled* with a mixture of sand and sawdust, or any similar material available, to make an equivalent weight to the bursting charge.

Shell so issued from Woolwich are stencilled "sand" in white, and are marked with a yellow band.

Length of projectile.

To insure good shooting the projectiles should at least be 2 calibres in length, the length of the projectile will be limited by the twist of the rifling; thus we find that the 12″ 25-ton gun, which has an exceptionally slow twist, is unable to fire as long projectiles as the other Woolwich guns, consequently its common and Shrapnel shell cannot be brought up to the weight of its Palliser projectiles.

§§ 1764, 2426.

All rifled common shell of 64 lbs. and upwards made between March, 1869, and Feburary, 1873, have unloading holes, closed with gun-metal plugs and papier mâché wads, in the head. These holes are used in unloading the shell when it is found difficult to extract the percussion fuze. Experience has shown that the unloading hole was very rarely of use, and to simplify manufacture it has been discontinued. If there is any difficulty in extracting the fuze, the shell may be fired from a gun, or thrown into the sea, or otherwise safely disposed of.

Blowing charges. § 2099.

80-pr. and 64-pr. .. .. .. 4 oz.

A calico bag made with a neck to fit the fuze-hole will be used, the shell is first to be filled with dry coal dust, leaving a space for the blowing charge; the bag will be inserted and filled with the proper charge of powder, and the wooden time fuze will then be driven firmly into the fuze-hole. Blowing charges for the larger natures of shell are given in § 2099, but the use of time fuzes has been discontinued with these shells.

---

* The black lacquer was discontinued as it was found to cause prematures. Vol. III, Extracts, pp. 143, 240.

† Serge bags are approved of for the 7″ B.L. 80-pr. and 64-pr. shells. At present for India only.

In all ammunition for rifled guns the pattern or mark is shown by a Roman numeral; this is important, as by quoting the numeral the nature of store in a fort or field battery can be identified.*

§§ 1126, 1162, 1545.

It is desirable when practicable to store Shrapnel shell under cover, they should never be kept standing on their bases when exposed to the weather.

Cl. 95, A.C., 1872.

Projectiles should not be stacked in contact with the ground, but a base should be formed of old shot or shell.

A plank is issued to facilitate stacking projectiles, it is made of elm, is 7 feet in length, and bevelled off at one end; it is bound near the ends by two iron bands. A tray for raising projectiles is issued to wharfs where there are cranes.

§ 2206. Cl. 58, A.C, 1872.

A pattern of this clip has been sealed to govern supplies. See cut.

The studs on the inside of the arms fit into the extracting holes in the head of the projectiles, and are retained in their place by the screw-bolt which passes through both arms.

Clip for lifting rifled M.L. projectiles from 7-in. to 12·5-inch (Mark I.). § 2418.

CLIP, LIFTING R.M.L. PROJECTILES.

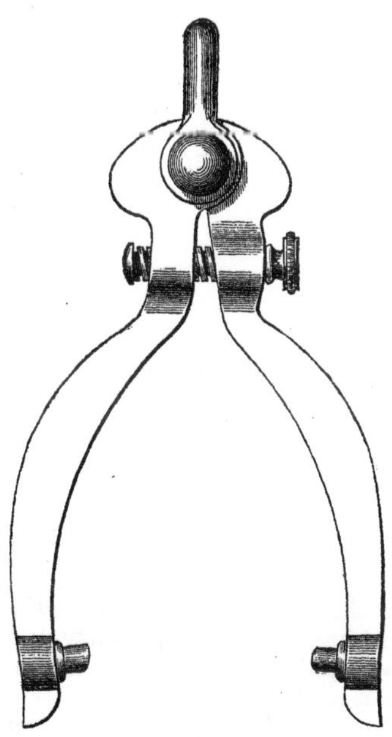

* The system of marking commenced in January 1866. In 1867 the word "mark" was directed to be substituted for "pattern." § 1545 states "consecutive "numerals will still be applied to stores on every change of manufacture, but these "are to be considered merely for the purpose of identification; and while therefore "they should be quoted in all reports having reference to any particular store, they "are not to be quoted in demands for stores, except in the few cases that may arise "which may call for special supply to suit the existing store at the station. It is to "be understood that no store becomes obsolete by the mere introduction of a later "pattern. The condemnation of stores as obsolete will always be followed by their "withdrawal on specific orders to that effect."

In some cases, however, marks are not interchangeable; thus, in demanding R.L. percussion fuzes for a 64-pr. R.M.L. gun, Mark II. should be demanded.

I 2

Strap, leather, slinging projectiles R.M.L. 12″·5 and 12″. §§ 3073, 3236.

A pattern of this article has been sealed to govern manufacture.

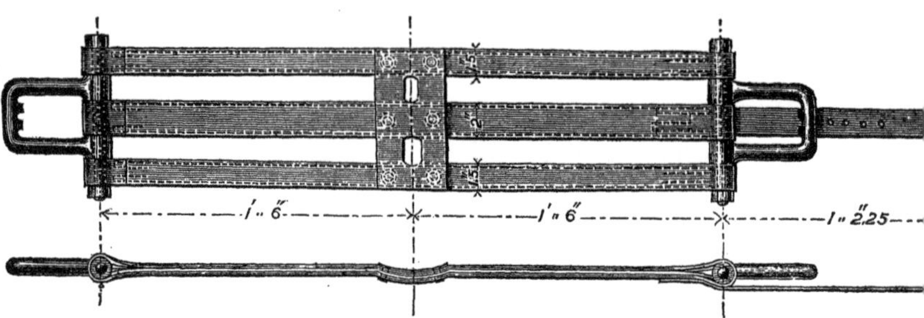


It consists of three three-fold lengths of leather, connected at each end by bolts passing through them, to which are attached shackles for lifting. A metal bearing plate, having two holes in it to receive studs on the projectiles, is secured across the centre of the strap. A strap and stud fasten the ends of the sling together to secure the projectile in it. It weighs 15 lbs. A similar strap is made for 12″, 35 ton, projectiles. It weighs 14 lb. 14 oz.

Experiments carried on at Shoeburyness in February, 1869, proved that filled and plugged shell were exploded when struck by a Palliser shell. (A 9″ Palliser shell was used with a striking velocity equivalent to that due to a battering charge at a range of 1,000 yards; the shell fired at were 9″ common shell.) Great destruction was caused in the casemate by the explosion.

The Committee concluded :—

(1.) That it is absolutely necessary to provide magazines proof against horizontal as well as vertical fire for storing shells.

(2.) That when shells are brought into the batteries for use, the greatest care should be taken to place them in positions where they may be secure from the danger which attends their exposure either to direct fire or to the impact of fragments of burst shells.

In May 1870 further experiments were carried on by firing 9″ Palliser shells, with a striking velocity equivalent to that due to a battering charge at a range of 200 yards, at filled Palliser shells and filled common shells fuzed with the Pettman G.S. fuze, placed behind a "Warrior" target (4½″ of iron, 18″ of wood). Both common and Palliser shell were exploded; in several cases the fuze of the exploded common shell was recovered and found not to have fired.

The Committee concluded :—

(1.) That the circumstance of filled shell being fuzed does not render them more liable to be exploded by a shell striking them than if they had been only filled and plugged.

(2.) That the best method of diminishing the havoc arising from a shell striking or bursting among rows of shell placed behind the bulwarks of an iron-clad ship is to place live and empty shell alternately, an arrangement which apparently confines the explosion to the shell actually struck.

The G.S. plug is used with all rifled shells except the B.L. field service segment and common shells; it is a conical plug with no shoulder, having a square hole in its head to take the G.S. keys, p. 69. A.G.S. plug with lanyard attached is issued for extracting field service R.M.L. shells at drill.

G.S. plug.

Drill plug, G.S., with lanyard. § 2090.

The B.L. field service plug is used only with the 6, 9, 12, and 20-pr. F.S. common and segment shells; it has a shoulder and may also be easily known by having a coarse left-handed screw thread. The 20-pr. plug has a loop* attached it. Leather collars are issued with these plugs, and 5 per cent. spare are issued to field batteries.

B.L. field service plug.

The Moorsom gauge may still be found in some B.L. shells. The large plug with a cylindrical body is easily recognised. The shells are to be converted to the G.S. gauge by fixing in an adapter, see p. 146.

Moorsom gauge plug.

*Primers, Shrapnel Shell.*—In Shrapnel shell a primer is used; it serves to convey the flash from the fuze to the powder, and also prevents the powder from working up into the fuze socket.

Primers, Shrapnel shell.

For method of fixing it in the shell, see p. 62.

Mark I. primer consisted of a metal cylinder tapped to screw into the pipe of the shell, the bottom was solid, pierced with a fire hole, the top was open. It contained mealed powder, driven and pierced like a tube.

§ 1395.

This construction was found defective as the flash had a tendency to pass upwards, and blind shell frequently occurred when it was used.

Mark II. primer was introduced to remedy the defect.

§ 2268.

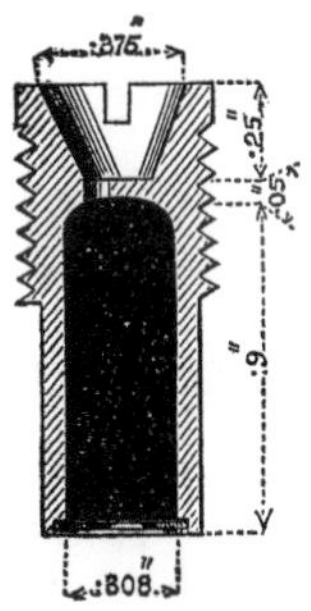


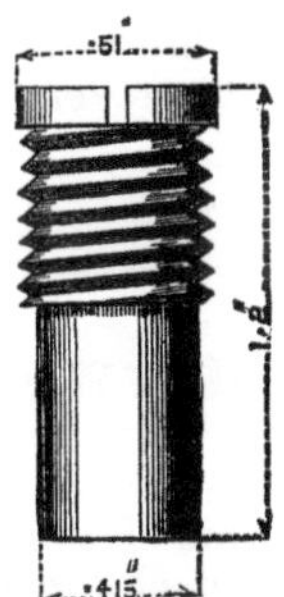


The top of this primer is solid, with a conical cup-shaped recess; the bottom of the cup is perforated with three small holes, communicating with loose powder, with which the body of the primer is filled. The bottom is closed by a thin annular disc of brass covered with shalloon.

There are two slots in the head for the screw driver. Mark I. had four slots, and in unscrewing a tight primer the head was apt to give under the screw-driver.

Mark II. is found to act well. Mark I. primers are to be returned except those fixed in shell.

§ 2521.

Primers are issued in a tin cylinder holding 10, closed by a tin band.

Issue.

* All the plugs had loops, but they were ordered to be cut off except the 20-pr.

§ 1975, and errata -/1/71.

Grummet wads. § 1973.

Grummet wads are supplied on special demand for use when firing rifled projectiles at angles of depression.

Wedge wads. § 2686.

Two sizes of wedge wads are issued. They both consist of two wooden wedges connected by a piece of cane, as shown in the cut.

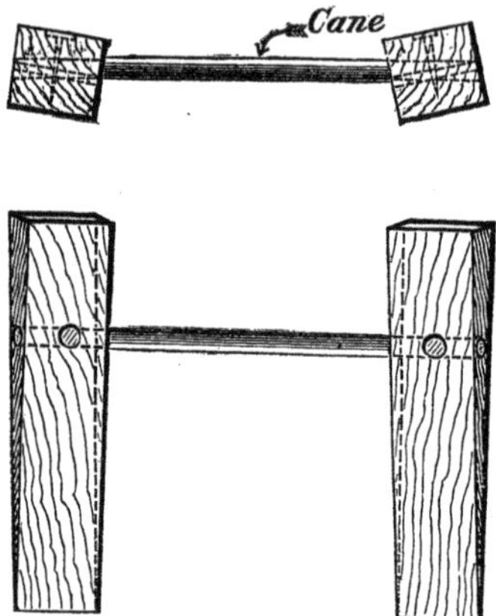


In the larger sized wad the length of cane is 7″·5 and length of wedge 7″. It is for use with 9″ guns and upwards.

In the smaller sized wad the length of cane is 6″·5 and length of wedge 5″·5. It is for use with 64-pr., 80-pr., 7″ and 8″ guns.

These wads are to be rammed home separately after the projectiles. Their use is to prevent the projectile from shifting when running the gun up.

## Gas-Checks.

Soon after the Woolwich guns were introduced, it became apparent that a great evil had to be combated. This was, that the heavy charges used, and the consequent rush of gas along the bore, especially at that portion of it immediately over the seat of the shot, seriously eroded the steel tubes of the guns. To such an extent does this erosion take place, that an impression of the rear portions of the bore of a heavy Woolwich gun that has been much fired resembles nothing so much as the bark of an elm tree in its rugosities. Obviously this erosion shortened the lives of the guns, and though they could be rendered serviceable again, for a time, by turning them upside down, then re-venting them, and plugging up the old vent, this entailed much labour and expense, could only be done at Woolwich, and was at best only a palliative, not a cure.

The earliest efforts to get rid of this erosive action were by means of papier mâché wads, cupped out to take the base of the projectile. and placed between the charge and the projectile. These were almost immediately superseded by "*Bolton wads*," which were made of pulp prepared from 75 per cent. of old rags, known as "tammies" or "woollens," and 25 per cent. of old tarred rope, formed in a mould and coated with a waterproof varnish when dry. Both the above were found to be useless, and were withdrawn from the service.

§ 1801.

§ 2353

In the meantime a Committee had been appointed "to investigate the question of adopting means for the prevention of scoring in the bores of heavy rifled M.L. guns." This Committee, after sending in progress reports, sent in a "Final Report," which was published in January, 1877.* It would be impracticable here to give even a tolerably full summary of this report, but a few of the remarks made therein may be quoted.

After recommending the disuse of the then service wads (papier mâché and Bolton), the Committee state,† that, "in their opinion, the "scoring action of the powder gases in heavy M.L. guns of the present "service pattern was not likely to be obviated by the employment of "any nature of wad, but that the desired result might be arrived at by "the introduction of some suitable form of gas-check. They con- "sidered the successful application of a metal gas-check was in great "measure dependent on the security with which it would attach itself "to the base of the shot on the explosion of the charge, and that a "metal gas-check, which would become detached, or which would fly "to pieces, would be inadmissible, at any rate for land service."

Various inventors submitted gas checks, and of these six were selected for trial in a 10″ R.M.L. gun. Besides these, a clay gas-check, to be made on the spot as wanted, was submitted by Major F. Bolton. This gas-check was tried in a 7″ R.M.L. gun, but the Committee decided that "no material softer than copper would stand "the rush of gas through the windage on the explosion of the charge." Moreover they considered that there were grave practical objections to the "adoption of an article of store which must be made as required, "and which would become useless if three days old."

The trial of the six selected metal gas-checks ended in the survival of the fittest, that proposed by Lt.-Col. Maitland, R.A., Assistant Superintendent R.G.F. This officer had proposed two designs. The successful one consisted of a copper disc or wide ring, with projections to fit the grooves of the rifling, and was cupped to the rear. On the base of the projectile was cast an undercut ring corresponding to a groove in the upper surface of the gas-check. The gas-check was automatic, *i.e.*, attached itself on firing to the base of the projectile. On firing the gun the pressure of the gas forced the inner portion of the gas-check to attach itself to the undercut ring, which held it firmly, while the flange of the outer portion was expanded so as to seal the windage.

Difficulties of manufacture arose in the preparation of projectiles for this mode of attachment, and it had the defect of not being applicable to the existing store of projectiles. Major Lyon, R.A., Assistant Superintendent R.L., then proposed a non-automatic gas-check, which consisted of a disc of copper, resembling one proposed by Lieut.-Col. Maitland in another design, attached to the projectile by means of a plug screwed into the base. This could be applied to existing projectiles, but required "to be fitted on the base of the projectile before "loading, instead of attaching itself automatically, and to facilitate "loading it became necessary to remove the projections fitting the "grooves of the gun; the gas appeared, however, to expand the meta "sufficiently to fill up the grooves."

* To this report the reader, who wishes to see full accounts of the experiments carried on from time to time, sundry interesting details, and plates of the principal gas-checks tried, is referred.

† This passage is quoted in the final report from a previous one rendered in July, 1872.

Experiments were successfully carried out with various modifications of this gas-check.

On 21st October, 1874, the Committee reported, "that a gas-check "could be produced which would adhere to the projectile throughout "its flight, and the use of which ensured a gain in length of range "and accuracy of shooting."* They further state, "that the erosion in "the bore of the gun would be mitigated, as the windage appeared to "be nearly, if not altogether, closed by the gas-check."

In June, 1875, the Committee on Explosives, who had been using Lyon gas-checks in a 12″·5 gun, reported that the use of such checks caused no material increase of pressure on the gun, and that a considerable gain in muzzle velocity would result from the use of such gas-checks.

So far the results were very promising, but the Navy specially desired an automatic gas-check which could be carried and loaded separately from the projectile, attaching itself to the latter when the gun was fired. The reason for this requirement was that, under existing arrangements, there was no room to stow their projectiles with attached gas-checks, and further, that the latter would be liable to injury from knocking about.

Three forms of automatic gas-checks were tried. The most successful one was that proposed by Lieut. Goold-Adams, R.A., slightly modified by the R.L., to facilitate manufacture. This gas-check generally resembled the one proposed by Lt.-Col. Maitland, and described above, but its automatic attachment was by means of a wrought-iron disc, with an undercut groove on its rear-face, and the disc was attached to the projectile by a screw plug. This form of gas-check possessed, moreover, the great merit of being applicable to existing projectiles.

Here may be pointed out the difficulties, which are still under consideration, attending the use of an automatic gas-check, as described above. If the flange be too narrow, the gas-check will not stand up in the bore when being rammed home, but will turn over and jam on account of the narrow bearing surface. The actual breadth of flange which will ensure the gas-check standing up in the bore is determined by the size of the bore, in other words, the requisite breadth of the flange varies with the diameter of the gas-check. If the flange be made deeper, so as to get over this difficulty, it is apt to have portions blown off it by the pressure of the gas. If again a narrow flange with projections to fit the grooves be tried, it will not go far enough home (owing to the termination of the rifling in the bore) when a full charge is used. In fact, in the use of the form of automatic gas-checks, we are met by many difficulties, but, as before mentioned, the subject is still under consideration.

Gas-checks, originally introduced to prevent erosion of the bore, have been found not only to fulfil this purpose, and, as mentioned above, to increase range and accuracy, but also to be capable of giving the requisite rotation to the projectile, so as to render the use of studs unnecessary. To effect this, the gas-checks must be capable of being very firmly attached to the projectile on discharge. This end is gained by having, in new projectiles, grooves, usually of a saw-shaped or undercut character on the base, in which the soft metal of the gas-check is forced by the gas, so that it becomes an integral and immovable part of the projectile.

* Note the importance of this collateral advantage.

A "ring" gas-check has been tried. It requires special fittings in the projectile, and may be described as somewhat resembling the leather collars used in the packings of the Bramah press, only it is of course made of metal. Its section looks something like an inverted ∩ with thick walls. The pressure of the gas forces it into an undercut recess running round the lower part of the projectile, and at the same time expands the hollow of the collar, so as to seal the windage; while on the base of the projectile are radial saw-shaped grooves, which cause the disc portion of the gas-check to "bite," and aid in rotating the projectile, or, in the case of unstudded projectiles to rotate them unaided. Brass gas-checks were tried as a substitute for copper, but without success, as they break up.

It is to be remembered that an *attached* gas-check, with projections to fit the grooves, must be capable of revolving independently to a certain extent, or the projections will jam when used with guns of increasing twist.

No gas-checks have, as yet, been formally adopted into the service, but two modifications of the Lyon gas-check have been issued to certain ships, and are daily used at the proof butts in the Royal Arsenal. We will proceed to describe these.

The first pattern consists of a flanged copper* disc of the same diameter as the shell, without any projections. It is slightly convex to the rear, so that the pressure of the powder gas flattens it on firing, and so helps to expand the metal of the flange, so as to seal the windage. It is attached to the base of projectiles by a gun-metal plug, with a solid hexagonal head, which screws into a hole in the base of the shell. This plug has the same pitch of thread as the ordinary Palliser base plugs, viz., nine threads to the inch. As in the case of existing projectiles the thread in the base is right-handed, there would be a tendency on the part of the plug to come unscrewed on firing. To guard against this, the end of the plug which goes into the base of the shell has a conical recess cut in it. The walls of this recess are split like the cap of a telescope, and into it fits a gun-metal plug with a coned shoulder, which does not reach to the bottom of the recess, and is suspended by a thin copper wire. On firing, the cone sets back, and expands the end of the plug, thus riveting it and preventing it from becoming unscrewed. In the case of new projectiles†, such as those for the 12″·5 gun, a left-handed screw would be used in the base, and the coned plug arrangement be thus rendered unnecessary.

The other pattern which has been issued, though not actually automatic, is so easily and quickly attached to the shell that it possesses most of the advantages of an automatic form, It differs from the one described above in that the base plug instead of having a solid head is prolonged and enlarged in diameter outside the shell. On this portion, which has a left-handed screw of 9-thread pitch upon it, screws a hexagonal wrought-iron cap.

The following instructions are issued for fitting these gas-checks to Palliser projectiles (having base plugs).

1. Remove the base plug by means of the "Wrench, removing, plug."

---

* A little zinc is added for convenience in turning.

† Or in preparing common or Shrapnel shell to take a gas-check. Existing heavy Shrapnel from the form of their bases will not take the gas-check. It is not proposed to use the gas-check with these shells under the 11″ of 25 tons calibre. Shrapnel for guns of 25 tons and upwards are in future to have their bases assimilated in form to those of other projectiles.

2. Insert a screw tap into the base plug-hole, screwing it up so as to ensure the whole of the screw thread in the plug hole being clear, then remove the screw tap.

N.B.—This operation (No. 2) must only be performed in empty shells.

3. Thoroughly clean the screw thread of the plug hole with a piece of cotton waste or rag.

4. The shell should now (if so required) be filled in the usual manner, as described in the instructions for use of fuze and shell implements.

5. The gas-check plug (from which the wrought-iron nut is to be removed) must now be well coated with a mixture of cocoa-nut oil and powdered chalk, or white lead and tallow if the former is not available, and then screwed firmly into the plug hole.

N.B.—The heads of the gas-check plugs will in future be stamped with the nature of projectile with which they are to be used, and they, with the wrought-iron nuts also, will each be stamped with an arrow to show the direction in which to turn, either when screwing in the gas-check plug, or when screwing on the wrought-iron nut.

The plugs and nuts, for shells of future manufacture which are to be fitted with gas-checks, will each have a left-handed screw thread, but the 12-inch and smaller natures of Palliser projectiles will at present be generally found fitted for a gas-check plug having a right-hand screw, with a wrought-iron nut having a left-hand screw, these may be known by the arrow on the plug pointing in a different direction to that on the nut: with this description of plug and nut, a few taps with a mallet or similar implement on the end of the wrench when screwing in the plug will ensure its not being loosened or unscrewed when fixing the wrought-iron nut.

6. Place the gas-check on the base of the projectile, with the concave or unpainted side next the base, and then screw the wrought-iron nut on to the end of the plug with the spanner till the nut binds against the gas-check.

7. Filled or empty shells intended to be stowed away in readiness to be fitted with gas-checks, will be fitted with the gas-check plug as described in § 5, and the wrought-iron nut well oiled, then screwed on to the plug without the gas-check.

The gas-check under trial for the common shell of the 6″·3 howitzer consists of a saucer-shaped piece of copper, having its concave surface to the rear, and attached by a solid-headed gun-metal screw plug. The screw is left-handed, and the plug 1″·3 in diameter.

On the base of the shell are eight saw-shaped radial grooves, into which the copper is forced on discharge into the grooves of the piece, the edges of the saucer being at the same time expanded and thereby rotation is given to the shell. The shell has no studs. Smaller natures will probably have gas-checks on the same principle.

## Short Notes on Projectiles for Rifled Guns.

The following remarks will be found again scattered through the subsequent pages of this work, but they are given here in a brief collective form, and as generally as possible. Details of the various projectiles will be found in the proper places, and as a rule only the *latest* patterns of projectiles will be dealt with here. The few notes here given are simply intended to aid the memories of those who already have a general knowledge of the subject.

All R.M.L. projectiles, as a rule, have two rings of gun-metal studs. The following are exceptions:— **Studs.**

The later patterns of projectiles for the 12″ gun of 35 tons,* and those for the 12″·5 gun have *three* rings of studs.

The projectiles for 64-pr. guns have *three* rings of pure copper studs (cut from copper rods) of small size, so as to fit both the shunt guns and those with "plain" grooves.

The studs for the 80-pr. projectiles are made of copper, with a little zinc added; though the gun is of uniform twist the projectiles have the front stud smaller than the rear one, so as to minimize the pressure requisite to force them in, and thus to diminish the strain on the metal of the shell.

The 25-pr. and 16-pr. common shells have gun-metal studs, but the Shrapnel shell have copper studs.

The projectiles for the 9-pr. and 7-pr. guns have copper studs.

**Palliser shot.** The Woolwich guns, up to the 10″ gun inclusive, have Palliser shot as well as shell; so also has the 12″ gun of 25 tons. The shot are lacquered inside.

**Palliser shell.** Palliser shell are used with all Woolwich guns.† They are lacquered inside.

**Common shell.** For R.M.L guns all common shell are bushed with gun-metal down to the 40-pr. inclusive.

All common shell for B.L. guns have gun-metal bushes; in the 20-pr. (L.S.), 12-pr. and 9-pr., the bush becomes a flanged socket. The 6-pr. B.L. has no common shell.

There is a countersink of ·2″ at the head of all R.M.L. common shell down to the 64-pr. inclusive, and of the 20-pr. (S.S.), 40-pr. and 7″ R.B.L. common shell. This is to allow of the naval wad being fitted in flush with the top of the shell.

The 80-pr. common shell has an interior strengthening belt running round the inside under the front stud.

All common shell are lacquered inside.

**Segment shell.** The segment shell for the 7″ and 40-pr. B.L. guns are bushed and countersunk like common shell. The interior is lacquered.

The 20, 12, and 9-prs. take a wrought-iron gaspipe burster.

The 6-pr. takes a cast-iron burster of peculiar shape, being larger at the top than at the bottom.

**Shrapnel shell.** Shrapnel shell for the 7″ R.M.L. guns and upwards contain iron sand shot. Lower natures contain mixed metal balls.

There are no Shrapnel shell made for the 20-pr. or 6-pr. B.L. guns. Patterns of Shrapnel for the 7″ and 40-pr. B.L. guns exist, but are not at present issued.

In 64-pr. and 80-pr. Shrapnel there is a thin wooden tube round the central iron tube.

Shrapnel for 64-pr. guns and upwards have longitudinal weakening grooves in the body as well as in the base; so has the 7-pr. Shrapnel, and the 9-pr. and 12-pr. Shrapnel for B.L. guns.

The 9-pr. M.L. has grooves in base only, the 16, 25, and 40-prs. have *no* grooves.

Up to the 40-pr. inclusive, the central tube is of gun-metal, and the socket is a composite one of gun-metal and tin.

Higher natures have an iron tube and a gun-metal socket fitting at its lower extremity into the central tube.

---

* Shrapnel excepted. † 7″ R.M.L. gun of 90 cwt. excepted.

Up to the 40-pr. inclusive, the "twisting pins" are of screw-wire. Other natures have plain rivets in two rows, the top set being called twisting pins.

Case shot.

Taking the latest patterns, there are practically only two classes of case shot in the service.

Up to the 7″ inclusive, the case shot have the body made of tin in three pieces soldered longitudinally together. The bottom is of tin, soldered to the body, and has an iron *ring* riveted on outside. The top end is fringed, and the fringes bent down and soldered on to a tinned iron top. For exceptions, see below.

Above the 7″ the case shot has the body made of one piece of tinned iron,* fringed at both ends. The bottom is an iron *disc*, the top is as above.

The 20, 12, 9, and 6-prs. B.L. have a tin top, covering a piece of wood with rounded edges.

The 9-pr. M.L. has a tin top, covering a thin disc of wood.

The 7-pr. has a plain tin top.

All case shot have their contents packed in clay and sand, and contain three loose wrought-iron segments as a lining, placed upon a loose wrought-iron disc at the bottom. To this rule there is one exception, viz., the 64 or 80-pr. case, which contains six segments.

For R.M.L. guns the case shot has mixed metal balls up to the 40-pr. inclusive; for R.B.L. guns up to the 20-pr. inclusive.

Higher natures contain 8 oz. sand shot.

Up to the 16-pr. inclusive, the case shot have no handle.<br>
From 25-pr. to 8″ inclusive " " one "<br>
Above the 8″ " " two handles.

The same case shot serves for 64 and 80-pr. guns, and for the 6″·3 howitzer, and there is only one 7″ case shot which is available for M.L. or B.L. guns. The same case shot serves for both 8″ gun and 8″ howitzer.

Rule for weight.

Up to 7″ guns inclusive, case shot weigh about ¾ weight of other projectiles; above that calibre, about the same weight as a spherical shot for the gun in question.

Rule for loading.

In loading case, the handle to be away from the charge.

Double shell.

Double shell are used with the 7″ and 7-pr. R.M.L. guns. The shell for the 7″ has three interior strengthening ribs.

Star shell.

The star shell is for use with the 7-pr., but may also be fired from the 9-pr. R.M.L. gun.

Extractor holes.

Up to the 40-pr. inclusive, the projectiles for R.M.L. guns have no extractor holes. The 64-pr. projectiles (case excepted which has none), have three extractor holes, those for higher natures two.

For obvious reasons, B.L. projectiles have no extractor holes.

Shells cast to gauge.

Certain shells will be in future "cast to gauge," *i.e.*, cast to final dimensions and not turned afterwards. This plan possesses the great recommendation of leaving the "skin" of the shell, which is the best and hardest part, untouched; and by care in manufacture the exterior can be made to come out of the mould as smooth as it could be made by the most careful turning. The following shells are ordered to be manufactured in this way:—

---

§ 2602.

* In future (March, 1874) the iron of which the envelops of the larger natures of case shot is made will be tinned. The term tin is limited to thin sheets of wrought-iron brought to a high polish and coated with tin. Tin sheet for case shot is obtained by contract, the thickness being regulated by the trade mark or brand. Tinned iron has not the brilliant surface of tinned plate.

| | | | |
|---|---|---|---|
| Common shell for the | 6″·3 howitzer | Has radial grooves in exterior base for gas-check to bite into. | § 3172. |
| ,, ,, | 64-pr. gun. | | |
| ,, ,, | 9-pr. ,, | | |
| Shrapnel shell for the | 64-pr. gun | Has internal weakening grooves. | |
| ,, ,, | 40-pr. ,, | Has no weakening grooves. | |
| ,, ,, | 25-pr. ,, | ,, ,, ,, | |
| ,, ,, | 16-pr. ,, | ,, ,, ,, | § 2898. |
| ,, ,, | 9-pr. ,, | Has weakening grooves in powder chamber only. | §§ 3172. |

Paint.

Projectiles for rifled guns are painted black. The heads, however, of the field service Shrapnel are painted red,* and the tips of the Palliser shell are painted white. The lead coating of B.L. projectiles is left unpainted, except about half-an-inch over the lead at either end of the projectile; for instructions as to painting, see Cl. 81, 115, Army Circulars, 1868, and table of paints, p. 321.

Issue.

As a general rule it will be found that common and segment shell of the garrison calibres are issued loose whether filled or empty, the F.S. calibres are issued loose when empty, except for transit to India, when they are boxed: and boxed when filled.†

Shrapnel shells are generally issued boxed.

Case shot are issued loose when made of iron, and boxed when made of tin plate, 7″ and 64-pr. excepted.

§§ 1349, 1880.

As a general rule all filled shell have the word "filled" stencilled on them in red. Shell which have their bursting charges contained in bags have the word "bag" stencilled on in addition to the word "filled." Shell filled in an arsenal would also have the date of filling and the monogram of the station. F.S. shells filled in an arsenal would be marked in accordance with the above rule, but when the shell are issued empty and filled by the battery this mark would not be found; all shells carried in the limbers and waggons are filled, so the mark is only of use when shell are returned into store.‡

Stencil plates filled shell. §§ 2204, 2648.

A stencil plate for marking shells "filled" has been sealed for issue. Also for marking shells "*bag*," to be used in addition to the above when required.

Filled Palliser shells can be recognised by the touch, as the key-hole of the base plug is filled up with red lead putty and painted over. Filled and fuzed common shells, from the 20-pr. B.L. and 64-pr. M.L. upwards in the naval service, can be recognized by touch, as they have the naval wad with loop on the top of the fuze hole. Similarly empty common shell would be recognised by the absence of a loop on the wad.

---

* The 40-pr. M.L. Shrapnel has a red head, the 40-pr. B.L. Shrapnel a black one, but doubtless if the latter were again manufactured in any quantity the colour would be red.

† Rifled projectiles used to be issued in jute bags, but this practice was discontinued in 1871. Grummets are to be attached to studded projectiles (7″ and upwards) issued to naval stations where there is no Ordnance Store Department. So far as present experience goes the studs do not suffer any material injury in transit when unprotected. § 2041.

‡ In 1868 experiments were carried out which showed that it was safer to carry the F.S. shell filled in the limbers than to carry the bursting charges separately in bags. The bags were therefore returned into store and the shell carried filled. Extracts, Vol. VI, p. 104, also § 1,581, Changes in War Stores.

# CHAPTER XI. — AMMUNITION FOR ARMSTRONG BREECH-LOADING GUNS.

SEGMENT, COMMON, AND SHRAPNEL SHELL.—GAUGES.—CASE, SOLID, AND DRILL SHOT.—CARTRIDGES, AND MISCELLANEOUS STORES CONNECTED WITH THEM.

---

THE systems of rifling now in the service are as follows:—*

1. Breech-loading, Armstrong, polygrooved.
2. Muzzle-loading, Armstrong, shunt.
3. „ Woolwich.
4. „ French.

## ARMSTRONG BREECH-LOADING SYSTEM.

The Armstrong breech-loading system adopted for the service in 1858, comprises the following guns: viz., 7″,† 40-pr., 20-pr., 12-pr., 9-pr., and 6-pr. In each of these the breech is closed by a vent piece supported by a breech screw.

Two natures of "wedge" guns, the 64-pr. and the 40-pr., were introduced into the service in 1864; but they have been withdrawn from the service; and though patterns of ammunition still exist for them, they differ in no essential particular from the ammunition for the other natures of B.L. guns. In fact, the ammunition for the 40-pr. wedge gun is identical with that for the 40-pr. breech-screw gun. No further mention, therefore, will be made in this treatise of them.

In the Armstrong B.L. system a soft-coated projectile is forced through a polygroove bore of such diameter that it can only pass through by the lands cutting their way into the soft coat. In other words, the diameter of the projectile is greater than that of the bore, except where the latter is enlarged towards the breech so as to allow of loading.

*Calibre of Armstrong guns.*

The calibres of Armstrong guns are the diameters taken at the smallest part of the bore immediately in front of the seat of the shot; this part is called the "grip," it extends a short distance up the bore, which is then enlarged by ·005 inch, so as to ease the projectile.

The dimensions of the powder chamber, shot chamber, grip, and barrel will be found in the "Treatise on the Construction, &c., of Ordnance," p. 166.

---

* See also foot-note, p. 128.

† The 7-inch gun was first called a 100-pr., then a 110-pr., lastly a 7-inch gun.

The calibres and twist of rifling of the guns are:—

| | Calibre. | Twist. | Diameters of Shells. Body. | Back end. |
|---|---|---|---|---|
| 7 inch .. | 7 inch | Uniform 1 in 37 cal. | 7·036 in. | 7·0875 in |
| 40 pr. .. | 4·75 ,, | ,, 1 in 36½ ,, | 4·78 ,, | 4·835 ,, |
| 20 ,, .. | 3·75 ,, | ,, 1 in 38 ,, | 3·78 ,, | 3·841 ,, |
| 12 ,, .. | 3·0 ,, | ,, 1 in 38 ,, | 3·029 ,, | 3·0705 ,, |
| 9 ,, .. | 3·0 ,, | ,, 1 in 38 ,, | 3·029 ,, | 3·0705 ,, |
| 6 ,, .. | 2·5 ,, | ,, 1 in 30 ,, | 2·531 ,, | 2·5705 ,, |

From the above dimensions it will be seen that the compression undergone by the shell at the grip varies from ·09 inch in the 20-pr. to about ·07 in the 12-pr.

The twist is in all cases uniform.

Windage is done away with, hence the accuracy is great: the projectile is detained in the bore until the force of the powder is more fully developed than where there is less initial resistance.

The absence of windage entails the necessity of a percussion arrangement for igniting the time fuzes; a considerable amount of force must be expended in overcoming friction in the bore, but there is a great advantage in the gun being preserved from "guttering," *i.e.*, the destructive action of the gas rushing over the projectile, and also in the shot being properly centered, which accounts to some extent for the great accuracy of B.L. guns.

As might have been expected, this system has been more successful with the smaller calibres than the larger, where the great difficulty of closing the breech effectually becomes aggravated by a system which saves powder at the expense of increased strain on the gun; this question has affected the ammunition considerably, for it has been found necessary in some cases to reduce the weight both of the projectile and charge.

It will be seen hereafter that the opposite plan has been adopted in the large muzzle-loading guns where the projectile is allowed to move directly forward at first, and rotation is gradually impressed on it after the gun has become relieved of the initial strain due to the inertia of the projectile, this being done at the expense of an increased charge of powder, since the gas presses on the shot in inverse ratio to the volume of the space in which it is generated (*i.e.*, the portion of the bore behind the shot), and the same object is still further carried out by the employment of pebble powder, which consumes more slowly than the powder used in the breech-loading guns.

There are certain advantages connected with the service of all breech-loading guns, which scarcely belong to this course, that of enabling any unconsumed matter left in the bore after firing to be readily seen, should, however, be mentioned; this especially applies to blank firing.

It is due to the breech-loaders to notice particularly the great accuracy mentioned above, which at the time of their introduction insured such good results in practice, that deficiencies or faults which might otherwise have been obvious to many were so little allowed to exist, that for some time one single projectile (the segment shell) was held to be efficient as shot, common shell, Shrapnel, and case.

The projectiles that are now in the service in connection with this system are as follows:—

A. Shell.
B. Shot.

A.—*Shell.**

1. Segment Shell.
2. Common ,,
3. Shrapnel ,,

Segment shell.

1. *Segment shell* (calibres, 7-inch, 40-pr., 20-pr., 12-pr., 9-pr., and 6-pr.) consists of a very thin cast-iron cylindro-conoidal shell, lined with cast-iron segments, built up in layers, having a cylindrical powder chamber in the centre. The base is closed with a cast-iron disc.

A thin coat of an alloy of 19 parts lead to 1 part antimony extends from base to shoulder; the alloy also flows in between the segments and lines the powder chamber, giving great weight and solidity. The lead is prevented from filling up the powder chamber by a core or mandril temporarily inserted. The alloy flows over a recess in the base of the iron disc which forms the bottom of the shell, thus retaining it in its place. The wide rim of lead at the base is the most certain way of distinguishing segment from common shell, which they closely resemble externally.

The head has been struck with various radii; the curve will be found generally more abrupt than that employed in common shell. Some are finished off with a nozzle.

The shell is strong against external pressure, while a small bursting charge opens it.

§ 543.

The coat (generally termed the lead coat) was at first made of the following alloy:—8 parts soft lead to 1 part antimonial lead, and 2 parts tin; but in June, 1862 this was superseded by an alloy of 19 parts lead to 1 of antimony on the score of economy.

Lead coating.

The lead coating is ·05″ deep over body, and ·1″ over base, a cannelure running round the shell to take any lead stripping off the front part.

The increased diameter at base is intended—

1st. To prevent windage.

2nd. To enable the projectile to be gripped simultaneously at shoulder and base on ramming home.

3rd. To retain the grip until the base leaves the muzzle.

Paint.

Uncoated portions (head and bottom) are painted black, the paint extending over the edge of the lead to prevent corrosion, &c.

The lead coat has been attached in three ways:—

Attachment.

1st. By tin solder and square cut groves in the shell. This stripped very much.

§ 271.

2nd. By mechanical means, viz., undercut groves. This was better.

§ 330.

3rd. By zinc solder, no grooves.

The zinc amalgamates sufficiently with the iron and lead to give a very complete attachment. To compensate for the absence of grooves on the outside of shells with zinc attachment, similar grooves are cast on the inside.

The 7-inch and 9-pr. have also grooves running round the outside edge

---

* Experiments are being carried on with lead-coated Palliser shells fired from the 7-inch B.L. gun, with a charge of 20lbs. P powder.

of the base and inside edge of the body; that is, in the surfaces of base and body which are in contact, for the lead to enter and seal up the joint.

Every segment shell has four longitudinal grooves in the interior of the head.

Segment shells are of two classes:— **Classes.**

1. Garrison or Naval.
2. Field or Boat.

*Garrison or Naval.*—7-inch and 40-pr. Gauge G.S. **Class I.**

Fuzes for L.S. 9 secs. and 20 secs. B.L. time. R.L. Percussion, Mark II.

Fuzes for S.S. 9 secs. and 20 secs. B.L. time. Pettman's G.S.*

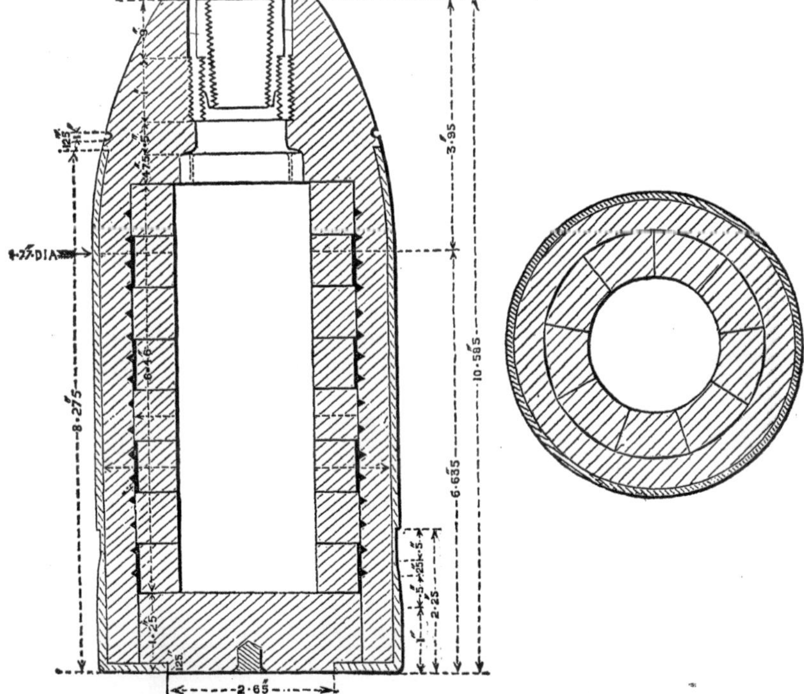


These shells have their powder chambers coated with red lacquer (see p. 322) in order to prevent premature explosion.† Originally these shells were made with a gun-metal bush of the Moorsom gauge, now obsolete; § 1120. § 2621.

---

* The Pettman G.S. fuze would answer well when the segment shell is used against wooden ships or against troops behind a thin wall or parapet, but for troops in the open, a fuze which acts on graze like the R.L. percussion, Mark II., is required. The latter fuze answers well with the B.L. guns, so far as some limited trials show.

† Black lacquer, consisting of coal tar, and pitch, was found to fail; it was generally rough, and apt to become detached from the shell, as the lacquer melted in hot weather, even when shell were exposed to a hot sun in England. For prematures ascribed to this cause, see p. 31, also Extracts, O.S.C., Vol. III, pp. 142 and 240. Many shell exist which have black lacquer. See Table of Patterns, p. 152. § 492.

§ 1427. they are now issued with an adapter screwed in and permanently secured,* converting them to the G.S. gauge, see p. 22.

For weight, bursting charge, dimensions of shell, see table, p. The length approximates to a little over two calibres.

For filling and securing shell, see p. 62.

Shells filled in the R.L. are marked as given on p. 141.

Issue. Empty, loose for garrison service.

§ 1089. Filled and fuzed with Pettman G.S.f uze and issued loose for S.S.

*Field or Boat.*—Calibres 20, 12, 9, and 6-pr.

Class 2. Gauge, Armstrong Field Service.†

Fuzes for L.S., B.L. plain percussion.

Fuzes for S.S., B.L. plain percussion, and "E." time.

§ 754. In the shells the powder "shell F.G." is contained in wrought-iron gas pipe bursters, the pipe is dropped into the powder chamber, which is of the same diameter as the fuze-hole, the brown paper cover in which the burster is issued is retained, the top end being torn off to allow

§ 1954. ignition, the ends of the burster are closed by serge and paper discs fastened to metal rings.‡ See plate, p. 355.

---

§ 1583. * "Apply a coating of red lead to the screw part of the adapter, and screw it home. After screwing the adapter into rifled shells, fill the space between the plain part of the adapter and the nose of the shell with the following composition :—

| | |
|---|---|
| Rosin .. .. .. .. | 12 lb. |
| Spanish brown .. .. | 2 ,, |
| Plaster of Paris .. .. | 1 ,, |
| Turpentine .. .. .. | ½ pint |

The composition should be poured in whilst very hot, a plug of wood being fitted into the adapter to prevent any of the composition running inside. No special tools are required."

† As the field-service gauge is tapped with a coarse, quick screw, the plugs are apt to unscrew from the jolting of the limber, care must be taken to examine the plugs frequently, and to screw up any that have become unscrewed. The thread is left-handed.

‡ The present pattern is Mark I.; the 6-pr. burster is of cast-iron, and smaller at the bottom to fit the chamber which contracts at the base. B.P. or B.C. on the burster shows that it is of a proper length for the brass percussion fuze, which used at first to be called concussion.

§ 1115. In 1865 the present pattern of bursters, Mark I., was introduced, and bursters of former patterns were ordered to be returned.

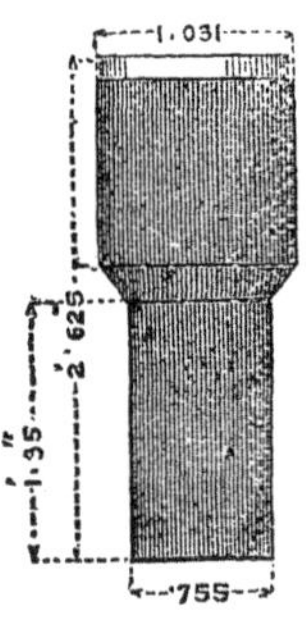


6-PR. BURSTER.

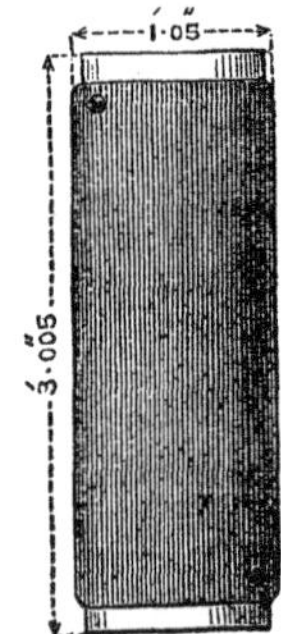


9-PR. BURSTER.

In 1868 it was ordered that the powder of the bursting charge in these bursters should not be shaken down as previously directed, it having been found to cake and

Field service shells are now carried filled, over the burster is placed a wooden plug covered with serge to keep it from shaking about, the B.L. field service gun-metal plug with leather collar (no loop at present, except 20-pr.) is then screwed in.

To prepare the shell for firing, the gun-metal plug is unscrewed, the serge covered plug taken out, and the fuze dropped in rim to the front; the safty pin being withdrawn (in most segment shells it is impossible to put the fuze in wrong, as the rim comes in the way), the gun-metal plug is again screwed in. Since the use of the time fuze has been discontinued it has been found necessary to insert a lead disc at the bottom of the shell to keep the percussion fuse from having play in the shell, owing to the screw thread in the gun-metal plugs being shorter than the time fuze. Shells which have these discs permanently fastened in are marked with the letter D on the lead coat. This does not apply to shells for S.S.

§ 1981. § 2028. § 2050.

These shells are about 2 calibres long. For weights, &c., see p. 152.

**Issue.**

1st. Filled and boxed for S.S.

2nd. Empty, loose for transit, or filled and boxed, for F.S. If issued empty, they are filled before placing them in the limbers.*

**Issue of bursters.**

Bursters are wrapped singly in brown paper, and packed in No. 6 packing case for Laboratory stores; in the following numbers, or less: 20-pr., 72; 12-pr., 88; 9-pr., 120; 6-pr., 154.

**Distinctive marks on B.L. projectiles.**

With reference to distinctive marks indicating the peculiarities of each pattern, the most important is the numeral. This is stamped on the coat of B.L. projectiles made since the commencement of 1866.

Very large numbers of shells, however, will be found without any numeral, having been made previous to this date, and many of a pattern previous to that identified by Mark I.

As a rule, shells made at Elswick are marked on the heads, and Laboratory made shells on the bases; both may carry marks on the coats.†

Shells for the service were at first made by the Elswick Ordnance Company. Such shells are generally marked E.O.C. on the head, and if passed by the Inspector of Artillery, I. ↑ A.‡ on the coat, besides the Marks agreeing with the sealed pattern.

§ 601.

Shells made in the Royal Laboratory have R.L. on the base, and generally Z, indicating zinc attachment. A few have Z. on the coat.

"U" upon a lead coat signifies the undercut method of attachment, but many of the undercut shells (especially 20-pr. segment) and all shells with tin attachment are without any mark of this kind.

§ 669.

---

sometimes fail to explode the shell; the bursters therefore are only to be filled loosely with F.G. shell powder.

The proper pattern is easily known from those made obsolete by having the ends secured by serge and paper, and by having the brass ends fastened in by punching three indentations in the iron tubing.

The bursters which had their ends closed with a wad were liable to cause blind shells.

Bursters are occasionally found to leak; this may be remedied as a temporary measure by inserting the burster in the shell, or, if already in the shell, by reversing it. The 6-pr. burster is about 2″·6 long, the 9-pr. 3″, the 12-pr. 4″·2, and the 20-pr. 5″·3.

* These shell must be assumed to be filled when carried in limbers. Batteries equipped at Woolwich would have the shell filled and marked in accordance with rule, p. 141. For transit to India the shell would be issued "Empty, boxed."

§ 1581.

† Some E.O.C. shell, re-leaded in the Royal Laboratory, carry Laboratory marks on their coats.

‡ Some shells of very early manufacture have ↑ only.

Shells with the lead and antimony coating, made in the Royal Laboratory, are marked (L / R ↑ L / A) in the cannelure, a few have been made at Elswick with the same coating, but there is no mark to indicate this.

*Marks which should be on the coats are frequently found deficient.*

All shells with I ↑ A on the coat may be assumed to have lead and tin coat (the attachment may be tin, undercut, or zinc).

Use.

Segment shell are very effective against troops in column, and should be burst close to them. A percussion fuze which acts on graze is necessary; the Dartmoor Experiments showed that a time fuze was nearly useless with a F.S. segment shell. When the ground is favourable and hard the effect would probably be very great. It is necessary to burst the shells close to the object, as they scatter very much, the shape, moreover, of the segments is unfavourable to prolonged flight. Segment shells have given good results when used against troops behind a thin wall, the shell bursting just as it passed through proved very destructive; the larger calibres would probably be effective against wooden ships. The distance up to which segment shell is effective will vary with the gun, the 12-pr. gun gave very good results at about 1,700 yards at Dartmoor (10 rounds gave 1,194 hits on targets, in column at 1,670 yards, bursting within 10 feet of the targets), and no doubt would be effective up to 2,000 yards. Beyond 2,000 yards their destructive effect would probably diminish rapidly; from garrison guns the effective range would of course be greater.

## Experience with Segment Shell.

*Shooting on Active Service.*

Most of the reports on the segment shell in active service have been favourable. See Appendix p. 297.

In new Zealand it was spoken of as very valuable in searching out men in rifle pits.

In China it was especially praised for its action when fired with percussion fuzes to burst through brickwork.

Its accuracy has been much commended, also its range, these being of course exhibited in a marked degree when compared with its spherical predecessors.

Some failures and dangerous accidents occurred with percussion fuzes, and some segment shells failed to open properly.

*Shooting, Home Service.*

Its action has been investigated by the Armstrong and Whitworth Ordnance Select and Dartmoor Committees, this being done in comparison with the Boxer Shrapnel shell.*

It would indeed be difficult to sum up concisely and impartially the results of all these experiments, or to select extracts from the proceedings of the committees; however, the following statements will

---

* *Vide* extracts from Proceedings of Ordnance Select Committee, Report of Armstrong and Whitworth Committee, and Report of a Select Committee on Ordnance, House of Commons, reprinted together in 1869.

hardly be contradicted. The Shrapnel bullets form a smaller "cone of dispersion," that is, carry much closer than the segments, and hence the Shrapnel is effective when burst at a much greater distance in front of the object* than the segment shell; it allows of wider limits of error in setting the fuze,† the bullets have also a decided advantage in penetration,‡ and it will hardly be denied that they are a better form for flight and ricochet.

---

* *Report of Armstrong and Whitworth Committee, given on page* VI.—After a careful consideration of this practice, the Committee have to report that when the shells are burst at only 10 yards in front of the targets, the concentration of the fragments with all the guns is greater than is desirable when fired against troops in line, almost the whole effect having been produced in each case upon a front of 18 feet. Where the shells were burst at 25 yards the distribution was better, and the effect of the three guns was about equal when fired against troops in line, but when burst at 50 or 100 yards in front of troops in line, Colonel Boxer's Shrapnel shell has a marked superiority, due to the bullets maintaining their velocity and direction better than the segments of Sir W. Armstrong's shell.

When fired with the time fuze at columns of troops, the greater penetration of the bullets of Colonel Boxer's Shrapnel shell make it a much more formidable projectile than Sir W. Armstrong's segment shell.

*Ordnance Select Committee Report, No.* 4, 190, *Minute* 18,928, 8/6/66.—"The result of this comparison (*i.e.*, a comparison of 12-pr. shells burst in air at rows of targets) is much in favour of Colonel Boxer's Shrapnel shell, in respect to the total number of balls and splinters striking the targets. This is in some measure accounted for by the actual number of segments being less than that of bullets in the proportion of 48 to 70, but making allowance for this, the proportional number is still greater at every distance.

"Reviewing the whole of the experiments with the 12-pr. gun, the Committee are led to the conclusion that if the segments in the segment shell were reduced in size and increased in number to equal the number of bullets in the Shrapnel shells, the latter would have no great advantage when burst in air by a time fuze at short distances from a line of troops, but would still possess a material advantage when burst between 20 and 100 yards in advance.

"The foregoing results (*i.e.*, of 64-pr. shells burst through screens placed at varying distances in front of rows of targets) exhibited very marked superiority in the Shrapnel shell when burst in the air by a time fuze at a suitable height and at any distance within 200 yards of the target."

† In the Dartmoor experiments on Friday, 18th June, 1869, three segment shells bursting from 40 to 45 yards in front of a row of targets, gave altogether only two throughs and one strike.

‡ *Ordnance Select Committee Report, No.* 4,190, *Minute* 18,928, 8/6/66.—In respect to penetrative powers, it appears, on examination, that the proportion of bullets, segments, and splinters which penetrated 2-inch deal targets, was as follows:—

| — | Shrapnel. | | | Segment. | | |
|---|---|---|---|---|---|---|
| | Total Effects. | Through. | Per Cent. | Total Effects. | Through. | Per Cent. |
| Bursts by passing through first row of targets .. .. 4 | 496 | 382 | 0·77 | 233 | (3 rounds) 110 | 0·47 |
| Burst by passing through the screen 15 yards in front of first row .. .. .. 5 | 913 | 696 | 0·76 | 549 | 298 | 0·54 |
| Do. 55 yards in front .. 5 | 453 | 314 | 0·69 | 128 | 60 | 0·47 |
| Do. 105 yards in front .. 5 | 258 | 144 | 0·56 | 45 | 14 | 0·31 |

Notwithstanding, therefore, the less weight of the bullets than of the segments, a arger proportion of the former penetrated 2-inch deal boards than of the latter.

On the other hand, when bursting quite close on the object, the wide spread of the segments tells well.*

In short, the action of the Boxer Shrapnel is rather adaptedto time fuzes, and that of the Armstrong segment to percussion fuzes, exploding close on the object, hence it will generally be found that the defenders of the segment shell advocate the use of the latter description of fuze, and those of the Shrapnel the former.

The very serious evil of premature explosion has frequently occurred in segment shells (as at Dartmoor). If the time fuze is not screwed home so as to press on the percussion fuze and burster, premature explosion may follow, and as such accidents occurred at Dartmoor particularly in independent firing, and especially with one battery, it is difficult to divest oneself of the idea that this may have been one cause, in one instance 9 and another 10 prematures occurring out of 27 rounds.

Since the above was written the Freeth and Dyer fuzes have been withdrawn, owing to their liability to prematures; the only fuze at present used with B.L. segment shell F.S. is the B.L. plain, which is very free from the defect of causing prematures.

**Stripping on active service.**

Complaints have been made of lead coated projectiles stripping in such a manner as to render it dangerous to fire over the heads of troops †; this objection, however, was made against shells with tin attachment, and can hardly apply to undercut or zinc attached shells.

With the zinc, although it is the best method, the coat has occasionally become detached in places where the lead has risen up into blisters from the formation of gas underneath it.

Such blisters are generally small, and may be pricked and then hammered down, when they will in no way affect the fitness of the shell for service. If left to develop themselves they have been known to attain a large size.‡

As has been noticed before, a place on a shell where the coat is detached is detected by the flat, dull, sound heard on tapping it,

---

* At Dartmoor, the most wonderful success of the 12-pr. segment was on 2nd July against targets in column, when 10 effective rounds gave 529 throughs, 234 lodges, and 431 strikes; total, 1,194. Ranges, 1,670 yards. On this occasion every effective shell burst within 10 yards in front of a target.

† Captain F. B. Seymour, R.N., gave the following evidence on this subject to a Special Committee of the House of Commons on Ordnance, 1862–63, see No. 2,300:—

On the 6th of March, 1862, " out of 14 rounds three shots stripped in such a way as to look more like canister falling on the water than anything I ever saw before. In March, 1861, while the force under General Pratt was attacking the Te Are Pa, near the Waitara River, in New Zealand, I was in the advanced sap, when a fire was opened over our heads from an Armstrong gun mounted in a redoubt about 600 yards to our rear; on that occasion also some of the shots stripped, the metal from which fell about the sap in a manner which might have endangered the men at work in the sap and in the trenches."

‡ Mr. Abel, Chemist to the War Department, has written a paper on this, with the title, " A Curious Instance of Electrolytic Action."

I gather the following from this paper and from information given me by his assistants, Mr. Brown and Mr. Dent, with reference to the chemical changes occurring on the performance of successive manufacturing operations:—

In order to get a clean metallic surface for the zinc to adhere to, the shell is dipped in sal-ammoniac solution, by which means any oxides of iron become converted into chlorides (say according to the following equation:—$FeO + NH_4Cl = FeCl + NH_3,HO$, such chloride of iron is probably left in the bath into which the shell is dipped; but any sal-ammoniac adhering to the shell may cause a similar formation of chloride of zinc, in the next operation, which consists in dipping the shell into molten zinc; this chloride has the property of retaining water at a very high temperature, hence a medium is provided for the commencement of voltaic action between the different

instead of the ring that comes from a sound spot; if the detached portion be encircled with a cut made with a chisel, dividing the coat through, it comes away from the shell showing in the case of a zinc attached shell, a yellowish green corroded surface beneath.

---

metals present (suppose the chemical change so brought about to be illustrated by the following equation:—Zn + ZnCl + HO = ZnO,ZnCl + H); the hydrogen so given off forms the blister, and this blister increases gradually if an escape be not provided for the gas.

Mr. Abel has in his possession two shells with curiously large blisters; one blister would nearly cover two half-crowns at the present time (September, 1870).

(Note by Captain C. O. Browne).

## SEGMENT SHELL.—B.L. GUNS.

| Calibre, and Mark of Pattern. | Date of Approval. | § Changes in War Stores. | Length, inches. | Diameter at Base. | Number and Nature of Segments. | *Weight, mean, empty. | Approximate bursting Charge. | Gauge of Fuze-Hole. | Attachment of Lead Coat. | Marks. Head. | Marks. Coat. | Marks. Base. | Remarks. |
|---|---|---|---|---|---|---|---|---|---|---|---|---|---|
| | | | | inches. | | lbs. oz. | Shell L.G. powder. lbs. ozs. | | | | | | |
| 7-inch (no Mark) ... | 21/2/61 | 211.1116 | 14·23 | 7·0875 | 112, 3·26 oz. each ... | 97 15 | 3 2 | † Moorsom | Tin ... | E. OC. ... | I. ↑ A. ... ... | E. OC. | Black lacquer. |
| ,, ,, ... | 30/12/61 | 405.1116 | 13·75 | ,, | ,, ... | 97 $13\frac{6}{16}$ | ,, | ,, | Undercut... | E. OC. A.... | ... ... | ... | ,, ,, |
| ,, Mark I. ... | 2/12/61 | 405.1116 | 14·48 | ,, | ,, ... | 98 $9\frac{11}{16}$ | ,, | ,, | Zinc ... | ... ... | Z., date, and I. | R.L. | Red ,, |
| 40-pr. (no Mark) ... | 21/2/61 | 211.1116 | 10·68 | 4·835 | 72, 2·47 oz. each ... | 39 9 | 0 13 | ,, | Tin ... | E. OC. C.... | ... ... | ... | Black ,, |
| ,, Mark I. ... | 2/12/61 | 405.1116 | 10·62 | ,, | ,, ... | 38 $9\frac{11}{16}$ | ,, | ,, | Zinc ... | ... ... | Z., I. ... ... | R.L. | Red ,, |
| ,, (no Mark) ... | 30/12/61 | 405.1116 | 10·58 | ,, | ,, ... | 38 15 | ,, | ,, | Undercut... | E. OC. C.... | ... ... | ... | Black ,, |
| | | | | | | | Shell F.G. | Armstrong | | | | | |
| 20-pr. (no Mark) ... | 17/12/61 | 405 | 8·16 | 3·841 | 56, 1·7 oz. each, and 14, 1·06 oz. each | 19 10½ | 700 grs. | Fd. Service | ,, ... | E. OC., ... probably | I. ↑ A. ... ... | — | |
| ,, Mark I. ... | 9/5/63 | 754 | 8·1 | ,, | ,, ... | 19 10 | ,, | ,, | Zinc ... | E. OC. Z. E. | I ... ... | ... | |
| 12-pr. (no Mark) ... | 13/4/60 | 90 | 6·82 | 3·0705 | 42, 1·32 oz. each ... | 10 9 | 550 grs. | ,, | Tin ... | ... ... | ... ... | ... | 12-pr. appd. 13/4/60, for practice only, beading low down and indistinct. Cannelure 1″·9 from base. |
| ,, ,, ... | 17/2/61 | 210 | 6·82 | ,, | 42, 1·32 oz. each, and 6, ·86 oz. each | 10 8 | ,, | ,, | Tin, probably | ... ... | ... ... | ... | |
| ,, ,, ... | 2/12/61 | 405 | 6·72 | ,, | ,, ... | 10 6¾ | ,, | ,, | Undercut... | E. OC. Q.... | — | — | |
| ,, ,, ... | 9/5/63 | 754 | 6·75 | ,, | ,, ... | 10 8 | ,, | ,, | Zinc ... | ,, | — | — | |
| ,, Mark I. ... | 28/10/64 | 1001 | 6·75 | ,, | ,, ... | 10 8⅔ | ,, | ,, | ,, ... | ... ... | (circle) L. R. ↑ L. A. | Z. R.L. | Rounded at base |
| 9-pr. (no Mark) ... | No record of approval | | 5·4 | ,, | 35, 1·5 oz. each, and 7, 1·2 oz. each | 8 3¾ | 300 grs. | ,, | Undercut... | E. OC. V. Some shell have T. | date, and I. ... ... | ... | For practice only. |
| ,, ,, ... | 22/5/62 | 542 | 5·35 | ,, | ,, ... | 8 3¾ | ,, | ,, | ,, ... | — | — | — | |
| ,, ,, ... | 9/5/63 | 754 | 5·4 | ,, | ,, ... | 8 3¾ | ,, | ,, | Zinc ... | E. OC. V. | — | — | |
| ,, Mark I. ... | 28/10/64 | 1001 | 5·4 | ,, | ,, ... | 8 3⅔ | ,, | ,, | ,, ... | ... ... | L. R. ↑ L. A. date & I. | R.L. Z. | Rounded at base. |
| 6-pr. (no Mark) ... | 3/7/61 | 344 | 5·0 | 2·5705 | 12, 1 oz. each, and 18, ½ oz. each | 5 9 | 200 grs. | ,, | Tin ... | — | — | — | |
| ,, ,, ... | 2/12/61 | 405 | 5·05 | ,, | ,, ... | 5 6 | ,, | ,, | Undercut... | — | — | — | |
| ,, Mark I. ... | 9/5/63 | 754 | 5·05 | ,, | ,, ... | 5 7 | ,, | ,, | Zinc ... | E. OC. E. Z. | I. ... ... | — | |

* By § 602 the following ± limits are allowed in the manufacture of segment shell, 7″ 2 lbs., 40-pr. 1¼ lbs., 20-pr. 9 oz., 12-pr. 4 oz., 9-pr. 2⅓ oz., 6-pr. 2 oz.

† § 1238 directs that existing shell with Moorsom gauge fuze-hole are to be fitted with G.S. adapters permanently screwed in. New shell to have the G.S. bush. See p. 146 for directions for fixing the adapters.

## 2. Common Shell.

Common shell are of two classes:—

1. Garrison or Naval.
2. Field.

Classes.

1. Garrison or Naval.

1. *Garrison or Naval.*—Calibres, 7 inch and 40-pr.; gauge, G.S.

| | | |
|---|---|---|
| Fuzes for L.S. | 9 secs. and 20 secs. B.L. time .. | Pettman's G.S. and R.L. percussion, Mark II. |
| " " S.S. | " " " " " " " .. | Pettman's G.S. percussion. |

They are cylindrical in shape, the chamber is lacquered as usual to protect the bursting charge. Many of the earlier patterns have black lacquer. (See p.155.)

Their length is generally about 2½ calibres.

For bursting charge, dimensions, &c., see table, p. 155.

It will be seen in the table that a light 7-inch shell has been introduced for S.S., the recoil of the heavy shell having proved inconveniently great on board ship. This shell may be met with in the L.S.; it is not suitable for use with the 7-inch gun of 72 cwt., charge 10 lbs., as the charge is not sufficient to ensure the lubricators breaking up properly when the light shell is used, the lubricant being blown out in lumps, instead of being evenly distributed along the bore; on an emergency, however, the shell might be used.

§§ 1038, 1119.

Revised equipment, 1876, p. 68.

The form of head adopted since 13/1/65 is ogival, struck with a radius of 1·5 diameters; prior to the above date various radii were used, and some of the shell are finished off with an abruptly curved nozzle.

2. *Field.*—Calibres, 20, 12, and 9-prs.; gauge, Armstrong field service.*

2. Field.

| | | |
|---|---|---|
| Fuzes for L.S., | B.L. plain percussion. | |
| " " S.S., (20-pr. only) | 9 secs. and 20 secs. B.L. time .. .. .. | Pettman's G.S. percussion. |

They have a flanged gun-metal socket to take the fuze (Marks I. and II., 9 and 12-prs., are unserviceable, the first having no socket and the socket in II. having been proved to be too weak), the socket contracts at the bottom to prevent the fuze being put in with the rim down. (See plate, p. 356.)

§ 2061.

These shell have the B.L. field service plug; a papier mâché wad fits into the hole of the flanged socked to prevent the powder working up. This wad does not want to be taken out, as the fuze will blow it in on acting; the wads are pressed in by a wooden drift, the recessed side of the wad goes uppermost.

§ 1708.

For preparation, see Instructions, p. 62.

Preparation.

For bursting charge, dimensions, &c., see table, p. 155.

Common shell would generally be used against material, fuzed with

Use.

* A 20-pr. shell with G.S. gauge has been introduced for S.S. Many 20-pr. shells have been converted from the Moorsom to the F.S. gauge by a flanged socket.

§§ 1345, 1426, 2843.

percussion fuzes, but time fuzes may be used if desired. It is well to remember that the fuzes with powder channels will generally act on impact, but the 20-seconds fuze is not to be depended on for the action.

**Issue.** Empty, loose, garrison service. Filled and fuzed with Pettman G.S. fuze, and issued loose for S.S.

*Field Service.*—Filled and boxed, or empty, loose, for transit.*

---

* For transit to India, the F.S. shell would be issued, empty, boxed.

## Common Shells.—B.L. Guns.

| Calibre, and Mark of Pattern. | Date of Approval. | § Changes in War Stores. | Length. | Weight, empty. | Approximate bursting charge, shell powder, L.G. | Gauge of Fuze-Hole. | Attachment of Lead Coat. | Marks. Head. | Marks. Coat. | Marks. Base. | Remarks. |
|---|---|---|---|---|---|---|---|---|---|---|---|
| | | | inches | lbs. oz. | lb. oz. | | | | | | |
| 7″ (no Mark) ... | 21/2/61 | 211 | 18·73 | 95 14 ± 2 lbs. | 7 10 | Moorsom* ... | Tin ... | E. OC. F.... | ... ... | ... | Projecting bottom with flat centre plug, nozzle round fuze-hole; black lacquer. |
| " " ... | 30/12/61 | 405 | 18·75 | 97 15 ± 2 lbs. | 7 10 | " ... ... | Undercut | E. OC. F.... | ... ... | ... | (as above) |
| " " ... | 10/5/62 | 541 | 18·53 | 98 2 ± 2 lbs. | 7 10 | " ... ... | Zinc ... | ... ... | Z, date ... ... ... ... | R.L. | Solid round bottom, nozzle; black lacquer. |
| " " ... | 26/3/63 | 753 | 18·53 | 98 0 ± 2 lbs. | 7 10 | " ... ... | " ... | | | | |
| " Mark I.... ... | 15/9/65 | 1119 | 15·8 | 83 0 ± 1½ lbs. | 6 8 | " ... ... | " ... | ... ... | L R ↑ L, date, I... ... ... A. | R. Z. Z. L. | Flat solid bottom; black lacquer For S.S. |
| " " II. ... | 11/5/67 | 1421 | 15·8 | 83 0 ± 1½ lbs. | 6 8 | General Service ... | " ... | ... ... | " " II. ... ... | R. ↑ L D. | Flat solid bottom; red lacquer. For S.S. |
| 40-pr. (no Mark) ... | 22/1/61 | 209 | 13·87 | 37 10 ± 1 lb. | 2 4 | Moorsom ... ... | Tin ... | E. OC. C.... | ... ... | ... | Projecting bottom with flat centr plug, nozzle; black lacquer. |
| " " ... | 30/12/61 | 405 | 13·87 | 38 7 11/16 ± 1 lb. | 2 4 | " ... ... | Undercut | E. OC. & C | ... ... | ... | (as above) |
| " Mark I. ... | 10/5/62 | 541 | 13·52 | 37 14 ± 1 lb. | 2 4 | " ... ... | Zinc ... | ... ... | Z, date, I. ... ... ... | R.L. | Flat solid bottom with flat centre plug, nozzle; black lacquer. |
| " II. ... | 11/5/67 | 1425 | 13·52 | 37 14 ± 1 lb. | 2 4 | General Service... | " ... | ... ... | (circle: II. R. ↑ L. A.) date, II.. ... | R. Z. L. | Flat solid bottom with flat centr plug, nozzle; red lacquer. |
| 20-pr. (no Mark) ... | 4/1/62 | 477 | 11·28 | 20 5 ± 9 oz. | 1 2 | Moorsom ... ... | Undercut | E. OC. ... | ... ... | ... | Projecting bottom with flat centre plug, nozzle; black lacquer. |
| " Mark I. ... | 4/1/62 | 477 | 10·91 | 20 9 ± 9 oz. | 1 2 | " ... ... | Zinc ... | ... ... | Z, date, I. ... ... ... | R.L. | Flat solid bottom, nozzle; black lacquer. |
| " " II. ... | 26/10/66 | 1342 | 10·875 | 20 8 ± 4 oz. | 1 2 | Field Service ... | " ... | ... ... | (circle: L. R. ↑ L. A.) date, II... ... | R. Z. L. | Flat solid bottom, flanged socket; red lacquer. |
| " " III. ... | 26/4/67 | 1426 | 10·5 | 20 8 ± 4 oz. | 1 2 | General Service... | " ... | ... ... | " III... ... | ... | Flat solid bottom, nozzle; red lacquer. For S.S. |
| 12-pr. Mark III. ... | 29/3/71 | 2061 | 8·35 | 10 12 ± 4 oz. | 0 8 | Field Service ... | " ... | ... ... | " " " ... ... | ... | Flanged socket; red lacquer. |
| 9-pr. Mark III. ... | 29/3/71 | 2061 | 6·30 | 8 2⅓ ± 2⅓ oz. | 0 6 | " ... ... | " ... | ... ... | " " " ... ... | ... | (as above) |

N.B.—Marks I. and II. 12-pr., (§§ 1087 and 1343), Mark I. 9-pr. (not published), and Mark II. 9-pr. (1344) may be regarded as unserviceable; Mark I. had no socket, and the socket in Mark II. was too weak.

* Existing shell with Moorsom gauge to have G.S. adapters screwed in § 1238. See p. 146, for directions for fixing. For diameters at base, see table of segment shell, p. 152.

### 3. *Shrapnel Shell.*

**Shrapnel shell.** *Shrapnel* shell are issued for the 12-pr. and 9-pr. guns only at present. Patterns exist for the 7″ and 40-pr. guns, but do not at present form any part of equipment. As, however, the general construction of all Shrapnel for rifled shell is much the same, we will give a description, first, of the larger calibres of B.L. Shrapnel, and then pass on to the F.S. calibres. The gauge of the 7″ and 40-pr. Shrapnel is G.S. Not being in equipment, it is unnecessary to give the fuzes.

**Construction.** *Each shell consists of a hollow body with a head lightly attached to it. The body is of cast-iron; it has a lead and antimony coat to take the grooving, in the case of the 40-pr. resembling exactly that of the segment shell Mark I. Externally the 12 and 9-pr. Shrapnel bodies nearly correspond with their respective segment and common shells.

The body is weakened internally by six longitudinal grooves running down the entire length of the interior, and forming lines of least resistance. The base is formed into a chamber to contain the bursting charge, the interior of the body is slightly conical, that is, it enlarges slightly towards the front, giving an increase of ·1″ in diameter in larger, and ·05″ in smaller calibres at the mouth; running round the mouth is a shoulder and groove forming a kind of recessed lip. A tin cup is placed in the chamber to contain the powder, and prevents iron fragments breaking into the charge on the shock of discharge, and so causing premature explosion. It also prevents any powder from working into the crevices round the diaphragm.

Over the mouth of the powder chamber rests a disc of wrought iron supported by a shoulder,† the disc is pierced in the centre and partly tapped to take a wrought-iron tube, which is screwed into it, this tube itself being tapped at the top to take a gun-metal primer (see page 133) employed to assist in carrying the flash of the fuze to the bursting charge in the chamber.

On the disc or diaphragm are placed bullets of lead and antimony, which are fixed by rosin being run in among them, brown paper being laid round the inside of the shell to prevent too firm adhesion of the rosin. Over the bullets and rosin is placed a kamptulicon or felt ring.

The head is made of elm covered with a light shell, formerly of wrought iron, latterly, since 8/66, of Bessemer metal, the wood being bored out to contain a tin socket fitting round the iron tube of the body and holding in its mouth a gun-metal bush of G.S. gauge tapped to take the G.S. screw plug; this bush projects slightly above the apex of the shell.

It will be seen that the chief functions of the head are to cut through

---

* In future manufacture of these shells probably they would be made similar to the later patterns of R.M.L. Shrapnel of large calibres. See p. 192.

† The cast-iron disc used in early patterns was found to break up, and was liable to cause prematures.

This disc is called a "diaphragm," after the similar arrangement used in the S.B. Shrapnel, used to prevent the bursting charge from being in contact with the bullets, &c.

the air in flight and contain the fuze; being very light it brings the centre of gravity of the shell, which would otherwise come too far forward, to its proper position, and its inertia being reduced, it is possible to hold it to the shell by a light method of attachment, viz., by means of steel rivets (12 in larger and 4 in smaller shells) and four steel twisting pins, these latter being rivets, sometimes fitted into slots instead of holes in the shell, so as to tend to prevent the head by its own inertia twisting away from the body, although in no way interfering with its liberation when blown to the front by the bursting charge.

The Bessemer metal head is flush with the exterior of the body in the 40-pr. and lower natures.

§ 2286.

All Shrapnel shell have weighed bursting charges of "service pistol powder," or "F.G.," as may be most convenient.*

For bursting charge, dimensions, &c., see table, p. 160; and for instructions for filling, see p. 62.

**Field service Shrapnel.**

*Field Service Shrapnel.*—Calibres 12-pr. and 9-pr., gauge G.S.

Fuzes for L.S.—5 seconds and 9 seconds B.L. time, and R.L. Percussion, Mark II.

Fuzes for S.S.—Shrapnel not issued for these guns to the Navy.

The above description of garrison Shrapnel applies to the early patterns of F.S. shell. Experience showed that the attachment of the head in the smaller natures was too weak, the jolting motion of the limbers was found to loosen the heads, in some cases they came off, and frequently the rivets were loosened, occasionally the rim of cast iron to which the Bessemer metal head is attached broke off, the rosin was found to work up between the iron pipe and the tin socket, thus getting above the primer, and causing a liability to blind shell. Some of the primers were found to be so tightly fastened by rust to the iron pipe, that it was impossible to unscrew them when the shells had to be emptied.

§ 2210.

In order to remedy these defects, the rim of cast-iron was thickened while the Bessemer metal was turned down to a corresponding amount at the junction; the latter metal being so tough as to have a large margin of strength. The head is attached to the body by screws as well as rivets. A gun-metal tube with an enlarged head is substituted for the iron one, thus preventing the primer becoming fixed by corrosion, and to the head of the tube a tin socket is soldered, so as to prevent the rosin from working up. The felt or kamptulicon washer is soaked in kit composition, so as to keep the rosin better in its place.

For bursting charge, dimensions, &c., see p. 160.

For instructions for filling, &c., see p. 62.

**Issue.**

Filled and boxed, primers and plug screwed in. Sometimes issued empty, loose, and filled by the battery.†

**Use of Shrapnel.**

The effect of shrapnel, like that of segment greatly depends on the correct estimate of the results that are being produced, and in most cases on the judgment displayed in the constant efforts to improve on the shooting: when used intelligently the effect is most excellent.

It is possible generally from the gun to estimate the line and the height of the burst of the shell, but not the distance at which it occurs, and bad practice commonly arises from a too sanguine estimate of

---

* R.F.G. may be substituted.

† For transit to India the shell would be issued empty, boxed.

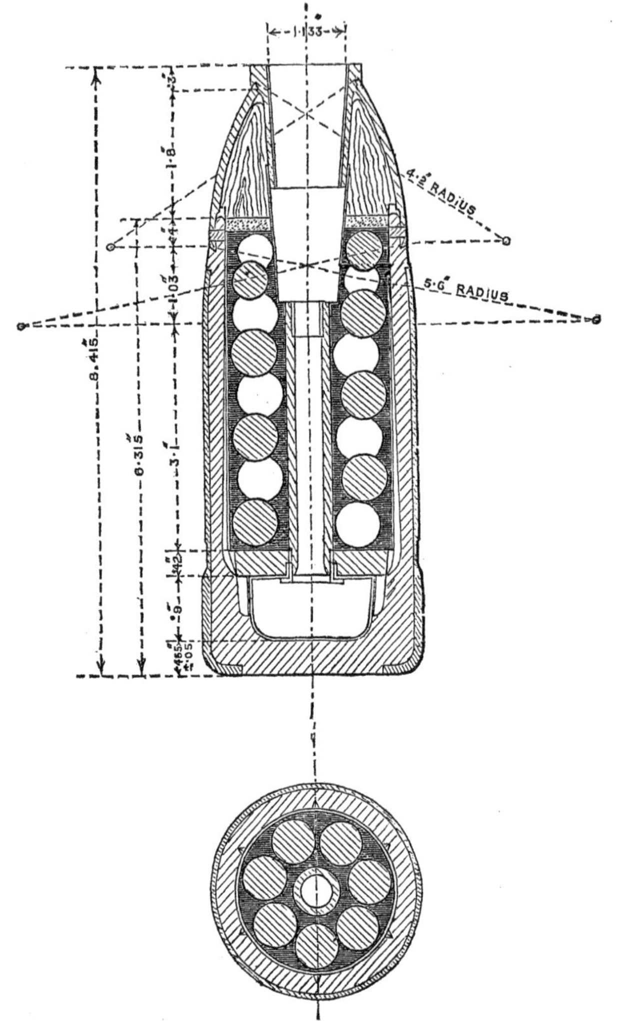


12-PR. SHRAPNEL, MARK I.

N.B.—In this pattern an *iron* central tube is shown.

effects judging from the appearance of the smoke of the burst alone; particular attention should therefore be paid to any visible marks of the bullets grazing. On water, splashes will be seen, on dry ground, puffs of dust, on ice, very distinctly scored marks, on wet or boggy ground nothing is commonly visible.

Shrapnel should be burst closer to compact masses of troops than to more open formations.

F.S. Shrapnel should be burst well within 100 yards of the object; as a rule, about 50 yards off, and about from 10 to 15 feet above the plane; these distances will, however, vary with the object

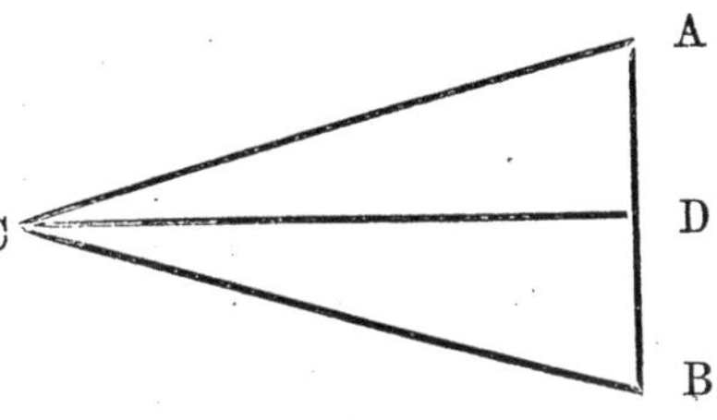


aimed at; these shells should be burst nearer to a column than to a line. In the Report of the Committee on Indian Field Equipment, page 53, it is stated that at 1,000 yards the dimensions of the cone of dispersion are as follows: A.B.$=\frac{1}{3}$ C,D., the centre of the cone D. falls $\frac{1}{12}$ of C. D.

Suppose the object aimed at to show a front of 60 feet, then A.B. = 60 feet, C.D., 180 feet, or 60 yards, the fall will be $\frac{180}{12}$ feet = 15 feet, therefore, in the supposed case firing at a front of 60 feet, the range being 1,000 yards, we ought to burst the shell 60 yards off and 15 feet above the plane.

In order to get the best results from F.S. Shrapnel, it is necessary to have a time fuze reading to very short intervals of time. We may say that the shell from a 12-pr. B.L. gun travels at the rate of 300 yards in a second; this would give 150 yards to half a second, and 75 yards to a quarter second, thus, even the five seconds fuze would hardly read close enough, and this fuze is available only up to about 1,600 yards from a B.L. gun; above this range the nine seconds fuze must be used which reads to half seconds, corresponding to intervals of about 150 yards in range; with this fuze much of the effect of Shrapnel must be lost.

The necessity for a finely divided fuze is even greater for M.L. guns, which have a higher velocity. As before pointed out, this object might probably be attained by making four powder channels in the fuzes, so causing the five seconds fuze to read $\frac{1}{8}$ second, and the nine seconds fuze, $\frac{1}{4}$ second.* See also page 23.

Until the last few years Shrapnel were employed exclusively with time fuzes, but recent experiments have shown that where local effects are desired (when firing against dense masses of men, or perhaps against artillery), percussion fuzes may be employed with advantage.† Very good results have been obtained in firing Shrapnel with the R.L. percussion fuze from the 16-pr. R.M.L., and from the 9-pr. R.M.L. At long ranges the percussion fuze does not answer well, as the shell descending at a sharp angle has a tendency to bury itself in the ground.

It is impossible to lay down the exact effective range of Shrapnel from field guns; probably with B.L. field guns they would be effective up to about 1,800 or 1,900 yards.‡

---

* Six nine-seconds fuzes, having four powder channels, were fired at Shoeburyness, they acted well.

† Up to about 2,000 yards.

‡ Two thousand metres is given as the limit in "Recherches sur les Fusées par Romberg."

The report of the Special Committee on Shrapnel *v.* Segment Shells gives the result of a large amount of practice; the practice was generally carried on at ranges well under 2,000 yards. On page 31 the result is given of firing Shrapnel at a range of 2,000 yards, the effect was very slight. The best results with B.L. Shrapnel, Dartmoor Committee, were—

| Gun. | Range. | No. of Rounds. | Total hits. | Hits per round. |
|---|---|---|---|---|
| 12-pr. B.L. | 1,670 yards. | 9 | 652 | 72 |

The targets were 54″ × 6″; 6 rows, 14′ apart.

SHRAPNEL SHELL B.L. GUNS.

| Calibre, and Mark of Pattern. | Date of Approval. | § Changes in War Stores. | Length, inches. | Number and Nature of Balls contained. | Weight of Shell empty, ± 1·5 per cent. lbs. | oz. | Weighed bursting charge, Pistol, R.F.G. or F.G. powder. lbs. | oz. | Total Weight of Shell. lbs. | oz. | Gauge of Fuze-hole. | Remarks. |
|---|---|---|---|---|---|---|---|---|---|---|---|---|
| 7″ Mark II.* .. | 4/4/68 | 1,609 | 15·98 | 305, lead and antimony, 14 per lb. | 97 | 0 | 0 | 8 | 97 | 8 | General Service | L.S. only. Not now issued. |
| 40-pr. „ I. .. | 7/7/69 | 1,807 | 12·4 | 162, „ 16 per lb. | 39 | 0 | 0 | 3 | 39 | 3 | „ | Not now issued. |
| 12-pr. „ I.† .. | 20/10/70 | 1,978 | 8·415 | 56, „ {42, 18 per lb. / 14, 34 „} | 10 | 11 | 0 | $0\frac{3}{4}$ | 10 | $11\frac{3}{4}$ | „ | |
| „ „ II. .. | 15/2/72 | 2,210 | „ | „ „ „ | „ | „ | „ | „ | „ | „ | „ | Junction of head and body stronger, and secured by screw as well as by plain rivets, centre tube of gun metal instead of iron. |
| 9-pr. „ I.† .. | 20/10/70 | 1,979 | 7·15 | 42, „ {21, 18 per lb. / 21, 34 „} | 8 | 11 | „ | „ | 8 | $11\frac{3}{4}$ | „ | |
| „ „ II. .. | 15/2/72 | 2,210 | „ | „ „ „ | „ | „ | „ | „ | „ | „ | „ | Same remarks as to 12-pr. II. |

* Mark I., 7″ (§ 1422) with cast-iron diaphragm, and without tin cup, is ordered to be broken up.

† Shells with stronger sockets, sealed to govern future manufacture, without a change of pattern, marked "P.S." (plain socket) on the lead coat, § 2062.

A few Mark I. shells have been made and issued which have the junction of head and body stronger (*vide* remarks), and some with all the improvements, strengthened head and gun metal tube, the former are distinguished by one small (*) and the latter, those with all the improvements, by two small stars viz. (**), on the lead coat, § 2210.

Diameters at base, same as segment and common shell.

All Shrapnel have the zinc attachment.

The various natures of shell for F.S. are readily distinguished, one from another, by the following marks:—The Shrapnel have their heads painted red and the gun-metal socket projects; the common shells have the gun-metal of their flanged sockets showing as a ring round their fuze-hole; while the segment has simply a fuze-hole tapped in the metal of the shell.

Distinctive features of B.L. shell.

The garrison service shells are more difficult to distinguish; the Shrapnel shells are known by their projecting socket, but it is sometimes hard to distinguish between common and segment shells; as a general rule the common are longer and have a more gradually curved head than the segment; however, the specially light 7″ common shell is but little longer than the segment, but by looking at the base of the shell it can be seen that the lead coating extends farther in the segment than in the common shell, as the base is secured by the lead coating. By unscrewing the plug, and feeling for the grooves in the head with a bent wire, the segment may be recognized, as every segment shell has four longitudinal grooves in the head.

The lead coating at the bottom of field service shells is rounded off to prevent their getting enlarged at the base by jolting in the limbers.

### Gauges, Iron, Ring, Shell or Shot, B.L. Guns.

* High and low gauges are issued to stations of inspection, the former passing over the projectile, the latter resting at the back end behind the cannelure.

§ 1313.

The high gauge passing over secures the shell loading; should the lead be a little set up, it can easily be filed down to the correct dimensions. This gauge is issued to Field and Garrison Artillery, and to the Navy.

The dimensions of the gauges are as follows:—

| | High, Diameter. | | Low, Diameter |
|---|---|---|---|
| | Inches. | | Inches. |
| 7 inch - - | 7·095 | .. | 7·08 |
| 40-pr. - - | 4·855 | .. | 4·83 |
| 20-pr. - - | 3·85 | .. | 3·836 |
| 12 and 9-pr. - | 3·10 | .. | 3·067 |
| 6-pr. - - | 2·60 | .. | 2·567 |

### Shot.

Are of the following classes:—

1. Case Shot.
2. Solid ,,
3. Drill ,,

---

* The high gauge was formerly called a "test ring."

§ 1017.

In manufacture, five gauges are used, viz.:—

1 for commencement of taper near point.
1 high and 1 low for body.
1 ,, and 1 ,, for back end.

In gauging a B.L. projectile standing on a table, the high back end should descend over the entire projectile and lie on the table; the low back end gauge should stop at the commencement of the back end behind the cannelure; the high body gauge should descend and rest on the last-mentioned one.

Gauging a B.L. projectile.

The proper position for the low body and commencement of taper gauges to rest is not so well defined, but the former should not pass over what is obviously the full diameter of the body, and the latter should stop at a point a little above it.

**1. Case shot.** The making of case shot for rifled guns is not such a simple business as to make case for S.B. guns. In the case of the S.B. guns all that is required is to make an envelope for the sand shot, &c., which is strong enough to resist the shock of discharge without smashing up altogether in the bore. It must, of course, in all cases be weak enough to burst and liberate its contents on leaving the gun. Injury to the bore of bronze S.B. guns was guarded against, as previously mentioned, by the use of wood bottoms to their case shot.

Now, in the case of a rifled gun, we are met in the outset by an anomaly. We want to fire a projectile from a rifled gun, and yet not allow it to take the rifling. Did the case shot leave the bore with the spin of other projectiles, the dispersion of the balls laterally would be very great, and their *direct* range to the front very small. In fact, for firing case shot we want to use the rifled gun as a smooth-bore. Then in the case of rifled guns, especially in the case of those having wrought iron barrels, as the service B.L. guns, we must, as much as possible, guard against injury to the bore, and yet insure the case breaking up properly on leaving the gun. In fact, the case shot must be strong enough in its construction:

1. Not to set up on discharge, so as to take the rifling;
2. Not to be easily injured by travelling, or the ordinary knocking about unavoidable on service;

while it must be weak enough, as above mentioned, to release its contents on leaving the gun.

These two considerations are obviously conflicting, and as in so many similar cases, it is a question of hitting the golden mean between the two.

It must be recollected that the weight of the envelope and of expedients used to strengthen it, as well as the weight of the packing in which the balls are placed, are necessary evils. The actually useful part of the projectile is the assemblage of balls within the case. The remaining constituent parts of the projectile are practically useless in a projectile sense. In the old S.B. case the envelope averaged about $\frac{1}{9}$ the weight of the projectile; in the case for rifled guns the envelope and accessories sometimes amount to nearly as much as (in one or two instances even more than) the weight of the balls.

§ 1241. The earlier patterns of case shot for B.L. guns were made after a design proposed by Lieut. Reeves, R.A. The bullets were imbedded in discs of wood, and the whole enclosed in an iron or tin case. The manufacture of these has long been discontinued.

§ 1611. The R.L. pattern of case is now adopted. For a description of the latest patterns, see p. 140.

The general construction of B.L. and R.M.L. case is similar, see p. 140, but the former have solder studs* to prevent their being rammed too far into the bore, *not* to take the rifling. The rule for the weight of case shot has been previously given (p. ). This rule was laid down in 1864,† but is not strictly adhered to. For B.L. guns, which have a small charge of powder compared to the old S.B. guns, it is evidently desirable to keep down the weight of the case shot.‡ The rule is less important for R.M.L. guns, but even their charges are less than those of S.B. guns of corresponding calibre.

---

§ 2508. * Some B.L. F.S. case shot have segmental rings instead of studs. They were abandoned.

† See Extracts, Vol. IV, p. 397.

‡ The 42-pr. S.B. had a charge of $10\frac{1}{2}$ lbs. while the 40-pr. B.L. has only a 5 lbs. charge, if the B.L. case weighed 40 lbs. its effect would be slight.

The balls of the earlier patterns of case shot are packed in coal dust, but equal parts of clay and sand are now used, the case thus made are found to give better results than those in which the coal dust was used.

*Garrison Case Shot, 7″ and 40-pr.*—For general description, see p. 140; for details, see table, p. 165.

Garrison case shot.

The 7″ case shot, both Reeves' and Laboratory pattern, is made to suit both M.L. and B.L. 7″ guns by having three large solder studs at one end, which cause it to stop in the shot chamber when ramming home in a B.L. gun, but find room in the three large grooves of the Woolwich gun.

Issue.

7″ loose, 40-pr. boxed.

Field case shot.

*Field Service Case Shot.*—20-pr., 12-pr., 9-pr. and 6-pr For construction, see p. 165; for details, see table, p. 165.

*12-pr. Case Shot. Mark IV.*

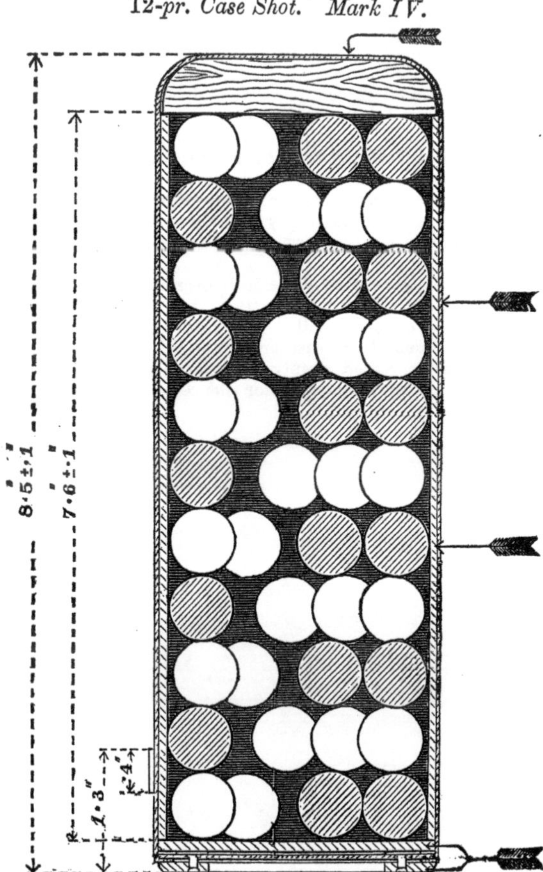


The lead and antimony balls, being much heavier than iron balls of a corresponding size, keep up their velocity better. Sand shot are used for economy in the garrison case shot.

Issue.

Packed in boxes for transit.

Use of case shot.

The general use of case is when firing at troops at about 300 yards; when the ground is hard and level it is not necessary to give elevation from heavy guns at short ranges, as the bullets run along the ground, on soft or uneven ground the effect of case is much diminished, as the destructive effects depends mainly on ricochet; over such ground elevation should be given according to distance.* Generally about 1° of elevation may be given from field guns at 300 yards.

**Loading.**

In loading B.L. case the studs go towards the rear, thus bringing the strongest part of the case next to the charge. The case shot should be pressed home gently, if much force is used the case may be rammed past the shot chamber, the soft studs being shorn off.

F.S. case occasionally become damaged in the limbers, they can readily be repaired by the battery artificers; a little solder is generally all that is wanting.

---

* Captain C. O. Browne states as follows:—

The effect of the case and grape fired on the English advancing on the Redan on September 8th, 1855, in the siege of Sebastopol, was such as will not readily be forgotten by those who saw it; the ground was fairly level, hard, and very dry. I happened to be in a battery well placed for view, and being ordered to discontinue my fire as the troops advanced I had full opportunity of watching the effects; each gun flash from the flanks of the work was followed by puffs of dust commencing perhaps 50 yards from the muzzle, and running along the ground in an accumulating shower which died away at some hundreds yards distance, some stray grape of large size even ringing on the stony ground and entering batteries 1,000 yards from the Russian works. A few rounds fired nearly simultaneously sufficed to sweep and regularly dust the ground, and gradually as the attack continued it became dotted with bodies of men.

The effects of the grape and case on June 18th over the same ground were very much greater, both the generals leading the attacks, viz., Sir J. Campbell and Yea, and a large number of the men, being killed in a few minutes; the column led by the former officer suffered chiefly from case and grape. I missed seeing this attack, but conclude the effect was increased simply from the fact that there were a larger number of Russian guns neither dismounted nor silenced.

# Case Shot.—B.L. Guns.

| Calibre, and Mark of Pattern. | Date of Approval. | § Changes in War Stores. | Length, inches. | Diameters, inches. | | Number and Nature of Balls contained. | Approximate weight of balls. | Approximate weight of case, linings, coal dust, or clay and sand | Total weight. | Marks on the top of the Case. | Number of Solder Studs at Base. | Remarks. |
|---|---|---|---|---|---|---|---|---|---|---|---|---|
| | | | | Over body. | Over studs. | | lbs. oz. | lbs. oz. | lbs. oz. | | | |
| 7-inch, Mark III. | 24/3/68 | 1611 & 2444 | 10·25 ± ·1 | 6·89 ± ·03 | 7·097 + ·03 − ·02 | 74 8-oz. sand shot ... | 36 5 | 30 11 | 67 0 ± 2 lbs. | III. W. ↑ D. 7-in- M.L. or B.L. | 3 | Balls packed in coal dust, surplus stores to be returned to Woolwich for alteration to IV. |
| ,, ,, IV. | 25/1/72 | 2188 | ,, | ,, | ,, | 70 ,, ... | 35 0 | 34 0 | 69 0 ± 2½ lbs. | IV. ,, ,, | ,, | Balls packed in clay and sand. |
| ,, ,, V. | 13/3/76 | 2924 | ,, | ,, | 7·11 ± ·005 | 71 ,, ... | 35 8 | 32 10½ | 68 2½ ± 2½ lbs. | V. ,, ,, | ,, | Do. Case in 3 pieces, a ring at bottom instead of a disc. Studs 3·5 inches from base. |
| 40-pr. ,, I. | 24/3/68 | 1611 | 10·15 ± ·1 | 4·715 ± ·015 | 4·885 ± ·015 | 37 ,, ... | 18 8 | 11 8 | 30 0 ± 1 lb. | I. ,, 40-pr. B.L. | 24 | Balls packed in coal dust. |
| ,, ,, II. | 30/4/72 | 2264 | ,, | ,, | ,, | 35 ,, ... | 17 8 | 14 0 | 31 8 ± 1 lb. | II. ,, ,, | 3 | Balls packed in clay and sand, tin case in 3 pieces. |
| 20-pr. ,, I. | 25/9/66 | 1299 | 8·4 ± ·05 | 3·705 ± ·015 | — | 55 lead and antimony, 4 oz. | 13 12 | 1 2 | 14 14 | I. ,, 20-pr. B.L. | Studs | Reeves' pattern. |
| ,, ,, II. | 22/3/68 | 1611 | 9·3 ± ·1 | ,, | 3·841 ± ·015 | 41 4 oz. sand shot ... | 9 12½ | 5 3½ | 15 0 ± 12 oz. | II. ,, ,, | Divided ring | Balls packed in coal dust. |
| ,, ,, III. | 1/11/75 | 2841 | ,, | ,, | 3·841 ± ·005 | 239 lead and antimony bullets, 16½ per lb. | 14 10½ | 5 11 | 20 5½ ± 8 oz. | III. ,, ,, | 3 studs | Balls packed in clay and sand, case in 3 pieces, ring at bottom instead of a disc. |
| 12 or 9-pr. ,, I. | 9/8/66 | 1242 1900 | 10·5 | 2·97 | 3·074 | 77 lead and antimony, 7 per lb. | 11 0 | 1 8½ | 12 8½ | I. ,, 12 or 9-pr. B.L. | 12 ,, | Reeves' pattern withdrawn from L.S., issued to S.S. only. |
| 12-pr. ,, II. | 9/8/67 | 1467 1900 | 6⅝ | — | — | 70 lead and antimony, 1⅓ oz. each | 6 9 | 2 7 | 9 0 | II. ,, ,, (marked wrongly) | 12 ,, | Balls packed in coal dust, withdrawn from L.S., issued to S.S. only. |
| ,, ,, III. | 24/3/68 | 1611 2058 | 8·5 ± ·1 | 2·95 ± ·015 | 3·12 ± ·015 | 48 2 oz. sand shot ... | 6 0 | 3 0 | 9 0 ± 8 oz. | III. ,, 12-pr. B.L. | Divided ring | ,, ,, |
| ,, ,, IV. | 15/2/71 | 2058 | ,, | 2·96 ± ·01 | 3·117 ± ·002 | 132 lead and antimony bullets, 16½ per lb. | 8 1½ | 3 6½ | 11 8 ± 6 oz. | IV. ,, ,, | 3 studs | Balls packed in clay and sand, case in 3 pieces, a ring at the bottom instead of a disc. |
| 9 pr. ,, II. | 29/1/70 | 1867 2058 | 6·85 ± ·1 | 2·95 ± ·01 | 3·07 + ·004 − ·003 | 35 2 oz. sand shot ... | 4 6 | 2 3½ | 6 9½ | II. ,, 9-pr. B.L. | Divided ring | Balls packed in coal dust, withdrawn from L.S., issued to S.S. only. |
| ,, ,, III. | 15/2/71 | 2059 | ,, | 2·87 ± ·08 | 3·117 ± ·002 | 101 lead and antimony bullets, 16½ per lb. | 6 3 | 2 13 | 9 0 ± 6 oz. | III. , ,, | 3 studs | Balls packed in clay and sand, case in 3 pieces, a ring at the bottom instead of a disc. |
| 6-pr. ,, II. | 25/9/66 Provly. | 1299 | 5·8 ± ·05 | 2·465 ± ·015 | — | 61 lead and antimony bullets, 16 per lb. and one do. of ½ oz. | 3 13¼ | 0 6¾ | 4 4 ± 2½ | II. ,, 6-pr. B.L. | 12 ,, | Reeves' pattern balls packed in sawdust, wood lining. |
| ,, ,, III. | 28/12/75 | 2899 | 6·1 ± ·01 | ,, | 2·57 ± ·004 | 64 lead and antimony bullets, 16¼ per lb. | 3 15 | 1 10 | 5 9 ± 14 oz. | III. ,, ,, | 3 ,, | R.L. pattern, balls packed in clay and sand. |

7″ Marks I. and II. §§ 1241 and 1611, are by § 2444 ordered to be returned to Woolwich to be broken up.
A few 6-pr. Mark I. were issued to Jamaica on urgent demand in 12/65 as Mark I. but no circular was promulgated, and no sealed pattern exists.

Solid shot.

*Solid Shot.*—Calibres 40-pr.,* 20-pr.,† 12-pr., 9-pr., and 6-pr.

They are only used for practice.

They have much the same form and length as segment shells. They approximate closely to the nominal weight: various patterns exist having tin, undercut, or zinc attachment.

Issue.

They are issued loose, but might be packed in boxes if convenient for transit or for India.‡

Drill shot. §§ 942, 1088, 1742.

*Drill Shot.*—For the larger guns, down to the 20-pr. inclusive, recovered shells, with their coating turned down, are used. For the 9 and 12-prs., special shot are used with the lead coating extending over the head, to avoid injury to the copper bush. The 6-pr. has a special drill shot.

The shell or shot used for drill shot may be known by the absence of cannelures.

## Cartridges, and Miscellaneous Stores connected with them.

The weight of the charge is ⅛ of the weight of the projectile, except for the 7″ guns. The 7″ of 82 cwt. has a charge of 11 lbs.; the 7′ of 72 cwt. has a charge of 10 lbs.§ In the L.S. the 98-lb. shell is used with both guns.

The powder employed is L.G., under the rule given on page 8, but R.L.G. will be resorted to when the stock of L.G. is used up.

The cartridge is made of serge, the bottom is formed of a circular piece, and the cylindrical part from a rectangular piece; the seams are made to overlap and are sewn with three rows of stitches, except the seam forming the junction of the body and bottom in the 12-pr., and under, where two rows of stitches are used. The cartridges are hooped with blue braid; the empty cartridges are issued with braid inserted; one end of the braid has a loop through which the loose end is passed, and a single bend is formed with it on to the loop. Care is necessary, as men are apt to make the knot off the loop or a wrong knot, which slips and the hooping is rendered useless.

Lubricators. § 1486.

‖A lubricator, consisting of two thin cups of tinned iron soldered together, containing a mixture of equal parts of tallow and linseed oil, attached to a wad of felt, backed by millboard (the wad being coated with beeswax),¶ is inserted into the cartridge. It is put on the top of the powder, just above the top hoop in all cases, except the 40-pr. S.S.

---

§§ 1038, 1815, 1347.

* The 7-inch shot were withdrawn, as they were too heavy for the guns; hollow and hollow-headed shot may also be regarded as obsolete.

§ 2907.

† The string loops on 20-pr. B.L. solid shot being no longer necessary, the original pattern, Mark I., has been resealed to govern future manufacture. All shot of Mark II. in store will be issued without loops.

‡ The dimensions, &c., of the solid shot will be found in the Changes in War Stores, 40-pr. § 209, 405; 20-pr. § 341, 405, 1,298; 12-pr. § 162, 270, 405, 477; 9-pr. § 600; 6-pr. § 344, 405, 477.

§ 1038.

§ These guns had at first heavier charges, but it was found necessary to reduce them to diminish the strain on the guns, hence the necessity for paper cylinders to enable the cartridge to be brought up to the required length to fill the chamber. By using an equivalent charge of 14 lbs. of P powder instead of 11 lbs. of R.L.G., the strain on the vent-piece was found to be reduced from 17·4 tons per square inch to 5 tons. Extracts XII, page 153.

§ 1510.

‖ The edge of the millboard wad of every inside lubricator is rounded off.

§ 2970.

¶ Previous to August, 1876, the edges of the wad only were coated with beeswax. The uncovered portion was found liable to become moth-eaten, hence the change. No change in pattern.

and 7″, which have their lubricators detached to save room in the magazine.

Cartridges carried in the limbers of field batteries have suffered from grains of powder finding their way up between the edge of the lubricator and the cartridge; and grains so situated cause lumps which wear the serge through and form holes, through which the powder may gradually work out. For this reason it is desirable that the lower edge of the lubricator should be as nearly opposite the upper edge of the first braid hoop as possible.

The lubricator proper is fastened to the wads by the stalk in the case of a detached lubricator, and by a copper wire in the case of inside lubricators, the wire being fastened to the closing disc.

**Use.**

The use of the lubricator is to prevent the guns from leading; the cups being crushed by the discharge, the lubricant is squeezed out, and the wad following wipes and polishes the bore.

**Packing and issue.**

Lubricators are generally packed in "cases for Laboratory stores," as follows:—

| | | |
|---|---|---|
| 7″ | .. | 24 in No. 2 case. |
| 40-pr. S.S. | .. | 48 in No. 3 " |
| 40-pr. L.S. | .. | 66 in No. 3 " |
| 20-pr. | .. | 99 in No. 4 " |
| 12 or 9-pr. | .. | 168 in No. 4 " |
| 6-pr. | .. | 224 in No. 4 " |

**Sockets, wood, B.L. cartridges. §§ 1939, 2170.**

With the 40-pr. S.S. and 7-inch, a wooden socket is choked into the neck of the cartridge, on to which the lubricator screws.* All B.L. service cartridges are choked with twine; the 7″ L.S. and S.S., and the 40-pr. S.S. have a running string in the mouth for choking in the socket for the lubricator. The chokes of all, except the 7″, are cut to a length of one inch; the 7″ choke is cut to $1\frac{1}{2}$″.

**Paper cylinders, B.L. cartridges. § 792.**

With the 7″, 40,† and 20-pr. a paper cylinder is used to bring the cartridge up to length, the cartridge is half-filled with powder, the cylinder is next inserted, so as to be in the centre of the cartridge, and then the rest of the charge. One end is formed by choking up the

*Cylinder for 7-inch 10 lbs. Charge.*

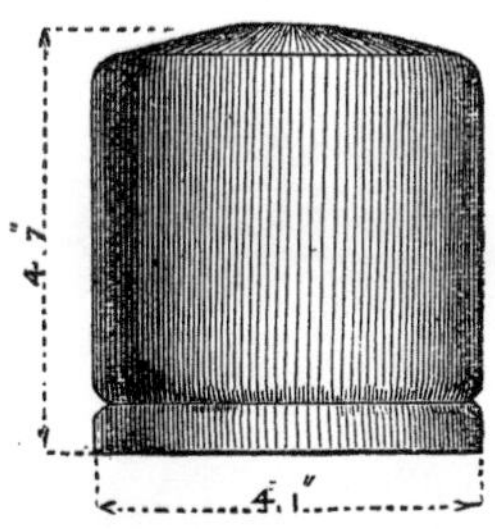


* A paper socket was used at first, but was abandoned as it was found to absorb moisture. The socket has a groove around its neck, smeared with kit composition and Spanish brown, by means of which it is securely fixed in its place. §§ 406, 422.

† The same cylinder is used for the 40-pr. and the 11 lbs. charge of the 7″, length 4″·5, diameter 3″·25. For 20-pr. the cylinder is 4″·5 long, diameter 2″·5. An early pattern existed which was found too weak, it was liable to be crushed in the cartridge, and portions of them were found to remain in the gun after discharge so as to impede loading for the next round. The weak pattern may be known by both ends being closed by paper discs. § 792.

cylinder on a former and closing the hollow by a paper plug; the other end has a groove choked into it, so as to support a cardboard disc, which is glued in. The end formed of the cylinder itself is the strongest, and is placed downwards in the cartridge. These cylinders are issued separately in vats containing 220 for the 10 lbs. cartridge 7″ gun, 390 for the 11 lb. cartridge 7″ gun, or for the 40-pr., and 700 for the 20-pr. cartridge.

Issue.

§ 1939.
Cl. 97, A.C. 1870.

It was found necessary to varnish the paper, as otherwise it absorbed moisture and damaged the powder, for this reason also the paper socket for the lubricator was abandoned, as paper is very apt to absorb moisture* when unvarnished; it is never desirable to have it in contact with powder. See page 11.

§ 1998.

The cartridges are marked B.L. in addition to the nature of gun and weight of charge.

Issue of cartridges.

For packing of filled cartridges, see Table, p. 172, and also p. 126†.

For issue of empty cartridge see pp. 126, 331.

Tin cups.

*Tin Cups* are used for all B.L. guns to prevent any escape of gas; they have a rim ·32″ deep, which is pressed back by the explosion of the powder against the sides of the bore, thus preventing the gas from getting behind them. The central hole allows the flash from the tube or primer to reach the cartridge.

§ 1794, Cl.197 A.C. /71 and errata A.C. December /71.

The tin cups are only used for practice and exercise with field guns and the 40-pr., as they have got a copper bush which stands the escape of gas better than iron, but 7″ guns require them also as a service store, as these guns have no copper bush.

For the 7″ one tin cup per cartridge is to be issued for service, and as many as required are to be used at practice.

Issue.

They are packed in packing cases for Laboratory stores.

See table, p. 169.

---

* See Extracts, Vol. VIII, p. 80, where an account is given of the deterioration of B.L. cartridges having unvarnished cylinders in magazines where M.L. cartridges kept well. Army Circular, July, 1870, directs that the unvarnished cylinders be replaced by varnished ones, and the paper socket to be replaced by a wooden one. The varnish used is made of 4 lbs. of shellac to 1 gallon of methylated spirit.

† The 12 or 9.pr. B.L. gun carries four filled service cartridges in the axletree boxes, packed in copper cylinders.

## DETAILS OF PATTERNS OF TIN CUPS FOR B.L. GUNS.

| Nature of Gun for which intended. | Date of Approval. | § of Changes in War Stores | Diameter. | Marks. | Remarks. |
|---|---|---|---|---|---|
| 7" .. .. .. .. | 11/1/62 | 489 | ins. 7·253 | 7-in. B.L. 7·253. I. L.G. R.L. W. ↑ | Low gauge cup to suit greater rigidity of the wrought-iron bush ring, also used for copper bushes at first. |
| All 7" except the above has become obsolete by § 1815, published 7/9/69. | | | | | |
| 40-pr. .. .. .. | 6/7/65 | 1163 | 4·96 | I. 40-pr. 4·96. R.L. W. ↑ D.* | Approved for general issue for broadside guns, 21/2/66, vide § 1208. |
| 20-pr. .. .. .. | 6/7/65 | 1163 | 3·94 | I. 20-pr. R.L. W. ↑ D.* .. .. | Approved for general service for broadside guns, 21/2/66, vide § 1208. |
| 12 or 9-pr. .. .. .. | 6/7/65 | 1163 | 3·197 | I. 12 or 9-pr. R.L. W. ↑ D.* .. | — |
| 6-pr.. .. .. .. | 6/7/65 | 1163 | 2·622 | 6-pr. 2"·622. R.L. W. ↑ D.* .. | — |

* These cups may be found with the letters M.G. (for medium gauge) stamped on them.

N.B.—§ 736 approves of 40, 20, and 6-pr. high and low gauge cups, both of tin and copper, for testing endurance, but these are not service articles and need not be particularly noticed.

§ 1014 directs that tin cups are to be employed in firing guns for the proof of vent pieces, approved 8/2/64.

Primers, vent-piece. § 943.

Issue.

§ 913.

*Primers* are used for the 7″ and 40 prs.; they consist of tubes of leather paper about $2\frac{1}{2}$″ long, driven with mealed powder and pierced like a tube, having strands of red worsted attached, which keep the primer in the hole in the vent piece. These are the only special primers made; they are painted black, and packed 25 in a tin cylinder. Field batteries, however, carry 15 blank muzzle-loading small-arm cartridges, in a blue serge bag, with each sub-division; they are only to be used if the friction tubes are found to fail to fire the charge. The powder is poured into the hole in the face of the vent piece, and the paper of the cartridge placed on top, to prevent its falling out: it is very seldom necessary to use them.*

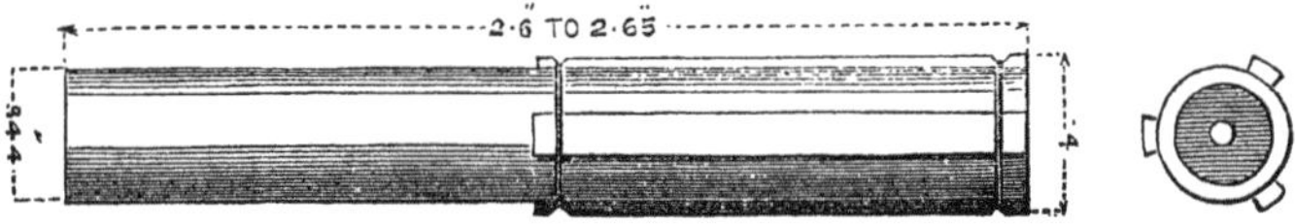


Saluting charges.

*Saluting, or Exercise Cartridges*, are generally similar to the service cartridge, except that they contain less powder, have no lubricators or paper cylinders, and are choked with worsted, as twine is liable to carry fire, they have a becket of braid sown on behind to facilitate unloading if necessary; the braid is sewn down for 1″ at each side on the edge to prevent jamming with the vent piece. The 9 and 12-prs. have worsted instead of braid, which was found to obstruct the flash.

The charges are 7″, 7 lbs.; 40-pr., 3 lbs.; 20-pr., $1\frac{1}{2}$ lbs.; 12-pr. or 9-pr., 1 lb.; 6-pr., 10 oz. Exercise L.G. or R.L.G. being used.

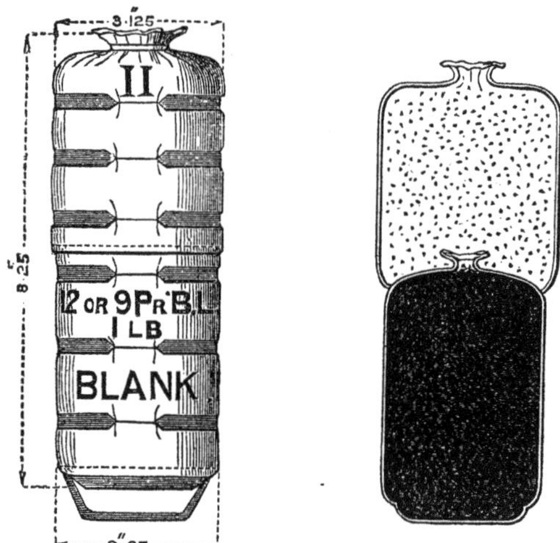


§ 165.

§§ 611, 1815.

* Primers for 12-pr. gun were withdrawn in 1860, a change in the vent-piece rendering them unnecessary.

A primer made like that described in the text, but made of parchment and shorter, was introduced for the 7′ and 40-prs. in 1862, but was ordered to be withdrawn in 1869.

Both parchment and leather paper consume in the gun, but the latter is the cheapest; it is manufactured from scraps of leather instead of the ordinary pulp.

† Silk cloth cartridges will shortly be introduced for saluting and exercise with all natures of B.L. guns. Patterns have been made but are not yet sealed.

The 12-pr. or 9-pr. has a sawdust cartridge stitched on in front to lengthen it. The 6-pr. has the same arrangement. § 1293.

The cartridges are marked with the nature of gun, weight of charge, and also with the word "Blank," thus 7″ B.L., 7 lbs. Blank.

*Packing and Issue.*

Packing and issue.

Blank cartridges are packed and issued in the same manner as the service cartridges. See table, p. 331, and p. 126.

## Table of Filled Cannon Cartridges.—B.L. Rifled Ordnance.

| Nature. | Description. | Date of Approval. | § Of Changes in War Stores. | Charge. | Length. | Diameter. | Paper Cylinder. | | No. and Nature of Braid Hoops. | Marks. | Number Packed and Weight of Package in Pounds. | | | | | | | | | | | | | | | | | | | |
|---|---|---|---|---|---|---|---|---|---|---|---|---|---|---|---|---|---|---|---|---|---|---|---|---|---|---|---|---|---|---|
| | | | | | | | | | | | Cases, Powder. | | | | | | | | | | | | | | | | | | | |
| | | | | | | | | | | | Copper-lined. | | | | | | Metal or Brass. | | | | | | | | | | | | | |
| | | | | | | | | | | | | | | | | | Pentagon. | | | | Rectangular. | | | | | | | | | |
| | | | | | | | | | | | Whole | | Half. | | Quarter. | | Whole | | Sectional | | Plain. | | Corrugated. | | | | | | | |
| | | | | | | | | | | | | | | | | | | | | | | | A | | B | | C | | D | |
| | | | | | | | Length. | Diameter. | | | Number. | Weight. | Number. | Weight. | Number. | Weight. | Number. | Weight. | Number. | Weight. | Number. | Weight. | Number. | Weight. | Number. | Weight. | Number. | Weight. | Number. | Weight. |
| | | | | lbs. | Inches. | Inches. | Ins. | Ins. | Narrow. | | | | | | | | | | | | | | | | | | | | | |
| 7-inch | Full, I. | 20/1/65 | 1038–9 | 11* L.G. | 10 to 11 | 7·03 | 4·5 | 3·25 | 5 | I. 7-in. B.L. 11 lbs. ... ... | 8 | 141 | 3 | 65 | — | — | 7 | 142 | — | — | 9 | 174 | 9 | 152 | – | – | – | – | – | – |
| | | 12/11/62 | 662 | 10† ,, | 10 ,, 11 | ,, | ,, | 4·1 | ,, | ,, 10 ,, ... ... | 8 | 133 | 3 | 62 | — | — | 7 | 135 | — | — | 9 | 165 | 9 | 143 | – | – | – | – | – | – |
| | Blank, I. ... | 27/6/67 | 1415 | 7 ,, | 6 ,, 6·5 | ,, | — | — | 2 | I. 7-in. B.L. 7 lbs. Blank ... | — | — | — | — | — | — | — | — | — | — | — | — | — | — | – | – | – | – | – | – |
| 40-pr. | Full, I. S.S. | 17/12/61 | — | 5 ,, | 10 ,, 10·75 | 4·77 | 4·5 | 3·25 | 5 | I. 40-pr. B.L. 5 lbs. ... ... | 15 | 131 | 5 | 57 | 2 | 29 | 14 | 138 | 5 | 70 | 17 | 163 | 20 | 157 | – | – | – | – | – | – |
| | L.S. | 1/8/62 | — | 5 ,, | 12·25 ,, 12·75 | ,, | ,, | ,, | 6 | ,, ,, | 12 | 123 | 5 | 61 | — | — | 12 | 136 | 4 | 67 | 13 | 151 | 16 | 151 | – | – | – | – | – | – |
| | Blank, II. ... | 6/2/66 | 1195 | 3 ,, | 5·75 ,, 6·25 | ,, | — | — | 2 | II. 40-pr. B.L. 3 lbs. Blank ... | — | — | — | — | — | — | — | — | — | — | — | — | — | — | – | – | – | – | – | – |
| 20-pr. | Full, I... ... | 5/11/61 | 234–408 | 2½ ,, | 10 ,, 10·75 | 3·77 | 4·25 | 2·5 | 5 | I. 20-pr. B.L. 2 lbs. 8 oz. ... | 26 | 134 | 12 | 69 | 4 | 31 | 22 | 133 | 11 | 79 | 30 | 168 | 30 | 145 | – | – | – | – | – | – |
| | Blank, I. ... | 6/2/66 | 1195 | 1½ ,, | 4·75 ,, 5·25 | ,, | — | — | 2 | II. ,, 1 lb. 8 oz. Blank | — | — | — | — | — | — | — | — | — | — | — | — | — | — | – | – | – | – | – | – |
| 12-pr. ... | Full, I.... ... | 5/11/61 | 409 | 1½ ,, | 8 ,, 8·5 | 3·02 | — | — | 4 | I. 12-pr. B.L. 1 lb. 8 oz. ... | 50 | 142 | 25 | 77 | 9 | 35 | 50 | 155 | 23 | 86 | 58 | 178 | 64 | 166 | – | – | – | – | – | – |
| 9-pr. ... | Full, I.... ... | 5/11/61 | 410 | 1⅛ ,, | 6 ,, 6·5 | 3·07 | — | — | 3 | ,, 1 lb. 2 oz. ... | 70 | 154 | 32 | 78 | 12 | 36 | 63 | 156 | 28 | 85 | 72 | 178 | 80 | 166 | – | – | – | – | – | – |
| 12 or 9-pr. | Blank, II. ... | 17/9/66 | 1293 | 1 ,, | 8·25 | 3·125 & 2·87 | — | — | 6 | II. 12 or 9-pr. B.L. 1 lb. Blank | — | — | — | — | — | — | — | — | — | — | — | — | — | — | – | – | – | – | – | – |
| 6-pr. | Full, I.... ... | 5/11/66 | 411 | 12 oz. ,, | 6·25 ,, 6·5 | 2·53 | — | — | 3 | I. 6-pr. B.L. 12 oz. ... ... | 100 | 150 | 48 | 79 | 18 | 36 | 93 | 155 | 46 | 89 | 108 | 179 | 129 | 178 | – | – | – | – | – | – |
| | Blank, I. ... | 20/7/67 | 1415 | 10 oz. ,, | 6·25 ,, 6·5 | ,, | — | — | 3 | ,, 10 oz. Blank ... | — | — | — | — | — | — | — | — | — | — | — | — | — | — | – | – | – | – | – | – |

* For 7-inch, 82 cwt. † For 7-inch, 72 cwt.

N.B.—Army Circular 7/74 directs the use of L.G. powder with B.L. rifled guns (L.G. in red paint being marked on the cartridge, §1998). Blank or exercise R.L.G. and L.G. is used for Blank, and Blank or Exercise R.F.G. and F.G. may also, if considered advisable, where there is a surplus store, be used for blank charges.

## B.L. Drill or Dummy Cartridges.

| Nature and Numeral. | Date of Approval. | § of Changes in War Stores. | Length. | Material of Outside Cover. | No. of Hoops. | Lubricator, inside or detached. | Marks. | Remarks. |
|---|---|---|---|---|---|---|---|---|
| 7″, Mark I.* .. .. | 3/2/64 | 876 | Length not laid down, but should correspond to that of the service cartridge represented by each. | Blue serge .. | 5 | Detached | I. A. 110-pr. Rifle Gun, 12 lbs. | |
| „ „ II. .. .. | 14/3/67 | 1384 | | Leather .. | 5 | Detached | II. 7-in. B.L., 11 lbs. | |
| 40-pr., for L.S., Mark I.*.. | 3/2/64 | 876 | | Blue serge .. | 6 | Inside .. | I. A. 40-pr. Rifle Gun, 5 lbs. | |
| „ „ „ II. .. | 14/3/67 | 1384 | | Leather .. | 6 | Inside .. | II. 40-pr. B.L., 5 lbs. | |
| „ S.S. „ I.* .. | 3/2/64 | 876 | | Blue serge .. | 5 | Detached | I. A. 40-pr. Rifle Gun, 5 lbs. | |
| 20-pr., Mark I.* .. .. | 3/2/64 | 876 | | Blue serge .. | 5 | Inside .. | I. A. 20-pr. Rifle Gun, 2 lbs. 8 oz. | |
| „ II. .. .. | 14/3/67 | 1384 | | Leather .. | 5 | Inside .. | II. 20-pr. B.L., 2 lbs. 8 oz. | |
| 12-pr., Mark I.* .. .. | 3/2/64 | 876 | | Blue serge .. | 4 | Inside .. | I. A. 12-pr. Rifle Gun, 1 lb. 8 oz. | |
| „ „ II. .. .. | 14/3/67 | 1384 | | Leather .. | 4 | Inside .. | II. 12-pr. B.L., 1 lb. 8 oz. | |
| 9-pr., Mark I.* .. .. | 3/2/64 | 876 | | Blue serge .. | 3 | Inside .. | I. A. 9-pr. Rifle Gun, 1 lb. 2 oz. | |
| „ „ II. .. .. | 14/3/67 | 1384 | | Leather .. | 3 | Inside .. | II. 9-pr. B.L , 1 lb. 2 oz. | |
| 6-pr., Mark I.* .. .. | 3/2/64 | 876 | | Blue serge .. | 3 | Inside .. | I. A. 6-pr. Rifle Gun, 12 oz. | |
| „ „ II. .. .. | 14/3/67 | 1384 | | Leather .. | 3 | Inside .. | II. 6-pr. B.L., 12 oz. | |

* Cartridges similar to these, except that the diameter was larger (which caused inconvenience), were approved in May, 1863, § 761, but no patterns of them now exist.

Drill cartridges. § 1384.

*Drill Cartridges* are made of a wood cylinder covered with felt and placed in a leather case, the base of the cartridge is shod with copper, and they have dummy lubricators, the detached lubricators having gun-metal stalks screwing into gun-metal sockets in the cartridges. The cartridge is marked in black, same as the service cartridge, viz., with the nature of gun and weight of charge.

Issue.

Loose, in numbers as demanded.

### Gauges for B.L. Rifle Cartridges.

Gauges are of two sorts, viz.:—

1st, Brass Ring for diameter.
2nd, Wood Sliding for length.

Gauges filled cartridges brass, ring, rifled gun B.L.

(1.) Brass ring gauges for diameter,—

High gauges only are necessary. They consist of rings of gun-metal with straight handles; they are marked on and near the handle with the designation and numeral, also the diameter, and the words "FILLED CARTRIDGE."

| Designation and Numeral. | Date of Approval. | § of Changes in War Stores, | Diameter. |
|---|---|---|---|
| 7″ B.L., II.* .. .. .. .. | 1/9/68 | 1695 | 7·05 |
| 40-pr. B.L., II.* .. .. .. .. | | | 4·84 |
| 20-pr. B.L., II.* .. .. .. .. | | | 3·84 |
| 12 or 9-pr. B.L., II.* .. .. .. | | | 3·07 |
| 6-pr. B.L., II.* .. .. .. .. | | | 2·57 |

*Issue.*

Issue.

Gauges are issued loose, in numbers according to demand.

(2.) Wood sliding for length,—

Gauges, filled, cartridges, wood length, rifled B.L. § 2849.

These have not been sealed for any calibre except 7″, but unsealed patterns exist for all; they are used in making up cartridges; each one consists of an open frame with a cross piece, with certain play (for which *vide* below) allowed as limit of error. The cartridge is passed beneath to test its length.

| | | |
|---|---|---|
| 7″ (for 11, or 10 lbs. cartridge) - | Limits | 10″ to 11″. |
| 40-pr., 5 lbs. L.S. (formerly for 64-pr., 8 lbs.) | ,, | 12″·25 to 12″·75. |
| 40-pr., 5 lbs. S.S., and 20-pr., 2 lbs. 8 oz. - | ,, | 10″ to 10″·75. |
| 12-pr., 1 lb. 8 oz. - - - - | ,, | 8″ to 8″·5. |
| 9-pr., 1 lb. 2 oz. - - - - | ,, | 6″ to 6″·5. |
| 6-pr., 12 oz. - - - - | ,, | 6″·25 to 6″·5. |

*Issue.*

Issue.

Loose, in numbers according to demand.

* Gauge I. is destroyed

# CHAPTER XII. — AMMUNITION FOR R.M.L. GUNS OF 7″ AND UPWARDS.

INTRODUCTORY REMARKS—COMMON, DOUBLE, AND SHRAPNEL SHELL.—PALLISER SHELL AND SHOT.—CASE SHOT.—GAUGES.—CARTRIDGES, AND STORES CONNECTED WITH THEM.

## THE WOOLWICH SYSTEM.

Woolwich system.

THE Woolwich system, so called, is nearly identical with one adopted by the French.* It owes its designation to the fact that some name of general signification was thought desirable, and "Woolwich" was chosen on the precedent of the Enfield rifle, which was called after its place of manufacture, and was not connected with the name of any individual.

Increasing and uniform spiral.

The system embraces both uniform and increasing twists; in both cases the rifling is given by studs moving in three or more grooves.

The increasing twist is preferred † chiefly because the strain on the

* The 7-inch experimental gun, rifled on the French system, tried in competition with guns with Commander Scott's, Messrs. Jeffrey and Britten, and Mr. Lancaster's rifling, had three grooves ·25″ deep, 2″·09 wide, with a twist increasing from zero up to one turn in 259″, or 37 calibres.—Extracts, Vol. II, pp. 289, 290.

† The Ordnance Select Committee say—

"The supposed advantage of the increasing spiral is that the projectile having at the moment of leaving its seat only to move directly forward gets away from the powder charged more readily than when, as in a gun with a uniform spiral, it is forced simultaneously into a rotatory motion; somewhat more of the powder is therefore consumed before the shot moves in the latter than in the former case, and therefore the total force acting on the projectile during its passage through the bore is somewhat greater, and hence a somewhat greater velocity at the muzzle. . . . .

"The blow struck by the one shot on leaving the muzzle would be to that struck by the other as 100 to 103.

"*Recoil.*—It was found that these guns would run off the ordinary platforms if fired without compressors; it was therefore impracticable to compare their recoils, because any measure of actual distance run back when compressors are used is only a measure of the goodness of the compressor or the strength with which it is hove up. . . . .

"*Easiness of loading.*—Both guns were loaded with ease, and there was no difference between them in this respect.

"*Condition of vent.*—The gutta-percha impressions show a small but decided difference between the wear of the two vents. That of the gun with uniform spiral is most enlarged.

"*Cost.*—There is no difference in the cost of rifling these guns, but there is 8 oz. less metal in the studs for the gun with increasing spiral as now made, and each of its shot will cost 10*d.* less than those for the other gun. It is possible that the leading stud of the gun with uniform spiral might bear an equal reduction without affecting the shooting.

"*General results.*—This difference of cost is too small to be worth consideration if any other reason determined the choice, and as the guns are practically equal in range and accuracy, the question is narrowed to that of whether it is worth while to

gun and consequent wear* is less; it also has a slight advantage in accuracy of fire, although a rather lower initial velocity. In accounting for these effects with the increasing twist, it may be well to consider very briefly the entire question.

The object is to get the maximum velocity that can be given to the shot without imposing an intolerable strain on the gun. Now, as noticed before, with large calibres the main difficulty is to keep this strain, and consequent wear and tear, on the gun under control; that is, to make the piece last a reasonable time. Hence it is desirable to attain this end, even at the sacrifice of powder, and this is precisely what is effected by the adoption of the increasing twist, instead of the uniform; for on the latter system the projectile can only move forward after rotation has been imparted to it by the indirect resistance of the driving edges of the grooves against the studs; in fact, after resistance to rotation as well as inertia has been overcome, while with an increasing twist the projectile moves forward as soon as its inertia is overcome, and the rotation is imparted gradually during its passage through the bore.

Capt. A. Noble has demonstrated that the substitution of increasing for uniform twist in the Woolwich guns decreases the maximum pressure on the studs by about one-half, a matter of much importance. For instance in a 10″ Woolwich gun, rifled with a uniform twist of 1 in 40 calibres, the total pressure on the studs varies from 68·5 tons, when the shot has just started, to 9·1 tons when the shot is leaving the gun. In the case of the same gun rifled with a twist increasing from 1 in 100 to 1 in 40 calibres the total pressure varies from 31·2 tons at starting to 36·3 tons on leaving the bore.†

To compare the two systems it is advisable to take the case of projectiles leaving both guns with the same velocity, when, it stands to reason, both shot having the same "work done" on them, both guns must also have the same "work done" on them, and the entire value

---

sacrifice 3 per cent. of the *vis viva* at the muzzle for the sake of a small relief to the powder chamber. Sir W. Armstrong has always insisted much upon the necessity for relieving the breech of the guns in heavy natures, but he attains his end by distributing the powder charge, and as the Committee are persuaded that the theoretical reasons in favour of an increasing twist are sound ones, and in some measure confirmed by the practice now reported, although it has not evinced a superiority in accuracy, they adhere to the opinion expressed in their former report.

"Colonel Younghusband records his dissent from the opinion of the majority of the Committee, and does not consider the adoption of an increasing spiral as warranted by the present comparison."

*N.B.*—The mean velocities were 1,440·7 feet and 1,465 feet.

* "It will be seen that in point of accuracy of fire, taken on a general average, the increasing spiral has the advantage, the uniform spiral gun has a higher initial velocity and consequent length of range, but the difference is not sufficient to be of any practical importance. Being still persuaded that the theoretical reasons in favour of an increasing twist are sound ones, that a system of rifling which admits of the projectile moving directly forward from its seat at the moment of ignition of the charge must be more favourable to endurance than one which, by impeding the first movement of the shot in the bore, narrows the space for the expanding gas, and consequently brings a greater pressure on the breech of the gun, the Committee as a body have no hesitation in recommending that it be adopted for 8-inch guns of 9 tons, and combined with four grooves of the Woolwich form, viz., width, 1″·5; depth, ·18″. Mean initial velocities, increasing 1303·3 feet a second. Uniform, 1338·6 feet."—Extracts, Vol. IV, p. 253, 2/7/66. *Also*, opinion repeated, Vol. III, p. 325, and Vol. V, p. 18.

† For a detailed discussion on this point and explanations of the formulæ used, see "Treatise on the Construction of Ordnance, 1877," pp. 43 *et seq.* The short notice given above is extracted therefrom.

of the increasing system depends on the shape in which this work is performed. Now it appears that "work" may be done on the particles of the mass among each other, or equally on the whole as a rigid body; thus, when the force acts too quickly to develop any recoil, it is obvious that it has acted on particles among themselves, which may be supposed to be vibrating from elasticity, or else permanently crushed, and this may be considered generally to be the nature of destructive effect; but when a force acts more gradually on a gun, the whole mass recoils more rigidly, and less destructive effect takes place; in short, the "work done" on a gun is shown by recoil and destructive effect; and the more gradually the work is performed, in plain words, the more it becomes a push rather than a blow, the more recoil and the less destructive effect is produced.

This, then, is the nature generally of the work developed by an increasing as compared with a uniform twist, and it is reasonable to conclude that where comparison was made in the cases of two projectiles of equal initial velocity, the gun with the increasing twist would have the greater recoil.

This method of performing the work gradually, by increasing spiral, however, necessitates the expenditure of more powder, because, the shot moving more readily, less gas is generated before it commences to move, and since the pressure of this gas is in inverse proportion to the space it occupies, it acts with less and less power as the space behind the shot gets larger and larger, and in addition to this, the shot being already in motion, the pressure becomes dependent on the difference between the velocity of the gas itself and that of the projectile in front of it.

**Uniformly increasing spiral.**

Every increasing spiral in the service is what is termed "uniformly gaining"; that is, suppose the bore to be cut across into any number of equal lengths, each successive one as the shot passes through it gives the same addition to the twist.

Before entering into details as to the studs and grooves, it may be remarked that the work they perform may be limited to imparting rotation to the projectile, or it may include giving it direction also, according to whether there is more windage over the top of the stud in the groove, or over the body of the projectile in the bore. In the former case the shot lies in the bore, and is spun by its studs; in the latter it lies and moves through the bore on its studs, whose faces are the medium through which direction is given, while their driving edges impart rotation.

**Action of studs.**

In all Woolwich guns both the direction and twist are given by the bearing of the studs on the grooves, the body of the shot never being intended to come into contact with the bore, so that, in fact, the action of the latter is merely to confine the gas and keep it behind the projectile, which is made with a windage of ·08" in all calibres.

Before entering on the details of the ammunition, it is necessary to point out the meaning of a few of the terms used in connection with studded projectiles.

**Windage.**

It is important to clearly distinguish between "windage" and "clearance."

*Windage* is the difference between the diameters of the body of the projectile and the diameter of the bore taken between the "lands"; while, as is shown at more length below, *clearance* is the difference between the height of a stud above the body of a projectile, and the depth of the groove into which that stud fits.

**Clearance.**

*Clearance.*—If the depth of the groove is less than the projection of the stud beyond the body of the shot there will be "clearance," that is, there will be an interval between the body of the shot and the bore of the gun, and consequently the shot will rest entirely on the studs, the amount of clearance will be the difference between the depth of groove and the projection of the stud, thus in some of the Woolwich guns the depth of groove is ·18″, the projection of stud is ·195″, therefore there is a clearance of ·015″, this is the mean clearance, and is liable to be reduced by manufacturing limits. With iron guns it is not generally considered of much importance to have clearance, but where bronze is used it becomes essential, as the iron projectile would injure the softer metal of the gun. If, on the other hand, the depth of the grooves was greater than the projection of the stud, the metal of the projectile would be in contact with the bore of the gun; in this latter case the studs only serve to twist the shot, whereas in the former they give it both direction and twist.

Centring.

The projectile in loading bears on one side of the stud, which is called the "loading edge"; on firing, it bears against the other side, which is called the "driving edge"; * if the sides of the studs be perpendicular to the projectile, or to speak more correctly, normal to its surface, and the groove be of a corresponding shape, the stud will remain at the bottom of the groove, but if the side of the stud be made an inclined plane, and the side of the groove corresponding in shape, the stud on being pressed against the groove will have a tendency to run up the inclined plane, and by so doing will centre the shot in the bore, that is, will bring the axis of the shot to correspond with the axis of the gun; the windage will in this case be evenly distributed all round the shot, and the shot will not be in contact with the bore, hence with bronze guns centring is desirable, as even with a groove of a depth equal to a height of the stud there would be no contact.

Studs.

In the Woolwich system the rotation is given by studs made of an alloy of copper and tin, 10 parts of copper to 1 of tin; they are secured to the projectiles by being pressed into undercut holes.†

The depth of the groove is ·18″ in the later patterns of guns up to the 9″; in the larger guns, the depth is ·2″, the bottom of the groove is circular, but struck with a radius of 3″, so in no case is the bottom of the grooves concentric with the bottom of the gun (this entails the necessity of planing the face of the studs instead of the simpler operation of turning), the edges of the grooves are circular, struck with radius of from ·3″ to ·25″, according to the nature of the gun; the width of the groove is 1″·5, the studs are formed with faces and shoulders corresponding to the grooves, they project ·195″, up to the 9″, and ·215″ in the larger shells, beyond the body. The width of the rear stud is 1″·42, that of the front stud is limited by the amount of twist of rifling at the muzzle. For Palliser projectiles the figures differ slightly; for

* The driving edge of the studs of all service projectiles is the *left hand* side of the stud as seen when the projectile stands on its base. The right hand is of course the loading edge.

Extracts, Vol. IV, p. 389, Vol. V, p. 153. § 1980.

† Studs were formerly screwed into the shells, but pressing them is more economical, and though it entailed using softer metal, the shooting was found to be equally accurate. At first an alloy of seven of copper to one of tin was employed, but the softer alloy was found necessary in the 7-inch, where pressing in the stud occasionally split the shell; its use was extended to Palliser projectiles in 1870, and subsequently (8/3/72) was extended to all projectiles from 7-inch, inclusive, upwards. A pressure of 26·79 tons is required to press the stud into the 7″ shell, and the rear studs into 8″ shells and upwards, and 22·77 tons for the top stud of projectiles for 8″ guns and upwards.

these the mean heights of the studs are ·1925″ and ·2125″, giving a mean clearance of ·0125″. The mean windage over the body is ·075″.

From the above dimensions of grooves and studs it will be seen that there is a clearance of ·015″, practically there is no centring, the play between the side of stud and side of groove is ·08″, the windage is ·08″ over the body and ·05″ over the studs.

As to the number of studs, two to each groove seem to be necessary to carry out the above principles, and the fact that surprisingly good results have been obtained with only one, simply shows how little practically depends on the centring principle,* and on the decrease of ·03″ of windage, where gun-metal bearings perform such sudden and violent work; for a shot with a single stud which is not opposite its centre of gravity must lie partly on the rear stud in the groove, and partly on the front portion of the body in the bore, the rear portion, having ·05″ windage and a gun-metal bearing on iron, and the front ·08″ windage and iron bearings. This then is its condition of rest, but when it commences to move the forces become complicated with shot of the service form, for the metal in the ogival head is placed so as to tell more on the balance of the shot, that is, on the position of its centre of gravity, than a similar quantity of metal in the base, while being placed nearer the axis of the piece it gives less resistance to rotation, hence the part where any force applied to spin the shot would meet with equal resistance on each side of it would not be opposite to the centre of gravity.

In point of fact, four centres may be taken in the longer axis of a projectile on different principles:—

1st. A point bisecting the longer axis, commonly called the centre of the figure.

2nd. The centre of gravity on which the shot would balance.

3rd. Such a point that a plane section passing through it at right angles to the longer axis would divide the metal into equal volumes or weights.

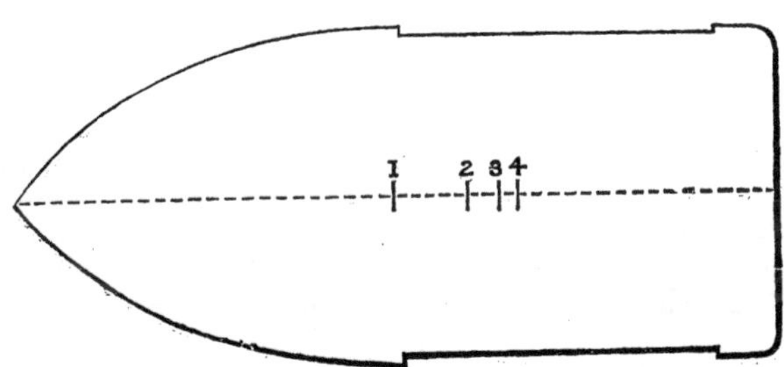


4th. Such a point that a plane section passing through it at right angles to the longer axis divides the projectile so that the "work done" in giving rotation is the same on each portion.

These points will probably be found in the above order, reading from point to base in service projectiles, and unless the base be made symmetrical with the head, they can never all coincide.

It has been said that practically projectiles shoot well with one ring of studs only, but a small number of rounds as yet have been fired, and

* Extracts, Vol. VII, pp. 141, 243.

it remains to be seen whether the bore suffers from this treatment. The advantage of firing projectiles with a single ring of studs would be great in some cases, for the projectile would be available for any pitch of groove. The front studs in all the service projectiles project the same amount as the rear ones, so as to keep the axis of the projectile parallel to, although, as has been noticed, not mathematically, coincident with, that of the bore; then as to rotation, with a uniform twist the front studs exactly resemble the rear ones, and both together drive the shot throughout its passage up the bore, but with the increasing spiral this is impossible,*

Front studs with increasing twist.

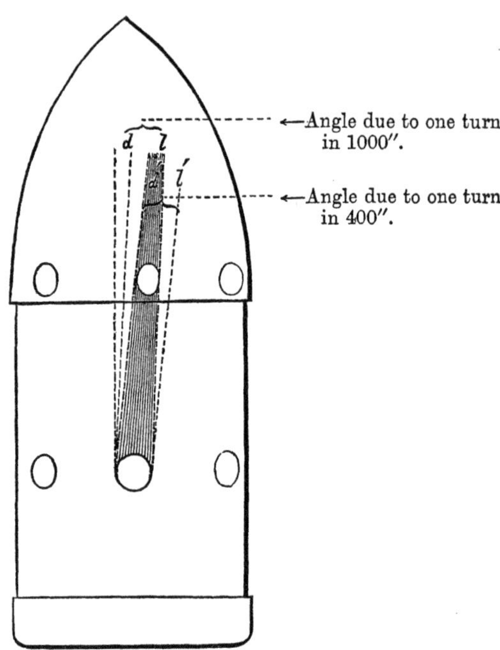


for the groove runs nearly longitudinally along the projectile at the bottom of the bore, while it crosses it at almost the angle due to the full rate of rotation before the front studs clear the grooves at the muzzle, hence it follows that there is only a very much narrower strip of the projectile that comes opposite to the groove in both its extreme positions, which are roughly shown in the annexed figure. The rear stud being of the dimensions given above, it will be seen that the front stud must be reduced to remain opposite to the groove in all positions, and it is therefore necessarily limited in its breadth to enable the shot to pass down the bore; for to enter the grooves *d′ l′* at the muzzle the front stud must not extend to the left of the edge *d′*, while to admit of being rammed home it must not extend to the right of the edge *l*. The exact breadth will of course vary with the circumstances of each case, and cannot therefore be laid down here.

Thus the front stud touches the driving edge *d′* on entering the bore, and the loading edge *l*, when well home; and the reverse action occur-

* In the French system of increasing twist the front stud was the full size and the rear one reduced, but this appears inferior to the opposite arrangement for more than one reason.

ring in firing the share it takes in the work of rotation is very small, for until the driving edge meets it, the whole pressure is on the rear stud; indeed it is only owing to the wear of the latter that the former comes in contact with the driving edge of the groove before it reaches the muzzle, and it has even been doubted whether it actually does so, but an inspection of the front stud* of recovered shot bears out the statement that a certain part of the work of rotation falls on it, although the chief use appears to be to steady the projectile.†

Two rings of studs have not been found sufficient for the Palliser projectiles of the 12-inch 35-ton gun, in some cases the front ring of studs was shorn off; this has been remedied by inserting a second ring midway between the front and rear ring. It has also been found advisable to add a third ring of studs to the common shell for those guns, as two were found insufficient when battering charges were used. The projectiles for the 12″·5 gun also have three rings of studs.

§ 2490.

§ 2557.

Longitudinal position of studs.

The longitudinal position of the studs has now been fixed as follows: The centre of the rear stud is to be not less than four inches from the base of all projectiles.‡

The following cuts show the way in which the required position and

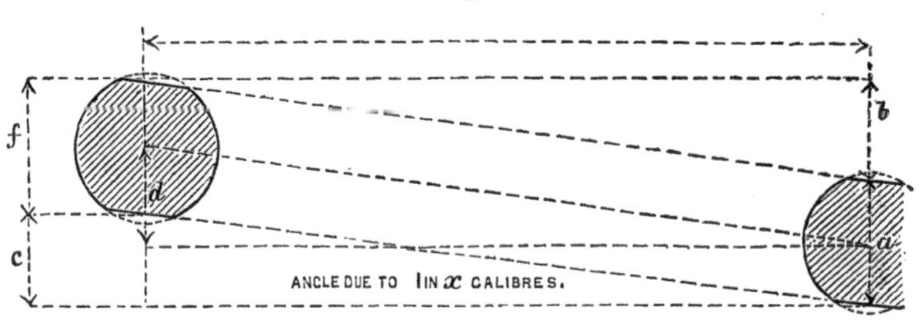


1st. For a uniform spiral of 1 turn in $x$ calibres

$$c=\frac{\pi e}{x} \quad d=c \quad f=a \quad b=c.$$

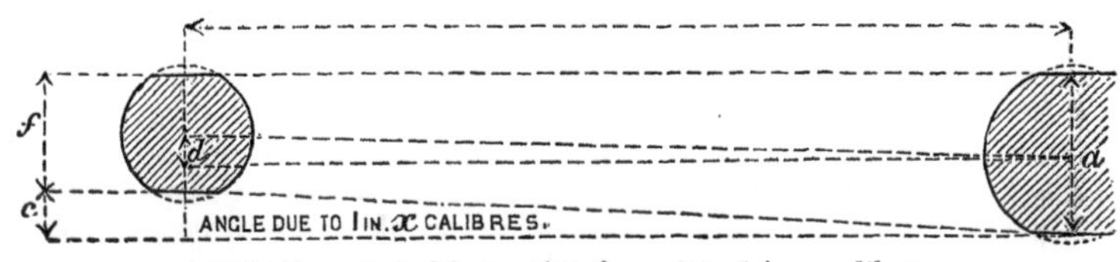


2nd. For a spiral increasing from 0 to 1 in $x$ calibres.

$$c=\frac{\pi e}{x} \quad d=\frac{c}{2} \quad f=a-c \quad b=o.$$

---

* The Superintendent (Shoeburyness) remarks that "on examination of the shell, after being recovered, the front stud was found in every instance to have been worn away on the driving side, showing that it had come into bearing." The same effect was seen even when the front stud was "very much reduced in size."—Extracts, Vol. III, p. 156, 5/6/65 and 7/6/65.

† On experiment with front studs still further reduced in size. "On this occasion the reduction in the size of the front studs was effectual, those on the recovered shell showing no symptoms of having touched the driving side of the groove. The practice is very nearly as good as that reported in June, and inclines the Committee to the conclusion that the use of those studs is to steady the shot in the bore and not to rotate them."—Extracts, Vol. III, p. 256, 24/7/65.

‡ Fixed by a letter from Ordnance Select Committee to Superintendent Royal Laboratory, dated 12/9/67. In the Palliser 7-in the distance is 3″·77.

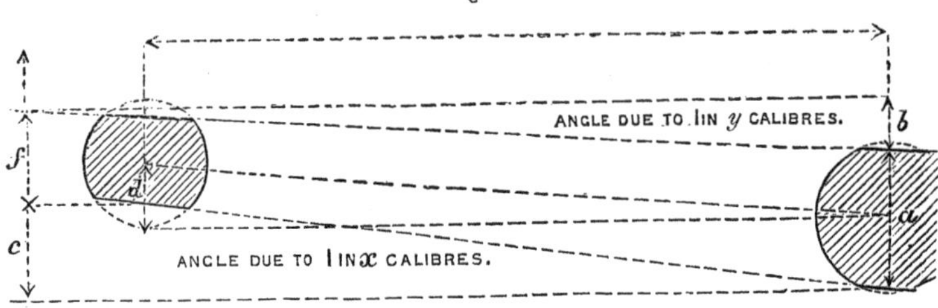


3rd. For a spiral increasing from 1 turn in $y$ calibres to 1 in $x$ calibres

$$c=\frac{\pi s}{x}\quad b=\frac{\pi s}{y}\quad d=\frac{b+c}{2}\quad f=(a+b)-c.$$

shape are given to the studs in the manufacture; in the cases of guns having respectively the three modes of rifling, viz.:—

1. Uniform twist. *E.g.* 7″ gun.
2. Twist increasing from 0 to 1 turn in $x$ calibres. *E.g.* 8″ gun.
3. Twist increasing from 1 turn in $y$ to 1 turn in $x$ calibres. *E.g.* 10″ gun.

The front and rear studs are to be put at equal distances in front and rear of the centre of gravity of the projectile, except in the case of Palliser projectiles, and are to be the same distance apart on all projectiles for the same gun.*

The rear stud has been moved forward to suit the Navy† who use a canvas bearer in loading, which does not bring the projectile up to the muzzle of the gun (like the land service bearer), with its studs opposite the grooves, but in an accidental position; it is therefore convenient to rest that part of the base which is clear of studs in the bore, and thus relieved of the weight of the projectile, to turn it round until its studs come into proper position.‡

The question arises of how this advanced position of the rear stud affects the shooting; this has of course been tested, and it has been found by practice that no harm results from it, indeed, as to direction, it stands to reason that the large stud in rear should be as near, if not nearer, to the centre of gravity than the front one, while as to rotation with the increasing twist it seems still a question whether the rear studs, which perform the main part of this work, would not still better effect this purpose if placed much more nearly opposite the centre of gravity.§

---

* In all the projectiles of the same calibre, the front and rear stud shall be placed (as far as practicable) at one uniform distance from each other. § 1518.

† A letter from Capt. Hood, R.N., H.M.S. "Excellent," 21/3/67, forwarded to the Ordnance Select Committee, says, "That the bearing actually necessary to sustain the weight of the shell in the muzzle when lodged for the purpose of turning it for the entry of the studs into the grooves is, in the case of the 7-inch gun, 1″·5, and in that of the 9-inch, 1″·75." He therefore considers that these distances are "the least that can be allowed from bottom of projectile to studs."—*Ordnance Select Committee, Min.* 21, 430, *subject No* 2424 ( *Ordnance Select Committee Proceedings*).

‡ It has been thought advantageous to lift the shot and enter its base in the bore and turn it to its required position in preference to hooking the shot bearer on the muzzle in serving these guns on land as well as on board ship.

§ There is an advantage in having the studs a considerable distance apart, as any unequal wear on the bearing surfaces will alter the position of the shot less than when the studs are nearer together.

Placing the studs at a uniform* distance apart on all projectiles of the same calibre simplifies gauging, which otherwise becomes complicated in the case of an increasing twist. This plan entails one gauge only to test the edges of the studs of each calibre (*vide* page 213). § 1518.

### *Length.*

**Length of projectiles.**

This necessarily varies in the different descriptions of projectiles for the same gun, inasmuch as it is to some extent subordinate to the consideration of bringing them all (with certain exceptions) to the same weight, but it has been decided that a length of two calibres† at least is necessary for very accurate shooting, and it is desirable for good "vis viva" or destructive effect on impact at any but very short ranges to have the weight great in proportion to the calibre, or in fact to the surface of resistance, and of course this is favoured by an increased length of projectile.

The question of hollow shot in place of solid connects itself with this, indeed a solid shot of two calibres long would in some cases exceed the desired weight for each gun, but besides the consideration of external form, the hollow shot, having its weight distributed further from its axis, has a slight advantage in having a longer radius of gyration and greater power of keeping up its rotation, though entailing a slightly increased strain on the gun. It will be seen, however, that this question only arises with reference to Palliser shot, case being the only other description of shot for Woolwich guns.

**Form of head.**

The form of head is governed by two considerations, flight and pene-

---

* As nearly as possible equidistant from centre of gravity, and at the same distance from one another in the same calibre so as to allow one gauge to be used for all.—*See Report* (*Ordnance Select Committee Proceedings*), 4,666, *Minute*, 22,679, date 30/7/67.

† Captain (now Admiral) Key, of H.M.S. "Excellent," in a letter, says, "He has fired both shell and shot from a 64-pr. muzzle-loading gun without detecting any difference between them as to accuracy or range; the short solid shot were very inferior to both."—Extracts, Vol. III, p. 35.

"He (Professor Bashworth) desires to call the Committee's attention to the superior accuracy of shell over shot in the trials with guns of various degrees of twist, from which it appears that the increased length of projectile tended to increased steadiness."—Extracts, Vol. III, p. 70.

"Experience has led the Committee to conclude that to attain very accurate shooting the projectiles of M.L. guns must not be less than two calibres in length. It has been observed that the shooting of common shell is almost invariably better than that of segment shell or common shot. This difference is attributed partly to their length and partly to the favourable conditions of the hollow form, by which the centre of gyration is thrown further from the axis of the projectile. The one condition points to elongating the shot, and the other to making it hollow, and inasmuch as cast-iron hollow shot of moderate thickness are strong enough for all battering purposes for which cast-iron projectiles are likely to be used, while they have the further advantage of being easily convertible into battering shells, the Committee propose the introduction of hollow instead of solid shot, and to make these shot of such weight as will give a length favourable to accurate shooting. This weight will also give a greater *vis viva* at a given distance than a lighter and consequently a shorter shot fired with a larger charge. There will therefore be economy in the stowage and quantity of gunpowder, combined with increased efficiency, in a system based on the foregoing principle."—Extracts, Vol. III, p. 128.

"Recent experience has led the Committee to conclude that to attain very accurate shooting the projectiles of muzzle-loading rifle guns must not be less than about two calibres in length, and inasmuch as a solid shot of a weight consistent with the safety of the gun will not fulfil this condition, it became necessary to look for a more favourable form."—Extracts, Vol. III, p. 128.

tration, and the latter, which gives different* forms in different instances will be discussed in detail. The question of flight affects all equally, and on this experiments† have been made which resulted in the adoption of what is termed an ogival head, struck with a radius or 1·5 diameters for common shell of all calibres.‡ §

---

* A letter from the Ordnance Select Committee, 13/11/65, to the Superintendent Royal Laboratory, gives forms for the heads of shells and shot for Woolwich guns nearly approaching those now in the service.

† Of the forms tested as to the resistance they meet with flight, viz., No. 1 nearly parabolical (service form), No. 2, hemispherical, No. 3, parabolical, focus $\frac{1}{10}$ diameter; 4, equilateral cone; 5, ogival, radius two diameters. The Ordnance Select Committee report "that the conoidal or ogival form is superior to either of the above, and the fronts of any new shot introduced into the service should approximate to this form as nearly as is compatible with suitability for the service in other respects."—Extracts, Vol. II, p. 305.

Much information on this subject will be found in the Reports on Experiments with the Bashforth Chronograph, 1865–1870, p. 10, 166, and also in Professor Bashforth's Treatise on the Motion of Projectiles. On p. 30 it is stated that the resistance of the air to the hemispheroidal and to the ogival heads (struck with radii of 1 and 2 diameters) varied so little that it was plain that any of these forms most serviceable in other respects might be adopted; the slight variations in the resistances lead to the conclusion that the amount of resistance offered by the air to the motion of elongated shot is little affected by the more or less pointed apex, but depends chiefly on the form of the head near its junction with the cylindrical body of the shot. The resistance to a hemispherical headed shot was much greater than to the above form. See also Tables of Velocity, Time of Flight and Energy, published by Spon, Charing Cross. On p. 12 it is stated that at a velocity of 1,100 f.s., the loss of velocity of a flat headed shot is to that of an ogival headed shot as 80 to 63.

‡ The construction of ogival heads of radii of 1, 1¼ and 1½ diameters respectively may be seen in the figs. 1, 2 and 3 below.

Fig. 1. Fig. 2. Fig. 3.

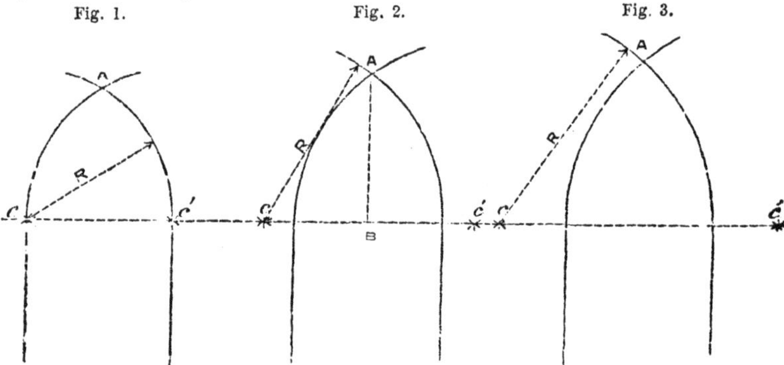


C and C′ being the centres, and R. the length, of the radii, in each case.

§ Captain Browne gives from integration the volumes of the three forms of head respectively, as—

1 — $V$ = ·395592 D³ (where D = diameter of projectile).
2 — $V$ = ·44765 D³.
3 — $V$ = ·49425 D³.

He also furnishes, with the assistance of Professor Bashforth, a general formula:—

$$R^2 = (x - a^2) + (y - b)^2.$$

$$\text{Vol.} = \pi \int_0^{\sqrt{(R^2 - b^2)}} (\sqrt{R^2 - x^2} - b)^2 \, dx.$$

$$= \pi \left\{ R^2 x - \frac{x^3}{3} - b\left( x\sqrt{R^2 - x^2} + R^2 \sin^{-1} \frac{x}{R}\right) + b^2 x \right\}$$

The origin is taken at B. See fig. 2.

In all three cases $a = o, b = R - \dfrac{D}{2}$

Form of base. § 1599. § 1765.

To facilitate loading at sea with the canvas bearer noticed above, the bottoms of the more recent patterns of all projectiles, case excepted, are rounded, all, except Shrapnel, with a circle of 1 inch radius; Shrapnel vary in this respect, as noticed in details hereafter given.

Before dealing individually with each projectile, it may be well to observe such distinctive features as necessarily spring from the rifling of the gun, and are therefore common to all the projectiles taking the grooves of each gun, in other words, all except case shot; it will be sufficient to notice these once for all, and will save vexatious repetition.

The characteristics then of each gun, to which their relative projectiles necessarily correspond, are as follows: * —

12″ gun and 12″·5 gun (35 and 38 ton) 9 grooves; twist; increasing from 0 to 1 in 35 calibres at muzzle.
12″ gun, 25 ton, 9 grooves; twist; increasing from 1 in 100 to 1 in 50 calibres.
11″ gun, 9 grooves, twist; increasing from 0 to 1 in 35 calibres at muzzle.
10″ gun, 7 grooves; twist; increasing from 1 in 100 to 1 in 40 calibres at muzzle.
9″ gun, 6 grooves; twist; increasing from 0 to 1 in 45 calibres at muzzle.
8″ gun,† 4 grooves; twist; increasing from 0 to 1 in 40 calibres at muzzle.
7″ gun, 3 grooves; twist; uniform 1 in 35 calibres.

The 7″ gun has a uniform twist, because the Admiralty at the time of its introduction preferred the uniform to the increasing spiral.‡

The following properties are common to the projectiles of each calibre, and may be assumed where the reverse is not stated, and as they vary with the calibre, it would cause confusing and vexatious interruptions to introduce them all under the head of each description of projectile:—

The various projectiles for each calibre of gun are generally brought to the same weight (except case shot and double shell.) It has not, however, been found practicable to do this in all cases. The 12-inch 25-ton gun has an exceptionally slow twist, therefore a long projectile cannot be fired from it, hence the weights of the common and Shrapnel shell are much less than those of the Palliser projectile.

The 12-inch 35-ton gun has a twist as rapid as the majority of Woolwich guns, but even with this twist it would be impossible to lengthen the common shell or Shrapnel, so as to bring them up to the weight of the Palliser projectiles, which is exceptionally heavy for a gun of that calibre. By reference to the table, p. 190, it will be seen that the common shell is nearly three calibres in length, beyond which it is impossible to go without injuring the shooting qualities of the

---

The special formula for the service head of 1½ diameters is obtained by substitution in the above general formula, and becomes $\left(R = \frac{3D}{2}\right)$

$$\text{Vol} = \pi\left\{\frac{9}{4}D^2x - \frac{x^3}{3} - 2D\left(\frac{x}{2}\sqrt{\frac{9}{4}D^2 - x^2} + \frac{9}{8}D^2\sin^{-1}\frac{2x}{3D}\right) + D^2x\right\}$$

* 13-inch gun, now obsolete, 10 grooves; twist; uniform; 1 in 55 calibres. Only four made, and only two of these remain serviceable.

† For S.S. only.

‡ O.S.C., Vol. III, p. 227, Minute 16,625.

shell.* The weights of the Palliser projectiles are given below, the weights of the other projectiles (except double shell and case shot), approximate closely to those of the Palliser projectiles, except for the guns of 12-inch calibre.

The weights are as follows †:—

| | | |
|---|---|---|
| 12·5 inch | .. .. .. | 800 lbs. |
| 12 ,, 35-ton | .. .. | 700 ,, |
| 12 ,, 25-ton | .. .. | 600 ,, |
| 11 ,, | .. .. .. | 530 ,, |
| 10 ,, | .. .. .. | 400 ,, |
| 9 ,, | .. .. .. | 250 ,, |
| 8 ,, | .. .. .. | 180 ,, |
| 7 ,, | .. .. .. | 115 ,, |

The constant distance between the front and rear stud adopted for all the projectiles of each calibre in the more recent patterns, as mentioned at page 183, is as follows :—

| | |
|---|---|
| For the 12″·5 gun .. .. | it is 7″ |
| ,, 12″ 35 and 25 ton guns | ,, 7″ |
| ,, 11″ .. .. .. | ,, 8″ |
| ,, 10″ .. .. .. | ,, 8″ |
| ,, 9″ .. .. .. | ,, 6″ |
| ,, 8″ .. .. .. | ,, 5″ |
| ,, 7″ .. .. .. | ,, 4″·6 |

Table of the distance of the centre of gravity from the base in Palliser projectiles :—

| Nature. | Shot. | Shell. |
|---|---|---|
| | inches. | inches. |
| 12″·5 gun .. .. | — | 14·35 |
| 12″ (35-ton) gun .. | — | 13·50 |
| 12″ (25-ton) ,, .. | 11·9 | 12·45 |
| 11″ gun .. .. | — | 12·30 |
| 10″ ,, .. .. | 11·3 | 11·40 |
| 9″ ,, .. .. | 8·9 | 9·20 |
| 8″ ,, .. .. | 8·0 | 8·30 |
| 7″ ,, .. .. | 6·9 | 7·10 |

The projectiles for each gun may be recognised by the calibre being cast on the base, the weight of gun is also given for the 12-inch pro-

---

* This applies to guns with such an amount of twist as is given to them in our service, the length of shell depends directly upon the amount of twist, and we may say generally that the sharper the twist the longer the projectile may be.

† The 13-inch projectiles approximated in weight to 600 lbs., they are ordered to be broken up.

jectiles. Shell cast prior to 1873 are not thus marked, but they may be known by the number of studs in each ring, except in the 11-inch, 12-inch, and 12·5 inch, which have the same number of studs. The number of studs in each ring is as follows :—

| | | |
|---|---|---|
| 12·5 inch .. .. | } | 9 |
| 12 „ 35-ton* | | |
| 12 „ 25 „ | | |
| 11 „ .. .. | | |
| 10 „ .. .. | .. | 7 |
| 9 „ .. .. | .. | 6 |
| 8 „ .. .. | .. | 4 |
| 7 „ .. .. | .. | 3 |

The projectiles now in the service for Woolwich guns are as follows :—

*a.* Shell (viz., common, double, Boxer Shrapnel, and Palliser).

*b.* Shot (viz., Palliser and case).†

N.B.—Besides these, steel shot and shell (Alderson's), although obsolete, might possibly be met with.‡

1. Shell.

1. Common.
2. Double.
3. Boxer Shrapnel.
4. Palliser's.

1. *Common Shell.* (*Calibres* 12″·5,§ 12″, 11″, 10″, 9″, 8″, and 7″, gauge G.S.)

Common shell. Calibres.

Fuzes for L.S. Pettman's G.S. percussion.
„ S.S.‖ „ „ „

See plate p. 357.

They are about three calibres long, except that for the 12-inch 25-ton gun, which is 2½ calibres long, for the reason given above. All the shells have two extractor holes in the head; their interior is lacquered like other shells. They are to be completely filled with powder, but bags are invariably to be used. For method of filling, see p. 63.

The capacity, dimensions, &c., of the various patterns will be found in the table, p. 190.

The thickness of the walls varies from about $\frac{1}{5}$ in the larger, to about $\frac{1}{6}$ of the diameter in the smaller shells. In the shell for the 12-inch 35-ton gun, the thickness of the walls increase slightly towards the base, the same plan is adopted in the shell for the 11-inch gun, and in Mark II. 10-inch shell, in the other shells the walls are of even thickness throughout.

---

* Any 12″ 35-ton projectiles issued before 3/73 will have "35-ton" stamped on the rear studs. § 2489.

† A 90-cwt. 7-inch gun has been introduced for S.S. The ammunition is the same as that for the heavier gun, except that it is not to be fired with Palliser projectiles and battering charges. This gun does not fire double shell.

‡ *Vide* §§ 1297, 1118. Steel shot and shell had heads painted white, that of the shell had a red tip, the head was unscrewed to enter the bursting charge. Steel heads latterly were stamped S.

Experiments are being carried on with a view to the possible introduction of steel shell of Palliser form.

§ Not yet sealed.

‖ The 9 and 20-seconds M.L. time fuzes may be used by the Navy when firing 14lbs. charges from the 7″ gun.

All shell of late manufacture have the bases rounded off to facilitate loading.

The gun-metal bush is countersunk about 0·2 inch in all, but in a few of the earlier patterns the countersink is only 0·1. The latter do perfectly well for L.S. but cannot be issued for S.S., which require a deep countersink to take the wad, see p. 72.

**Issue.** §§ 2370, 2413, 2421.

For S.S. loose.

1. Filled and fuzed with Pettman's G.S. fuze, over which is cemented in the wad, papier mâché fuze hole, Naval, with loop.
2. Empty with G.S. plug and wad, papier mâché, Naval, plain.

For L.S. empty, loose, with G.S. plug.

**Use.**

Would be used against material generally; they would be most destructive against wooden ships.

It will be remarked that only a percussion fuze, which acts on direct impact, is supplied. It is assumed in the L.S. equipment that the Woolwich guns will for the present only be found on sea fronts, therefore a fuze which will not act on striking water is necessary.*

*Experience.*

**Experience.**

The risk of prematures, their importance, and the method of preventing them which has been found effectual, has been fully given on p. 30.

The studs have stood well, and the shooting of the common shell has generally been good. In some cases the practice has not been very good from the 9-inch gun when using full charges,† possibly the rotation is not rapid enough with the smaller charge.

The 10-inch shell altered its form, when fired, rather more than the other calibres, and experiments carried on against a target representing the side of a wooden ship indicated that a stronger shell was desirable;‡ the walls were consequently thickened, and the capacity reduced. See table, pp. 190–191.

The effect produced by the pressure on the sides of a R.M.L. common shell from the gas rushing past is found to be most remarkable, the shell having a tendency to assume a form approaching that of a dumb-bell. The base being solid transversely, is not appreciably compressed, but the end of the body close to the commencement of the head, although receiving considerable support from the latter, is acted on to some extent, while the body about the middle, where it is weakest, is forced inwards, and decreased in diameter to an extent which would hardly be credited.§

The lower side of the shell in the gun appears to be convex, and the upper concave; in the 10-inch the concavity has been found to be as much at 0·1 inch, and in one instance the shell in front of the rear stud was reduced from its original diameter of 9·92 inch to 9·84 inch.||

The shell have occasionally marks of scoring showing that they have been in contact with the iron of the gun.

---

* Possibly there may be cases where it is desirable to burst these shells on graze, in bombarding a fortress the effect would be lost except where a shell struck some solid obstacle directly; in silencing guns it would probably be advantageous to burst shells when they graze.

† See Extracts, Vol. X, p. 199.

‡ See Extracts, Vol. X, p. 302.

§ For example, the diameter of a 9″ shell has been reduced from 8″·92 to 8″·55 about the middle.

|| For alterations of form of the 11-inch common shell, see Extracts, Vol. X, p. 185.

2. *Double Shell.* Calibre 7-inch, gauge G.S.

2. Double shell.

Fuzes for L.S. Pettman's G.S. percussion.
" S.S. " " , *

See plate p. 358.

It is a shell nearly four calibres long, strengthened by three longitudinal ribs internally, otherwise resembling the common shell.

A bag is used to contain the bursting charge as given for common shell.

For dimensions, capacity, &c., of the different patterns, see table, pp. 190–191.

Issue.

As given for common shell, p. 358.

The double shell is chiefly intended for use against wooden ships;† owing to its great length,‡ it is inaccurate at long ranges, but at 2,000 yards it has given good results.

*Bags, Serge, bursting charge R.M.L., Common shell, Mark I.*

Use. § 2493.

The following table shows the dimensions of the bags:—

| —— | Width. | Length to Neck. | Neck. | | | No. Packed in Bale. |
|---|---|---|---|---|---|---|
| | | | Length. | Width. | | |
| | | | | Top. | Bottom. | |
| | inches. | inches. | inches. | inches. | inches. | |
| 12′ 35-ton .. .. | 12·25 | 32·5 | 6·5 | 4·25 | 2 | 300 |
| 12′ 25-ton .. .. | 13 | 28·5 | 6·5 | 4·25 | 2 | 300 |
| 11″ .. .. .. | 10 | 31·25 | 6·5 | 4·25 | 2 | 300 |
| 10″ .. .. .. | 10·5 | 30 | 6·5 | 4·24 | 2 | 300 |
| 9″ .. .. .. | 10 | 25·75 | 4·75 | 3·5 | 2 | 400 |
| 8″ Gun or Howitzer .. | 8·5 | 23·5 | 4·75 | 3·5 | 2 | 500 |
| 7″ " .. .. .. | 7·5 | 20·5 | 4·75 | 3·5 | 2 | 500 |
| ′ double .. .. | 10 | 26 | 4·75 | 3·5 | 2 | 400 |

Issue.

In bales, same as cartridges, see table above.

* The 9-seconds M.L. fuze might be used by the Navy when firing 14lbs. charges.

† Admiral Key, R.N., reports in a letter, 22/2/66, on practice conducted against the "America," at 700 yards range; that "the destructive effect was unusually great, although in no instance did they set the ship on fire. . . . One burst about six feet inboard on the lower deck, entirely destroying one-half of a main deck beam with about 8 feet square of planking, and cutting severely into the planks and a beam of the lower deck."

"Another struck the chain cable on starboard quarter, by which the ship was slung, and burst before passing through, making a clean hole inside about 26″ square, and laying eight planks on the side open to the extent of 28 feet by 8 feet; the damage in this case was very close to the water line These shells have been fired over 700 yards with astonishing accuracy. . . . The real value of these shells has been more prominent in the late trials against the 'America,' where the effect of the large burster is so apparent."—Vide Extracts, Vol. IV, p. 28.

‡ The Ordnance Select Committee report, "although these shell roll considerably, owing to their great length and low velocity, yet their accuracy is satisfactory up to more than 2,000 yards, a range at which they would probably never be used."—Vide *Ordnance Select Committee Report*, 3,858, 4/8/65.

| Calibre, and Mark of Pattern. | Date of Approval. | § Changes in War Stores. | Length. ± $\frac{1}{10}$ in. per foot. | Diameters. | | Thickness of Metal. | | | Weight empty. |
|---|---|---|---|---|---|---|---|---|---|
| | | | | | | Walls. | | | |
| | | | | Body. | Studs. | Top. | Bottom. | Base. | |
| | | | ins. | ins. ± ·01 | ins. ± ·005 | ins. | ins. | ins. | lbs. oz. |
| 12″ 5, Mark ... ... | Not yet | sealed | 36·7 | 12·42 | 12·85 | 2·25 | 2·5 | 3·1 | 657 0 |
| 12″ 35-ton, Mark I. ... | 7/1/73 | 2419 | 34·45 | 11·92 | 12·35 | 2·2 | 2·4 | 3·0 | 575 0 |
| ″ ″ II. ... | 15/10/73 | 2557 2655 | 34·45 | 11·92 | 12·35 | 2·2 | 2·4 | 3·0 | 575 3 |
| 12″ 25-ton, Mark I. ... | 20/3/69 | 1766 2655 | 30·0 | 11·92 | 12·35 | 1·96 | 1·96 | 3·0 | 459 6 |
| 11″ Mark I. ... ... | 3/9/72 | 2378 2655 | 34·2 | 10·92 | 11·35 | 2·15 | 2·4 | 2·75 | 506 4 |
| 10″ Mark I. ... ... | 18/9/68 | 1678,1767, and 2356 | 32·5 | 9·92 | 10·35 | 1·85 | 1·85 | 2·5 | 373 12 |
| ″ ″ II. ... ... | 28/8/73 | 2524 2655 | 30·55 | 9·92 | 10·35 | 1·95 | 2·15 | 2·15 | 377 12 |
| 9″ Mark I. ... ... Afterwards altered to Mark II. | 19/5/66 | 1239 and 2356 | 26·6 | 8·92 | 9·31 | 1·5 | 1·5 | 2·25 | 232 0 |
| 9″ Mark III. ... ... | 26/11/66 | 1337 and 2356 | 26·6 | 8·92 | 9·31 | 1·5 | 1·5 | 2·25 | 232 0 |
| ″ ″ IV. ... ... | 21/10/67 | 1518 and 2356 | 26·6 | 8·92 | 9·31 | 1·5 | 1·5 | 2·25 | 232 0 |
| ″ ″ V. ... ... | 9/12/68 | 1765 and 2655 | 26·75 | 8·92 | 9·31 | 1·5 | 1·5 | 2·25 | 230 9 |
| 8″ Mark I. ... ... | 15/10/66 | 1338 and 2356 | 24·0 | 7·92 | 8·31 | 1·335 | 1·335 | 2·0 | 167 0 |
| ″ ″ II. ... ... | 21/10/67 | 1518 and 2356 | 24·0 | 7·92 | 8·31 | 1·335 | 1·335 | 2·0 | 167 0 |
| ″ ″ III. ... ... | 9/12/68 | 1765 and 2655 | 24·17 | 7·92 | 8·31 | 1·335 | 1·335 | 2·0 | 166 0 |
| 7″ Mark I. ... ... | 18/1/66 | 1188 and 2356 | 20·05 | 6·92 | 7·31 | 1·15 | 1·15 | 1·75 | 106 12 |
| ″ ″ II. ... ... | 19/5/66 | 1240 and 2356 | 20·05 | 6·92 | 7·31 | 1·15 | 1·15 | 1·75 | 106 12 |
| ″ ″ III. ... ... | 24/10/66 | 1341 and 2356 | 20·05 | 6·92 | 7·31 | 1·15 | 1·15 | 1·75 | 106 12 |
| ″ ″ IV. ... ... | 21/10/67 | 1518 and 2356 | 20·1 | 6·92 | 7·31 | 1·15 | 1·15 | 1·75 | 106 12 |
| ″ ″ V. ... ... | 9/12/68 | 1765 and 2655 | 20·4 | 6·92 | 7·31 | 1·15 | 1·15 | 1·75 | 106 14 |
| ″ Double, Mark I. ... | 9/10/66 | 1339 and 2356 | 27·2 | 6·92 | 7·31 | 1·0 | 1·0 | 2·0 | 146 12 |
| ″ ″ ″ II. ... | 21/10/67 | 1518 and 2356 | 27·2 | 6·92 | 7·31 | 1·0 | 1·0 | 2·0 | 146 12 |
| ″ ″ ″ III. ... | 9/12/68 | 1765 2655 | 27·2 | 6·92 | 7·31 | 1·0 | 1·0 | 2·0 | 145 6 |

N.B.—× indicates that the studs are formed to correspond to the curve of the groove instead of being con that they are of hard alloy, viz.: 7 of copper to 1 of tin. Soft alloy, viz.: 10 of copper to 1 of tin was approved for above order.

On 20/12/72 it was ordered that all common shell for the Woolwich guns should be marked on the base with the being marked on the stud as usual.

Unloading holes discontinued, 27/1/73, § 2426, without a change of pattern.

The weights given above will not be found in every case to strictly correspond with those given in "Changes." the high and low limits of manufacture. See also § 2655.

Some Mark IV. 9″ shell and Mark II. 7″ double shell were made with rounded base, and advance numeral

| Approximate bursting charge, Shell Powder, L.G. | Weight of Filled Shell. Limits of Error ±1·5 per cent. | Studs screwed in or swedged into Undercut Holes. | Distance between centres of Front and Rear Studs. | Edge of Bottom rounded. | Marks in Front of One Front Stud. | Marks on One Rear Stud. The date refers to the manufacture of each individual shell. | Remarks. |
|---|---|---|---|---|---|---|---|
| lbs. oz. | lbs. oz. | | ins. | | | | |
| 43 0 | 700 0 | Swedged | 7 | — | — | — | Three rings of studs. |
| 38 4 | 613 4 | ,, | 7 | Yes | R. I. ↑ L. | Date and × ... ("35 ton" on every alternate stud) | Two rings of studs. |
| 38 4 | 613 7 | ,, | 7 | ,, | R. II. ↑ L. | Date and × ("35 ton" on every alternate stud) | Three rings of studs. |
| 37 12 | 497 2 | ,, | 7 | ,, | R. I. ↑ L. | Date and × | |
| 29 12 | 536 0 | ,, | 8 | ,, | R. I. ↑ L. | Date and × | |
| 27 6 | 401 2 | ,, | 8 | ,, | R. I. ↑ L. | Date and × | |
| 20 4 | 398 0 | ,, | 8 | ,, | R. II. ↑ L. | Date and × ... | Thicker walls. |
| 20 0 | 252 0 | Screwed | 8 | No | R. II. ↑ L. | Nil ... ... ... | Groove round head for O.P. extractor countersink ·1″. |
| 20 0 | 252 0 | ,, | 8 | ,, | R. III. ↑ L. | Nil ... ... ... | Two extractor holes for N.P. extractor. |
| 20 0 | 252 0 | Swedged | 6 | ,, | R. IV ↑ L. | Date and × ... | Studs swedged in and placed at a constant distance apart. |
| 19 0 | 249 9 | ,, | 6 | Yes | R. V. ↑ L. | Date and × ... | Rounded base, unloading hole. |
| 14 9 | 181 9 | Screwed | 7 | No | R. I. ↑ L. | Nil. | |
| 14 9 | 181 9 | Swedged | 5 | ,, | R. II. ↑ L. | Date and × ... | Studs swedged in and placed at a constant distance apart. |
| 14 8 | 180 8 | ,, | 5 | Yes | R. III. ↑ L. | Date and × ... | Rounded base, unloading hole. |
| 9 4 | 116 0 | Screwed | 9 | No | R. I. ↑ L. | Nil ... ... ... | Three extractor holes, studs of copper 9 lbs., tin 1 lb., zinc 1 oz. This shell has the Moorsom bush, and requires a G.S. adapter. |
| 9 4 | 116 0 | ,, | 9 | ,, | R. II. ↑ L. | Nil ... ... ... | Three extractor holes, countersink ·1″, and G.S. bush. |
| 9 4 | 116 0 | ,, | 9 | ,, | R. III. ↑ L. | Nil ... ... ... | Two extractor holes and countersink ·2″. |
| 9 4 | 116 0 | Swedged | 4·6 | ,, | R. IV. ↑ L. | Date and × ... | Studs swedged in and placed at constant distance of 4″·6. |
| 8 12 | 115 10 | ,, | 4·6 | Yes | R. V. ↑ L. | Date and × ... | Rounded base and unloading hole. |
| 13 3 | 159 15 | Screwed | 12·0 | No | R. ↑ L. | | |
| 13 3 | 159 15 | Swedged | 4·6 | ,, | R. II. ↑ L. | Date and × ... | Studs swedged in and placed at constant distance of 4″·6. |
| 10 12 | 156 2 | ,, | 4·6 | Yes | R. III. ↑ L. | Date and × ... | Rounded base and unloading hole. |

centric with the projectiles as in previous patterns, and in all but 7″ projectiles previous to 8/3/72; it also indicates all natures on the above date. Soft studs are indicated by the date marked on them being subsequent to the

day and month of casting, and calibre; the 12″ being followed with "25 ton," or "35 ton," the date when finished.

Slight modifications have in some cases taken place, and the weights given above are those of the means between

stamped on a front stud. § 1776.

3. Shrapnel shell.

3. *Shrapnel Shell.*—Calibres: 12-inch, of 35 tons, and 12-inch of 25 tons; 11-inch, 10-inch, 9-inch, 8-inch, 7-inch. Gauge G.S.
Fuzes for L.S.—These shell not issued to L.S. See below.
" " S.S.—9 secs. M.L. time.*
See plate, page 359. For dimensions, &c., see table, p. 194.

These shells have been withdrawn from the L.S. equipment. The Woolwich guns, as before stated, are only mounted on sea fronts;† it is considered preferable to trust to common shell and Palliser projectiles for use against ships. They are still however, issued to the Navy, and would be available for L.S. if specially required.

Some of the early patterns gave indications of weakness, the 10-inch occasionally broke up,‡ so it was considered desirable to make them all stronger; this was done by increasing the thickness of the walls, the number of sand shot was diminished, and the bursting charge was increased,

Construction.

The construction generally resembles that of the B.L. Shrapnel given on page 156, but differs in several details.§

The body of the shell is nearly as thick as that of the common shell, which it resembles in studding and external dimensions.

The walls of the shell increase slightly in thickness from the top to the base,|| and are weakened by six longitudinal grooves or "lines of least resistance."

At the base the shell contracts, forming a chamber for the powder; a slightly recessed groove runs round the top of the shell near the exterior circumference to receive the Bessemer metal head.

A tin cup, coned at the top to facilitate unloading, fits the powder chamber; above this is a wrought iron disc, also coned to suit the cup, and screwed to receive a wrought iron pipe, lacquered internally, and 1½ inch in diameter, which occupies the centre of the shell. The fuze socket fits into the upper portion of this pipe; the diameter of the pipe is much larger than in the first patterns to facilitate loading and unloading.

The sides of the shell are lined with brown paper to prevent the rosin adhering too firmly to the iron.

The interior of the shell is filled with sand-shot,¶ secured in their place by rosin (which is poured in hot). Above the balls is placed a ring of felt or kamptulicon saturated with kit composition.

§ 2491.

The head is made of Bessemer metal, struck with a radius of one diameter and lined with wood, having a gun-metal socket soldered into the head, which serves to take the fuze. The Bessemer metal is bent down, as shown in the cut, so as to form a shoulder for the socket, which is made flush with the head of the shell (the projecting socket in former patterns being liable to be injured in travelling).

---

* Trial of 10″ and 12″ (25-ton Shrapnel), with 9-second fuzes, Mark I., bored to ·6; 25 rounds of each fired, no prematures. Committee report that these fuzes may be safely used when bored not shorter than ·6 = 900 yards. Under this range case shot may be used.—Extracts, XII, p. 60.

† It is thought that a boat attack could be repelled by case shot, which acts up to about 600 yards, while against ships the other shells would be more effective.

‡ See Extracts, Vol. IX, p. 127, 10-inch shell were fired with bursting charge and plugged, and broke up in the gun.

§ The construction given is that of the latest pattern.

|| The increase varies from ·2 inch in the larger calibres to ·1 inch in the smaller.

¶ Iron balls are used simply from economy; lead and antimony balls have been proved to be more effective.

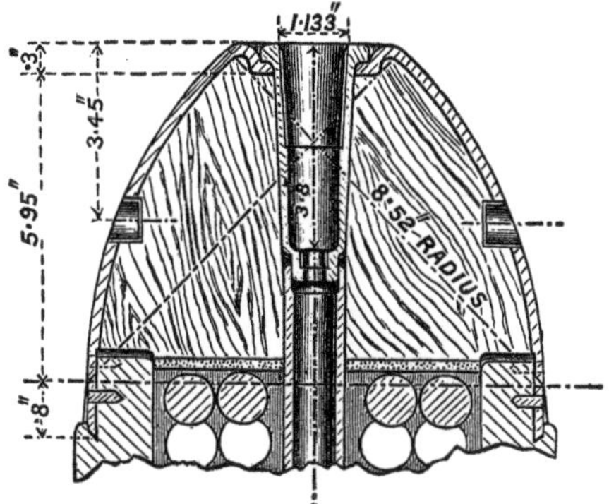


The socket is tapped at the base to receive the primer, see p. 133, and at the top to receive the G.S. plug.

The head is secured to the body by means of rivets and twisting-pins; the latter are only intended to prevent the head from twisting off in flight, and the holes in which they are inserted are close to the top of the cast-iron rim, so as to offer but little resistance in a longitudinal direction when the head is blown off by the action of the bursting charge. A band of solder prevents the pins and rivets from dropping out.

Holes are bored through the head to take small tin sockets, which are fitted into the wooden lining and serve as extractor holes.

**Issue.** §§ 2357, 2406, 2480.

Filled for S.S., with primer inserted and boxed; one per box.

If required for L.S. they would be issued, empty, loose.

**Use.**

Would be available against troops, and, owing to the great distance to which the shell keeps up its velocity, would probably be useful up to about 3,200 yards; this being the limit of range at which the 9-seconds fuze is available. The shell gives a good cone of dispersion if burst within 300 yards of the object; but should the troops be in close order it ought to be burst much nearer, and would be probably more effective if burst about 100 yards from the object.*

---

* Extracts from Proceedings of Ordnance Select Committee—Report of Armstrong and Whitworth Committee—Report of a Select Committee on Ordnance, House of Commons, 1862, &c., &c., relative to the merits of segment as compared with Shrapnel for field guns, published in a Blue Book for 1869, gives the results of the trials of a large number of heavy Shrapnel shell. On page 19 it is stated, with regard to the trial of 9″ and 7″ M.L., 7″ B.L., and 64-pr. R.M.L. Shrapnel, that "The results of this practice are most satisfactory and afford conclusive evidence of the formidable nature of Shrapnel shell. It is apparently most effective when burst within 100 yards of the target, and at about 10 feet above the plane; but its efficiency in the larger natures is still retained, even when the burst takes place at so great a distance as 300 yards short of the object, a condition which the Committee believe would not be realised by projectiles constructed on the segment principle."

"The pattern Shrapnel shell for 7″, 8″, and 9″ M.L. guns which have been provisionally approved to meet urgent supplies for land service, are composed of balls of mixed metal, viz., lead and antimony, and such was the construction of the Shrapnel tried by the Committee in February last. They have, however, since tried, at Colonel Boxer's request, some Shrapnel made up with iron balls. Although inferior in general effect to the shell with balls of mixed metal, they appear to be still efficient, but it should be observed that the total weight of these projectiles was in each case below that of the service projectile, and that in consequence of the substitution of iron for lead the number of balls in each shell was materially diminished."

On page 23 will be found some remarkably good practice. The shell were fired against three rows of targets, 20 yards apart, each row having a front of 9 feet by

## Shrapnel Shell, R.M.L. Woolwich Guns.

| Calibre, and Mark of Pattern. | Date of Approval. | § Changes in War Stores. | Length. ± $\frac{1}{10}$ in. per foot. | Diameters. | | Thickness of Metal. | | | Number and Nature of Balls contained. | Weight of Shell, empty. | Weighed Bursting Charge. Pistol F.G., or R.F.G. Powder. | Distance between Centres of Front and Rear Studs. | Bottom flat, or rounded at Edge, with a radius of | Remarks. |
|---|---|---|---|---|---|---|---|---|---|---|---|---|---|---|
| | | | | Body. | Studs. | Walls. Top. | Walls. Bottom. | Base. | | | | | | |
| | | | ins. | ins. ± ·01 | ins. ± ·005 | ins. | ins. | ins. | | lbs. oz. ± ·15 per cent. | lbs. oz. | ins. | ins. | |
| 12″ 35 ton, Mark I. | 25/7/73 | 2522 | 33·55 | 11·92 | 12·35 | 2·2 | 2·4 | 2·25 | 368, 4 oz., or 453, 3¼ oz. sand shot | 610 6 | 1 15 | 7 | 3·96 | N.B.—In future manufacture the base of Shrapnel shell, for guns of 25 tons and upwards, will have their bottoms rounded, similar to other shell, *i.e.*, with a radius of 1″. |
| 12″ 25 ton, Mark I. | 17/2/70 | 1873 | 30·75 | 11·92 | 12·35 | — | — | — | 530, 4 oz. sand shot ... | 495 0 | 1 2 | 7 | 3·96 | |
| ,, ,, ,, II. | 31/10/72 | 2381 | 29·75 | 11·92 | 12·35 | 2·04 | 2·24 | 2·25 | 276, 4 oz. sand shot ... | 495 0 | 1 15 | 7 | 3·96 | Thicker sides, gun metal socket, larger centre tube, thicker diaphragm, and diaphragm and tin cup coned. |
| ,, ,, ,, III. | 11/7/73 | 2523 | 29·05 | 11·92 | 12·35 | 2·04 | 2·24 | 2·25 | 276, 4 oz., or 339, 3¼ oz. sand shot | 494 13 | 1 15 | 7 | 3·96 | Flush socket. |
| 11″ Mark I. ... ... | 30/6/75 | 2780 | 31·8 | 10·92 | 11·35 | 1·78 | 2·24 | 2·25 | 378, 4 oz., or 465, 3¼ oz. sand shot | 529 10 | 1 12 | 8 | 3·5 | |
| 10″ Mark II.* ... | 28/3/72 | 2366 | 31·88 | 9·92 | 10·35 | 1·59 | 1·84 | 1·88 | 306, 4 oz. sand shot ... | 403 0 | 1 9 | 8 | 3·96 | Same improvements as 12″ 25 ton II. |
| ,, ,, III. ... | 11/7/73 | 2523 | 31·13 | 9·92 | 10·35 | 1·59 | 1·84 | 1·88 | 306, 4 oz., or 376, 3¼ oz. sand shot | 403 0 | 1 9 | 8 | 3·96 | ,, ,, III. |
| 9″ Mark I.† ... ... | 21/3/68 | 1609 | 26·25 | 8·92 | 9·31 | — | — | — | 564, 2 oz. sand shot ... | 250 0 | 0 12 | 6 | 3 | |
| ,, ,, II. ... ... | 28/3/72 | 2366 | 26·25 | 8·92 | 9·31 | 1·34 | 1·52 | 1·7 | 374 ,, ,, ... | 254 0 | 1 5 | 6 | 3 | ,, ,, II. |
| ,, ,, III. ... | 11/7/73 | 2491 | 25·35 | 8·92 | 9·31 | 1·34 | 1·52 | 1·7 | 374 ,, ,, ... | 254 0 | 1 5 | 6 | 3 | ,, ,, III. |
| 8″ Mark I.† ... ... | 21/3/68 | 1609 | 23·25 | 7·92 | 8·31 | — | — | — | 376 ,, ,, ... | 174 0 | 0 10 | 5 | 2·65 | |
| ,, ,, II. ... ... | 28/3/72 | 2366 | 23·23 | 7·92 | 8·31 | 1·11 | 1·27 | 1·5 | 300 ,, ,, ... | 180 0 | 1 0 | 5 | 2·65 | ,, ,, II. |
| ,, ,, III. ... | 11/7/73 | 2491 | 22·55 | 7·92 | 8·31 | 1·11 | 1·27 | 1·5 | 300 ,, ,, ... | 179 8 | 1 0 | 5 | 2·65 | ,, ,, III. |
| 7″ Mark II.‡ ... | 21/3/68 | 1609 | 19·72 | 6·92 | 7·31 | — | — | — | 227 ,, ,, ... | 112 0 | 0 8 | 4·6 | 2·0 | |
| ,, ,, III. ... | 28/3/72 | 2366 | 19·72 | 6·92 | 7·31 | ·9 | 1·04 | 1·25 | 192 ,, ,, ... | 117 4 | 0 12 | 4·6 | 1·0 | ,, ,, II. |
| ,, ,, IV. ... | 11/7/73 | 2491 | 19·0 | 6·92 | 7·31 | ·9 | 1·04 | 1·25 | 192 ,, ,, ... | 115 10 | 0 12 | 4·6 | 1·0 | ,, ,, III. |

* 10″ Mark I., described in § 1806, declared obsolete, § 2420.
† The patterns of 9″ and 8″ approved provisionally §§ 1388, 1389, never having governed any supplies for service were withdrawn, and those approved in § 1609 substituted as Mark I.
‡ 7″ Mark I. § 1390, with cast-iron diaphragm and no tin cup, is declared unserviceable and is broken up on return to Woolwich.
N.B.—The Marks on the rear stud, and the Marks in front of one front stud of Shrapnel are the same, and have the same signification as those in the common shell, table, p. 190.
The weights given above will not be found in every case to strictly correspond with those given in "Changes." Slight modifications have in some cases taken place, and the weights given above are those of the means between the high and low limits of manufacture. See also § 2655.

## Palliser Projectiles.

It seems desirable, before giving a description of the chilled projectiles now in the service, to enter on such an explanation of their character as will show the reasons for the qualities which they possess, but it does not seem necessary to give any lengthened account of their introduction, probably the following will suffice for most readers:—

*In 1863, Major Palliser proposed a projectile of iron cast in a metal chill to render it hard for the penetration of armour, providing for the evil effects of brittleness by the form of head he gave to it which was an "elongated point."

These projectiles, as fired in 1864, were chilled to a considerable depth, but not throughout;† they proved so successful that their manufacture was afterwards carried on in the Royal Laboratory, until Mr. Davidson, the manager of that department, so worked out the selection and trial of various samples of iron, and the method of conducting the manufacture in concert with Major Palliser, as to bring the projectiles to the state of efficiency in which they now exist in the service.

In those now manufactured the heads are cast in metal and the bodies in sand, the samples of iron being such that the heads are chilled white nearly to the centre, the bodies being an even mottle throughout.

In some degree the powers of these projectiles excel those of the original Palliser shot, but the principle advocated by the inventor remains, viz., that a hard iron, chilled white, is used to punch, its deficiency in tenacity being met by the form of head employed, which belongs to one of the classes proposed by Major Palliser in the first instance.

As is well known, the function of Palliser projectiles is to pierce armour plates. The successful attainment of this result depends on—

1. The metal used.
2. The mode of casting that metal.
3. The form of the projectile, or the arrangement of the metal.

*First*, as to the metal used. It is a particular kind of cast-iron made from carefully selected varieties, and generally speaking is a "white"

---

54 feet. The shells were not of the present pattern, but the following result is worth quoting, as showing the possible effect of a Shrapnel shell fired under favourable conditions. A 9-inch Shrapnel shell, weight 255 lbs., having 374 3 oz. balls and a bursting charge of 12 oz., was burst 136 yards short of the first target and 15·5 feet above the plane, the range to the first target being 1,200 yards. The total hits out of six rounds were 3,199, and of that number 3,038 went through the targets, the average number of hits per round being 533. This shows that many of the shot must have penetrated two targets, as the number of hits is considerably greater than the number of balls in the shell. By looking through the practice the enormous results obtained by a few rounds of Shrapnel shell, properly directed, will be seen. No doubt when the Woolwich guns are mounted on land fronts such shells will again become desirable for them.

Trials of 10″ Shrapnel containing 4 oz. and 3¼ oz. balls will be found in Extracts, Vol. XI, p. 156. A shell containing 376 3¼ oz. sand shot gave 646 hits on 3 rows of targets at a range of 1,050 yards; at 2,050 yards it gave 312 hits.

* For a more full account, *vide* a paper written by Captain V. D. Majendie—*Proceedings of Royal Artillery Institution*, Vol. V, Part 7, January, 1867.

† Captain Browne states:—I picked up fragments myself to investigate the truth of some remarks of Dr. Percy,‡ and tested them in a rough way for the chemical condition of the carbon and the iron, which went to prove the above, viz., that the interior was in a very different state from the exterior.

‡ Percy *Metallurgy*, "Iron and Steel," p. 815.

iron. Without going at all deeply into the subject, which would here be out of place, we will briefly explain what is meant by this term.

Pure iron is unknown as a commercial product. Even piano wire, which is the nearest approach to pure iron, contains minute quantities of other substances, notably of carbon. Wrought-iron, of which piano wire is an example, can hardly be melted under any circumstances, speaking in a commercial sense. It is produced, however, in the Bessemer process, but Dr. Percy in his work on iron and steel mentions his having at first doubted that it was actually obtained in that process.

When, however, carbon is present, iron can be readily melted, and combines with a portion of the carbon to form a carbide of iron, which, in the absence of any extended knowledge as to the exact proportions in which carbon combines with iron to form definite chemical compounds, we may call $Fe_xC_y$.* This non-malleable, fusible, form of iron we call cast-iron; it is distinguished from wrought-iron in that it is fusible in an ordinary blast furnace or cupola, and from both wrought-iron and steel in that it cannot be rolled or forged. The different varieties of cast-iron depend on the condition in which the carbon exists in the cooled product. If it is all, or very nearly all, chemically combined with the iron, the product is intensely hard, brittle, and of a silvery fracture, and comparatively hard to fuse; this variety is known as white cast-iron. If, on the other hand, the carbon is chiefly mechanically mixed in a free state with the iron in the form of graphite or "kish" we obtain a darker product, comparatively soft and easily fused, and known as *grey* cast-iron. A mixture of these two varieties results in an iron known as "mottled" iron from its appearance. These three classes of cast-iron shade into each other, and no hard and fast line can be drawn between them.

There is a curious indication of the character of the metal, even in a molten mass of cast-iron. Grey iron is known by a small breaking or cracking, running rapidly in the film which forms on the surface, the form being sometimes zigzag, in cold blast grey, or small stars or spider shaped cracks, as in hot blast grey. White iron may be recognized by the film breaking across in coarse broad cracks rather sluggishly compared with the grey iron.

Without stating dogmatically what is the exact action of a blast furnace, which is liable to vary with circumstances, and which is diffidently discussed, even by metallurgists or manufacturers, it may be generally said, in spite of the fact that some samples of iron will be white and others grey, however they may be cooled, that any tendency of iron and carbon to separate is encouraged by slow cooling, and thus certain irons, or mixtures of irons, which would be mottled if slowly cooled, will be white if chilled or suddenly cooled. This is the kind of iron used in the manufacture of Palliser projectiles.

*Secondly*, as to the mode of casting this metal. The projectiles are cast head downwards so as to ensure density and soundness there. The head is cast in an iron "chill" or mould in which, by virtue of the conducting power of the metal, the molten mass is rapidly solidified, and all, or nearly all the carbon remains chemically combined with the iron, so

---

* Of course the actual quantity of either free or combined carbon in any given sample of iron can always be found by analysis, but the exact proportions of carbon requisite to form definite chemical combinations with iron are not well known. "Iron will take up nearly 6 per cent. of carbon, and this corresponds nearly to the formula $Fe_4C$. The result is a white, brilliant, brittle compound. Bloxam's Chemistry, page 306.

that the head is composed of white iron. The high conductivity of metals as compared with other substances is so well known that it is needless to dwell upon it here.

The effect therefore of a larger mass of metal such as a chill on molten iron, may at once be seen to be very great as compared with the comparatively non-conducting sand used in ordinary moulds, nor is this question greatly affected by the metal being warmed up to 80 or 100 degrees temperature, which is nothing compared to the heat of molten iron, in fact, metal is a rather better conductor when warm than when quite cold, and so chills rather more efficiently and immediately.

The iron thus rendered white possesses generally the qualities of white iron, that is, intense hardness and crushing strength and considerable brittleness, but it is rather denser,* and appears much finer and more silky in its fracture.

Now this being the opposite extreme to annealing, which renders metal uniform and even throughout its mass, it is not surprising that its particles should be in an unnatural and constrained condition, and that on slight provocation such molecular action should take place as would cause the projectile to split, and this has actually occurred under certain circumstances (*vide* page 198), and the chilling effect in manufacture is carefully carried out with a view to meet this in ways which need not here be noticed.†

The Palliser projectiles then possess the following properties in an eminent degree :—

1 Intense hardness.
2. Crushing strength.
3. Brittleness or want of tenacity.
4. Increased density.

Hardness prevents the point from flattening in penetration, so that while the points even of steel shot are commonly found flattened after impact, those of chilled shot, however broken, retain their sharp form.‡

Crushing strength may appear at first to be very much the same as hardness, but while the latter relates to the rigidity of actual particles, the former applies to the rigid connection or building up of particles so as to resist their being forced in upon each other.

To compare glass and iron, the former has greater hardness, and the latter crushing strength, thus the former will scratch the latter but the latter will crush the former.

It seems probable that crushing strength is a very important element in Palliser projectiles, and it may be observed that in punching ordinary sheet iron with a steel punch, when sufficient force is applied, either the punch must crush or the part it presses must be torn out and separated from the rest of the sheet, thus the two opposing forces are

---

* Captain Browne states :—" Some limited experiments I made with the assistance of the Chemical Department as to this, gave an increase of density to mottled metal on casting in chills of about 3½ per cent., and large as this is, I have reason to think it is likely to be correct."

† Details as to the manufacture of Palliser projectiles, &c., are given at length in a paper by Major Barlow, R.A., in R.A.I. Proc., Vol. IX, No. 3.

‡ There is in the R.L. pattern room a portion of the head of a Palliser shell fired from the 12″·5 gun of 38 tons. It penetrated three 6½-inch iron plates, and the wood packing between them. The *point* is as sharp and unbroken as the day it left the foundry.

the crushing strength of the punch and the tensile strength of the sheet iron.

It seems probable that this reasoning may hold good to some extent in the question of piercing armour plates, although it must not be applied without reserve to such a sudden performance of work as must here take place.

The form of head of Palliser projectiles is admirably suited to enable such metal to do its work in penetration when its surface is being pressed from front and sides towards one centre.

The fact that it actually resists crushing seems to be established by the circumstance that the fragments which are picked up are *remarkably cool* as compared with pieces of plate or of ordinary cast iron-shot, for it is well known that material crushed or compressed becomes hot in the operation.

As to brittleness, it can hardly affect the front portion of the shot, pressed in, as noticed above, towards its centre; it is hard to say how much of its momentum the rear portion lends to the blow before it shivers away from it, but it is obvious that it is better here to have the iron in its rather less brittle, that is its mottled condition.

At the same time, even in the body of the projectile, it is necessary to have a hard unyielding metal, though not quite so hard as, and more tenacious than, the head. If the body was of soft metal while the head was hard, there would be considerable loss of power on impact, owing to the setting up of the soft metal. The reason of this is as follows: A projectile from a gun has a certain amount of energy stored up in it. This energy may be converted into mechanical work, or into heat. Now, when a projectile strikes an iron plate, we wish as much of the energy as possible to be devoted to penetration. What is used up in the setting up or distortion of the projectile itself is so much dead loss; hence we try to produce a projectile that will change its form as little as possible.

We have said above, that the body, though hard, need not be so very hard as, but requires more tenacity than, the head. This point is gained by casting the bodies in sand. Sand being a bad conductor, the bodies have time to cool gradually, and the carbon in the iron has time to partly separate in a free condition.

The result is mottled iron. Hence, if we examine a Palliser shell fractured longitudinally, we find the head and part of the shoulder to be composed of white iron, while the body is more or less mottled, though all the iron came out of the same ladle at the same time. Previously, however, to March, 1870, the whole of the projectiles were chilled throughout. Besides the advantage of greater tenacity gained in the body by the present system of casting, another most important advantage is gained. The projectile is rendered far less subject to the action of molecular forces, which, as noted above, may either split in store or crack it so as to cause rupture in the bore of the gun.

*Thirdly*, as to the arrangement of the metal. As noticed above, the form of the head is arranged so as to get the maximum penetrating effects. The ogival of $1\frac{1}{2}$ diameters radius has been selected as the best form for this purpose (see p. 183).

Major Palliser has very naturally tried to get the metal forward as far as possible, so as to impress its momentum on the plate without acting through the medium of the sides or walls of the shell which must be made thicker as the base of the shell becomes heavier, and thus diminish the interior capacity and consequently the bursting charge.

Some premature explosions of these shells in the bores of guns have been attributed to the base being made so light as not to support the bush of the filling hole under the pressure of the firing charge of the guns, and hence a slight increase in metal was made in more recent patterns.

It will be seen that the shells at first made for certain calibres (9″ and 8″) had thicker walls and less capacity for powder than those since manufactured, the momentum of both shell being equal, the thinner shell being slightly increased in length to make it up in weight;* the relative advantages are increased explosive force in favour of the large capacity shells, and resistance to crushing in favour of those of small capacity, the former have been adopted, Palliser shot being associated with them. **§ 1518.**

Projectiles on striking iron plates are more or less heated, any force converted into heat must be lost so far as useful effect goes. If the amount of heat generated could be exactly ascertained, the loss of force could be calculated by making use of the mechanical equivalent of heat. Sir. W. Armstrong carefully measured the amount of heat generated by concussion in the following metals, and obtained an approximate estimate of the amount of work lost by conversion into heat.

With hard tempered steel shot the force expended in heating the projectile was about $\frac{1}{10}$ of the power stored up on striking; with softer steel the loss was about $\frac{2}{10}$, and with wrought-iron the loss was more than half. Cast-iron has hitherto eluded observation on account of the difficulty of collecting the fragments and measuring the amount of heat in them.

We may, however, show theoretically how much power is lost by this heating. Take the case of a 400-lb. projectile, and suppose it to be raised 100° F. on impact, neglecting the heat imparted to the plate. The specific heat of the iron employed in Palliser projectiles may be take at ·12. Then the units of heat required to raise 400 lbs. of iron 100° F. $= 400 \times 100 \times \cdot 12 = 4800$.

Now one unit of heat $= 772$ foot-pounds.

Therefore the work expended in heating the projectile $= \frac{4800 \times 772}{2240} =$ 1654 foot-tons.

This is equivalent in the case of the 10″ projectile of 400 lbs. to an energy per inch of shot's circumference of 53·08 tons, and this again, using Major W. H. Noble's rule, is equivalent to a penetration of 6″·701. It must be remembered that this represents the *actual* penetration due to this energy, not that which would be attained could we prevent any of the power from being wasted as heat. Again, the energy of a 10″ Palliser shell fired with the service battering charge is at 1,000 yards range 4,185 foot-tons, and at this range it will penetrate 12 inches of iron. The whole of this power converted into heat would only raise the temperature of the projectile 253° F. This is easily proved from the data given above.

**Studs.** The studs of Palliser projectiles have in all cases been attached by swedging into undercut holes cast in the projectile.

**Penetration.** The penetration of Palliser projectiles has increased considerably since their first introduction, owing to improvements in the manufacture of the shell and in the powder. The old rule which used to be given was to add an inch to the calibre of the projectile for its penetration

---

* It must not be assumed from the wording of the paragraph in "Changes in War Stores" referred to, that all Mark I. shells were of small capacity, for this is by no means the case.

into iron plates at short ranges. Supposing the projectile to be fired with a battering charge at a range of 200 yards, the penetration will be found approximately by adding 2 inches to the calibre. This rule must not, however, be applied to guns which are specially powerful for their calibre; the 11-inch gun will penetrate about 14 inches, while the 12-inch 35-ton gun has penetrated 17 inches of iron, backed by 12 inches of teak and an iron skin $1\frac{1}{4}$ inches thick.*

The following empirical formulæ give good approximate results, not widely differing from each other:—

Captain A. Noble of Elswick gives $T=\left(\frac{e}{3\cdot 138}\right)^{\frac{1}{1\cdot 5}}$.†

Major W. H. Noble, R.A., from further experience gives $T=\left(\frac{e}{2.53}\right)^{\frac{1}{1\cdot 6}}$.

Major Maitland, R.A., gives $T=\frac{E}{14\cdot 87r^2}$.

In the above formulæ, T=thickness of iron perforated, $e$=energy per inch of shot's circumference, E=total energy of shot, and $r$=radius of shot.‡

## Palliser Shells.

### *General Description.*

Calibres 12″·5; 12″, 35-ton; 12″, 25-ton; 11″; 10″; 9″; 8″; and 7″. The form of the body in every case is cylindrical, the head being ogival struck with a radius of $1\frac{1}{2}$ diameters. See plate, p. 360.

The total length varies between a little over 2 calibres and a little over $2\frac{1}{2}$ calibres, as may be seen in the table p. 205.

The bottom is flat, in the more recent patterns it is rounded at the edge to facilitate loading.

---

* Extracts, Vol. X, p. 174.

Experiments are now being carried on with steel shell for armour-piercing purposes. If they can be manufactured successfully the advantages expected to be derived from their introduction are:—

1. They will not break against iron on impact with direct fire.
2. They will more often penetrate with oblique fire, owing to their being stronger, and thus less liable to break up on being brought up at right angles to a plate on the point biting.

† This rule may be applied to ron plates varying from from 4″ to 15″ in thickness.

‡ It is to be remembered that these formulæ are calculated to show the thickness of plate that can be *perforated*, not the depth to which a projectile will *penetrate* in a plate too thick for it to perforate.

To give an instance of the application of these formulæ, we may take the case of the projectile for the 80-ton gun, bearing in mind, however, that in this case the target was not completely perforated. On 4/5/77 this gun, having a bore of 16″, chambered to 18″, was fired at a target consisting of four 8″ plates, having 5″ teak between each plate. The projectile pierced through three of the plates and cracked and bulged out the fourth plate, so that the unbroken point of the projectile could be easily felt with the finger from the back of the target. The projectile weighed 1700 lbs., the charge was 425 lbs. P². The I.V. was 1600′ and the striking velocity 1585′. The stored energy (E) was 29615 foot-tons, the energy per inch of shot's circumference was 589·17 foot-tons.

By Captain A. Noble's formula this corresponds to a perforation of 32″·79;
by Major W. H. Noble's formula it corresponds to „ 30″·16;
and by Major Maitland's formula to 31″·12;

(in all cases the projectile is taken as having a 16″ diameter, which is slightly in excess of the truth).

Tables showing the energy, whence the penetration is easily calculated, are given for the various Woolwich guns, in "Treatise on Construction of Ordnance," pp. 349–50.

In the centre of the bottom is a filling hole bushed with cast-iron* and closed with a gun-metal screw plug which should fit tightly and evenly, and is therefore selected for each shell, and is not intended to be interchanged with that of another.

The cast-iron bush is cast into the shell, being placed for this purpose on the core spindle; it has four longitudinal grooves, and also horizontal steps, so as to be well gripped by the metal cast round it. There is an undercut annular groove in the shell's base. Into the groove is hammered a lead ring so as closely to seal the joint between the bush and the metal of the shell.

The mottled iron employed in casting Palliser shell is too hard to admit of tool work, even when the metal is cast in sand, hence the necessity for introducing the cast-iron bush in the base to allow a screw thread to be tapped in it to receive the gun-metal plug. The holes for the studs and extractor holes are cast in the projectile. The shell are brought as nearly as possible to their final dimensions in casting and ground down to their proper size by revolving grindstones.†

All Palliser shell are lacquered internally with red lacquer.‡ The bursting charge is further protected by a bag; the dimensions of the bags will be found on p. 203.§

One point has escaped without such definite marks as could be wished, and this must be specially noticed, viz., the difference between 9-inch and 8-inch shells of large and small capacity. By reference to the table, p. 205, it will be seen that all the patterns except I. are of large capacity, containing bursting charges of 5 lbs. 14 oz. and 4 lbs. 6 oz. respectively, all shells then of II. and subsequent patterns are of large capacity. The 11″ Mark I. is however of larger capacity than Mark II., which was originally a shot.

The table gives Mark I., shells of small capacity, viz., the 9-inch to contain 2 lbs. 13 oz. and the 8-inch 2 lbs. of powder, but unfortunately shells of large capacity were manufactured without changing the numeral, and therefore a Mark I. shell may be either a small capacity shell, as given in the table, or it may be one of the large capacity, and the only method of identifying them, without testing their capacity, is a careful measurement of the length, those of large capacity being the same length as Mark II. shell given in the table, but in numeral and other external marks resembling Mark I.‖

Palliser projectiles are tested by water pressure of 100 lbs. on the square inch, and by hammering the base all over with a pointed hammer, to detect any weak or porous portion. Shell so examined used to be marked E. on rear stud, but as this examination is now invariably carried out in manufacture it has not been thought necessary to continue marking them. Shell made between 1870 and August 1872 will be found to have E. on the stud, as well as any examined at

---

* Previously to 29/10/74 a tinned wrought-iron bush was used, and is the one shown in the plate. The cast-iron bush adheres more intimately to the metal of the base, it stands blows from a steam hammer (imitating the percussive action of the charge) better, and is rather cheaper. No change of pattern. § 2864.

† Formerly the bodies were cast slightly smaller, having bands at the shoulder and base; these bands were ground down to the final dimensions.

‡ Black lacquer was tried in 1870, but did not answer.

§ The use of bags was recommended in 1870. The trial of various kinds of bags and the comparative cost of bags and lacquering will be found in Extracts, Vol. VIII, p. 145. The question of discontinuing the lacquer was raised, but it was decided to continue it. Extracts, Vol. VIII, p. 227.

‖ The shells of large capacity are more than an inch longer than those of small capacity.

out-stations, (see below,) but shells made subsequent to August 1872, although examined, will not be marked with E.*

Paint.

Palliser shell were formerly painted black with a white ring and a red tip, but are now painted black with white tip, the studs being in all cases left unpainted as with other projectiles.

Marks.

For Marks, see table, p. 205. All shell made since 1873 inclusive have the word "shell" cast on the base and also the calibre. Where there are two guns of the same calibre the weight of the gun is given.

Issue.

Loose:—

1. Filled for S.S. wrench hole in base plug filled with red lead putty to enable a filled shell to be distinguished by touch.
2. Empty for L.S.

*Use and Experience.*

Use and experience.

As has been noticed above, these shells are intended for use againsf armour-clad vessels; it is very uncertain if they would explode if fired against iron vessels not armour-plated, and they would probably altogether fail to burst against wooden ships.†

The discovery of porous places under the surface of the bases of Palliser projectiles in the spring of 1870 led to a searching examination of those projectiles in store at Woolwich, and this defect was found to exist to a sufficient extent to render it expedient to extend the examination to certain out-stations, the cavities being detected by searching by slight blows with a pointed hammer; any flaws so discovered should be probed with a sharp wire, where the metal seems crumbling or globular in appearance, it will generally come away, and perhaps be followed by dust-like particles till a sound bottom is reached, a cavity of about the diameter and one-fourth the depth of an ordinary thimble may exist near the circumference of a shell or shot bottom without in the least weakening it, but it is ordered (in the instructions for examination) that all shells which may appear "doubtful" are to be returned to Woolwich with those which appear "unserviceable."‡

The tip or point of a chilled projectile is occasionally broken off by the impact of a shell or shot rolled or struck obliquely against it; for, strange as it may appear, the point which may penetrate directly through many inches of armour without injury may be fractured by a very slight transverse blow.

A. C. Cl. 134, 1877.

Plaster of Paris models, labelled respectively A and B, are issued to inspectors of warlike stores, to guide them in the examination of Palliser projectiles.

Any Palliser projectile that may be found with their points broken to a greater extent than what is shown in model A will be sent to the Royal Arsenal, Woolwich.

Projectiles with points broken similarly to model A, or to a greater extent than is shown in model B, up to A, may be retained at out-stations for practice if so required. If not so required they are to be sent to Woolwich.

---

* The base of an examined shell shows the marks of the sharp pointed hammer.

† Lieut. Boxer, R.N., advocated the employment of a percussion fuze in the base of a Palliser shell to ensure the explosion, in the event of a mistake as to the character of an adversary.

‡ On carrying out the examination a considerable number of Palliser shells were found unserviceable. Extracts, Vol. VIII, pp. 222, 224, 225.

Projectiles with points broken similarly to, or to a less extent than is shown in model B, will be considered serviceable.*

Clause 6, A.C., 1876, is cancelled.

12-inch 25-ton and 10-inch Palliser shells, Mark I., were found not to be satisfactory; they are broken up on return to Woolwich.†

The service lacquer which holds very well to other projectiles is apt to chip away from the interior of Palliser shell; this arises from the fact that the projectile is not re-heated for lacquering, but undergoes this operation in the stage of cooling most nearly approaching to the proper temperature, hence an exact degree of heat is not to be attained, and consequently the lacquer does not always hold well to the metal.

At the same time, another possible cause of premature explosion in Palliser shell was discovered in the presence in these shells of particles cut from the wrought-iron bush in the operation of tapping, and either from the warmth generated in the operation, adhering to the lacquer or remaining in the curve of the interior of the shell so as to escape removal. In consequence of this it has been ordered that any powder removed from Palliser shell is to be treated as unserviceable, and "at once thoroughly wetted and destroyed." Under these circumstances it has been decided to introduce serge bags to contain the bursting charge as an additional precaution against premature explosion; these bags are made bottle-shaped, and are introduced through the filling hole. All Palliser shells are completely filled with "shell powder L.G." See page 63.

§ 1992.

*Bags, Serge, Bursting Charge, Palliser Shell.*

The following table shows the dimensions of the bags:—

| Calibre, Charge, Mark. | § Changes in War Stores. | Greatest Width. | Length to neck. | Neck. | | Number packed in a Bale. | Remarks. |
|---|---|---|---|---|---|---|---|
| | | | | Length of. | Width of. | | |
| | | ins. | ins. | ins. | ins. | | |
| 12"·5 11 lbs. 12 oz., I. ... | 2889 | — | — | — | — | 500 | |
| 12" 35 ton, 9 lbs. 14 oz., I. ... | 2459 | 9·0 | 19·6 | 5·0 | 2·7 | 500 | |
| 12" 25 ton, 14 lbs., I. ... | 1970 | 10·1 | 23·0 | 4·1 | 2·7 | 500 | |
| 11" 9 lbs. 4 oz., I, ... ... | 2372 | 8·4 | 20·0 | 5·0 | 2·7 | 500 | For 11" shell, Mark I. of which, though not approved, some issues were made. § 2429. |
| 11" 6 lbs. 7 oz., II. ... ... | 2422 | 8·4 | 16·0 | 5·0 | 2·7 | 500 | |
| 10" 6 lbs. 14 oz., I. ... ... | 1970 | 7·2 | 19·5 | 4·0 | 2·7 | 500 | |
| 9" 5 lbs. 8 oz., I. ... ... | 1970 | 7·7 | 15·6 | 3·6 | 2·7 | 500 | |
| 9" 2 lbs. 13 oz., I. ... .. | 1970 | 6·1 | 12·6 | 3·6 | 2·7 | 500 | Small capacity shell. |
| 8" 4 lbs. 8 oz., I. ... ... | 1970 | 6·9 | 15·0 | 3·4 | 2·7 | 500 | |
| 8" 2 lbs., I. ... ... ... | 1970 | 5·3 | 12·0 | 3·0 | 2·7 | 500 | ,, ,, |
| 7" 2 lbs. 8 oz., I. ... .. | 1970 | 6·0 | 12·0 | 3·2 | 2·7 | 500 | |

* Experiments at Shoeburyness in 1875 tend to show that projectiles with slightly damaged points are not unserviceable. The range and accuracy are not materially affected. Extracts, Vol. XII, p. 22.

† Only 152 12-inch and 708 10-inch, Mark I., were made.—Extracts, Vol. VIII, p. 65.

Issue.

Issue, see table above.

Different reasons have been assigned as to the cause of the explosion of the bursting charge in Palliser shells.* It is not due to the shell breaking up, as the same action takes place in Whitworth's hardened steel shell, which do not break up. The metal of the shell does not acquire a sufficiently high temperature on striking, so the explosion is not due to heat transmitted from the shell; the most probable cause appears to be the violent concussion of the powder on striking.

On firing, the powder sets back, forming a dense compact mass so hard that sometimes it cannot be cut by a copper tool; on the shell striking the plate, this hard mass of powder would be dashed forward and pressed into the contracted part of the shell, thus undergoing great friction, probably sufficient to tear the bag and thus cause the powder to explode.

The setting back of the powder on firing is necessarily much less violent from the cushioning action of the loose grains of powder.†

---

* See Captain W. H. Noble, R.A., on Ballistics, p. 93.

† Occasionally, to prevent the explosion taking place before the shell has penetrated sufficiently, two bags have been used in experimental practice.

§ 1648.

Palliser projectiles which have been fired and recovered will not be fired again, but will be returned into store to be condemned and re-cast, as they are liable to break up in the bore when fired a second time.

## Palliser Shell, R.M.L. Woolwich Guns.

| Calibre, and Mark of Pattern. | Date of Approval. | § Changes in War Stores. | Length in Inches, ± 1/10 in. per foot. | Diameters. Body.* | Diameters. Studs. | Thickness of Metal. Walls. | Thickness of Metal. Base. | Weight empty. | Approximate bursting charge, Shell Powder, L.G. | Weight of Filled Shell. | Studs, Hard or Soft. | Distance between Centres of Front and Rear Studs. | Edge of Bottom, rounded. | Bands round Shoulder and Base. | Bodies cast in Sand. | Marks on a Rear Stud. The Date referring to the Manufacture of each individual Shell. | Remarks. |
|---|---|---|---|---|---|---|---|---|---|---|---|---|---|---|---|---|---|
| | | | ins. | ins. ± ·01 | ins. ± ·005 | ins. | ins. ± ·05 | lbs. oz. | lbs. oz. | lbs. oz. ± 1·5 per cent. | | ins. | | | | | |
| 12"·5, I. ... | 24/1/76 | 2877 | 33·0 | 12·425 | 12·85 | 3·25 | 3·4 | 790 8 | 11 12 | 802 4 | Soft | 7 | Yes | No | Yes | Date, × and I. ... | Three rings of studs. |
| 12" 35 ton, I. ... | 10/10/72 Provly. | 2380 2459 | 31·3 | 11·925 ± ·015 | 12·35 | 3·16 | 3·25 | 691 2 | 9 14 | 701 0 | " | 7 | Yes | No | Yes | Date, × and I. ... "35 ton," on every alternate stud | Two rings of studs. |
| " " II.... | 25/4/73 | 2490 2655 | 31·3 | 11·925 | 12·35 | 3·16 | 3·25 | 688 5 | 9 14 | 698 3 | " | 7 | Yes | No | Yes | Date, × and II. ... "35 ton," on every alternate stud | Three rings of studs. |
| 12" 25 ton, I.†... | 11/11/68 | 1707 | 29·2 | 11·92 | 12·35 | 2·785 | 2·5 | 585 0 | 15 0 | 600 0 | Hard | 7 | Yes | Yes | No | Date and × | |
| " " II.... | 7/3/70 | 1872 2655 | 29·2 | 11·92 | 12·35 | 2·85 | 2·85 | 586 4 | 14 0 | 600 0 | " | 7 | Yes | No | Yes | Date, × and II. | |
| 11", Mark II. ... | 18/11/72 28/1/73 | 2106 2429 2655 | 28·3 | 10·925 ± ·015 | 11·35 | 2·9 | 3·0 | 529 8 | 6 7 | 535 15 | Soft | 8 | Yes | No | Yes | Date, × and II. ... | 11", Mark I., was not finally approved, a number issued, capacity 9 lbs. 4 oz. By § 2429 it was decided that only one description of Palliser projectile should be issued for the 11" gun, viz., that described in § 2106 as Shot, Palliser, 11", Mark I., which was then ordered to be called Shell, Palliser, 11", Mark II. |
| 10", Mark I.† ... | 18/9/68 | 1678 | 26·8 | 9·92 | 10·35 | 2·5 | 1·94 | 391 0 | 9 0 | 400 0 | Hard | 8 | Yes | Yes | No | Date and × | |
| " " II.... | 7/3/70 | 1872 2655 | 26·3 | 9·92 | 10·35 | 2·8 | 2·5 | 393 8 | 6 14 | 400 6 | " | 8 | Yes | No | Yes | Date, × and II. | |
| 9", Mark I.‡ ... | 14/1/67 Provly. | 1386 | 20·2 | 8·92 ± ·015 | 9·31 | 2·71 | 2·1 | 247 3 | 2 13 | 250 0 | Soft | 8·7 | No | No | No | Nil ... ... ... | Small capacity shell. |
| " " II. .. | — | Nil | 21·5 | 8·92 | 9·31 | — | — | 244 2 | 5 1' | 250 0 | Hard | 6 | No | Yes | No | Date and ×. | |
| " " III. | 9/12/68 | 1765 | 21·45 | 8·92 | 9·31 | — | — | 244 2 | 5 14 | 250 0 | " | 6 | Yes | Yes | No | Date and ×. | |
| " " IV. | 7/3/70 | 1872 2655 | 21·45 | 8·92 | 9·31 | 2·15 | 2·0 | 244 3 | 5 8 | 249 11 | " | 6 | Yes | No | Yes | Date, × and IV. | |
| 8", Mark I.‡ ... | 14/1/67 Provly. | 1387 | 18·4 | 7·92 ± ·015 | 8·31 | 2·36 | 1·8 | 178 0 | 2 0 | 180 0 | Soft | 8·05 | No | No | No | Nil ... ... ... | Small capacity shell. |
| " " II. ... | 0/11/67 | Nil | 19·4 | 7·92 | 8·31 | — | — | 175 10 | 4 6 | 180 0 | Hard | 5 | No | Yes | No | Date and ×. | |
| " " III. | 9/12/68 | 1765 | 19·6 | 7·92 | 8·31 | — | — | 175 9½ | 4 6½ | 180 0 | " | 5 | Yes | Yes | No | Date and ×. | |
| " " IV. | 7/3/70 | 1872 2655 | 19·25 | 7·92 | 8·31 | 1·92 | 2 0 | 174 12 | 4 8 | 179 4 | " | 5 | Yes | No | Yes | Date, × and IV. | |
| 7", Mark I. ... | 0/11/67 | Nil | 16·4 | — | — | — | — | 112 8 | 2 8 | 115 0 | Soft | 4·6 | No | Yes | No | Date and ×. | |
| " " II. . | 9/12/68 | 1765 | 16·5 | 6·92 | 7·31 | — | — | 112 8 | 2 8 | 115 0 | " | 4·6 | Yes | Yes | No | Date and ×. | |
| " " III. | 7/3/70 | 1872 2655 | 16·5 | 6·92 | 7·31 | 1·685 | 1·6 | 112 1 | 2 8 | 114 9 | " | 4·6 | Yes | No | Yes | Date, × and III. | |

* Limits of error in diameter over the body of Palliser projectiles increased to ± ·015, § 1899.

† Ordered to be broken up.

‡ Many 9" and 8" shells of large capacity were made as Mark I., these may be recognised from Mark I. small capacity shells by being 1"·3 and 1"·0 longer. To facilitate identification, Palliser shell with sand cast bodies have their distinguishing numeral stamped on a rear stud in addition to being cast on the base. § 1872. "E" on a rear stud denotes that the base has been tested for porous places; this mark was discontinued 1/8/72. × and the date on a rear stud has the same signification as in common shell, table, p. 190. On the base of Palliser shell manufactured since 1872, will be found the mark Palr shell, followed by the calibre, R.L., date, numeral indicating pattern, and letters indicating the nature of iron used. Thus a 10" Palliser shell cast on the 1st January would be marked: Palr shell, 10 in., R.L., 1/1 R.C. II. The letters R.C. standing for Ridsdale and Cwmbran.

The weights given above will not be found in every case to strictly correspond with those given in "Changes." Slight modifications have in some cases taken place, and the weights given above are those of the means between the high and low limits of manufacture.

*Palliser Shot.*

*Calibres.*—12″ 25 ton, 10″, 9″, 8″, 7″:—Palliser shot resemble the shell very closely,* indeed the difference between the present pattern is so slight that it has not been found necessary to introduce more than one projectile for the 12·5 inch, 12-inch 35 ton and 11-inch guns. (See plate, p.361).

In the earlier patterns there were, however, some differences, a solid 7-inch shot was introduced, but it is withdrawn as a service projectile, though still issued for practice.† The earlier shot were made with small cores, and were a little shorter than the shells.

§ 2040.

The hollow in the centre is an advantage in manufacture, as it is very difficult to cast such a dense metal well when quite solid, even when apparently sound they were liable to break up. Shot of the earlier patterns may be found with their bases closed by a wrought iron plug hammered in, this was found objectionable as the action of the powder tended to drive the plug in, and split the shot when it was fired.* Therefore, all Palliser shot of patterns antecedent to those approved 7/3/70, were ordered to be returned to Woolwich and to have their

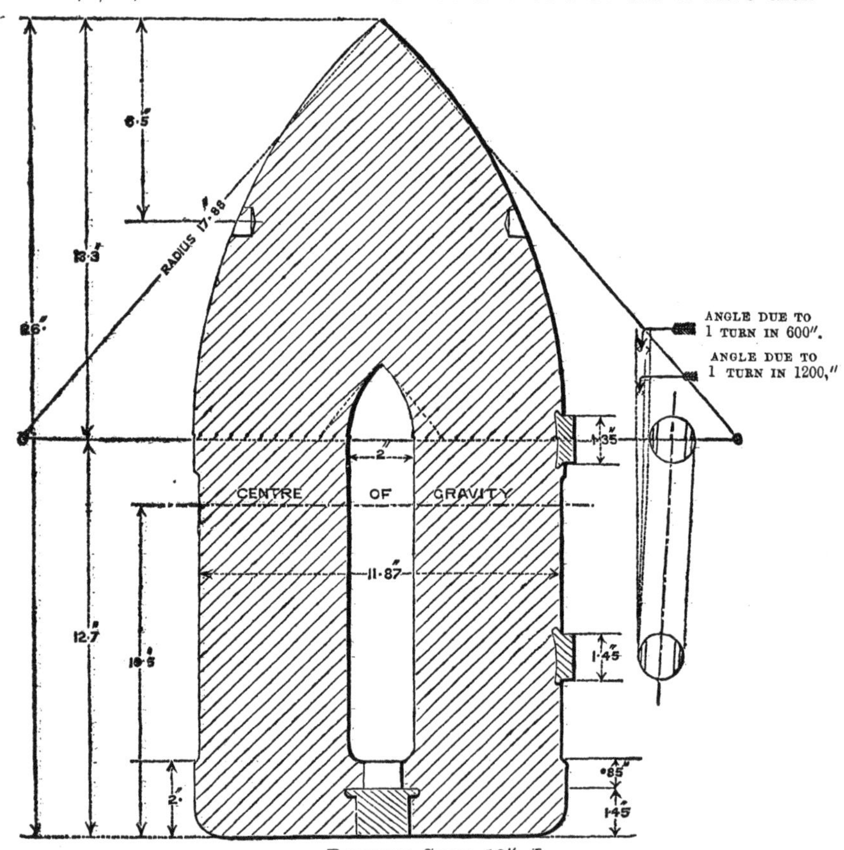


Palliser Shot, 12″, I.

(Base closed by a plug hammered in.)

---

* So slight is the difference that the 11-inch Palliser shell, Mark II., was originally sealed as a shot, and subsequently its name was changed to shell.

§ 2106.
§ 2429.

† Solid shot do not shoot as well as cored shot.—Extracts, Vol. VIII, p. 65.

bases strengthened, this was done by tapping the wrought-iron plug, so as to take a wrought-iron screw plug with a mushroom-shaped head which projects over and covers the weak part of the shot. Shot so strengthened can be easily recognised by the projecting plug.

The succeeding pattern had a wrought-iron bush cast into the base of the shot similar to that used with the shell, into this a *wrought-iron* plug was screwed.

Finally, experiments were carried on to ascertain whether the efficiency of the shot would be lessened by increasing the hollow in the interior. The main advantage claimed for shot over shell was that they possessed greater penetrative power in oblique firing than shell.† It was found desirable to increase the capacity of the shot to get rid of defects in the manufacture due to the form of the cavity.‡

Experiments were therefore carried on against iron plates to try the comparative effects of the small cored, and the proposed large cored shot, when fired against iron plates at angles of incidence varying from 60° to 66°. The result was satisfactory, and the large cored shot were approved.§

In order that these shot may be used as shell with a bursting charge, a gun-metal plug similar to that used with the shell is employed to close the base. This use of the shot would be exceptional, therefore no bags are included in the equipment: if it is desired to fire them with bursting charges, bags should be specially demanded.‖

Palliser shot manufactured since 1873 inclusive, are marked on the base with the word shot and with the calibre of the gun, the weight of the gun being added when there are two guns of the same calibre.

For dimensions, &c., *vide* table p. 208.

**Paint.** Black.¶

**Issue.** Loose.

**Use.** Their use, as above pointed out, is similar to that of the shells, but they are considered to penetrate better when firing at oblique angles.

By comparing the drawings of the shell and last pattern shot, it will be seen that the only differences between them are, that the shot is a very little shorter, and that the cavity of the shot contracts more rapidly towards the point than that of the shell.**

---

* Extracts, Vol. VIII, p. 65.

† Extracts, Vol. VIII, p. 65.

‡ Extracts, Vol. VIII, p. 224. Superintendent R.L. reports a serious defect in the present pattern of cored shot of all calibres, namely, the form of the interior cavity. He has caused a large number to be broken up in course of manufacture, and in every case fissures have been found extending from the rear end of the cavity towards the outside of the projectile. Experience shows that this defect is due to the unnecessary thickness of the shot towards the rear end, to the unequal effect of cooling the front end when it is cast in chill, and the rear end which is cast in sand, and the consequent unequal force exerted in the contraction of the metal. He finds that by enlarging the cavity and assimilating the form to that of the shell, this tendency to the production of fissures is apparently obviated.

§ Extracts, Vol. IX, p. 257.

‖ Extracts, Vol. IX, pp. 178–183. Under conditions where the target when struck direct is about a match for the gun opposed to it, Palliser live shell have more penetrative power than shot.

With shot of the form recently introduced, viz., large cored shot which are capable of holding a very fair bursting charge, the effect produced in direct fire appears to be at least equal to that of Palliser shell.

Much information as to the penetration of projectiles will be found in the Extracts above quoted.

¶ Those first issued had a white ring painted round the head.

** In the 9-inch the difference in length between the shot and the shell is only ·6″, and the difference between the bursting charges is 1¾ lb.

## Palliser Shot, R.M.L.

| Calibre and Mark of Pattern. | Date of Approval. | § Changes in War Stores. | Length, ± 1/10 in. per foot. | Diameters. | | Thickness of metal. | | Weight empty. § | Approximate capacity for bursting charge, shell powder, L.G. |
|---|---|---|---|---|---|---|---|---|---|
| | | | | Body.* | Studs. | Walls (minimum). | Base. | | |
| | | | ins. | ins. ± ·01 | ins. ± ·005 | ins. | ins. | lbs. oz. ± ¾ per cent. | lbs. oz. |
| 12″ 25 ton, Mark I. ... | 11/11/68 | 1707 | 26·0 | 11·92 | 12·35 | 4·935 | 2·3 | 600 0 | — |
| " " II. ... | 7/3/70 | 1872 | 26·2 | 11·92 | 12·35 | 4·935 | 2·3 | 600 0 | — |
| " " III. ... | 14/12/70 Provly. | 2038 | 28·15 | 11·925 ± ·015 | 12·35 | 3·0 | 2·85 | 600 9 | — |
| " " IV. ... | 6/4/72 | 2263 | 28·15 | 11·925 ± ·015 | 12·35 | 3·0 | 2·85 | 596 4 | 7 12 |
| 10″ Mark I. ... ... | 18/9/68 Provly. | 1678 | 24·5 | 9·92 | 10·35 | 3·935 | 2·3 | 400 0 | — |
| " " II. ... ... | 7/3/70 | 1872 | 24·5 | 9·92 | 10·35 | 3·935 | 2·3 | 400 0 | — |
| " " III. ... ... | 10/1/71 Provly. | 2039 | 25·8 | 9·925 ± ·015 | 10·35 | 2·8 | 2·5 | 400 0 | — |
| " " IV. ... ... | 11/8/71 | 2107 | 25·8 | 9·925 ± ·015 | 10·35 | 2·8 | 2·5 | 400 8 | 4 0 |
| 9″ Mark I. ... ... | 21/10/67 | 1518 | 18·8 | 8·92 | 9·31 | — | — | 250 0 | — |
| " " II. ... ... | 9/12/68 | 1765 | 19·0 | 8·92 | 9·31 | — | — | 250 0 | — |
| " " III. ... ... | 7/3/70 | 1872 | 19·1 | 8·92 | 9·31 | — | — | 250 0 | — |
| " " IV.† ... ... | — | — | 19·1 | 8·92 | 9·31 | — | — | 250 0 | — |
| " " V. ... ... | 16/12/71 | 2222 | 20·85 | 8·925 ± ·015 | 9·31 | 2·15 | 2·0 | 244 8 | 3 12 |
| 8″ Mark I. ... ... | 21/10/67 | 1518 | 17·15 | 7·92 | 8·31 | — | — | 180 0 | — |
| " " II. ... ... | 9/12/68 | 1765 | 17·39 | 7·92 | 8·31 | — | — | 180 0 | — |
| " " III. ... ... | 7/3/70 | 1872 | 17·4 | 7·92 | 8·31 | — | — | 180 0 | — |
| " " IV. ... ... | 16/12/71 | 2222 | 18·8 | 7·925 ± ·015 | 8·31 | 1·92 | 2·0 | 176 0 | 2 10 |
| 7″ Mark I‡. ... ... | — | — | — | — | — | — | — | — | — |
| " " II. ... ... | 18/12/66 Provly. | 1340 | 14·0 | 6·92 | 7·31 | — | — | 115 0 | — |
| " " III. ... ... | 21/10/67 | 1518 | 14·55 | 6·92 | 7·31 | — | — | 115 0 | — |
| " " IV. ... ... | 9/12/68 | 1765 | 14·7 | 6·92 | 7·31 | — | — | 115 0 | — |
| " " V. ... ... | 7/3/70 | 1872 | 14·7 | 6·92 | 7·31 | — | — | 115 0 | — |
| " " VI. ... ... | 16/12/71 | 2222 | 16·1 | 6·925 ± ·015 | 7·31 | 1·685 | 1·6 | 112 0 | 1 10 |

* Limits of error in diameter over the body of Palliser projectiles increased to ± ·015″, § 1899.
† Does not appear in "Changes in War Stores."
‡ A few experimental issues, no pattern extant.
§ The weight given above will not be found in every case to strictly correspond with those given in "Changes." Slight modifications have in some cases taken place, and the weights given above are those of the means between the high and low limits of manufacture.

## Woolwich Guns.

| Radius of Head. | Studs, hard or soft. | Distance between Centres of Front and of Rear Studs. | Edge of bottom rounded. | Band round shoulder and base. | Bodies cast in sand. | Marks on a Rear Stud, the Date referring to the manufacture of each individual Shot. | Remarks. |
|---|---|---|---|---|---|---|---|
| diameter. | | ins. | | | | | |
| 1½ | Hard | 7 | Yes | Yes | No | Date and × ... | Base closed with a wrought iron plug swedged into an undercut, requires strengthening plug. § 2040. |
| 1½ | Hard | 7 | Yes | No | Yes | Date, ×, and II. ... | Base closed with a bush and screw plug of wrought iron. No keyhole in plug. |
| 1½ | Soft | 7 | Yes | No | Yes | Date, ×, and III. ... | Do., and enlarged core |
| 1½ | Soft | 7 | Yes | No | Yes | Date, ×, and IV. ... | Do., do., and gun-metal plug with keyhole. Red lacquer. |
| 1½ | Hard | 8 | Yes | Yes | No | Date and ×. ... | Base closed, same as 12″, I. |
| 1½ | Hard | 8 | Yes | No | Yes | Date, ×, and II. ... | ,, ,, II. |
| 1½ | Soft | 8 | Yes | No | Yes | Date, ×, and III. ... | ,, ,, III. |
| 1½ | Soft | 8 | Yes | No | Yes | Date, ×, and IV. ... | ,, ,, IV. |
| 1¼ | Hard | 6 | No | Yes | No | Date and × ... | ,, ,, I. |
| 1¼ | Hard | 6 | Yes | Yes | No | Date and × ... | ,, ,, I. |
| 1¼ | Hard | 6 | Yes | No | Yès | Date, ×, and III. ... | ,, ,, II. |
| 1½ | — | 6 | Yes | No | Yes | — | ,, ,, II. |
| 1½ | Soft | 6 | Yes | No | Yes | Date, ×, and V. ... | ,, ,, IV. |
| 1¼ | Hard | 5 | No | Yes | No | Date and × ... | ,, ,, I. |
| 1¼ | Hard | 5 | Yes | Yes | No | Date and × ... | ,, ,, I. |
| 1¼ | Hard | 5 | Yes | No | Yes | Date, ×, and III. ... | ,, ,, II. |
| 1½ | Soft | 5 | Yes | No | Yes | Date, ×, and IV. ... | ,, ,, IV. |
| — | — | — | — | — | — | — | — |
| 1 | Soft | 6·1 | No | No | No | Nil ... ... ... | Solid shot, Practice only. |
| 1¼ | Soft | 4·6 | No | Yes | No | Date and × ... | Base closed, same as 12″, I. |
| 1¼ | Soft | 4·6 | Yes | Yes | No | Date and × ... | ,, ,, I. |
| 1¼ | Soft | 4·6 | Yes | No | Yes | Date, ×, and V. ... | ,, ,, II. |
| 1½ | Soft | 4·6 | Yes | No | Yes | Date, ×, and VI. ... | ,, ,, IV. |

N.B.—The marks on the rear stud are similar to those on the Palliser shell, table, p. , and have the same meaning. On the base of Palliser shot manufactured since 1872 will be found the mark, Palliser shot, followed by the calibre, R.L., date, numeral indicating pattern, and letters indicating the nature of iron used. Thus a 10″ Palliser shot cast on the 1st January would be marked: Pal[r] shot, 10 in., R.L. $\frac{1}{1}$. R.C. IV The letters R.C. standing for Ridsdale and Cwmbran.

## *Case Shot.*

*Case Shot.*—Calibres 12″, 11″, 10″, 9″, 8″, 7″. For description see p. 140.

The rule for the weight given on p. 140 is followed, but not very closely, the 7-inch being about three-fourths the weight of the other service projectiles, the 8-inch and upwards being about the weight of a round shot of the same calibre.

The reasons for the rule will be found in Extracts, Vol. IV, p. 397. Some of them have become inapplicable, as there are no B.L. guns of calibres higher than the 7-inch in the service.

The reasons given are briefly:—

1st. To allow of the same case shot being used for B.L. and R.M.L. guns of corresponding calibre. One or two case shot being used with R.M.L. guns according to the range, and to the velocity required for the penetration of the balls.

2nd. To secure from rifled guns the same range and penetrative effect as when case are fired in S.B. guns.*

Experience.

The experiments carried on with case shot show that much better results are obtained by using two case shot than by using one, at moderate ranges.†

When the rule of using battering charges (see p. 215), when firing at an enemy from a casemate is followed, the effects of two case shot will be still better.

The case carries close, and is effective up to about 600 yards;‡ much depends on the ricochet of the balls; case will be more effective and range to a greater distance over smooth water than over rough. ~~As Shrapnel are withdrawn~~ from the L.S. equipment of Woolwich guns, case will be the only projectile available against boats. About 1° of elevation should be given when firing over water at a range of 500 yards, the elevation should be reduced for shorter ranges or when firing over sands.

The dimensions, weights, number of balls, &c., will be found in the table, page 212. 8oz. sand shot are employed in all for motives of economy. As before mentioned, much of the weight is expended on the envelope, which is not efficient as a projectile: the segmental linings

---

* The 68-pr. case shot, weight 48 lbs., was fired with a 16 lbs. charge from the 68-pr. (the proportion of charge to shot being about one-third). The 8-inch R.M.L. gun has a full charge of 20 lbs.; if the case was made the same weight as the other projectiles the proportion of the charge to the shot would be about one-ninth.

† Extracts, Vol. V, pp. 51, 137, 228, show that up to 450 yards double shotting is by far the most effective.

‡ Extracts, Vol. IX, p. 108. Trial of 9 and 10-inch case at two rows of targets showing a front of 63 × 9 feet, the first row 500 and the second 550 yards from the gun; 5 rounds fired from the 9-inch at 1° 30′ elevation gave 129 hits; 5 rounds from the 10-inch gun at 30′ elevation gave 406 hits. The 9-inch seemed to break up 100 yards from the gun and ranged about 1,000 yards, covering a front of about 50 yards at that range, p. 200. Trial of 11-inch case; 10 rounds were fired against targets arranged as before, at 30′ elevation, and gave 1,102 hits. Over smooth water case has been found effective up to 700 yards.

See also Vol. VIII, p. 229.

The indifferent results given by the 9-inch case are probably due to the elevation being too great, the other guns were fired at ½°.

The results do not compare favourably with the effects of Shrapnel, see p. 195.

had to be thickened, until they were found sufficient to protect the bore, and also to prevent the shot from setting up and taking the rifling.

Issue.

Loose.

Drill shot. § 2267.

*Drill Shot.*—Service projectiles are used at drill to accustom men to their weight, the wooden drill shot formerly issued are withdrawn.

INSTRUMENT FOR EXTRACTING PROJECTILES.

Instrument for extracting projectile, rifled M.L. gun. §§ 1206, 1266.

An instrument for the 9″ and 7″ projectiles, resembling the one described for the 64-pr. M.L. p. 23, was first adopted (24/1/66 and 8/5/66); that for the 9″ was only provisionally sealed. 9″ common shell were made with a groove round the head to enable this extractor to work with the grooves of the increasing twist.

Instrument, Extracting Projectiles, for Woolwich Guns.

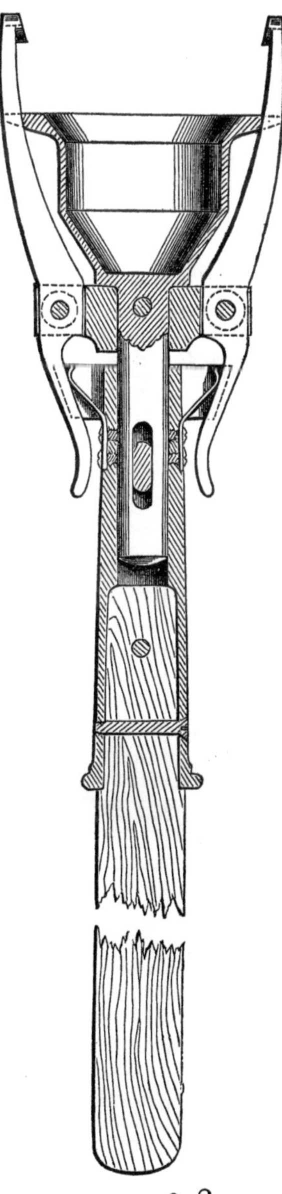

Instrument extracting 7″, 8″ and 9″, II. §§ 1363, 1081, 1712, 2850.

The extractor since made for every calibre of Woolwich projectile is constructed on a different principle, so as to act independently of the grooves of the gun.

It has two jaws and teeth for the two extractor holes of the projectile.

The jaws are closed by a strong spring acting on each counterjaw. The increasing twist prevents the use of guides. The extractor is pushed home until the edge of a cup in its head (*vide* figure) bears on the edge of the projectile. It is then turned round until the teeth spring into the holes, which are placed at a fixed distance from the ring on the head on which the cup edge will rest.

On withdrawing the staff the socket moves slightly in a slot in the shaft of the head, bringing two wedges, fixed on each side of each spring, to support each counterjaw.

This extractor has no means provided for releasing its hold while in the bore.

## Case Shot, R.M.L. Woolwich Guns.

| Calibre and Mark of Pattern. | Date of Approval. | § Changes in War Stores. | Length. | Diameter. | No. of Handles. | Number and Nature of Balls contained. | Approximate weight of balls. | Approximate weight of case, linings, coal dust, or clay and sand. | Total weight. | Marks on the top of the Case. | Remarks. |
|---|---|---|---|---|---|---|---|---|---|---|---|
| | | | ins. | ins. | | | lbs. oz. | lbs. oz. | lbs. oz. | | |
| 12″ 25 ton, Mark I. ... ... | 6/5/69 | 1777 | 11·35 | 11·88 ± ·05″ | 2 | 255 8 oz. sand shot ... | 127 8 | 102 8 | 230 0 ± 5 lbs. | I., R ↑ L. 12″ M.L. ... | Balls packed in coal dust. |
| ,, 25 or 35 ton, Mark II. ... | 30/4/72 5/10/72 | 2264 2382 | 11·35 | 11·88 ± ·05″ | 2 | 258 8 oz. sand shot ... | 129 0 | 117 0 | 246 0 ± 5 lbs. | II., ,, ,, ... | ,, half sand, half clay. |
| 11″ Mark I. ... ... ... | 15/4/72 | 2265 | 10·9 | 10·88 ± ·04″ | 2 | 210 8 oz. sand shot ... | 102 0 | 98 0 | 200 0 ± 6 lbs. | I., R ↑ L, 11″ M.L. ... | ,, ,, ,, |
| 10″ Mark I. ... ... ... | 27/1/68 Provly. | 1705 | 9·6 | 9·88 ± ·04″ | 2 | 136 8 oz. sand shot ... | 68 0 | 62 0 | 130 0 ± 4 lbs. | I., R ↑ L, 10″ M.L. ... | ,, coal dust. |
| ,, ,, II. ... ... ... | 25&31/1/72 | 2188 | 9·6 | 9·88 ± ·04″ | 2 | 139 8 oz. sand shot ... | 69 8 | 73 8 | 143 0 ± 4 lbs. | II., R ↑ L. 10″ M.L. ... | ,, half sand, half clay. |
| 9″ Mark III. ... ... ... | 24/3/68 | 1611 | 9·1 | 8·88 ± ·04″ | 2 | 113 8 oz. sand shot ... | 56 8 | 43 8 | 100 0 ± 3 lbs. | III., W ↑ D, 9″ M.L. ... | ,, coal dust. |
| ,, ,, IV. ... ... ... | 10/10/71 | 2115 | 9·1 | 8·88 ± ·04″ | 2 | 113 8 oz. sand shot ... | 56 8 | 50 8 | 107 0 ± 3 lbs. | IV., W ↑ D, 9″ M.L. ... | ,, half sand, half clay. |
| 8″ Mark I. ... ... ... | 24/3/68 | 1611 | 8·4 | 7·885 ± ·035″ | 1 | 75 8 oz. sand shot ... | 37 8 | 30 8 | 68 0 ± 2 lbs. | I., W ↑ D, 8″, M.L. ... | ,, coal dust. |
| 8″ (Gun or Howitzer) Mark II. | 25&31/1/72 & 1/5/75 | 2188 2742 | 8·4 | 7·885 ± ·035″ | 1 | 75 8 oz. sand shot ... | 37 8 | 36 8 | 74 0 ± 2½ lbs. | II., W ↑ D, 8″ M.L. ... | ,, half sand, half clay. |
| 7″ Mark III. ... ... ... | 24/3/68 | 1611 | 10·25 | 6·89 ± ·03″ | 1 | 74 8 oz. sand shot ... | 37 0 | 30 0 | 67 0 ± 2 lbs. | III., W ↑ D, 7″ M.L. or B.L. | ,, coal dust. |
| ,, ,, IV. ... ... ... | 25&31/1/72 | 2188 | 10·25 | 6·89 ± ·03″ | 1 | 70 8 oz. sand shot ... | 35 0 | 34 0 | 69 0 ± 2½ lbs. | IV., W ↑ D, 7″ M.L. or B.L. | ,, half sand, half clay. |
| ,, ,, V. ... ... ... | 13/3/76 | 2924 | 10·25 | 6·89 ± ·03″ | 1 | 71 8 oz. sand shot ... | 35 8 | 32 10½ | 68 2½ ± 2½ lbs. | V., W ↑ D, 7″ M.L. or B.L. | ,, ,, ,, Case of tin in 3 pieces. Ring at bottom. |

The 9″ and 7″, Marks I. and II. (§ 1241, 1611) are obsolete, being ordered to Woolwich to be broken up. The 12″, 10″, and 8″, Mark I., and the 9″ and 7″, Mark III., are to be returned to Woolwich for alteration to the present latest pattern, § 2444.

By § 2971 the high gauge for the diameters of R.M.L. case shot is to be the same as that for common and Shrapnel shell; and the limit between the high and low gauges will be as follows:—7″, ·07″; 8″, ·08″, 9″, ·09″; 10″ and upwards, ·1″.

The following patterns of extractors for heavy R.M.L. projectiles exist:—

For the 7″ gun three extractors have been made. Mark I. resembled the extractor for the 64-pr. to be hereafter described. It is only fitted for shells which have three extractor holes in the head, of which a few were issued. Mark II. has the present pattern head and a smooth stave. Mark III. differs from Mark II. in having a groove like a screw thread running up the stave to give a better grip to the hand.

§ 1206.
§ 1363.
§ 1712.

For the 8″ gun, Mark I. has a smooth stave, and Mark II. a grooved stave.

§§ 1363, 1712

For the 9″ gun Mark I. was made like Mark I. for the 7″ gun. The claws fitted into a groove cut round the head of the shell,* and will not fit the present pattern. Mark II. has the present pattern head and a smooth stave. Mark III. has a grooved stave.

§ 1266.
§§ 1363, 1712.

The 10″, 11″, and 12″ guns take the same extractor, which is Mark I., and has a grooved stave. The instrument for the 12″·5 gun differs only in length of stave. The lengths of the staves for the different extractors are as follows:—7″ gun, 9′ 6″; 8″ gun, 10′ 4″; 9″ gun, 10′ 6″; 10″, 11″, and 12″ guns, 11′ 3″; 12″·5 gun, 14′ 3″.

§ 1681.
§ 2850.

Bags for extracting instruments.
§§ 1734, 2912

There are two sizes of bags made of painted canvas, the mouth constructed to draw round the stave, and tied with small cord running through eyelet holes.

The larger size is for the 12″·5, 12″, 11″, and 10″ extractor; the smaller for the 9″, the 8″, the 7″, or the 64-pr. extractor.

Each bag is marked I. and R ⩚ L in white paint.

Holder, shell, Palliser.

Holder, shell, Palliser, *vide* Implements, page 63.

Gauges, iron, cylinder, shell or shot, rifled, muzzle-loading.
§§ 1546, 1697, 1714, 2000, 1313, 1547.

Cylinder gauges are issued to stations of inspection, and also low ring gauges over studs.

The former is important as, being a high gauge for both body and studs, a projectile which passes through it is sure to enter the bore of the gun; the diameters both of the cylinder and groove are ·045 inch less than the corresponding diameters in the gun.†

Guns with a uniform twist have the front and rear stud of their projectiles the same size, therefore only one set of grooves is required in the cylinder gauge corresponding to the grooves in the gun, the front stud in projectiles for guns having an increasing twist is smaller than the rear stud, hence for 8-inch guns and upwards a cylinder gauge is employed having a second set of grooves, which narrow up to the point so as to test the size of the front stud. A flange round the top of the gauge should rest on the front stud; a slot cut in the flange allows the front stud to be seen.

For dimensions, &c., see table p. 324.

§ 2177.
Ring gauges.

*Gauges, iron ring, shell or shot; R.M.L. body and studs.*

These gauges answer sufficiently well to test the shell as far as loading is concerned; they are issued to H.M. ships in commission, garrison batteries, and control officers, instead of cylinder gauges, the issue of which will be restricted to stations of inspection. They do not test the pitch of rifling, but are much more portable than the cylinder gauges.

---

* Such shell when returned to Woolwich have extractor holes of the present type cut in them.

† In manufacture the cylinder gauge has the advantage of detecting an eccentric stud which could not be found by ring gauges.

The 8-inch and upwards have two sets of grooves, one for the rear and one for the front stud.

Using.

Some skill is required to use a gauge properly; unless held quite fair it will not pass over the projectile.

For dimensions, &c., see table, p. 324.

RING GAUGE FOR 9-IN. RIFLED M.L. PROJECTILES.

§§ 2477, 2536.

Section.

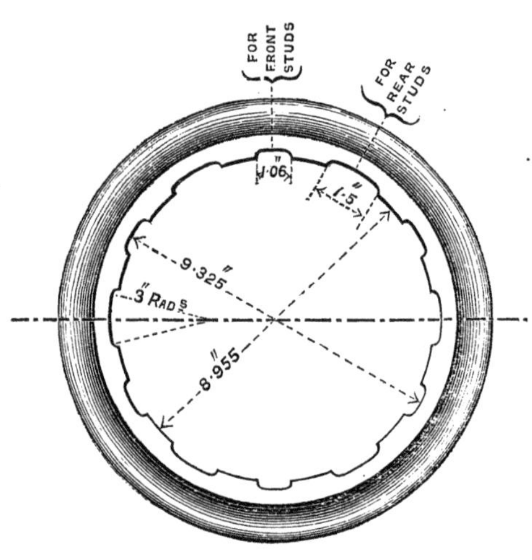


Plan.

## CARTRIDGES.

1st, Service ("battering" and "full").
2nd. "Reduced."
3rd. Drill or dummy.

The powders used are pebble for all battering charges (except the 12″·5 gun, which takes P²), and for full charges of 40 lbs. and upwards, L.S. and S.S.; L.G. in the L.S. and R.L.G. in the S.S. for full charges under 40 lbs.* See p. 6.

---

* General results connected with pressure established in Experiments by Committee on Explosives, 1872, p. 6:—

(1) If the powder be burned uniformly in the gun without indication of wave action, the pressure increases with the increase of the charge at first rapidly, but after 20 tons on the square inch has been exceeded, then very slowly. It may be remarked that in the whole course of the Committee's experiments a uniform pressure by the crusher gauge of 30 tons in the powder chamber has never been attained. This fact appears strongly to corroborate the experiments of Captain Noble, of Elswick, on the pressures produced by ignited powder in closed vessels,

There are two classes of service cartridges: battering and full.

The first would be used with Palliser projectiles, and only under special circumstances with common shell;* the second would be the ordinary charge used with common, double, and Shrapnel shells, and case shot. The reason for using pebble powder with heavy charges has been given on p. 6.

Cartridges for pebble† powder are made of two pieces of serge, one rectangular and the other circular; they are coned to fit the chamber of the gun. By reference to the table, p. 220, it will be seen that their diameter is larger when filled with P. than when filled with R.L.G., which latter may be substituted when no P. is available.‡

When R.L.G. powder is used, the cartridge is to be made up to the length given in the table by hooping the bags tightly.

All cartridges containing 20 lbs. charges and upwards have a becket over the choke, to enable them to be easily withdrawn from the cases. The becket is of the same material as the cartridge.

**Reduced charge.**

The reduced charges are only used for saluting, R.L.G. or L.G. Exercise powder being employed.

The dimensions, charges, &c., will be found in the table, p. 220. See also table, p. 329, for particulars as to the amount of materials used in making up cartridges.

The battering charges of P. powder will generally be found to vary between $\frac{1}{5}$th and $\frac{1}{6}$th the weight of the projectile; the charge for the 12-inch 25-ton gun is exceptionally light, about $\frac{1}{7}$th, and that of the 7-inch is exceptionally heavy, about $\frac{1}{4}$th.

See tables, giving packing, p. 220.

**Issue of cartridges.**

With reference to the cases in which they are packed, for land service barrels and boxes would not be used, special zinc cylinders being made for this purpose. See pp. 76, 77.

**Gauges**

The diameter of the cartridge is tested by brass ring gauges, the length or diameter by the universal gauge.

Brass ring gauges for diameter:—

---

which indicated that the maximum pressure produced by ignited powder in a perfectly closed space is somewhat less than 40 tons to the square inch.

(2) When a charge of any description of powder is increased beyond a certain limit, wave or local pressures are set up, which strain the gun unduly without affording an equivalent of useful effect on the projectile.

(3) Provided the battering charge is not exceeded the pressure in the gun increases steadily with the increase of weight of the projectile up to a certain point; beyond this point no material increase of pressure can be obtained by increasing the weight of the projectile. This again corroborates Captain A. Noble's experiments.

* When actually engaging an enemy from a casemate battery, it is advisable to use battering charges with all projectiles; the recoil being insufficient when the full charge is used—.Extracts, Vol. XI, pp. 100, 103. See also Extracts, Vol. XI, p. 165, where it is stated that common shell may be fired with battering charges when it is desirable to do so, but the 12″ 35-ton gun common shell is not to be fired with the battering charge, § 2618.

† Committee on Explosives, 1872, p. 3.

The course pursued by the Committee in order to determine the battering charge is "to increase the charge gradually until distinct wave pressures are exhibited, the "highest charge which can be employed without these local pressures appearing should "then be accepted as the battering charge."

‡ This change in diameter arose from the fact of the corrugated powder cases being made to hold cartridges filled with R.L.G. In order that the larger amount of P. powder should fit in, it was necessary to increase the diameter, so as to keep down the length.

Some cartridges approved on the 11/1/71 were marked for P. powder only. See table, p. 222.

Gauges, filled, cartridges, brass ring, rifled gun (M.L.).

High gauges are only necessary. They consist of rings of gun-metal, with straight handles; they are marked on and near the handle with the designation and numeral, also the diameter, and the words "FILLED CARTRIDGE," and further where the gauge does not apply to all the cartridges of a gun, as in the case of the 9″ and 8″, which have two gauges each; the weight of the charges of those for which it is intended is stamped on the gauge.

§ 3043. Ring gauge, P² powder for 12″·5 gun, diameter 12″. Mark I.

§ 2330. Ring gauges, P. powder:—

| | | |
|---|---|---|
| For 12-inch guns, diameter | 11″·5 | Mark I. |
| " 11 " " " | 10″·5 | |
| " 10 " " " | 9″·5 | |
| " 9 " " " | 8″·5 | |
| " 8 " " " | 7″·5 | |
| " 7 " " " | 6″·5 | |

§§ 1695, 1849. Ring gauges, R.L.G. powder:—

| | | | | |
|---|---|---|---|---|
| For 12-inch uns | .. .. | diameter | 11″·0 | Not yet sealed or published in "Changes." |
| " 10 " | .. .. | " | 9″·0 | Mark I. |
| " 9 " " | (43 and 30 lbs.) | " | 8″·2 | Mark II. Mark I. broken up. |
| " 9 " " | (15 lbs.) | " | 6″·8 | " " " |
| " 8 " " | (30 and 20 lbs.) | " | 7″·3 | " " " |
| " 8 " " | (12 lbs.) | " | 6″·26 | " " " |
| " 7 " " | .. .. | " | 6″·4 | " " " |

Issue. Loose, in numbers as demanded.

Gauge, cartridge, wood, length, universal, § 2074.

The length can be tested by the universal gauge, which has been provided for filled R.M.L. cartridges, having the lengths marked on one side and the diameters on the other. There is an arrangement enabling the proper limit to be given in measuring the length.

It is simply an adaptation of the sliding callipers, having two movable arms, *a* and *b*, which can be placed in position on the slide or scale, on either side of the fixed stop *d;* the scale on that part on which the arm *b* moves is marked on one side with lines corresponding to the "*low gauges*" of the lengths of all muzzle-loading rifled cartridges, and on the other side with lines representing the diameters of all cartridges; the opposite end of the scale on which the arm *a* moves is marked with lines which represent the difference or "limit" between the "*high*" and "*low gauges*" of length.

*Directions for using the Gauge.*

1st. For gauging diameters; fix the arm *b* with the inner edge coinciding with the diameter required, and the arm *a* close up to the fixed stop *d*, by means of the sliding stop *c*.

2nd. For gauging lengths; fix the arm *b* with the inner edge coinciding with the line representing the "low gauge" for length, and fix the sliding stop *c* with the inner edge coinciding with the line representing the difference or limit between the "*high*" and "*low gauges*" of length, leaving the arm *a* free to move between the stops *c* and *d;* the low gauge is now obtained by holding the arm *a* against the stop *d*, and the high gauge by holding the arm against the stop *c*.

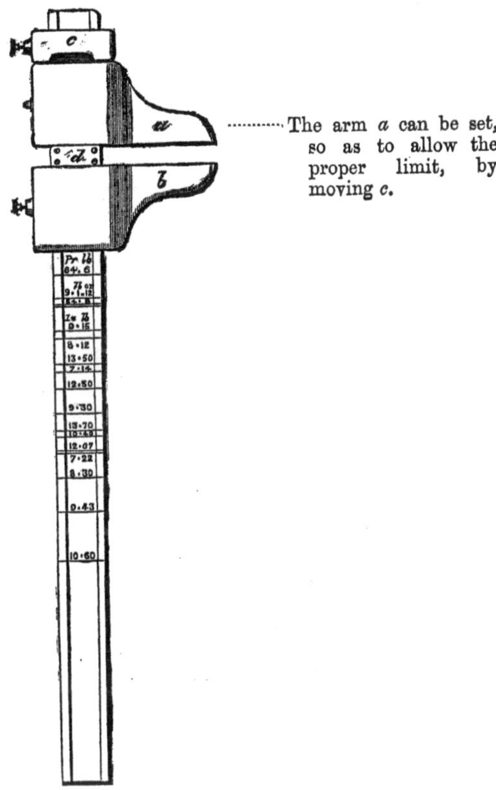


*Reduced Cartridges.**—As before mentioned, these cartridges are only used for saluting purposes; they are issued to guns of 9-inch calibre and under, and would rarely or ever be used in the L.S.

**Reduced or saluting cartridges.**

As to manufacture and construction, reduced cartridges differ only from the "battering" and "full" in their charges and dimensions, and further in the case of the 7-inch of 10 lbs., Mark I., in the choke being cut off to a length of two inches instead of three; they, like the smaller service cartridges, have no beckets over their choked ends. Blank, or Exercise, R.L.G. or L.G. powder is used.

It may be noticed that the diameters of the 9″ and 8″ reduced cartridges are less than those of the service cartridges; this is necessary in order to lengthen the cartridge, and ensure its extending far enough up the bore to reach past the vent, which in the case of the 9-inch gun is 9·7 inches, and in the case of the 8-inch 9·2 inches from the bottom of the bore.†

It will be seen in the table, p. 222, that the 32-pr. S.B. 10 lbs. cartridge has been approved as a reduced or saluting charge for the 7″ R.M.L. gun.

---

* The name was introduced because it was originally intended to use these cartridges with case shot, but this use was abandoned.

† The long quill friction tube is issued to fire these cartridges.

**Packing and issue.** For packing and issue, see pp. 220, 222.

**Drill cartridge. §§ 2358, 2533.** *Drill Cartridge.**—Is made of a hollow wood block covered with hide, and the nature of gun and charge marked on it.

**Issue.** Loose in numbers as demanded.

They are now made of the same size as P. powder cartridges, and without the imitation choke in order to admit of their being packed in zinc cylinders. For details, &c., see table, p. 219. The 12″ III., 10″ III., 9″ and 7″ IV., will supersede those of previous pattern when the existing store is used up.

**Proof cartridges.** Proof cartridges hardly come into a Laboratory course. The following occurs in Extracts, Vol. XI, p. 80:—

"It has been found difficult to fix a proof charge which will ensure a heavier pressure on the gun without unduly straining it. Powder chosen within the proper limits strain the gun to very different extent when large charges are used, *e.g.*, one brand of Pigou and Wilks "P." gave pressures ranging from 20 to 24 tons per square inch, with a charge of 58 lbs., projectile 250 lbs., while another brand gave a pressure of from 26 to 44 tons, with the same charge."

The following proof charges have been decided on:—

| | |
|---|---|
| 10-inch | tons, 75 lbs., Pebble powder. |
| 9 ,, | 58 ,, ,, |

The dimensions of the empty cartridges are †:—

| | Length. | Diameter. |
|---|---|---|
| 10″ .. .. .. | 33″·75 .. | 9″·5 |
| 9″ .. .. .. | 42″·75 .. | 8″·5 |

---

* Mark I. drill cartridges, 7″, 8″, and 9″, were stuffed with junk, and were found to set up on ramming home, and so to allow the rear ring of studs to get beyond the rifling into the powder chamber where the rifling ceases, and so cause the shot to jam.

† Brown braid is used in making up these cartridges. Flat-headed cylinders are fired at proof, as the service ogival-headed projectiles would be dangerous to fire at the butts.

## Drill, or Dummy Cartridges. Woolwich, R.M.L. Guns.

| Calibre and mark of Pattern. | Date of Approval. | § Changes in War Stores. | Length, Inches. From | Length, Inches. To | Diameter, Inches. From | Diameter, Inches. To | Diameter decreased at Bottom to Inches. From | Diameter decreased at Bottom to Inches. To | Imitation Choke on Front End. | Marks. | Remarks. |
|---|---|---|---|---|---|---|---|---|---|---|---|
| 12″, 35 ton, Mark I... ... | 2/9/73 | 2516 | 27·5 | 28·5 | 10·8 | 11·1 | 9·1 | 9·4 | No | I, 12-in., M.L. 110 lb. R. ↑ L. ... ... | |
| ,, 25 ton, Mark I... | 6/8/68 | 1659 | 17·0 | 18·0 | 10·8 | 11·1* | — | — | Yes | I, 12-in., 54 lb. R. ↑ L. ... ... ... | Contains hollow wood block; it has two crossed rope beckets at each end. |
| ,, ,, II. ... | 8/3/70 | 1878 | 17·0 | 18·0 | 10·8 | 11·1 | 9·1 | 9·4 | Yes | II, 12-in., M.L., 54 lb., R. ↑ L. ... ... | Differs from I. in being conical at base. |
| ,, ,, III. .. | 27/9/72 | 2358 | 20·0 | 21·0 | 10·8 | 11·1 | 9·1 | 9·4 | No | III, 12-in., M.L. 85 lb., R. ↑ L. ... ... | Dimensions increased to those of P. powder cartridges, and made without a choke to admit of being packed in zinc cylinders. |
| 11″, Mark I. ... ... ... | 27/9/72 | 2358 | 25·0 | 26·0 | 9·9 | 10·2 | 8·4 | 8·7 | No | I, 11-in., M.L., 85 lb., R. ↑ L. | |
| 10″, Mark I. ... ... ... | 27/11/68 | 1706 | 17·5 | 18·5 | 8·9 | 9·2 | — | — | Yes | I, 10-in., M.L., 40 lb., R. ↑ L. ... ... | Contains hollow wood block, it has two crossed rope beckets at each end. |
| ,, ,, II... ... ... | Provly. 8/3/70 | 1878 | 17·5 | 18·5 | 8·9 | 9·2* | 7·5 | 7·8 | Yes | II, 10-in., M.L., 40 lb., R. ↑ L. ... ... | Differs from I. in being conical at base. |
| ,, ,, III. ... ... | 27/9/72 | 2358 | 25·0 | 26·0 | 8·9 | 9·2 | 7·5 | 7·8 | No | III, 10-in., M.L., 70 lb., R. ↑ L.... ... | Dimensions increased to those of P. powder cartridges, and made without a choke to admit of being packed in zinc cylinders. |
| 9″, Mark I. ... ... .. | 23/10/67 | 1508 | 16·0 | 17·0 | 8·15 | 8·25 | — | — | No | I, 9-in., 30 lb., R. ↑ L. ... ... ... | The inside is formed of junk, it has two crossed rope beckets at each end. |
| ,, ,, II... ... ... | 4/2/69 | 1763 | 16·0 | 17·0 | 8·0 | 8·3 | — | — | Yes | II, 9-in., M.L., 30 lb., R. ↑ L. ... ... | Contains hollow wood block, it has two crossed rope beckets at each end |
| ,, ,, III. ... ... | 8/3/70 | 1878 | 16·0 | 17·0 | 8·0 | 8·3* | 6·8 | 7·1 | Yes | III, 9-in., M.L., 30 lb., R. ↑ L..... ... | Differs from II. in being conical at base. |
| ,, ,, IV. ... ... | 27/9/72 | 2358 | 21·0 | 22·0 | 8·0 | 8·3* | 6·8 | 7·1 | No | IV, 9-in., M.L., 50 lb., R. ↑ L. ... ... | Dimensions increased to those of P. powder cartridges, and made without a choke to admit of being packed in zinc cylinders. |
| 8″, Mark I. ... ... ... | 23/10/67 | 1508 | 15·2 | 15·7 | 7·25 | 7·35 | — | — | No | I. ... ... ... ... ... ... | The inside is formed of junk, it has two crossed rope beckets at each end. |
| ,, ,, II... ... ... | 4/2/69 | 1763 | 15·2 | 15·7 | 7·1 | 7·4 | — | — | Yes | II, 8-in., M.L., 20 lb., R. ↑ L. ... ... | Contains hollow wood block, it has two crossed rope beckets at each end. |
| ,, ,, III. ... ... | 8/3/70 | 1878 | 15·2 | 15·7 | 7·1 | 7·4 | 6·1 | 6·4 | Yes | III, 8-in., M.L., 20 lb., R. ↑ L. ... ... | Differs from II. in being conical at base. |
| 7″, Mark I. ... ... ... | 23/10/67 | 1508 | 13·5 | 14·3 | 6·35 | 6·45 | — | — | No | I, 7-in., 14 lb., R. ↑ L. ... ... ... | The inside is formed of junk, it has two crossed rope beckets at each end. |
| ,, ,, II... ... ... | 4/2/69 | 1763 | 13·5 | 14·3 | 6·2 | 6·5 | — | — | Yes | II, 7-in., M.L. 14 lb., R. ↑ L. ... ... | Contains hollow wood block, it has two crossed rope beckets at each end. |
| ,, ,, III. ... ... | 8/3/70 | 1878 | 13·5 | 14·3 | 6·2 | 6·5* | 5·2 | 5·5 | Yes | III, 7-in., M.L. 14 lb., R. ↑ L. ... ... | Differs from II. in being conical at base. |
| ,, ,, IV. ... ... | 27/9/72 | 2358 | 22·5 | 23·5 | 6·2 | 6·5 | 5·2 | 5·5 | No | IV, 7-in., M.L., 30 lb., R. ↑ L. ... ... | Dimensions increased to those of P. powder cartridges, and made without a choke to admit of being packed in zinc cylinders. |

By § 1839, Changes in War Stores, the following alterations were ordered to be made in future manufacture, without a change of pattern, viz., the rope becket to be removed from the base end of cartridges, and the becket at the front end set nearer the circumference to make it easier for a wad hook to catch them. Commanding officers were permitted to make these alterations, at their discretion, to such cartridges as had been issued to them, and also a rope's end with a tuft of loose yarn as an imitation choke.

* These limits in diameter were approved 8/3/70, § 1878, Changes in War Stores, to supersede those given in § 1839, besides those given at the time of sealing each pattern.

## CARTRIDGES (SILK CLOTH),

| Calibre, Nature, and Mark of Pattern | Date of Approval. | § Changes in War Stores. | Charge lbs., and Nature of Powder. | Length when filled. From | Length when filled. To | Diameter when filled. Body | Diameter when filled. Bottom | Number and description of Hoops. | Marks. |
|---|---|---|---|---|---|---|---|---|---|
| | | | | ins. | ins. | ins. | ins. | | |
| 12″·5 Battering, I.† ... | 1/1/77 | 3040 | 130, $P^2$. | 29·75 | 31·25 | 12·0 | 10·4 | 17 Broad. | I., 12″·5 M.L., 130 lb. $P^2$. |
| 12″, 35 ton, Battering, I. | 8/3/75 24/7/75 and 29/10/75 | 2838 | 110, P. | 27·5 | 28·5 | 11·5 | 9·9 | 17 ,, | I., 12″ M.L., 110 lb. P. |
| 12″, 35 ton, Full I. ... ... | ,, | ,, | 85, P., or<br>67, R.L.G. | 21·5<br>18·4 | 22·5<br>19·4 | 11·5<br>11·0 | 9·9<br>9·9 | 17 ,, | I., 12″ M.L., 85 lb. P. |
| 12″, 25 ton, Battering, I. ... | ,, | ,, | 85, P., or<br>67, R.L.G. | 21·5<br>18·4 | 22·5<br>19·4 | 11·5<br>11·0 | 9·9<br>9·9 | 12 ,, | I., 12″ M.L., 85 lb. P. |
| 12″, 25 ton, Full, I. ... ... | ,, | ,, | 55, P., or<br>50, R.L.G. | 14·25<br>14·5 | 15.25<br>15·5 | 11·5<br>11·0 | 9·9<br>9·9 | 7 ,, | I., 12″ M.L., 55 lb. P., 50 lb. R.L.G. |
| 11″, Battering, I. ... ... | ,, | ,, | 85, P., or<br>70, R.L.G. | 25·0<br>24·0 | 26·0<br>25·0 | 10·5<br>10·0 | 9·3<br>9·3 | 14 ,, | I., 11″ M.L., 85 lb. P., 70 lb. R.L.G. |
| 11″, Full, I. ... ... ... | ,, | ,, | 60, P., or<br>50, R.L.G. | 18·0<br>18·0 | 19·0<br>19·0 | 10·5<br>10·0 | 9·3<br>9·3 | 10 ,, | I., 11″ M.L., 60 lb. P., 50 lb. R.L.G. |
| 10″, Battering, I. ... ... | ,, | ,, | 70, P., or<br>60, R.L.G. | 25·0<br>25·0 | 26·0<br>26·0 | 9·5<br>9·0 | 8·25<br>8·25 | 13 ,, | I., 10″ M.L., 70 lb. P., 60 lb. R.L.G. |
| 10″, Full, I. ... ... ... | ,, | ,, | 44, P., or<br>40, R.L.G. | 16·3<br>17·5 | 17·8<br>18·5 | 9·5<br>9·0 | 8·25<br>8·25 | 8 ,, | I., 10″ M.L., 44 lb. P., 40 lb. R.L.G. |
| 9″, Battering, I. ... ... | ,, | ,, | 50, P., or<br>43, R.L.G. | 22·5<br>22·0 | 23·5<br>23·0 | 8·5<br>8·2 | 7·3<br>7·3 | 11 ,, | I., 9″ M.L., 50 lb. P., 43 lb. R.L.G. |
| 9″, Full, I. ... ... ... | ,, | ,, | 30, R.L.G. | 16·0 | 17·0 | 8·2 | ... | 7 Narrow. | I., 9″ M.L., 30 lb. |
| 9″, Reduced, I. ... ... | ,, | ,, | 15, R.L.G., or L.G., Blank | 11·0 | 12·0 | 6·8 | ... | 4 ,, | I., 9″ M.L., 15 lb. |
| 8″, Battering, I. ... ... | ,, | ,, | 35, P., or<br>30, R.L.G. | 21·0<br>20·0 | 22·0<br>21·0 | 7·5<br>7·3 | 6·43<br>6·43 | 11 ,, | I., 8″ M.L., 35 lb. P., 30 lb. R.L.G. |
| 8″, Full, I. ... ... ... | ,, | ,, | 20, R.L.G. | 15·2 | 15·7 | 7.3 | ... | 6 ,, | I., 8″ M.L., 20 lb. |
| 8″, Reduced, I. ... ... | ,, | ,, | 12, R.L.G., or L.G., Blank | 11·5 | 12·0 | 6·26 | ... | 4 ,, | I., 8″ M.L., 12 lb. |
| 7″, Battering, 1. ... ... | ,, | ,, | 30, P., or<br>22, R.L.G. | 22·5<br>18·3 | 23·5<br>19·3 | 6·5<br>6·4 | 5·57<br>5·57 | 12 ,, | I., 7″ M.L., 30 lb. P., 22 lb. R.L.G. |
| 7″, Full, I. ... ... ... | ,, | ,, | 14, R.L.G. | 13·5 | 14·3 | 6·4 | ... | 5 ,, | I., 7″ M.L., 14 lb. |
| 7″, Reduced, I. ... ... | ,, | ,, | 10, R.L.G., or L.G., Blank | 9·25 | 9·75 | 6·4 | ... | 3 ,, | I., 7″ M.L., 10 lb. |
| Bags, Spare Powder: 15 lbs. $P^2$ ... ... | — | — | — | — | — | — | — | — | — — |
| 15 ,, P. ... ... | — | — | — | — | — | — | — | — | — — |
| 10 ,, P. ... ... | — | — | — | — | — | — | — | — | — — |
| 15 ,, R.L.G. ... | — | — | — | — | — | — | — | — | — — |
| 10 ,, R.L.G. ... | — | — | — | — | — | — | — | — | — — |
| 15 ,, L.G. ... | — | — | — | — | — | — | — | — | — — |
| 10 ,, L.G. ... | — | — | — | — | — | — | — | — | — — |
| Weight of Case, empty ... | — | — | — | — | — | — | — | — | — — |

* Not yet sealed. † Only temporarily approved. ‡ 11 when hooped with worsted.

The above mode of packing is for § S. S. The L. S. cartridges for the Woolwich guns are packed in zinc cylinders. L. G. powder is used in the L. S. for full charges under the 10″, A. C. 7/74, and § 2838. The letters "L. G." to be

| Number Packed, and Weight of Package in Pounds. Cases, Powder. | | | | | | | | | | | | | | | | | | | | | | | | | |
|---|---|---|---|---|---|---|---|---|---|---|---|---|---|---|---|---|---|---|---|---|---|---|---|---|---|
| Metal-lined. | | | | | | Brass. | | | | | | | | | | | | | | | | | | | |
| Whole. | | Half. | | Quarter. | | Pentagon. | | | | Rectangular. | | | | | | | | | | | | | | | |
| | | | | | | Whole. | | Sectional. | | Plain. | | Corrugated. | | | | | | | | | | | | | |
| | | | | | | | | | | | | A | | B | | C | | D | | E | | F* | |
| Number. | Weight. | Number. | Weight. | Number. | Weight. | Number. | Weight. | Number. | Weight. | Number. | Weight. | Number. | Weight. | Number. | Weight. | Number. | Weight. | Number. | Weight. | Number. | Weight. | Number. | Weight. | Number. | Weight. |
| — | — | — | — | — | — | — | — | — | — | — | — | — | — | — | — | — | — | — | — | — | — | — | — | 2 | 337 |
| — | — | — | — | — | — | — | — | — | — | — | — | — | — | — | — | — | — | — | — | 2 | 294 | — | — | 2 | 297 |
| — | — | — | — | — | — | — | — | — | — | — | — | — | — | — | — | — | — | 2 | 228 | 3 | 329 | — | — | 3 | 332 |
| — | — | — | — | — | — | — | — | — | — | — | — | — | — | — | — | — | — | 2 | 192 | — | — | — | — | — | — |
| — | — | — | — | — | — | — | — | — | — | — | — | — | — | — | — | — | — | 2 | 228 | — | — | — | — | — | — |
| — | — | — | — | — | — | — | — | — | — | | — | — | — | — | — | — | — | 2 | 192 | — | — | — | — | — | — |
| — | — | — | — | — | — | — | — | — | — | — | — | — | — | — | — | — | — | 3 | 222 | — | — | — | — | — | — |
| — | — | — | — | — | — | — | — | — | — | — | — | — | — | — | — | — | — | 3 | 208 | — | — | — | — | — | — |
| — | — | — | — | — | — | — | — | — | — | — | — | — | — | — | — | — | — | — | — | — | — | 2 | 229 | — | — |
| — | — | — | — | — | — | — | — | — | — | — | — | — | — | — | — | — | — | 2 | 198 | — | — | 2 | 199 | — | — |
| — | — | — | — | — | — | — | — | — | — | — | — | — | — | — | — | — | — | — | — | — | — | 3 | 239 | — | — |
| — | — | — | — | — | — | — | — | — | — | — | — | — | — | — | — | — | — | 2 | 158 | — | — | 3 | 209 | — | — |
| — | — | — | — | — | — | — | — | — | — | — | — | — | — | — | — | 2 | 193 | — | — | — | — | — | — | — | — |
| — | — | — | — | — | — | — | — | — | — | — | — | — | — | — | — | 2 | 173 | — | — | — | — | — | — | — | — |
| — | — | — | — | — | — | — | — | — | — | — | — | — | — | — | — | 3 | 185 | — | — | — | — | — | — | — | — |
| — | — | — | — | — | — | — | — | — | — | — | — | — | — | — | — | 3 | 173 | — | — | — | — | — | — | — | — |
| — | — | — | — | — | — | — | — | — | — | — | — | — | — | 3 | 198 | — | — | — | — | — | — | — | — | — | — |
| — | — | — | — | — | — | — | — | — | — | — | — | 1 | 113 | 3 | 177 | — | — | — | — | — | — | — | — | — | — |
| 3 | 140 | — | — | — | — | — | — | — | — | — | — | 3 | 160 | 4 | 168 | — | — | — | — | — | — | — | — | — | — |
| 7 | 155 | — | — | — | — | 7 | 168 | — | — | — | — | 7 | 176 | 8 | 168 | — | — | — | — | — | — | — | — | — | — |
| — | — | — | — | — | — | — | — | — | — | — | — | — | — | 3 | 152 | — | — | — | — | — | — | — | — | — | — |
| 2 | 109 | — | — | — | — | 2 | 122 | — | — | — | — | 3 | 160 | 3 | 137 | — | — | — | — | — | — | — | — | — | — |
| 5 | 150 | — | — | — | — | 5 | 163 | — | — | — | — | 5 | 171 | 5 | 148 | — | — | — | — | — | — | — | — | — | — |
| 9 | 158 | 4 | 79 | — | — | 9 | 171 | — | — | — | — | 9 | 179 | 10 | 168 | — | — | — | — | — | — | — | — | — | — |
| — | — | — | — | — | — | — | — | — | — | — | — | 3 | 160 | 4 | 168 | — | — | — | — | — | — | — | — | — | — |
| 5 | 160 | — | — | — | — | 4 | 151 | — | — | — | — | 5 | 181 | 5 | 158 | — | — | — | — | — | — | — | — | — | — |
| 7 | 147 | 3 | 73 | — | — | 7 | 160 | — | — | — | — | 7 | 168 | 8 | 160 | — | — | — | — | — | — | — | — | — | — |
| 11 | 160 | 4 | 71 | — | — | 10‡ | 162 | — | — | — | — | 12 | 191 | 14 | 188 | — | — | — | — | — | — | — | — | — | — |
| 10 | 201 | 4 | 91 | — | — | 9 | 198 | — | — | 10 | 222 | 11 | 214 | 14 | 265 | 16 | 300 | 25 | 453 | 17 | 317 | 26 | 472 | | |
| 10 | 201 | 4 | 91 | — | — | 9 | 198 | — | — | 10 | 222 | 12 | 229 | 14 | 265 | 16 | 300 | 26 | 469 | 18 | 332 | 27 | 487 | | |
| 15 | 201 | — | — | — | — | 13×10 1×5 | 198 | — | — | — | — | — | — | — | — | — | — | — | — | — | — | — | — | — | — |
| 8 | 170 | 4 | 91 | — | — | 8 | 183 | — | — | 9 | 207 | 10 | 199 | 13 | 249 | 15 | 285 | 22 | 408 | 15 | 287 | 23 | 427 | | |
| 12 | 170 | — | — | — | — | 12 | 183 | — | — | — | — | — | — | — | — | — | — | — | — | — | — | — | — | — | — |
| 8 | 170 | 4 | 91 | — | — | 8 | 183 | — | — | 9 | 207 | 10 | 199 | 13 | 249 | 15 | 285 | 22 | 408 | 15 | 287 | 23 | 427 | | |
| 12 | 170 | — | — | — | — | 12 | 183 | — | — | — | — | — | — | — | — | — | — | — | — | — | — | — | — | — | — |
| — | 48 | — | 30 | — | 18 | — | 61 | — | 43 | — | 69 | — | 46 | — | 51 | — | 56 | — | 72 | — | 57 | — | 75 | | |

see p. 77.
marked on the cartridges in red, § 1998.

CARTRIDGES (SERGE),

| Calibre, Nature and Mark of Pattern. | Date of Approval. | § Changes in War Stores. | Charge, lbs. and Nature of Powder. | Length when filled. From. | To. |
|---|---|---|---|---|---|
| | | | | ins. | ins. |
| 12″, 35 Ton, Battering, I. ... ... | 10/10/72 | 2373 2607 | 110, P. ... ... | 27·5 | 28·5 |
| 12″ 35 Ton, Full, or ... | 28/8/71 | 2103 | 85, P., or ... | 21·5 | 22·5 |
| 12″, 25 Ton, Battering, I. ... | 23/4/72 | 2517 | 67, R.L.G. ... | 18·4 | 19·4 |
| 12″, 25 Ton ,, I. ... ... | 16/10/68 and 9/1/69 | 1740 | 67, R.L.G. ... | 18·4 | 19·4 |
| 12″, 25 Ton ,, I. ... ... | 11/1/71 | — | 85, P. ... ... | 21·5 | 22·5 |
| 12″, 25 Ton, Full, I. ... ... ... | 16/10/68 and 9/1/69 | 1740 | 50, R.L.G. ... | 14·5 | 15·5 |
| 12″ 25 Ton ,, I. ... ... ... | 28/8/71 | 2103 | 55, P., or ... | 14·25 | 15·25 |
| | | | 50, R.L.G. ... | 14·5 | 15·5 |
| 11″, Battering, I. ... ... ... | 15/4/72 | 2269 | 85, P., or ... | 25·0 | 26·0 |
| | | | 70, R.L.G. ... | 24·0 | 25·0 |
| 11″, Full, I. ... ... ... ... | ,, | ,, | 60, P., or ... | 18·0 | 19·0 |
| | | | 50, R.L.G. ... | 18·0 | 19·0 |
| 10″, Battering, I. ... ... ... | 8/9/68 | 1673 | 60, R.L.G. ... | 25·0 | 26·0 |
| ,, ,, II. ... ... ... | 16/10/68 and 9/1/69 | 1740 | 60, R.L.G. ... | 25·0 | 26·0 |
| ,, ,, I. ... ... ... | 11/1/71 | — | 70, P. ... ... | 25·0 | 26·0 |
| ,, ,, I. ... ... ... | 28/8/71 | 2103 | 70, P., or ... | 25·0 | 26·0 |
| | | | 60, R.L.G. ... | 25·0 | 26·0 |
| ,, Full, I. ... ... ... ... | 8/9/68 | 1673 | 40, R.L.G. ... | 17·5 | 18·5 |
| ,, ,, II. ... ... ... ... | 16/10/68 and 9/1/69 | 1740 | 40, R.L.G. ... | 17·5 | 18·5 |
| ,, ,, I. ... ... ... ... | 23/8/71 | 2103 | 44, P., or ... | 16·3 | 17·3 |
| | | | 40, R.L.G. ... | 17·5 | 18·5 |
| 9″, Battering, I. ... ... ... | 20/3/66 | 1194 | 43, R.L.G. ... | 22·0 | 23·0 |
| ,, ,, II. ... ... ... | 16/10/68 and 9/1/69 | 1740 | 43, R.L.G. ... | 22·0 | 23·0 |
| ,, ,, I. ... ... ... | 11/1/71 | — | 50, P. ... ... | 22·5 | 23·5 |
| ,, ,, I. ... ... ... | 28/8/71 | 2103 | 50, P., or ... | 22·5 | 23·5 |
| | | | 43, R.L.G. ... | 22·0 | 23·0 |
| 9″, Full, I. ... ... ... ... | 20/3/66 | 1194 | 30, R.L.G. ... | 16·0 | 17·0 |
| 9″, Reduced, I. ... ... ... ... | 20/3/66 | 1194 | 15, R.L.G. or L.G., Blank | 11·0 | 12·0 |
| 8″, Battering, I. ... ... ... | 18/1/66 | 1292 | 30, R.L.G. ... | 20·0 | 21·0 |
| ,, ,, I. ... ... ... | 17/1/71 | — | 35, P. ... ... | 21·0 | 22·0 |
| ,, ,, I. .. ... ... | 28/8/71 | 2103 | 35, P., or ... | 21·0 | 22·0 |
| | | | 30, R.L.G. ... | 20·0 | 21·0 |
| 8″, Full, I. ... ... ... ... | 24/8/66 | 1292 | 20, R.L.G. ... | 15·2 | 15·7 |
| 8″ Reduced, I. ... ... ... ... | 24/8/66 | 1292 | 12, R.L.G., or... L.G. Blank | 11·5 | 12·0 |
| 7″, Battering, I. ... ... ... | 18/1/66 | 1188 | 22, R.L.G. ... | 18·3 | 19·3 |
| ,, ,, I. ... ... ... | 17/1/71 | — | 30, P. ... ... | 22·5 | 23·5 |
| ,, ,, I. ... ... ... | 28/8/71 | 2103 | 30, P., or ... | 22·5 | 23·5 |
| | | | 22, R.L.G. ... | 18·3 | 19·3 |
| 7″, Full, I. ... ... ... ... | 18/1/66 | 1188 | 14, R.L.G. ... | 13·5 | 14·3 |
| 7″, Reduced, I. ... ... ... ... | 18/1/66 | 1188 | 10, R.L.G. or ... L.G. Blank | 9·25 | 9·75 |
| 7″, Reduced, or 32 pr. S.B., I. ... | 25/10/70 | 1977 | 10, R.L.G. or L.G. Blank | 10·75 | — |

For packing, see Table p. 220.

## Woolwich Guns.

| Diameters when filled. | | No. and Description of Braid Hoops. | Marks. | Remarks. |
|---|---|---|---|---|
| Body. | Bottom. | | | |
| ins. | ins. | | | |
| 11·5 | 9·9 | 16 double width | I., 12 in. M.L., 110 lb. P. | Previous to 28/8/71 there were separate patterns of cartridges for R.L.G. and P., viz., of a cylindrical form with rounded bottoms for R.L.G. charges, and with a conical and flat bottom for P. By § 2103 it was decided that the cartridges for all battering charges and full charges of 40 lbs. and over should in future be of the shape for P. powder to suit the chambers of the guns, and marked for R.L.G. as well as for P., so that they might be used for P, when R.L.G. is not available, the cartridges being made up to the length specified in the table for charges of R.L.G. by hooping them tightly, they being of larger diameter than those heretofore used for similar charges of R.L.G. When filled with P., the marking for R.L.G. will be obliterated and *vice versâ*.<br>As many of the earlier patterns may still exist, they are shown in the table. The cylindrical cartridge with rounded bottom is still the pattern for full charges under 40 lbs. for which P. is not used.<br>All new cartridges for R.M.L guns (7 pr. 4 oz. excepted) will be manufactured of silk cloth.<br>Filled and empty R.M.L. cartridges made of serge, now in store, with the exception of empty carriages of 85 lbs. and upwards, will be issued or used up, as a rule, before any store of silk cloth is made use of. The empty cartridges for 85 lbs. and upwards will be exchanged and returned for conversion, and all future issues will be of silk cloth. § 2838. Appendix 10/75. |
| 11·5 | 9·9 | 12 ,, ,, | I., ,, ,, 85 lb. P. | |
| 11·0 | 9·9 | 12 ,, ,, | | |
| 11·0 | — | 9 ,, ,, | I., ,, ,, 67 lb. | |
| 11·5 | 9·9 | 12 ,, ,, | I., ,, ,, 85 lb. P. | |
| 11·0 | — | 7 ,, ,, | I., ,, ,, 50 lb. | |
| 11·5 | 9·9 | 7 ,, ,, | I., ,, ,, 55 lb. P. 50 R.L.G. | |
| 11·0 | 9·9 | 7 ,, ,, | | |
| 10·5 | 9·3 | 14 ,, ,, | I., 11 in. M.L., 85 P., 70 R.L.G. | |
| 10·0 | 9·3 | 14 ,, ,, | | |
| 10·5 | 9·3 | 10 ,, ,, | I., 11 in. M.L., 60 P., 50 R.L.G. | |
| 10·0 | 9·3 | 10 ,, ,, | | |
| 9·0 | — | 12 single width | I., 10 in. M.L., 60 lb. | |
| 9·0 | — | 12 double width | II., ,, ,, 60 lb. | |
| 9·5 | 8·25 | 12 ,, ,, | I., ,, ,, 70 lb. | |
| 9·5 | 8·25 | 13 ,, ,, | I., ,, ,, 70 P., 60 R.L.G. | |
| 9·0 | 8·25 | 13 ,, ,, | | |
| 9·0 | — | 7 single width | I., 10 in. M.L., 40 lb. | |
| 9·0 | — | 7 double width | II., ,, ,, 40 lb. | |
| 9·5 | 8·25 | 8 ,, ,, | I., ,, ,, 44 P., 40 R.L.G. | |
| 9·0 | 8·25 | 8 ,, ,, | | |
| 8·2 | — | 11 single width | I., 9 in. M.L., 43 lb. | |
| 8·2 | — | 11 double width | II., 9 in. M.L., 43 lb. | |
| 8·5 | 7·3 | 10 ,, ,, | I., 9 in. M.L., 50 lb. | |
| 8·5 | 7·3 | 11 ,, ,, | I., ,, ,, 50 P., 43 R.L.G. | |
| 8·2 | 7·3 | 11 ,, ,, | | |
| 8·2 | — | 7 single width | I., 9 in. M.L., 30 lb. | |
| 6·8 | — | 4 ,, ,, | I., ,, ,, 15 lb. | |
| 7·3 | — | 9 ,, ,, | I., 8 in. ,, 30 lb. | |
| 7·5 | 6·43 | 10 ,, ,, | I., ,, ,, 35 lb. P. | |
| 7·5 | 6·43 | 11 ,, ,, | I., ,, ,, 35 P., 30 R.L.G. | |
| 7·3 | 6·43 | 11 ,, ,, | | |
| 7·3 | — | 6 ,, ,, | I., 8 in. M.L., 20 lb. | |
| 6·26 | — | 4 ,, ,, | I., ,, ,, 12 lb. | |
| 6·4 | — | 8 ,, ,, | I., 7 in. ,, 22 lb. | |
| 6·5 | 5·57 | 11 ,, ,, | I., ,, ,, 30 lb. P. | |
| 6·5 | 5·57 | 12 ,, ,, | I., ,, ,, 30 P., 22 R.L.G. | |
| 6·4 | 5·57 | 12 ,, ,, | | |
| 6·4 | — | 5 ,, ,, | I., 7 in. M.L., 14 lb. | |
| 6·4 | — | 3 ,, ,, | I., ,, ,, 10 lb. | |
| 6·0 | — | 3 worsted | I., 32 pr. 58 or 56 cwt. D. or 7 in. M.L., 10 lb. | |

# CHAPTER XIII.—AMMUNITION FOR 8-INCH AND 6·3-INCH R.M.L. HOWITZERS; 80-PR., 64-PR., 40-PR., AND 25-PR. R.M.L. GUNS.

COMMON AND SHRAPNEL SHELL.—CASE SHOT.—GAUGES.—CARTRIDGES AND STORES CONNECTED WITH THEM.

THE pieces of ordnance enumerated above, with the exception of the 80-pr., form the heavy and light units of a siege train,* and thus it will be convenient to bring them under one heading, and the 80-pr., though not a siege gun, may also be classed with them as being associated in its ammunition to a certain extent with the 64-pr.

Besides the two howitzers treated of in this chapter there is a 10-inch howitzer of 6 tons; but it is not yet in the service, nor is any ammunition sealed for it. It will, therefore, not receive further notice in this work.

The projectiles fired from the ordnance treated of in this chapter are common and Shrapnel shell, and case shot. Patterns of all these projectiles have not yet been sealed for the howitzers, in fact, no projectiles are as yet formally approved of for the 6″·3 howitzer, except the case shot which is the same as that for the 64 and 80-pr. guns.

*8-inch Howitzer.*

Projectiles for 8″ howitzer.

Common shell and Shrapnel shell have been tried, but no pattern of Shrapnel has been yet sealed.†

Common shell. § 2539.

The small charges used render the construction of a percussion fuze difficult. The R.L. percussion fuze, Mark II., requires a 7 lb. charge to ensure its action from the 8″ Howitzer. A special 30-seconds time fuze has been tried. See Extracts, Vol. XI, p. 53. The "sensitive" fuze is for use with this piece.

*Common Shell.*—This is the same as the shell for the 8″ gun, except as to the studs, which are made to suit the quick uniform twist of the howitzer (the twist is 1 turn in 16 calibres). As the twist is uniform,

* The ordnance composing a Heavy Siege Train Unit consists of 8 64-prs. 8 40-prs., and 14 8-inch howitzers; that composing a Light Siege Unit consists of 10 40-prs., 10 25-prs., and 10 6·3-inch howitzers.

With each unit are associated 6 7-pr. guns and 300 24-pr. Hale rockets. The ammunition, &c., for the 7-pr. guns will be found in the next chapter. For the rockets, see chapter

† A Shrapnel shell studded to suit the quick uniform twist of the howitzer, but otherwise resembling the 8″ gun Shrapnel has been made, but is not sealed.

the front and rear studs are of the same size, so the shell can be easily distinguished from that for the gun.*

For capacity, &c., see table, p. 234.

*Common Shell, 8″ Howitzer.*

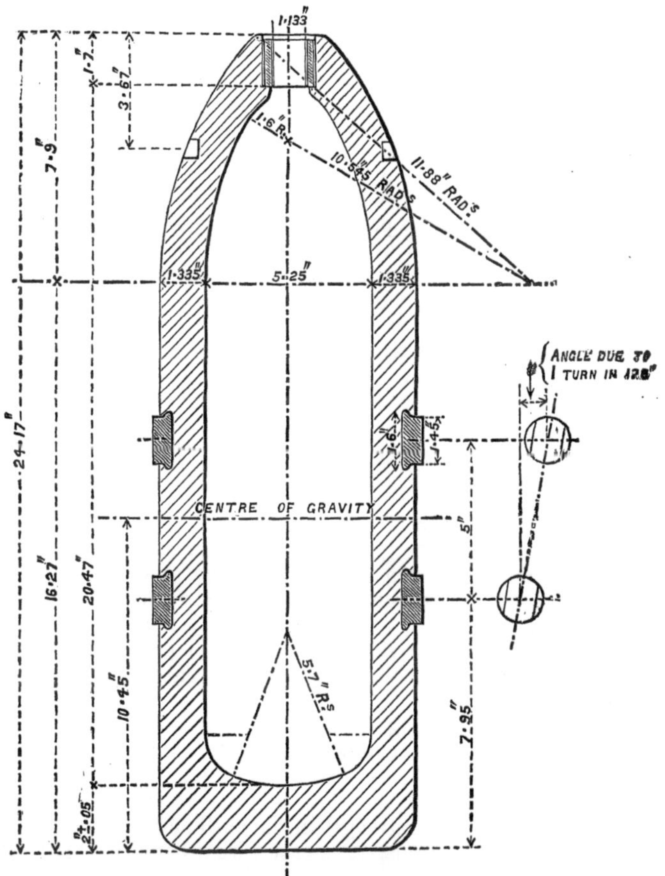


This shell has evidently a large margin of strength, as it has only to stand a charge of 10 lbs. in the howitzer, while it is strong enough to stand a 35 lb. charge. If desired, its capacity could be increased without unduly weakening the shell.

Star shell and carcass shell have been tried with this howitzer, and will probably be introduced, but no pattern is as yet sealed.

The star shell somewhat resembles a parachute light ball externally.

---

* A shell weighing 146 lbs. has been recommended. Various shell have been tried, the general object being to get a shell of less weight and more capacity. See Extracts, Vol. XII, pp. 202, 218, &c. It is not however probable that the pattern will be changed at present.

A battering shell has been tried for the 8″ howitzer. It is of Palliser form and arranged so as to take a fuze in the base. The head is slightly chilled, but the metal is not so hard as that used for Palliser projectiles.

It is spherical in form, and contains a number of stars of magnesium light composition in paper cylinders. It is very much the same in principle as the shells used in firework displays.

The carcass is like an ordinary shell, but as there is no fuze hole requisite, the head resembles that of a Palliser projectile. It is filled with carcass composition, and in general construction is like the S.B. carcass, only elongated instead of spherical.

**Case shot.**
**§ 2742.**

The 8″ gun case shot is to be fired from the 8″ howitzer with a 10 lb. charge.

### 6·3-*inch Howitzer.*

Projectiles for the 6″·3 howitzer.

This howitzer has no projectiles as yet actually sealed for it. It is polygrooved, having a twist increasing from 1 in 100 to 1 in 35 calibres, and the rotation to the projectiles will be given by a cup-shaped gas-check, there being no studs. The common shell to be used will probably be about 16″ long, contain a bursting charge of 7 lbs., and have a total weight, when filled, including gas-check, of 68 lbs.

Common shell.

Star shell and carcass shell resembling those for the 8″ howitzer have been tried, but no patterns are yet sealed.

No Shrapnel shell has as yet been decided on for this howitzer.

Case Shot.

The case shot is the same as that for the 80-pr. or 64-pr. guns.

### 80-*pr. R.M.L.* (*Converted*) *Gun.**

This gun is converted on Major Palliser's principle from the 68-pr. of 95 cwt. and weighs 5 tons. It has three grooves of the same form as those in the Woolwich guns, but of smaller dimensions. The twist is uniform, and is 1 turn in 40 calibres. The studs in the projectiles are made of copper, to which a little zinc (·214 per cent.) is added. This soft alloy is used as the shells are greatly strained in pressing in the studs. There are two rings of studs, 3 in each ring, the front studs are much smaller than the rear ones, to enable them to be fixed with less pressure.†

Common shell.
§ 2278.

Common shell. Gauge G.S.
Fuzes for L.S. Pettman G.S. and R.L. Mark II.
Fuzes for S.S. (This is a L.S. gun only.)

The construction of the shell is peculiar (the form is given in the woodcut); in order to strengthen the shell sufficiently to bear the strain of pressing in the studs, it is cast with a band of increased thickness

---

* The calibre of the gun is 6″·29, the diameter of gun and projectile are as follows:—

| | inch. | | inch. |
|---|---|---|---|
| Diameter of gun .. .. | 6·29 | Width of groove .. .. | 1·3 |
| " shell .. .. | 6·22 | " rear stud .. .. | 1·22 |
| Windage over body .. | ·07 | Play of stud .. .. | 0·08 |
| Diameter of gun + depth of grooves .. .. .. | 6·58 | | |
| Diameter of shell + height of studs .. .. .. | 6·53 | Height of studs.. .. .. | 0·155 |
| | | Depth of grooves .. .. | 0·145 |
| Windage over studs .. | 0·05 | Clearance .. .. .. | 0·01 |

† Pressure on bottom stud 16·52 tons, on top stud 9·82 tons.

80-pr. Common Shell. Mark I.

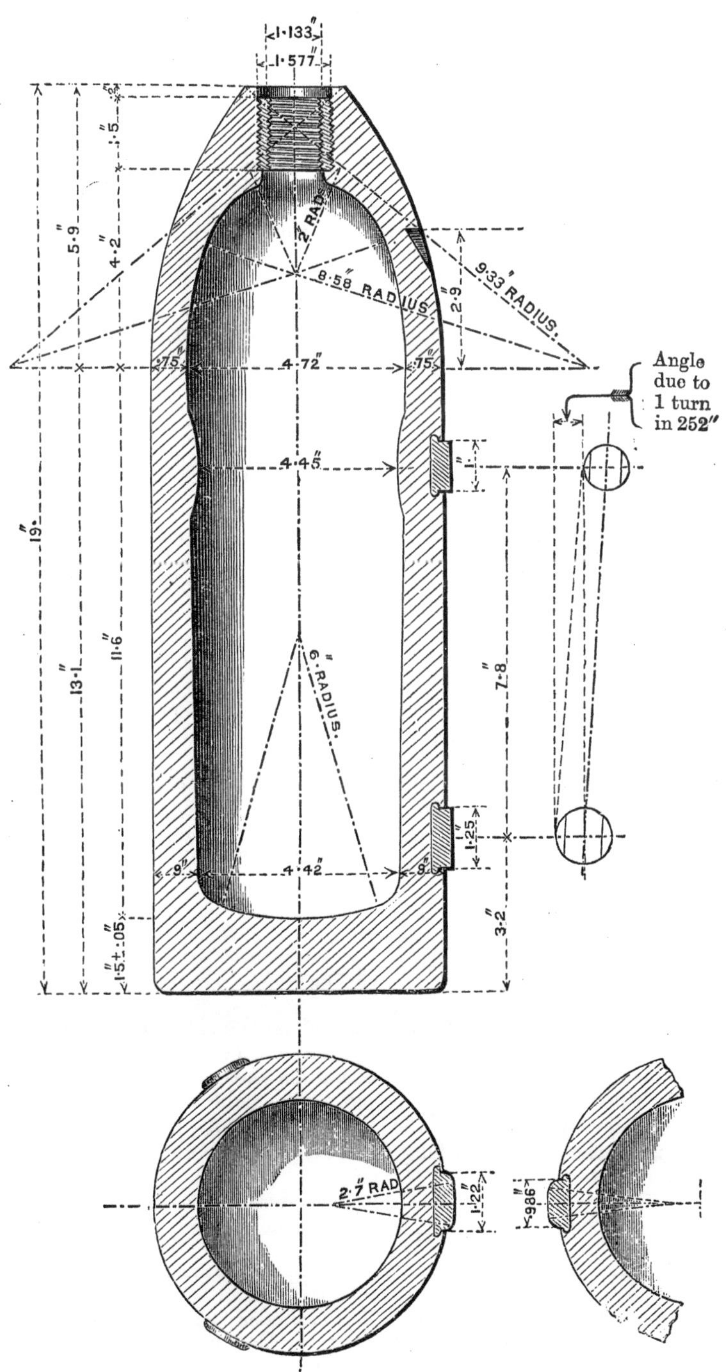

under the front stud; the thickness of the shell increases gradually towards the base.

For dimensions, bursting charge, patterns, and other details, *vide* table, p. 234.

Issue.

For L.S. empty, loose.

Shrapnel shell.

*Shrapnel Shell.* Gauge, G.S.

Fuzes for L.S. 5 and 9 sec. M.L. time.

Fuzes for S.S. (This is a L.S. gun only.)

These shell generally resemble in construction the Shrapnel for guns of higher calibres, but the balls are of mixed metal (*i.e.* lead and antimony), and round the central iron tube is a thin wooden tube, which is introduced to fill up space and bring the shell to the proper length, without unduly increasing the weight. Two longitudinal slots are cut at each end, so as to allow the rosin to flow in and bind the tubes together. The slots at the bottom are cut at right angles to those at the top, so as to enable them all to extend half the length of the tube without actually dividing the latter. The Shrapnel for the 64-pr. has a similar wooden tube. See cut of the latter shell, p. 231.

The first pattern of this shell had a projecting socket, but this was found to render the head liable to injury in transit, or on grazing when fired. Mark II. had a flush socket, and the head resembled in construction the cut given on p. .

Mark III. has also a flush socket, and the head is as given in the cut of the 64-pr. Mark V., p. .

For dimensions, bursting charge, &c., see table, p. 235.

Issue.

For L.S., empty, loose.

Case shot.

The present pattern of case shot is that described on p. 140, but it has the peculiarity of having six iron segmental linings instead of three. The patterns made with three did not break up satisfactorily, hence the change. The same projectile serves for the 64-pr. and 6″·3 howitzer.

For details, dimensions, &c., see table, p. 236.

Issue.

Loose.

Extractors.
§ 2850.

The extractors for the 8″ howitzer is the same as that for the 8″ gun, differing only in the length of the stave, which is 6′ long.

No extractor is yet sealed for the 6″·3 howitzer. It will probably resemble the others described previously.

§ 2362.

The extractor for the 80-pr. is on the same principle as those for Woolwich guns. Length of stave, 7′ 3″.

## 64-pr. Guns.

The manufacture of 64-pr. guns* on the shunt principle has ceased but a large number of S.B. guns have been converted to 64-pr. rifled guns; these take the same ammunition as the shunt gun, although they are rifled on a different principle. The calibre of the 64-pr. guns are very nearly the same, the shunt being 6·3″, and the converted gun 6·29″; the twist of rifling is the same in both guns, viz., 1 turn in 40 calibres.

*The Shunt System* has been considered inferior to the Woolwich, because, besides being unnecessarily complicated, the grooves which are

* The 64-pr. will take 32-pr. S.B. ammunition; 32-pr. solid shot is fired from this gun at practice by the reserve forces, but 32-pr. case is not used with this gun, as it scores the bore.—Extracts, Vol. X, p. 169.

cut with abrupt sharp angles weaken the gun.* The principle is as follows:—The studs of the shell bear on the deep side of a double groove on entering, and on the shallow side on leaving the bore, thus loading esaily, yet leaving the muzzle tightly gripped with a stable axis. This may be easily explained:—A rifled projectile, both on entering and leaving the bore, is driven by a force acting along its axis, and rotation is given by the stud coming against the spiral formed by the edge of the groove, thus in figure, ifthe projectile were pushed base first, the stud would move against the loading edge C C of the groove, while, if pressure were applied to the base so as to move it point first, the studs would meet the driving edge D D and work along it. In the shunt gun, however, in addition to the mere fact of the driving side of the double groove being shallow, and the loading side deep, the two grooves join in one deep one at 2 feet 7·5 inches from the muzzle, each stud thus running down the deep side into the common groove, and returning up the common groove and shallow side, just as a railway engine shunts from one line to another by backing on a common piece, into which both are made to lead. The shallow groove does not break with an abrupt step into the single groove, but by a gradual incline, the exact details of which belong to the Gun Factory. It may be noticed also that at the bottom of the bore, one groove is made to narrow slightly so that after running along the loading edge, each stud is brought to touch the driving edge just at the bottom of the bore; this saves it from bounding on to it with a blow on the first impulse from the charge.

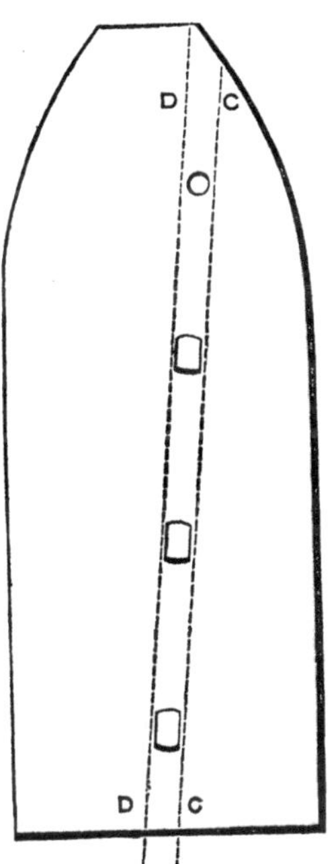


The studs are of copper; there were formerly five, but now three only to each groove; they are attached by swedging into undercut holes; their form is cylindrical, with the sides bevelled off to a certain extent. The pressure required to swedge them in is 7·14 tons.

The 64-pr. R.M.L. projectiles may be easily recognised by their having three rings of copper studs, three in each ring, whilst formerly they had five rings of studs, three in each ring. It is necessary to employ a soft metal to allow the studs to be compressed when firing from the shunt gun, and to prevent shearing, a larger bearing surface is desirable, this latter point is attained by having three rings of studs.

---

* In the case of the two guns which burst, the steel tubes split in the direction of the sharp angles of the grooves, and the Committee are inclined to think that the angular form of the groove in the shunt system, for the greater part of the length, may possibly have promoted the ultimate disruption of the tubes. The Woolwich form of groove is free from the defect due to sharp angles. The Superintendent, Royal Gun Factories, in a letter to the Ordnance Select Committee, attributes the failure of the 13″ gun, among other causes, to the gun having been rifled on the shunt system.—*Vide* Extracts, Vol. V, p. 99.

The dimensions of grooves and studs are as follows:—A section of the deep groove gives a depth of ·11″, width at bottom ·6″. The incline of the edges is given by tangents drawn from each side of the groove to a circle concentric with the bore 1″·85 radius. The shallow or "pinching" groove whose driving edge is in the production of that of the deep groove (*i.e.*, at a spiral of 1 in 40 calibres), decreases in depth by ·025″ in the first 12″·6 incline, and by ·005″ the next 12″·6, from which it runs at a constant depth of ·08″ to the muzzle, where the deep groove has so far shunted from it as to give a step or shallow groove ·4″ inches wide. The width of the stud is ·55 inches, so that it must always partly extend over the deep groove. The height ·125″ shows that for any part to enter the shallow groove it must undergo a compression or wearing down of ·01 inch; supposing the compression to be equal on all the studs, it would be ·005″ on each.

The windage over the body of all shunt projectiles is ·08″, over the studs ·04″, the clearance is ·015″, and the play of the stud is ·18″.

In the 64-pr. converted gun the plain groove is employed, the edges are angular and slightly rounded off.*

All shells have extractor holes, three in number, in the same line as the studs. The mark showing the pattern of the shell will be found between an extractor hole and the row of studs.

Common shell.

*Common Shell.* Gauge G.S.
Fuzes L.S. Pettman's G.S. and R.L. Mark II.†
Fuzes S.S. 9 secs. and 20 secs. M.L. time. Pettman's G.S.
For details see table p. 234.

Issue.

Generally filled for siege train or S.S., in the latter case fuzed with Pettman's G.S. fuze. Loose. For other L.S. empty, loose.

Battering shell.

Trials are now being made with a battering shell for the 64-pr. It is of Palliser form, but only slightly chilled, and the metal is not so hard as that used for Palliser projectiles. It is arranged to take a fuze in the base.

Shrapnel shell.

*Shrapnel Shell.*—Gauge G.S.
Fuzes for L.S. 5 sec. and 9 sec. M.L. time.
Fuzes for S.S. 9 secs. M.L. time.

The construction of this shell (see cut) is precisely similar, except in dimensions and in the position and number of the extractor holes, to the

---

* For particulars and drawings of groove of both shunt and converted guns, see "Treatise on Construction of Ordnance," pp. 265, 267.

The following dimensions in the converted 64-pr. show the clearance, play of studs, and windage over body and studs:—

| | Inches. | | Inches |
|---|---|---|---|
| Calibre of gun .. .. .. | 6·29 | Calibre of gun + depth of grooves .. .. .. | 6·52 |
| Diameter of shell .. .. | 6·22 | Diameter of shell + height of studs.. .. .. .. | 6·47 |
| Windage over body .. | ·07 | Windage over studs .. | ·05 |
| Height of stud .. .. | ·125 | Width of groove .. .. | ·60 |
| Depth of groove .. .. | ·115 | " stud .. .. | ·55 |
| Clearance .. .. | ·010 | Play of stud .. .. | ·05 |

† There will doubtless be time fuzes for the 64-pr. guns used in the siege train.

80-pr. Shrapnel. Earlier patterns had a projecting socket, which was found to render the head liable to injury in transit, or on grazing when fired. A strong gun-metal socket replaces the composite one of tin and gun-metal which serves to receive the fuze. The bottom of the socket is screwed to receive the brass primer, the iron central tube fits on to the end of the fuze socket, and is enlarged in diameter to facilitate loading, being $1\frac{1}{2}''$ wide; and the under surface of the diaphragm and the top of the tin cup are coned to facilitate unloading.

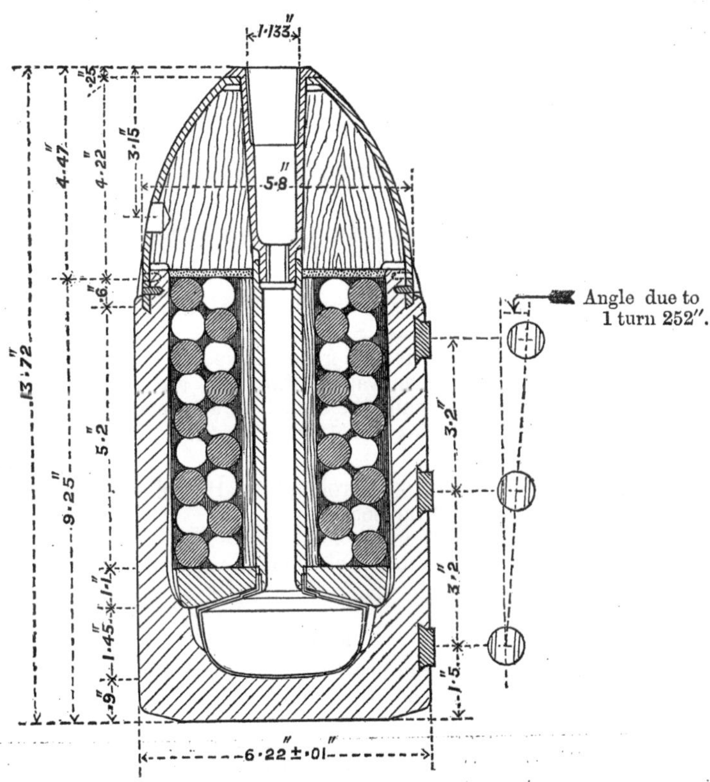


For S.S. filled and boxed, primer fixed and marked as usual. Empty, loose for L.S.

**Issue.** § 2357.

For dimensions, bursting charges, &c., *vide* table, p. 235.

**Case shot.**

The case shot for the 64-pr. is the same projectile as that used for the 80-pr., and it is marked for both guns in the later patterns

**Issue.**

Loose.

For dimensions, &c., see table, p. 236.

**Extractor.** §§ 1205, 1712.

The extractor has three guides, which keep in the grooves and bring

the teeth of the jaws over the extractor holes; by turning the handle the collar, A, attached to the counter-jaws, recedes or advances, opening or closing the jaws by means of pins attached to the collar, which work in slots in the counter-jaws.

### 40-*pr. R.M.L. Gun.**

40 pr. R. M.L.

The calibre of this gun is the same as that of the B.L. gun, viz., 4″·75; it is rifled with 3 grooves of the Woolwich form, and the rifling is uniform, 1 turn in 35 calibres.

The studs on the shells are made of gun-metal, there are two rings of studs, with three studs in each ring.

Common shell

*Common Shell.*—Gauge, G.S.

Fuzes for L.S. 9 secs. and 20 secs., M.L., Marks II. and III.—R.L. percussion, Mark II.

Fuzes for S.S. (this is a L.S. gun only).

This shell is the lowest one that has a gun-metal bush. The latter is not countersunk. For dimensions, &c., see table, p. 234.

Issue.

For siege train, filled and boxed.

Shrapnel shell

*Shrapnel Shell.*—Gauge, G.S.

Fuzes for L.S. 5 secs. and 9 secs. M.L.—R.L. percussion, Mark II.

Fuzes for S.S. (this is a L.S. gun only).

Mark II.
§ 3172.

The construction of this shell generally resembles that of the Shrapnel for larger natures, but the walls of the shell are thin, and the central tube is of gun-metal, while the socket is a composite one of tin and gun-metal. It resembles the 16-pr. Shrapnel, of which a cut is given on p. 244.

Mark I.
§ 2504.

This shell is ordered to be broken up. It had a narrow central iron tube over which fitted the socket.

Issue.

For siege train, filled and boxed.

Case shot.

*Case Shot.*—The case shot is of the R.L. pattern. For dimensions, &c., see table, p. 236.

Issue.

Boxed.

### 25-*pr. Gun.*†

§ 2673.

The calibre of this gun is 4″.00. It is rifled with three grooves of the Woolwich form, and the rifling is uniform, being one turn in 35 calibres.

| | | | |
|---|---|---|---|
| *Calibre of gun .. .. | 4″·75 | Calibre of gun + depth of grooves | 4″·95 |
| Diameter of shell .. .. | 4″·69 | Diameter of shell + height of studs | 4″·93 |
| Windage over body .. | 0″·06 | Windage over studs .. .. | 0″·02 |
| Height of stud .. .. | 0″·12 | Width of groove .. .. .. | 0″·80 |
| Depth of groove .. .. | 0″·10 | „ stud .. .. .. | 0″·76 |
| Clearance .. .. | 0″·02 | Play of stud .. .. .. | 0″·04 |

| | | | |
|---|---|---|---|
| † Calibre .. .. .. | 4″·00 | Calibre of gun + depth of grooves | 4″·20 |
| Diameter of shell .. | 3″·94 | Diameter of shell + height of studs | 4″·18 |
| Windage over body .. | 0″·06 | Windage over studs .. .. | 0″·02 |
| Height of stud .. .. | 0″·12 | Width of groove .. .. .. | 0″·80 |
| Depth of groove .. .. | 0″·10 | „ stud .. .. .. | 0″·74 |
| Clearance .. .. | 0″·02 | Play of stud .. .. .. | 0″·06 |

The studs of the common shell are gun-metal, those of Shrapnel are copper, so as to have as little pressure as possible on the thin Shrapnel shell.

Common shell.

*Common Shell* of ordinary form but unbushed. Gauge, G.S.

Fuzes for L.S. 9 secs. and 20 secs. M.L., Marks II. and III.—R.L. percussion, Mark II.

Fuzes for S.S. (this gun is a L.S. gun only).

For dimensions, &c., see table p. 234.

Issue.

Filled and boxed for siege train.

Shrapnel shell

*Shrapnel Shell.*—Gauge, G.S.

Fuzes for L.S. 5 secs. and 9 secs. M.L.—R.L. percussion, Mark II.

Fuzes for S.S. (this gun is for L.S. only).

The construction generally resembles the Shrapnel for the larger natures of guns, but the walls of the shell are thin, and the central tube is of gun-metal, while the socket is a composite one of tin and gun-metal. It resembles the 16-pr. Shrapnel, of which a cut is given on p. 244.

For dimensions, &c., see table, p. 235.

Issue.

Filled and boxed for siege train.

Case shot.

*Case Shot* of R.L. pattern. For dimensions, &c., see table, p. 236.

Issue.

Boxed.

For the gauges used in testing R.M.L. projectiles, see table, p. 324.

For directions for filling shell, preparing fuzes, &c., see pp. 62, 64, 65.

COMMON SHELL, 8″, and 6″·3 HOWITZERS, and 80, 64, 40 and 25-pr. R.M.L. GUNS.

| Calibre, and Mark of Pattern. | Date of Approval. | § Changes in War Stores. | Length ± $\frac{1}{10}$″ per foot. | Diameter. | | Thickness of Metal. | | | Weight empty. | Approximate Bursting Charge Shell Powder L.G. | Weight of Filled Shell. | Number of Rings of Studs. | Gauge of Fuze-Hole. | Remarks. |
|---|---|---|---|---|---|---|---|---|---|---|---|---|---|---|
| | | | | | | Walls. | | | | | | | | |
| | | | | Body. | Studs. | Top. | Bottom. | Base. | | | | | | |
| | | | inches. | inches. ±·01″ | inches. ±·005 | inches. | inches. | inches. ±·05″ | lbs. oz. | lbs. oz. | lbs. oz. | | | |
| 8″ Howitzer Mark I. .. | 23/5/73 | 2539 | 24·17 | 7·92 | 8·31 | 1·335 | 1·335 | 2·0 | 166 0 | 14 8 | 180 8 | 2 | General Service. | |
| 6″·3 Howitzer .. 80-pr., Mark I. .. | 16/5/72 | 2278 | 19·0 | 6·22 | 6·53 | ·75 | ·9 | 1·5 | 71 1 | 8 13 | 79 14 | 2 | General Service. | |
| 64-pr., Mark I. .. | 3/12/64 Provly. | 1042 | 14·71 | 6·22 | 6·47 | ·76 | ·76 | 1·3 | 59 8 | 4 8 | 64 0 | 5 | Moorsom | Black lacquer. Issued for practice only. |
| 64-pr., Mark II... | 4/1/67 Provly. | 1393 | 16·0 | 6·22 | 6·47 | ·76 | ·76 | 1·3 | 57 9 | 7 0 | 64 9 | 3 | " | Larger capacity, 3 rings of studs. Black lacquer. |
| 64-pr., Mark III. | 18/2/67 | 1394 | 16·0 | 6·22 | 6·47 | ·76 | ·76 | 1·3 | 57 9 | 7 0 | 64 9 | 3 | General Service. | |
| 64-pr., Mark IV. | 17/2/69 | 1768 | 16·0 | 6·22 | 6·47 | ·76 | ·76 | 1·3 | 56 14 | 7 2 | 64 0 | 3 | " | Unloading hole. |
| 64-pr., Mark V... | 6·77 | 3172 | 16·0 | 6·22 | 6·47 | ·76 | ·76 | 1·3 | 57 6 | 7 2 | 65 8 | 3 | " | Cast to gauge. |
| 40-pr., Mark I. .. | 23/5/73 | 2503 | 13·53 | 4·69 | 4·93 | ·8 | ·8 | 1·2 | 35 9½ | 2 8 | 38 1½ | 2 | " | Bushed, not countersunk. |
| 25-pr., Mark I. .. | 5/3/75 | 2725 | 12·85 | 3·94 | 4·18 | ·69 | ·69 | ·95 | 23 3½ | 1 12 | 24 15½ | 2 | " | Unbushed. |

In front of one front stud in the 8″ howitzer and 80 and 64-prs. and below the stud in the 40 and 25-prs. will be found the mark of pattern and R↑L.
Shell having the Moorsom gauge require conversion by an adapter.
See foot note § p. 208.

## Shrapnel Shell, 80, 64, 40, and 25-pr. R.M.L. Guns.

| Calibre, and Mark of Pattern. | Date of Approval. | § Changes in War Stores. | Length ± 1/10″ per foot. | Diameters. Body. | Diameters. Studs | Thickness of Metal. Walls. Top. | Thickness of Metal. Walls. Bottom. | Thickness of Metal. Base. | Number and Nature of Balls contained.‡ | Weight of Shell empty. | Weight of Bursting Charge, Pistol, R.F.G. or F.G. Powder. | Gauge of Fuze-Hole. | Number of Rings of Studs. | Remarks. |
|---|---|---|---|---|---|---|---|---|---|---|---|---|---|---|
| | | | inches | inches ± ·01 | inches ± ·005 | inches | inches | inches | | lbs. oz. | lbs. oz. | | | |
| 80-pr., Mark I. | 4/9/72 | 2379 | 15·8 | 6·22 | 6·53 | ·8 | ·9 | 1·0 | 288 lead and anti-mony, 14 lb. | 77 7 | 0 9 | GS. | 2 | |
| 80-pr., Mark II. | 11/7/73 | 2491 | 15·13 | 6·22 | 6·53 | ·8 | ·9 | 1·0 | Do. do. do. | 77 6 | 0 9 | „ | „ | Flush socket. |
| 80-pr., Mark III. | 10/12/75 | 2878 | 15·13 | 6·22 | 6·53 | ·8 | ·9 | 1·0 | Do. do. do. | 77 6 | 0 9 | „ | „ | Improved form of head. |
| 64-pr., Mark II.* | 21/3/68 | 1609 | 14·02 | 6·22 | 6·47 | ·76 | ·81 | ·9 | 234 do. do. | 64 3 | 0 5 | „ | 3 | |
| 64-pr., Mark III. | 5/2/72 | 2209 | 14·02 | 6·22 | 6·47 | ·76 | ·81 | ·9 | Do. do. do. | 64 7 | 0 9 | „ | „ | Metal socket, larger tube, diaphragm and cup coned. |
| 64-pr., Mark IV. | 11/7/73 | 2491 | 13·55 | 6·22 | 6·47 | ·76 | ·81 | ·9 | Do. do. do. | 65 10 | 0 9 | „ | „ | Flush socket. |
| 64-pr., Mark V. | 20/10/75 | 2852 | 13·72 | 6·22 | 6·47 | .. | .. | ·9 | Do. do. do. | 66 0 | 0 9 | „ | „ | Improved form of head. |
| 64-pr., Mark VI. | 6/77 | 3172 | 13·72 | 6·22 | 6·47 | ·76 | ·86 | ·9 | Do. do. do. | 66 0 | 0 9 | „ | „ | Cast to gauge. |
| 40-pr., Mark I.† | 23/5/73 | 2504 | 13·53 | 4·69 | 4·93 | ·595 | ·62 | ·89 | 180 do. at 18 per lb. | 39 5½ | 0 5 | „ | 2 | Flush socket. |
| 40-pr., Mark II. | 6/77 | 3172 | 13·03 | 4·69 | 4·93 | ·69 | ·69 | ·9 | Do. do. do. | 41 9 | 0 4½ | „ | „ | Cast to gauge, no internal grooves, gun-metal tube, and composite socket. |
| 25-pr., Mark I. | 5/3/75 | 2526 | 11·4 | 3·94 | 4·18 | ·50 | ·50 | ·65 | 45 at 34 and 113 at 20 per lb. | 25 0 | 0 3 | „ | „ | Flush socket. |
| 25-pr., Mark II. | 6/77 | 3172 | 11·4 | 3·94 | 4·18 | ·50 | ·50 | ·65 | Do. do. do. | 25 0 | 0 3 | „ | „ | Cast to gauge, no internal grooves. |

* 64-pr., Mark I. (1424), with cast-iron diaphragm, and without a tin cup, is unserviceable, and when returned to Woolwich is broken up.

† To be broken up, § 3172.

‡ 4 parts of lead to 1 part of antimony.

The mark of pattern, and R↑L will be found under one front stud.

See foot note §, p. 208.

Case Shot, 8″ and 6″·3 Howitzers and 80-pr., 64-pr., 40-pr., and 25-pr. R.M.L. Guns.

| Calibre, and Mark of Pattern. | Date of Approval. | § Changes in War Stores. | Length in inches. | Diameter, inches. | Number and Nature of Balls contained. | Approximate weight of Balls. | Approximate weight of Case, Linings, Clay, and Sand, &c. | Total Weight. | Marks on the top of the Case. | Remarks. |
|---|---|---|---|---|---|---|---|---|---|---|
| | | | | | | lbs. oz. | lbs. oz. | lbs. oz. | | |
| 8″ Howitzer<br>6″·3 Howitzer<br>64-pr., Mark I. | Same as<br>24/3/68 | 64 or<br>1611 | 80-pr.<br>9·6<br>±·1 | Mark IV.,<br>6·2±·03 | which see<br>56 8-oz. sand shot | 27 6 | 20 10 | 48 0<br>± 1·12 | I. W. ↑ D. 64-pr., M. L. | Balls packed in coal dust. |
| 64-pr. or 80-pr., Mark II. | 25 & 31/1/70 & 29/4/72 | 2188<br>2266 | 9·6<br>±·1 | „ | 49 8-oz. sand shot | 24 8 | 26 8 | 51 0<br>± 2 lb. | II. W. ↑ D. 64 or 80-pr. M.L. | Balls packed in clay and sand. |
| 64-pr. or 80-pr., Mark III. | 4/2/75 | 2716 | 9·55<br>±·1 | „ | 51 8-oz. sand shot | 25 8 | 25 0½ | 50 8½<br>± 1¾lbs. | „ „ | Balls packed in clay and sand. Outer case of tin in 3 segments soldered together. |
| 64-pr. or 80-pr., Mark IV. | 1/3/76 | 2925 | 9·6<br>±·1 | „ | 50 8-oz. sand shot | 25 0 | 24 14½ | 49 14½<br>± 2 lbs. | „ „ | Balls packed in clay and sand. Outer case of sheet iron (tinned) in 3 segments, soldered together. Inside lining of 6 segments.* |
| 40-pr. Mark I. | 23/5/73 | 2479 | 9·75 | 4·69±·025 | 405 mixed metal balls, 16½ per lb.† | 24 8 | 13 14 | 38 6<br>± 1 lb. | I. W. ↑ D. 40-pr. M.L. | Balls packed in clay and sand. |
| 25-pr., Mark I. | 22/5/74 | 2675 | 9·75 | 3·94±·02 | 245 mixed metal balls, 16½ per lb.† | 14 14 | 9 6 | 24 4<br>± 12 oz. | I. W. ↑ D. 25-pr., M.L. | „ „ |

In future the high gauge for the diameter of Case Shot will be the same as that for Common and Shrapnel Shells, while the limit between the high and low gauges will be as follows: 8″·88, 80-pr. ·06″, 64-pr. ·06″, 40-pr. ·04″, 25-pr. ·03″, § 2971.

* Tin plate is now used instead of tinned iron. No change in pattern.

† 4 of lead to 1 of antimony. For future manufacture 16 per lb.

## Cartridges for R.M.L. 8″ and 6″·3 Howitzers; 80-pr., 64-pr., 40-pr., and 25-pr. Guns.

The powder used in all these cartridges is L.G., or R.L.G. The latter will alone be used when the present stock of L.G. is exhausted. The advantage of the R.L.G. has been pointed out. Exercise L.G. or R.L.G. is used for blank charges.

Serge cartridges exist for all these ordnance*, with the exception of the 6″·3 howitzer, for which no cartridges have yet been sealed. Silk cloth cartridges are sealed for all, with the exception at present of the 6″·3 howitzer. The blank cartridges for L.S. are of silk cloth.

The choke of all these cartridges is cut to a length of one inch. For dimensions, packing, &c., see table, p. 238.

The cartridges for the howitzers are made like those for B.L. guns, *i.e.*, they have a cylindrical body and a circular bottom.

The cartridges for the 80-prs. and 64-prs. are cylindro-conical in shape, the serge or silk being cut in one piece, and sewn with an overlap, and three rows of worsted or silk stitches down one side of the body of the cartridge, and round the conical end.

The cartridges for the 40-pr. and 25-pr. are like those for the howitzers.

The serge cartridges are choked with worsted, and hooped with worsted braid; the silk cloth cartridges are choked with silk and hooped with silk braid.

Empty cartridges are issued in bales. For numbers, see table p. 331.

**Drill cartridges.**

The *Drill Cartridges* generally resemble those described for the Woolwich guns, p. 218. For dimensions, &c., see table, p. 240.

**Brass ring gauges for cartridges.**

*Brass Ring Gauges* are issued to test the diameter of the cartridge. The gauges are as follows:—

| | Diameter. |
|---|---|
| 8″ howitzer (10 lbs.) | 6″·8 |
| ″ ″ (5 lbs.) | 6″·3 |
| ″ ″ (2½ lbs.) | 6″·0 |
| 6″·3 howitzer (4 lbs.) | |
| ″ ″ (3 lbs.) | |
| ″ ″ (2 lbs.) | |
| ″ ″ (1 lb.) | |
| 80-pr. or 64-pr. | 6″·0 |
| 40-pr. | 4″·3 |
| 25-pr. | 3″·6 |

**Issue.**

Loose, in numbers according to demand.

**Gauge, wood, length, universal. § 2074.**

The length of the cartridge can be tested by the gauge, length, universal, see p. 217, in the event of the length of a cartridge not being marked on it, it can be ascertained from the tables, p. 238, 239, and marked.

---

* After 1st April, 1878, no serge cartridges will be used in S.S. for marks I., II. III., 64-prs. This is in consequence of the vent being considerably forward in all these guns, which necessitates the employment of all available means to guard against smouldering fragments of the cartridge being left in the bore. As previously observed, silk cloth is not nearly so liable to carry fire as serge.

## SERVICE and BLANK CARTRIDGES (SILK) 8″ and 6″·3 HOWITZERS and 80-pr., 64-pr., 40-pr., and 25-pr. R.M.L. GUNS.

| Calibre and Mark of Pattern. | Date of Approval. | § Changes in War Stores. | Charge. | Length, inches. From ins. | Length, inches. To ins. | Diameter; Body; inches. | No. of Hoops (Silk). | Marks. | Metal lined. Whole. Number. | Metal lined. Whole. Weight. | Metal lined. Half. Number. | Metal lined. Half. Weight. | Metal lined. Quarter. Number. | Metal lined. Quarter. Weight. | Brass. Pentagon. Whole. Number. | Brass. Pentagon. Whole. Weight. | Brass. Pentagon. Sectional. Number. | Brass. Pentagon. Sectional. Weight. | Brass. Rectangular. Plain. Number. | Brass. Rectangular. Plain. Weight. | Rectangular, corrugated. A Number. | A Weight. | B Number. | B Weight. | C Number. | C Weight. | D Number. | D Weight. | E Number. | E Weight. | F* Number. | F* Weight. | Remarks. |
|---|---|---|---|---|---|---|---|---|---|---|---|---|---|---|---|---|---|---|---|---|---|---|---|---|---|---|---|---|---|---|---|---|---|
| 8″ How., Service, I. | 29/10/75 | 2838 | 10 lb. R.L.G. ... | 8·75 | 9·25 | 6·8 | 3 | I., 8″ M.L. How. 10 lb. ... ... | 11<br>11 | 161<br>160 | 4<br>5 | 71<br>81 | – | – | – | – | – | – | – | – | – | – | – | – | – | – | – | – | – | – | – | – | |
| | | | 5 lb. R.L.G. ... | 5·75 | 6·25 | 6·3 | 1 | " " 5 lb. ... ... ... | 21<br>21 | 157<br>155 | 9<br>10 | 77<br>81 | – | – | – | – | – | – | – | – | – | – | – | – | – | – | – | – | – | – | – | – | |
| | | | 2½ lb. R.L.G. ... | 4·0 | ... | 6·0 | ... | " " 2½ lb. ... ... | 42<br>47 | 159<br>169 | 18<br>20 | 77<br>83 | – | – | – | – | – | – | – | – | – | – | – | – | – | – | – | – | – | – | – | – | |
| | | | 10 lb. L.G. ... | 8·75 | 9·25 | 6·8 | 3 | " " 10 lb. ... ... | 11<br>11 | 161<br>160 | 4<br>4 | 71<br>71 | – | – | – | – | – | – | – | – | – | – | – | – | – | – | – | – | – | – | – | – | |
| | | | 5 lb. L.G. ... | 5·75 | 6·25 | 6·3 | 1 | " " 5 lb. ... ... | 21<br>21 | 157<br>155 | 9<br>10 | 77<br>81 | – | – | – | – | – | – | – | – | – | – | – | – | – | – | – | – | – | – | – | – | |
| | | | 2½ lb. L.G. ... | 4·0 | ... | 6·0 | ... | " " 2½ lb. ... ... | 41<br>47 | 157<br>169 | 18<br>20 | 77<br>83 | – | – | – | – | – | – | – | – | – | – | – | – | – | – | – | – | – | – | – | – | |
| 6″·3 How., Service* | ... | ... | 4 lb. R.L.G. ... | 6·0 | ... | 5·6 | 1 | ... ... ... ... ... ... | 27<br>27 | 161<br>158 | 12<br>13 | 80<br>83 | – | – | – | – | – | – | – | – | – | – | – | – | – | – | – | – | – | – | – | – | |
| | | | 3 lb. R.L.G. ... | 5·2 | ... | 5·2 | 1 | ... ... ... ... ... ... | 36<br>40 | 162<br>171 | 6<br>17 | 81<br>82 | – | – | – | – | – | – | – | – | – | – | – | – | – | – | – | – | – | – | – | – | |
| | | | 2 lb. R.L.G. ... | 4·5 | ... | 4·6 | 1 | ... ... ... ... ... ... | 50<br>57 | 155<br>166 | 21<br>25 | 75<br>82 | – | – | – | – | – | – | – | – | – | – | – | – | – | – | – | – | – | – | – | – | |
| | | | 1 lb. R.L.G. ... | 3·0 | ... | 3·5 | 1 | ... ... ... ... ... ... | 106<br>110 | 162<br>162 | 49<br>50 | 83<br>82 | – | – | – | – | – | – | – | – | – | – | – | – | – | – | – | – | – | – | – | – | |
| | | | 4 lb. L.G. ... | 6·0 | ... | 5·6 | 1 | ... ... ... ... ... ... | 27<br>27 | 161<br>158 | 12<br>13 | 80<br>83 | – | – | – | – | – | – | – | – | – | – | – | – | – | – | – | – | – | – | – | – | |
| | | | 3 lb. L.G. ... | 5·2 | ... | 5·2 | 1 | ... ... ... ... ... ... | 35<br>37 | 159<br>162 | 16<br>17 | 81<br>82 | – | – | – | – | – | – | – | – | – | – | – | – | – | – | – | – | – | – | – | – | |
| | | | 2 lb. L.G. ... | 4·5 | ... | 4·6 | 1 | ... ... ... ... ... ... | 49<br>55 | 153<br>161 | 21<br>25 | 75<br>82 | – | – | – | – | – | – | – | – | – | – | – | – | – | – | – | – | – | – | – | – | |
| | | | 1 lb. L.G. ... | 3·0 | ... | 3·5 | 1 | ... ... ... ... ... ... | 106<br>110 | 162<br>162 | 49<br>50 | 83<br>82 | – | – | – | – | – | – | – | – | – | – | – | – | – | – | – | – | – | – | – | – | *a* For exceptional use under certain circumstances, with 64-pr. 64 cwt. siege guns (Mark III. with *steel* tubes). § 2837. |
| 80-pr., Service, I.... | 29/10/75 | 2838 | 10 lb. R.L.G. or L.G. | 11·0 | 12·0 | 6·0 | 4 | I., 80-pr., or 64-pr. M.L. 10 lb. ... | 10<br>11 | 151<br>159 | 4<br>5 | 71<br>81 | ... | ... | 10 | 162 | ... | ... | 12 | 190 | 14 | 188 | 16 | 213 | 20 | 258 | 28 | 355 | 20 | 259 | 30 | 378 | |
| Blank, I. ... ... | 2/4/74 | 2633 | 5lb. Blank R.L.G. | 6·5 | ... | 5·8 | 2 | I., 32, 64, or 80-pr. M.L. 5 lb. ... | 22 | 162 | 10 | 82 | | | | | | | | | | | | | | | | | | | | | *b* Service charge for 64-pr. 64 cwt. (Mark III. *only*). § 2837. |
| Blank, II. ... ... | 29/10/75 | 2838 | or L.G. ... ... | 6·5 | ... | 6·0 | 2 | II., 80 or 64-pr. M.L. or 32-pr., 5 lb. | 22 | 160 | 10 | 81 | ... | ... | 22 | 173 | ... | ... | 25 | 196 | 28 | 188 | 34 | 224 | 40 | 259 | 58 | 367 | 41 | 265 | 61 | 385 | |
| 64-pr., Service, I.... | 29/10/75 | 2838 | 12 lb. R.L.G. or L.G. | 13·0 | 14·0 | 6·0 | 5 | I., 64-pr. M.L. 12 lb. ... ... ... | 8<br>9 | 146<br>157 | 4<br>4 | 79<br>79 | ... | ... | 9 | 170 | ... | ... | 10 | 190 | 10 | 167 | 14 | 221 | 16 | 250 | 24 | 363 | 17 | 263 | 25 | 378*a* | |
| | | | 10 lb. R.L.G. or L.G. | 11·0 | 12·0 | 6·0 | 4 | I., 80 or 64-pr. M.L. 10 lb. ... ... | 10<br>11 | 151<br>159 | 4<br>5 | 71<br>81 | ... | ... | 10 | 162 | ... | ... | 12 | 190 | 14 | 188 | 16 | 213 | 20 | 258 | 28 | 355 | 20 | 259 | 30 | 378*b* | *c* Also reduced charge Naval Service for 64-pr. 64 cwt. (Mark III. *only*). § 2837. |
| | | | 8 lb. R.L.G. or L.G. | 9·2 | 9·7 | 6·0 | 3 | I., 64-pr. M.L. 8 lb. ... ... | 13<br>14 | 155<br>162 | 6<br>6 | 80<br>79 | ... | ... | 14 | 175 | ... | ... | 16 | 199 | 18 | 192 | 20 | 214 | 24 | 251 | 35 | 356 | 24 | 252 | 35 | 359*c* | |
| Reduced, or Saluting, S.S., I. | 29/10/75 | 2838 | 6 lb. R.L.G. or L.G. | 7·5 | 8·5 | 6·0 | 2 | I., 64-pr. M.L. 6 lb. ... ... | 18<br>19 | 160<br>164 | 7<br>8 | 73<br>79 | ... | ... | 19 | 177 | ... | ... | 21 | 198 | 24 | 194 | 28 | 222 | 32 | 252 | 48 | 365 | 33 | 259 | 50 | 380 | |
| Blank, I. ... ... | 26/4/72 | 2283 | 5lb. Blank R.L.G. | 6·5 | ... | 5·8 | 2 | I., 32-pr. or 64-pr. M.L. 5 lb. ... | 22 | 162 | 10 | 82 | | | | | | | | | | | | | | | | | | | | | |
| Blank, II. ... ... | 29/10/75 | 1868<br>2838 | or L.G. ... ... | 6·5 | ... | 6·0 | 2 | II., 80 or 64-pr. M.L. or 32-pr., 5 lb. | 22 | 160 | 10 | 81 | ... | ... | 22 | 173 | ... | ... | 25 | 196 | 28 | 188 | 34 | 224 | 40 | 259 | 58 | 367 | 41 | 265 | 61 | 385 | |
| 40-pr., Service, I.... | 29/10/75 | 2838 | 7 lb. R.L.G. or L.G. | 13·75 | 14·25 | 4·3 | 7 | I., 40-pr. M.L. 7 lb. ... ... ... | 15<br>16 | 156<br>162 | 6<br>7 | 73<br>80 | – | – | – | – | – | – | – | – | – | – | – | – | – | – | – | – | – | – | – | – | |
| 25-pr., Service, I. | 29/10/75 | 2838 | 4 lb. R.L.G. or L.G. | 10·75 | 11·25 | 3·6 | 6 | I., 25-pr. M.L. 4 lb. ... ... ... | 26<br>27 | 156<br>158 | 12<br>12 | 80<br>79 | 5<br>5 | 39<br>38 | – | – | – | – | – | – | – | – | – | – | – | – | – | – | – | – | – | – | |

For bags containing spare powder, and weight of empty cases, see table p 200.

SERVICE and BLANK CARTRIDGES (SERGE), 8″ HOWITZER, and 80-pr., 64-pr., 40-pr., and 25-pr. R.M.L. GUNS.

| Calibre and Mark of Pattern. | | Date of Approval. | § Changes in War Stores. | Charge. | Length, inches. | | Diameter, inches. | No. and Description of Hoops. | Marks. | — |
|---|---|---|---|---|---|---|---|---|---|---|
| | | | | lbs. | From | To | | | | |
| 8-in. Howr. | 10 lbs., I. .. .. | 23/5/73 | 2518 | 10 | 8·75 | 9·25 | 6·8 | 3 braid .. | I. 8-in. M.L. Howr. 10lbs. | |
| | 5 ,, I. .. .. | 23/5/73 | 2518 | 5 | 5·75 | 6·25 | 6·25 | 1 braid .. | I. 8-in. M.L. Howr. 5lbs. | |
| | 2½ ,, I. .. .. | 23/5/73 | 2518 | 2½ | 4·0 | — | 6·0 | — | I. 8-in. M.L. Howr. 2lbs. 8oz. | |
| 80-pr. | (Service), Mark I. .. | 16/5/72 | 2281 | 10 | 11·25 | 11·75 | 6 | 4 braid | I. 80-pr. M.L. 10lbs. | |
| 64-pr. | (Service), Mark I. .. | 2/1/65 | 1044 | 8 | 9·2 | 9·7 | 6 | 3 braid | I. 64-pr. M.L. 8lbs. | |
| | (Reduced), Mark .. | 2/1/65 | 1044 | 6 | 7·5 | 8·0 | 6 | 2 braid | I. 64-pr. M.L. 6lbs. | S.S. only. |
| | (Reduced), Mark .. | 6/4/72 | 2282 and 2177 | 6 | 7·5 | — | 5·9 | 2 worsted .. | I. 32-pr. or 64-pr. M.L. 6lbs. | S.S. only. |
| 40-pr. | (Service), Mark I... | 23/5/73 | 2476 | 7 | 13·75 | 14·75 | 4·3 | 7 braid .. | I. 40-pr. M.L. 7lbs. .. .. | |
| 25-pr. | (Service), Mark I... | 3/3/75 | 2721 | 4 | 10·75 | 11·25 | 3·6 | 6 braid .. | I. 25-pr. M.L. 4lbs. .. .. | |

For packing, see p. 238.

DRILL, OR DUMMY CARTRIDGES, 8″ and 6″·3 HOWITZERS, 80-pr., 64-pr., 40-pr. and 25-pr., R.M.L. GUNS.

| Calibre and Mark of Pattern. | Date of Approval. | § Changes in War Stores. | Length, inches. | | Diameter, inches. | | Diameter, decreased at bottom, inches. | | Imitation choke on front end. | Marks. | Remarks. |
|---|---|---|---|---|---|---|---|---|---|---|---|
| | | | From | To | From | To | | | | | |
| 8″ Howr., Mark 1. | 23/5/73 | 2533 | 8·75 | 9·25 | 6·7 | 7·0 | 6·1 | 6·4 | Yes | I., 8-in. M.L., Howr. 10 lb. R.↑L. | |
| 6″·3 Howr. .. | | | | | | | | | | | |
| 80-pr., Mark I. .. | 27/9/72 | 2358 | 11·25 | 11·75 | 5·8 | 6·1 | 3·8 | 3·5 | ,, | I., 80-pr. M.L. 10 lbs .. | |
| 64-pr., Mark I. .. | 16/9/68 | 1674 | 9·2 | 9·7 | 5·95 | 6·05 | — | — | ,, | I., 64-pr. M.L. 8 lbs... | |
| 64-pr., Mark II... | 8/3/70 | 1878 | 9·2 | 9·7 | 6·1 | 5·8 | 4·0 | 3·7 | ,, | II., 64-pr. M.L. 8 lbs... | |
| 40-pr., Mark I. .. | 23/5/73 | 2534 | 13·75 | 14·25 | 4·2 | 4·4 | 3·6 | 3·8 | ,, | I., 40-pr. M.L. 7 lbs. | |
| 25-pr. .. .. | | | | | | | | | | | |

# CHAPTER XIV.—AMMUNITION FOR 16-PR., 9-PR., AND 7-PR. R.M.L. GUNS.

COMMON AND SHRAPNEL SHELL; 7-PR., DOUBLE AND STAR SHELL, CASE SHOT, GAUGES, CARTRIDGES, AND STORES CONNECTED WITH THEM.

THE French system of rifling has been adopted for the field guns. The bottom of the groove is concentric with the bore in the 7-pr. and eccentric in the 9 pr. and 16-pr.; the radius of the bottom of the groove in these being 1″·35,* the driving edge making an angle of 70° with a radius to the centre of the bore, and the loading edge an angle of 56°. The studs are made of a shape corresponding to the grooves. The driving edge of the groove is an easy inclined plane up which the stud mounts so as to centre the projectile. The twist is uniform in all these guns, and is 1 turn in 30 calibres for the 16-pr. and 9-pr., and 1 turn in 20 calibres for the 7-pr.

In the 16 and 9-prs. the windage over the body is ·06″, over the studs ·02″, the clearance before centring is ·02″, and after ·03″ the play of the stud in the groove is ·04″ in the earlier patterns of projectiles; it is ·06″ in the later patterns.

The depth of the groove in the 9 and 16-prs. is ·11″, the height of the stud is ·13″, the width of the groove is ·8″, that of the studs ·76″ in the earlier patterns, but, to prevent jamming,† the width is now reduced to ·74″.

The calibre of the 16-pr. is 3″·6, that of the shell 3″·54; the calibre of the 9 pr. is 3″, that of the shell is 2″·94.

In the 7-pr. the windage over the body is ·06″, over the studs ·02″, the clearance before centring is ·02″, and after ·03″ the play of the stud in the groove is ·05″.‡

The studs for the 16-pr.§ are made of gun-metal for common shell, and of copper for Shrapnel shell; those of the 9 and 7-prs. were made of zinc in order not to injure the bronze guns first introduced for these calibres; the manufacture of 9-pr. bronze guns has been discontinued, and the studs are now made of copper. Zinc is more liable to corrosion than gun-metal, hence it is desirable to store projectiles with zinc studs under cover where practicable.‖

---

* The system is slightly modified in the 16-pr. and 9-pr., the edge of the grooves being slightly rounded off at bottom.

† In case of jamming use a wet sponge if possible. See Extracts, XII, p. 45.

‡ The calibre of the 7-pr. is 3″, the diameter of the shell is 2″·94, the depth of the groove is ·1″, the height of the stud is ·12″. The width of the groove is ·9″, that of the stud is ·85″.

§ Pressure to attach studs—16-pr., 8·48 tons.
9-pr., 6·25 „
7-pr., 7·14 „

‖ Zinc does not corrode rapidly when not in contact with another metal, but when iron and zinc come together a galvanic action is set up. Projectiles may be seen with their zinc studs a mere white patch where they have been long exposed to weather.

The 7-pr. and 9-pr. have the same calibre, but the projectiles of the one gun cannot be fired from the other as the twist is different, the studs also on the 7-pr. are much wider than those on the 9-pr. projectiles.*

Projectiles.

The projectiles fired from the R.M.L. field guns are common and Shrapnel shell and case shot.† The 7-pr. has in addition a double shell and a star shell. The star shell can be fired out of the 9-pr. also.

The Shrapnel shell are the most important projectiles, about $\frac{2}{3}$ of all the projectiles in the equipment being Shrapnel.

As before pointed out, M.L. time fuzes, Marks II. or III., with increased priming, are used for these guns, and for small charges when firing at high angles of elevation, gun cotton priming is used, see p. 24.

§§ 2621, 2622.

The R.L. percussion fuze is used, Mark II. for the 16-pr., and Mark I. or II. for the 9-pr. and 7-pr. Mark II. will be issued for the 9 and 7-prs., when the present stock of Mark I. fuzes is exhausted.

The R.L. percussion fuze does not answer well with the 7-pr., and will not act with a charge under 10 oz. in the 9-pr.

There are several 7-prs. in the service.

§ 2498.

| | | | | |
|---|---|---|---|---|
| A steel gun | weighing | 150 lbs., | charge | 6 oz. |
| A bronze gun | „ | 200 „ | „ | 8 „ |
| A steel gun | „ | 200 „ | „ | 12 „ |

The 6 oz. charge sometimes fails to prepare the percussion fuze to act, the shock being too slight to ensure the guard always shearing the feathers.‡ The small charge also causes a difficulty with case shot, which frequently does not break up.

---

* The case shot would fit in, but the 9-pr. case would not break up in the 7-pr.

† The Indian Field Equipment Committee recommended equal proportions of Shrapnel and segment shells, and numerous comparative trials were made.

Numerous trials were also carried out with bullet shells, and some good results were obtained. The bullet shell differs from Shrapnel in having the body and head cast in one piece; the bottom of the shell consists of an iron disc secured in a somewhat similar manner to the bottom of a segment shell. In the interior of the shell is a wide iron pipe leading down to an iron chamber; this pipe and chamber contain the bursting charge; round the pipe and above the chamber are arranged bullets secured in rosin; the shell carried a heavier bursting charge and the bullets were more scattered than in Shrapnel shell.

Extracts, Vol. IX, p. 212.

The Special Committee on Rifled Shell Guns report that the results of experiments made with 9-pr. R.M.L. Shrapnel, segment, and bullet shells, show that the former is equal to the two latter when burst on graze, and is superior when used to burst with a time fuze.

Comparative trials of 16-pr. Shrapnel and bullet shell will be found in Extracts, Vol. X, pp. 45, 120, Vol. XI, p. 42. The Shrapnel shell shows a decided superiority over the bullet shell when used as a time shell. The trial of the bullet shell was suspended pending the results of a trial of 16-pr. common shell burst with picric powder. Extracts, Vol. XI, p. 46.

16-pr. common shell have been burst by gun-cotton in conjunction with a fulminate of mercury detonator. When an ounce of gun-cotton was used the shell burst into 262 pieces (about 1 lb. 4 oz. was broken too small to be found); when $\frac{1}{2}$ oz. of gun-cotton was used the shell burst into 186 pieces. In both cases the shells were filled up with water, which is a convenient and sufficiently incompressible medium for transmitting the force of explosion. Experiments have been carried on at Shoebury to ascertain whether gun-cotton arranged in conjunction with a fuze and detonator can be safely fired in a 16-pr. shell. So far the experiments have been satisfactory.

A trial of some 16-pr. common shell cast with spherical heads (as in the design of the new Russian shell) will be found in Extracts, Vol. XI, p. 115. The idea is that the spherical head when blown off will continue to ricochet. The targets were put up at 1,200 yards, the head ricochetted beyond the targets for about 1,500 yards. The heads of the shell after bursting ricochetted like round shot and fairly straight.

‡ *Vide* p. 55. The Mark II. fuze is very uncertain with an 8 oz. charge.
Letter of Assistant Superintendent R.L., 19/12/73.

The shell for all the guns have the G.S. gauge of fuze-hole, the shells for the 7-pr. had originally the common gauge, but this was changed to the G.S. All existing shell were altered to the G.S. gauge in 1871.*

§ 2097. Common shell.

*Common Shell.*—Gauge G.S.

Fuzes for L.S., 5 secs. and 9 secs., M.L. time, Marks II. or III., and R.L. Percussion Mark II. for 16-pr., Marks I. or II. for 9-pr. and 7-pr.

Fuze for S.S. (9-pr. and 7-pr. only), 9 sec. M.L. time, Marks II. or III and R.L. Percussion.

The sensitive fuze is for use with the 7-pr.

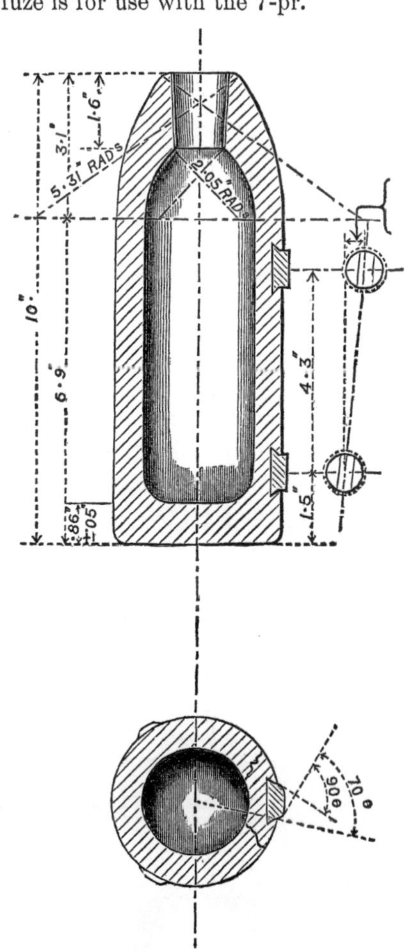


16-pr. Common Shell. Mark I.

The construction of the shell will be seen from the cut. They are all lacquered internally. The various patterns, dimensions, &c., are given in the table p. 249.

* 5 seconds, 10 seconds, and 15 seconds special fuzes were introduced for use with the 7-pr. shells having the common gauge fuze-hole. These have been withdrawn from the 7-pr. equipment, they may be used with S.B. common shells if required. These fuzes were all of one size, but the rate of burning was altered by using different compositions.

§ 2097.

Directions for filling, &c., will be found at p. 62.

These shell when filled have the G.S. wad, see p. 72, placed in the fuze-hole.

**Issue.**

1. Filled, wad G.S. inserted and boxed.
2. Empty, loose. For transit to India, empty, boxed.

The Indian Field Equipment Committee, p. xvii., state that the 9-pr. shell with its charge of $7\frac{1}{4}$ oz. gives considerable incendiary power and that its penetration and explosive force are sufficiently great for the destruction of the strongest masonry usually found in villages.*

**Shrapnel shell.**

*Shrapnel Shell.*—Gauge G.S.

Fuzes for L.S. 5-seconds and 9-seconds M.L. time, Marks II. or III., (for 7-pr. 5-seconds only) and R.L. percussion; Mark II. for 16-pr.; Marks I. or II. for 7-pr. and 9-pr.

Fuzes for S.S. (9 and 7-pr. only) 5-seconds M.L. time, Marks II. or III.

For dimensions, &c., see table p. 249.

The first pattern introduced for the 16-pr. proved too weak,† and is to be broken up. That now introduced has no grooves, and has the coned cup and diaphragm described at p. 192; the studs are of pure copper to avoid undue pressure on the shell in pressing them in: in other respects it resembles the 9-and 7-pr. Shrapnel, the later patterns of which also have copper studs. Mark III. Shrapnel is cast to gauge, and so is Mark VIII. 9-pr.

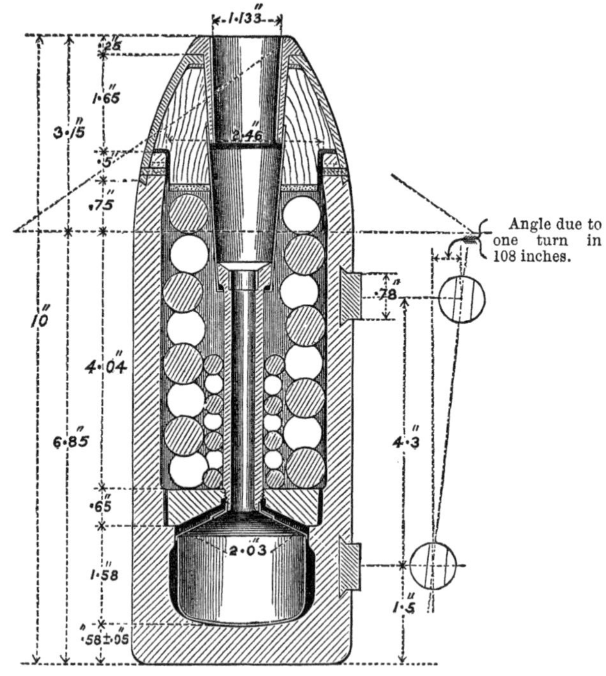


16-pr. Shrapnel, Mark III.

* Experiments are being carried on with a view to the introduction of common shell filled with incendiary composition.—See Extracts, Vol. XIV, p. 65.

† Extracts, Vol. X, p. 242.

The 9 and 7-pr. Shrapnel resemble the B.L.F.S. Shrapnel described at p. 157, except that they are studded, and the latest patterns have a flush instead of a projecting socket.

Some 9-pr. Shrapnel having failed at proof, Mark V. was introduced without any internal grooves, except those in the powder chamber.* § 2625.

The balls are of lead and antimony. It will be seen by the table, p. 250, that two sizes of balls are used so as to avoid wasting space in the shell.

Issue.

1. Filled, with primer, and boxed.
2. Empty, loose. For transit to India, empty, boxed.

Percussion fuzes have been used with good effect. The general result of experiments seem to show that the effect of percussion fuzes would be great when the shell are burst within 10 yards of troops in masses; at long ranges the effect of percussion fuzes is not so good, as the shell tends to bury itself in the earth.

The time fuze would have probably the best effect against open formations, the rule for bursting the shell being similar to that given on p. 159.†

*Experience with Shrapnel (Active Service).*‡

Experience with Shrapnel.

Active service.

This has been confined to the campaigns in Abyssinia and the Gold

---

* See Extracts, Vol. XI, p. 168. Some of the early pattern Shrapnel shell failed, the attachment of the head not being strong enough. The improvements which render the head firm are the same as those pointed out on p.     for B.L.F.S. Shrapnel.

The strengthened head is also an advantage when using a percussion fuze. An account of the defects will be found in Extracts, Vol. IX, p. 292; Vol. X, p. 137.

† 9-pr. Shrapnel fired with time fuzes against a column of troops represented by targets 54′ × 9′ (2 inches thick) in four ranks 20 yards apart made 48 hits *through* per round at 1,200 yards, 40 through at 1,600 yards, and 10 through at 2,000 yards.

The Committee report that Shrapnel loaded with the nose of the shell next the powder may be used as case shot in circumstances of any great emergency with good results up to about 100 yards.

Extracts, Vol. VIII, p. 85. At 800 yards 9-pr. Shrapnel fired against targets arranged as above, gave 82 hits per round when burst 50 yards in front and 10 feet above the plane; when burst 25 yards in front and 7 feet above the plane, 110 hits per round. The 16-pr. Shrapnel gave 144 hits per round when burst 50 yards in front and 9 feet above the plane; when burst 25 yards in front and 7 feet above plane 201 hits per round. Extracts, Vol, XI, p. 179.

Effects of Shrapnel.

9-pr. Shrapnel fired at a range of 600 yards, with a percussion fuze, gave very good results; two rows of targets, 54′ × 9′, were used 20 yards apart, the practice shows well the necessity of bursting the shell close up to the object; a shell burst 10 yards from the object gave 128 hits, one 25 yards off gave 54, one 45 yards off 49, and one 70 yards off gave 26.

Ten effective rounds of 9-pr, Shrapnel fired at a range of 800 yards with a percussion fuze against one row of targets, 54′ × 9′, gave an average of 54 hits per round.

Extracts, Vol. VIII, p. 224.

The results of some experimental segment shell will also be found in the Extracts.

7-pr. Shrapnel, Mark II., 12 oz. charge, range 1,200 yards; 5 rounds gave 97 hits on two rows of targets, 54′ × 9′.

‡ Much valuable information with regard to the power and capabilities of common and Shrapnel shell fired from 16-pr. and 9-pr. guns in the course of the Okehampton Experiments will be found in the report of the Special Committee on Rifled Field Guns and High Angle Fire on the Artillery Experiments at Okehampton. These experiments took place in August and September 1875. A précis of the report and a paper on it by Lieut.-Col. F. Lyon, R.A., late Assistant Superintendent R.L., is published in R.A.I. Proceedings, Vol. IX, No. 9.

Coast, and to the 7-pr. shell, which did well.* The gun employed was the 150 lb. steel gun, 6 oz. charge.

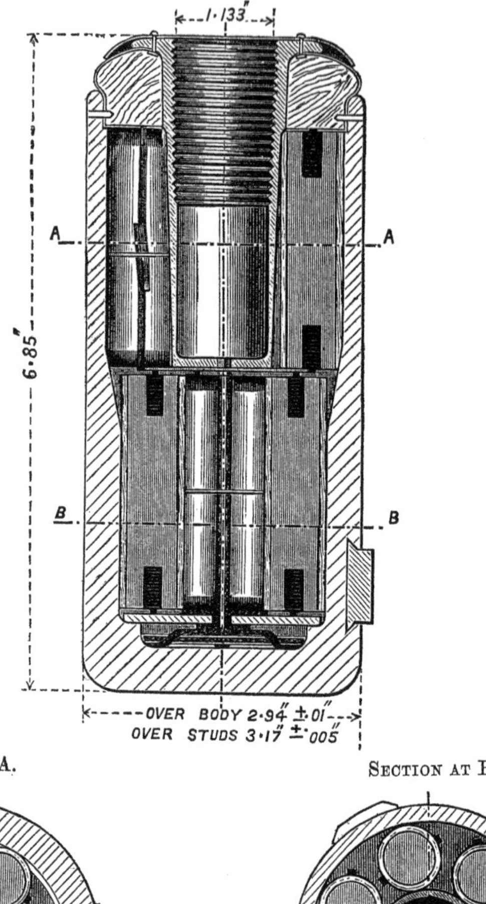


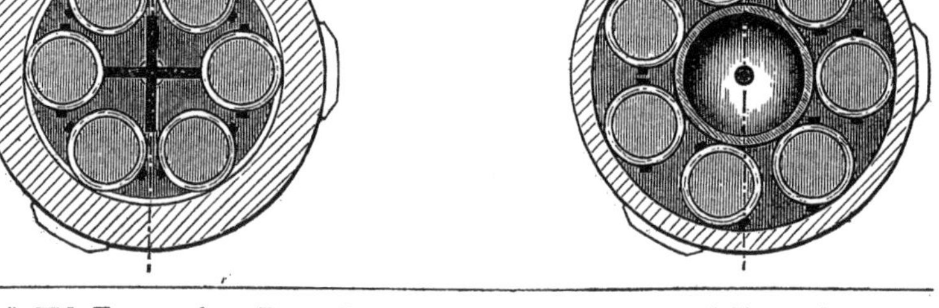


* *Vide* Extracts from Proceedings, &c., relative to segment and Shrapnel.

Extracts from Col. Milward's remarks on equipment of mountain batteries.—

"Common and Shrapnel shell good in all respects, but should be fitted for percussion fuzes."

Extracts from Lieut. Chapman's Journal of Practice:—

8/1/68.—"Two rounds Shrapnel at about 470 yards, 5 seconds fuze. Practice good."

10/1/68.—"Shrapnel at 700 yards was ineffective."

11/1/68.—"Two rounds Shrapnel at 450 yards very effective."

Action of 10/4/68.—"The Shrapnel and common shell took great effect at ranges from 500 to 1,500 yards, the effect of the Shrapnel was considered good. 55 rounds were fired."

For experience in the Gold Coast Expedition, see Appendix, p. 301.

*7-pr. Double Shell.*—Gauge, G.S. Fuzes for L.S. 9 sec. and 20 sec. M.L. time, Marks II. or III. Fuzes for S.S. 20 sec. M.L. time, Marks II. or III. This shell is intended for high angle firing with a low charge (4 oz.), its construction is similar to the common shell, it carries a large bursting charge (15 oz.), and would be affective against houses.

7-pr. double shell.

For dimensions, &c., see table, p. 249.

(1.) Filled and boxed. (2.) Empty, loose. For transit to India, empty, boxed.

Issue.

*Star Shell.*—Gauge, G.S. slightly reduced so as to cause the fuze to project farther, in order to ensure ignition with small charges. Fuze. 5 sec.; M.L., II. or III. This shell has been introduced for use with 7-pr. batteries. It consists of a thin iron shell, having a chamber in the base to take the bursting charge ($\frac{1}{2}$ dram of R.F.G. powder contained in a red shalloon bag); over the bursting charge is a wrought-iron disc with a hole in the centre.* The interior is filled with 13 stars, arranged as shown in the figure. Mark II. differs from the above only in having copper instead of zinc studs.

Star shell, Mark I. §§ 2492, 2526.

The stars are paper cylinders, filled with the composition given on p. 319; they gave a brilliant light and burn about 18 seconds. The top of the shell is made of wood covered with tin, and is lightly attached to the body by rivets, it contains a gun-metal socket, G.S. gauge to take the fuze, having a fire hole at the bottom. A kit plaster is placed over the head of the shell to secure it from damp. Quick match is wrapped round the stars as shown; it serves to ignite them and to convey the flash of the fuze to the burster. There are also loose strands of quick-match led down the walls of the shell and on to the wrought-iron disc.

The shells are issued in boxes, ready for use.

Issue.

To prepare them the kit plaster must be removed, the plug unscrewed, and a 5-seconds M.L. fuze, Mark II. or III., bored to the required length fixed in; gun-cotton priming is to be used.

Preparation.

The shell has been fired at high angles at 15° to 27°, and with from 4 oz. to 1 oz. of powder; ranges of from 400 to 600 yards have been obtained.†

As has already been mentioned this shell has only one row of studs, the front portion of the shell being too thin to allow the studs to be pressed in; the studs are small enough to fit into the 9-pr., from which the shell might be fired if necessary, They form no part, however, of the 9-pr. equipment.

*Case Shot.*—Of R.L. pattern. See page 140. The outer ring and the inner disc and segmental lining were in earlier patterns made of zinc to avoid injuring the bore of the bronze 9-pr. gun. The latter no longer being in the service, the ring, disc, &c., are now, like those in other case shot, of wrought-iron.

Case shot.

The 7-pr. case shot frequently does not break up with the 6 oz. charge from the light 7-pr. gun.‡

---

* The bursting charge is not intended to break the walls of the shell, only to blow the top off, and disperse the stars.

† 4 oz. charge, elevation 15°, fuze bored to 3, gave 600 yards range; 2 oz. charge, elevation 20°, fuze bored to 1·5, gave 500 yards range; 1 oz. charge, elevation 27°, fuze bored to 1, gave a range of 400 yards. The light was good.—Extracts, Vol. IX, p. 237.

‡ The average elevation of case shot from field guns is 1° for 300 yards. Increase or decrease the elevation by $\frac{1}{2}$° for every increase or decrease of 100 yards in range; this applies to such ground as the sands at Shoebury, over swampy ground the elevation might be slightly increased.

7-pr. case shot, fired with 12 oz. charge, all broke up, 1° of elevation did well at 200 yards.

For dimensions, &c., see table p. 251.

Issue. In boxes, 10 rounds per box for 7-prs., 8 for 9-pr., and 6 for 16-pr.

Gauges. Gauges are issued for use with the 16, 9, and 7-pr. projectiles. See table, page 324.

CARTRIDGES.

The cartridges for the 16-pr. and 9-pr. R.M.L. guns are cylindrical, made of two pieces, one rectangular, the other circular; those for the 7-pr. resemble S.B. cartridges, being made of one piece only. Both serge and silk cloth cartridges exist for these guns, with the exception of the 4-oz. charge used for high angle firing from the 7-pr. gun. This charge is made up in red shalloon, the thinner material being used to make the choke as small as possible, this is desirable as there is a chance of a small charge turning on being rammed home.

For charges, dimensions, &c., see table p. 252.

As previously pointed out, p. 242, there are three different service charges for the 7-pr. guns of different weights.

A.C. 74, Cl. 158. § 2708. R.L.G. is used with the 16-pr. and 9-pr. guns in possession of the R.H.A., R.A., and School of Gunnery. It gives better velocity and accuracy than L.G. powder. The guns are rather undersighted for L.G. For blank charges, exercise R.L.G. or L.G. is used.

§ 1998. F.G. or R.F.G. is used for the 7-pr. cartridges for the reasons given on p. 4.

Issue. Filled.

A.C. 77, Cl. 137. For numbers of empty cartridges packed in bales, see table, p. 331.

Blank charges. The blank charges, see table, p. 252, are made up in silk cloth, choked and hooped with silk. There is no blank charge for the 7-pr. gun.

Drill cartridges. §§ 2088, 2435, 3093. Drill cartridges of hollow wood blocks, covered with raw hide, are issued for these guns. See table, p. 252.

Brass ring gauges for cartridges. Brass ring gauges are issued to test the diameters of the cartridges, their dimensions are:—

| | |
|---|---|
| 16-pr. - - - | 3″·3 |
| 9-pr. - - - | 2″·6 |
| 7-pr. - - - | 2″·5 |

The gauge, "wood, length, universal," page 217, would be used for testing the length.

Issue. Loose in numbers as demanded.

## COMMON SHELL, 16, 9, AND 7-PR., AND DOUBLE SHELL, 7-PR. R.M.L. GUNS.

| Calibre and Mark of Pattern. | Date of Approval. | § Changes in War Stores. | Length $\pm \frac{1}{10}''$ per Foot. | Diameters. Body. | Diameters. Studs. | Thickness of Metal. Walls. | Thickness of Metal. Base. | Weight empty. | Approximate bursting Charge, Shell, Powder, L.G. | Weight of filled Shell. | Gauge of Fuze-Hole. | Remarks. |
|---|---|---|---|---|---|---|---|---|---|---|---|---|
| | | | inches | inches ± ·01 | inches ± ·005 | inches | inches ± ·05 | lbs. oz. | lbs. oz. | lbs. oz. ± 1·5 per cent. | | |
| 16-pr., Mark I. .. .. | 6/1/72 | 2190 | 10·0 | 3·54 | 3·8 | ·6 | ·86 | 15 1 | 1 0 | 16 1 | G.S. | |
| 16-pr., Mark II... .. | 25/11/75 | 2853 | 10·0 | 3·54 | 3·8 | ·6 | ·86 | 14 15½ | 1 0 | 15 15½ | ,, | Reduced Studs, viz., ·74. |
| 9-pr., Mark I. .. .. | 4/1/70 | 1921 | 7·93 | 2·94 | 3·2 | ·55 | ·6 | 8 8 | 0 8 | 9 0 | ,, | Studs, ·78″ in width, ordered to be reduced to ·76″. |
| 9-pr., Mark II... .. | 16/9/71 | 2104 | 7·93 | 2·94 | 3·2 | ·55 | ·6 | 8 9½ | 0 8 | 9 1½ | ,, | Reduced studs, viz., ·76″. |
| 9-pr., Mark III. .. | 24/3/75 | 2744 | 7·93 | 2·94 | 3·2 | ·55 | ·6 | 8 9½ | 0 7½ | 9 1 | ,, | Copper studs. |
| 9-pr., Mark IV. .. | 25/11/75 | 2853 | 7·93 | 2·94 | 3·2 | ·55 | ·6 | 8 7¼ | 0 7½ | 8 14¾ | ,, | Reduced studs, viz., ·74″. |
| 9-pr., Mark V... .. | 6/77 | 3172 | 7·93 | 2·94 | 3·2 | ·55 | ·6 | 8 7¼ | 0 7½ | 8 14¾ | ,, | Cast to gauge. |
| 7-pr., Mark I. .. .. | 4/12/65 | 1153 | 6·75 | 2·94 | 3·17 | ·5 | ·5 | 6 14 | 0 7 | 7 5 | Common. | Fuze-hole ordered to be altered to G.S. gauge, § 2097. |
| 7-pr., Mark II.... .. | 11/7/71 | 2097 | 6·75 | 2·94 | 3·17 | ·5 | ·5 | 6 15¾ | 0 7 | 7 6⅝ | G.S. | |
| 7-pr., Mark III. ... | 24/3/75 | 2744 | 6·75 | 2·94 | 3·17 | ·5 | ·5 | 6 14¼ | 0 6½ | 7 4¾ | ,, | Copper studs. |
| 7-pr., double, Mark II. | 30/12/68 | 1779 | 11·25 | 2·94 | 3·17 | ·5 | ·5 | 11 0 | 0 15 | 11 15 | Common. | Lighter shell, fuze-hole ordered to be altered to G.S. gauge, § 2097. |
| 7-pr., double, Mark III. | 11/7/71 | 2097 | 11·25 | 2·94 | 3·17 | ·5 | ·5 | 11 2⅜ | 0 15 | 12 1¾ | G.S. | |
| 7-pr., double, Mark IV. | 24/3/75 | 2744 | 11·25 | 2·94 | 3·17 | ·5 | ·5 | 11 2¼ | 0 15 | 12 1¼ | ,, | Copper studs. |

7-pr. double shell, Mark I. (§ 1520), obsolete, § 1779.
See foot note, § p. 208.

## SHRAPNEL SHELLS, 16, 9, AND 7-PR. R.M.L.

| Calibre and Mark of Pattern. | Date of Approval. | § Changes in War Stores. | Length ± $\frac{1}{10}$″ per foot. | Thickness of Metal. Walls. Top. | Thickness of Metal. Walls. Bottom. | Thickness of Metal. Base. | No. and Nature of Balls contained.‡ | Weight of Shell empty. | Weight, Bursting Charge, Pistol or F.G., from | Gauge of Fuze-Hole. | Remarks. |
|---|---|---|---|---|---|---|---|---|---|---|---|
| | | | ins. | ins. | ins. | ins. | | lbs. oz. ± 1·5 per cent. | lbs. oz. | | |
| 16-pr. Mark I. | 6/1/72 | 2189 | 10 | ·38 | ·4 | ·58 | 63, lead and antimony, 18 per lb.<br>56 ,, 84 ,, | 16 5 | 0 1½ | G.S. | This shell is too weak, and is to be broken up. Letter from Director of Artillery, 10/11/73. |
| ,, II. | — | — | ,, | ·45 | ·45 | ·58 | 72 ,, 18 ,,<br>56 ,, 84 ,, | 17 13 | 0 1½ | ,, | No internal grooves. |
| ,, III. | 25/11/75 | 2898 | ,, | ·45 | ·45 | ·58 | 72 ,, 18 ,,<br>56 ,, 84 ,, | 17 13 | 0 1½ | ,, | Copper studs, no internal grooves, and cast to gauge. |
| 9-pr. Mark I.* | 14/1/70 | 1921 | 7·93 | ·345 | ·37 | ·45 | 28 ,, 18 ,,<br>35 ,, 34 ,, | 9 4 | 0 0¾ | ,, | A pattern of this shell, with a stronger socket than those of a previous manufacture, was approved, 5/4/71, § 2062, without a change of pattern. P S. (plain socket) is marked on one of the studs. |
| ,, II.* | 16/9/71 | 2104 | ,, | ,, | ,, | ,, | ,, ,, ,, ,, | 9 4 | 0 0¾ | ,, | Studs reduced in width from ·78″ to ·76″, centre tube of gun metal instead of iron. |
| ,, III.† | 15/3/72 | 2210 | ,, | ,, | ,, | ,, | ,, ,, ,, ,, | 9 4 | 0 0¾ | ,, | Junction of head and body stronger, and secured by screw rivets as well as by plain rivets. |
| ,, IV. | 28/8/73 | 2525 | ,, | ,, | ,, | ,, | ,, ,, ,, ,, | 9 4 | 0 0¾ | ,, | Flush socket. |
| ,, V. | 14/4/74 | 2625 | ,, | ·375 | ·375 | ,, | ,, ,, ,, ,, | 9 12 | 0 0¾ | ,, | Do. No internal grooves (except in powder chamber.) |
| ,, VI. | 24/3/75 | 2744 | ,, | ,, | ,, | ,, | ,, ,, ,, ,, | 9 12 | 0 0¾ | ,, | Copper studs. |
| ,, VII. | 20/11/75 | 2853 | ,, | ,, | ,, | ,, | ,, ,, ,, ,, | 9 12 | 0 0¾ | ,, | Studs reduced in width from ·76″ to ·74″. |
| ,, VIII. | 6/77 | 3172 | ,, | ,, | ,, | ,, | ,, ,, ,, ,, | 9 12 | 0 0¾ | ,, | Cast to gauge. |
| 7-pr. ,, II. | Provly. 19/3/68 | 1600 | 6·5 | ·345 | ·37 | ,, | 21 ,, 18 ,,<br>,, ,, 34 ,, | 7 5 | 0 0½ | ,, | Wrought-iron diaphragm, cup, fuze-hole ordered to be altered to G.S. gauge, § 2097. |
| ,, III.† | 11/7/71 | 2097 | ,, | ,, | ,, | ,, | ,, ,, ,, ,, | 7 5 | 0 0½ | ,, | Fuze-hole, G.S. gauge. |
| ,, IV.† | 15/3/72 | 2210 | ,, | ,, | ,, | ,, | ,, ,, ,, ,, | 7 5 | 0 0½ | ,, | Junction of head and body stronger, and secured by screw rivets as well as by plain rivets, centre tube of gun-metal instead of iron. |
| ,, V. | 19/1/74 | 2577 | ,, | ,, | ,, | ,, | ,, ,, ,, ,, | 7 8⅛ | 0 0½ | ,, | Flush socket. |
| ,, VI. | 24/3/75 | 2744 | ,, | ,, | ,, | ,, | ,, ,, ,, ,, | 7 9⅜ | 0 0½ | ,, | Copper studs. |

* To be broken up, § 3046.

† A few Mark II. 9-pr. and Mark III. 7-pr. shell have been made and issued which have the junction of head and body stronger, and some with all the improvements; these are distinguished by one small star (*) or two small stars (**) on the studs. 7-pr., Mark I. (§ 1392), with cast-iron diaphragm, and without a tin cup, is unserviceable, and is broken up on return to Woolwich.

‡ 4 parts lead to 1 part of antimony.

N.B.—For diameters, see Table of Common Shell. See foot note §, p. 208.

## CASE SHOT, 16, 9, AND 7-PR. R.M.L. GUNS.

| Calibre and Mark of Pattern. | Date of Approval. | § Changes in War Stores. | Length. | Diameter. | No. and Nature of Balls* contained. | Approximate Weight of Balls. | Approximate Weight of Case Linings, Clay, and Sand, &c. | Total Weight. | Marks on the top of the Case. | Remarks. |
|---|---|---|---|---|---|---|---|---|---|---|
| | | | inches | inches | | lbs. oz. | lbs. oz. | lbs. oz. | | |
| 16-pr., Mark I. .. | 8/5/72 | 2279 | 7·1 ± ·05 | 3·54 ± ·02 | 176, lead and antimony, 16½ per lb.† | 10 9 | 4 10 | 15 3 ± 8 oz. | 16-pr., R.↑L., I. .. | Balls packed in clay and sand. |
| 9-pr., Mark I. .. | 4/7/70 | 1921 | 7·4 ± ·1 | 2·94 ± ·015 | 110, lead and antimony, 16½ per lb. | 6 12 | 3 0 | 9 12 ± 4 oz. | 9-pr., R.↑L., I. .. | Balls packed in equal parts of resin and sand. |
| 9-pr., Mark II... | 15/2/71 | 2060 | 7·4 ± ·1 | 2·94 ± ·015 | 108, lead and antimony, 16½ per lb.† | 6 11 | 3 2½ | 9 13½ ± 6 oz. | 9-pr., R.↑L., II. | Outer case in three parts, ring at bottom, instead of a disc, balls packed in clay and sand. |
| 9-pr., Mark III. | 16/9/71 | 2104 | 7·4 ± ·1 | 2·94 ± ·015 | " " " | 6 11 | 3 2½ | 9 13½ ± 6 oz. | 9-pr., R.↑L., III. | Knob removed. |
| 9-pr., Mark IV. | 13/10/75 | 2802 | 7·4 ± ·1 | 2·94 ± ·015 | " " " | 6 9¼ | 3 1 | 9 10¼ ± 6 oz. | 9-pr., R.↑L., IV. | Iron ring, disc, and ments, instead zin |
| 7-pr., Mark I. .. | 4/12/65 | 1153 | 5·6 | 2·94 ± ·01 | 48, lead and antimony | 6 0 | 1 8 | 7 8 ± 2½ oz. | I. .. .. .. | Balls loose. |
| 7-pr., Mark II... | Provly. 9/9/67 | 1468 | 5·2 ± ·05 | 2·94 ± ·015 | 36, 1½ oz. sand shot .. | 3 6½ | 1 13½ | 5 4 ± 4 oz. | W.↑D., 7-pr., M.L., II. III. | Balls packed in coal dust. |
| 7-pr., Mark III. | 31/7/69 | 1814 | 4·7 ± ·05 | 2·94 ± ·015 | 82, 1 oz. lead and antimony | 5 4 | 1 10 | 6 14 ± 4 oz. | R.↑L., 7-pr., M.L. IV. | " " |
| 7-pr., Mark IV. | 26/2/72 | 2280 | 4·7 ± ·05 | 2·94 ± ·015 | 70, lead and antimony, 16½ per lb.† | 4 4 | 2 0 | 6 4 ± 4 oz. | R.↑L., 7-pr., M.L. | Balls packed in clay and sand. Outer case in three parts. Ring at bottom. |

* Four parts lead to one part of antimony. † For future manufacture 16 per lb.

In future the high gauge for the diameter of Case Shot will be the same as that for Common and Shrapnel Shells, while the limit between the high and low gauges will be ·03″. § 2971.

## Service, Exercise or Saluting, and Drill Cartridges, 16, 9, and 7-pr. R.M.L. Guns.

| Calibre and Mark of Pattern. | Date of Approval. | § Changes in War Stores. | Charge, lbs. | Length, Inches. From | Length, Inches. To | Diameter, Inches. From | Diameter, Inches. To | No. of Hoops. | Material of Cartridge. | Marks. | Metal-lined Whole Number. | Metal-lined Whole Weight. | Metal-lined Half Number. | Metal-lined Half Weight. | Metal-lined Quarter Number. | Metal-lined Quarter Weight. | Pentagon Whole Number. | Pentagon Whole Weight. | Pentagon Sectional Number. | Pentagon Sectional Weight. | Rectangular, Plain Number. | Rectangular, Plain Weight. | Remarks. |
|---|---|---|---|---|---|---|---|---|---|---|---|---|---|---|---|---|---|---|---|---|---|---|---|
| 16-pr., Service, Mark I. ... | 12/3/72 | 2212 | 3 | 9·5 | 10·0 | 3·3 | — | 6 braid | Serge ... | I., 16-pr. M.L., 3 lbs. | *33 | 152 | 15 | 77 | 6 | 37 | | | | | | | To govern conversion of 68-pr., 16 lb. cartridges to 16-pr., 3 lbs. |
| ,, ,, ,, I. ... | 9/11/71 | ,, | ,, | ,, | ,, | 3·3 | — | 6 silk | Silk cloth ... | I., 16-pr., M.L., 3 lbs. | 33 | 149 | 15 | 76 | 6 | 36 | | | | | | | To govern supply of silk cartridges after the serge cartridges are exhausted. |
| ,, ,, ,, II. ... | 29/10/75 | 2838 | ,, | ,, | ,, | 3·3 | — | 6 silk | Silk cloth ... | II., 16-pr., M.L., 3 lbs. | | | | | | | | | | | | | |
| 16-pr., Exercise, Mark I. ... | 9/11/71 | 2212 | 1½ | 5·5 | 6·0 | 3·3 | — | 3 silk | Silk cloth ... | I., 16-pr., M.L., 1½ lbs. | *74 | 166 | 33 | 83 | 13 | 39 | | | | | | | |
| ,, ,, ,, II. ... | 29/10/75 | 2338 | 1½ | ,, | ,, | 3·3 | — | 3 silk | Silk cloth ... | II., 16-pr., M.L., 1½ lbs. | 75 | 165 | 35 | 85 | 13 | 39 | | | | | | | |
| 9-pr., Service, Mark I. ... | 14/2/70 | 1982 | 1¾ | 9·0 | 9·5 | 2·6 | — | 5 braid | Serge ... | I., 9-pr., M.L., 1 lb. 12 oz. | *60 | 160 | 27 | 80 | 10 | 37 | | | | | | | For 8 cwt. gun and 6 cwt. gun, L.S. |
| ,, ,, ,, II. ... | 29/10/75 | 2838 | 1¾ | 9·0 | 9·5 | 2·6 | — | 5 silk | Silk cloth ... | I., 9-pr., M.L., 1 lb. 12 oz. | 65 | 165 | 30 | 84 | 11 | 38 | | | | | | | |
| ,, ,, ,, I. ... | 25/10/70 | 2670 | 1½ | 8·25 | 8·75 | 2·6 | — | 5 braid | Serge ... | I., 9-pr., M.L., 1 lb. 8 oz. | *67 | 154 | 33 | 83 | 13 | 39 | | | | | | | For 6 cwt. gun, S.S. |
| ,, ,, ,, I. ... | 29/10/75 | 2838 | 1½ | 8·25 | 8·75 | 2·6 | — | 5 silk | Silk cloth ... | I., 9-pr., M.L., 1 lb. 8 oz. | 68 | 152 | 33 | 81 | 13 | 38 | 68 | 165 | 34 | 95 | | | |
| 9-pr., Exercise, Mark I. ... | 7/3/71 | 2047 | 1 | 5·75 | 6·25 | 2·6 | — | 3 silk | Silk cloth ... | I., 9-pr., M.L., 1 lb. | *101 | 155 | 44 | 77 | 20 | 39 | — | — | — | — | | | |
| ,, ,, ,, II. ... | 29/10/75 | 2838 | 1 | 5·75 | 6·25 | 2·6 | — | 3 silk | Silk cloth ... | I., 9-pr., M.L., 1 lb. | 120 | 172 | 54 | 86 | 21 | 40 | 107 | 171 | 51 | 96 | | | |
| 7-pr., 12 oz., Mark I. ... | — | — | 12 oz | 5·75 | 6·25 | 2·5 | — | 3 braid | Serge .. | I., 7-pr., M.L., 12 oz. | *135 | 157 | 65 | 83 | 27 | 40 | — | — | — | — | | | For the steel gun of 200 lbs. |
| ,, ,, ,, I. ... | 29/10/75 | 2838 | 12 oz. | 5·75 | 6·25 | 2·5 | — | 3 silk | Silk cloth ... | I., 7-pr., M.L., 12 oz. | 142 | 159 | 70 | 84 | 28 | 40 | 134 | 166 | 68 | 95 | | | |
| ,, 8 oz., Mark I. ... | 10/5/67 | 1414 | 8 oz. | 4·5 | — | 2·5 | — | 1 worsted | Serge ... | I., 7-pr., M.L., 8 oz. | *198 | 156 | 90 | 79 | 40 | 40 | — | — | — | — | | | For the bronze gun of 200 lbs. |
| ,, ,, ,, I. ... | 29/10/75 | 2838 | 8 oz. | 3·5 | 3·75 | 2·5 | — | 1 silk | Silk | I., 7-pr., M.L., 8 oz. | 214 | 162 | 102 | 84 | 42 | 40 | 204 | 169 | 100 | 96 | | | |
| ,, 6 oz., Mark I. ... | 17/9/67 | 1471 | 6 oz. | 3·5 | — | 2·5 | — | 1 worsted | Serge | I., 7-pr., M.L., 6 oz. | *260 | 154 | 116 | 78 | 46 | 37 | — | — | — | — | | | For the steel gun of 150 lbs. |
| ,, ,, ,, I. ... | 29/10/75 | 2838 | 6 oz. | 3·5 | — | 2·5 | — | 1 silk | Silk | I., 7-pr., M.L., 6 oz. | 290 | 162 | 133 | 82 | 56 | 40 | 266 | 165 | 126 | 92 | | | |
| ,, 4 oz., Mark I. ... | 14/2/69 | 1781 | 4 oz. | 2·7 | — | 2·2 | — | 1 worsted | Shalloon | I., 7-pr., M.L., 4 oz. | *380 | 153 | 170 | 78 | 72 | 38 | — | — | — | — | | | For use with double shell with both the steel and the bronze guns. |
| | | | | | | | | | | | 425 | 159 | 200 | 82 | 82 | 40 | 414 | 169 | 200 | 95 | | | |
| **Drill Cartridges.** | | | | | | | | | | | | | | | | | | | | | | | |
| 16-pr., Mark I. ... ... | 10/3/73 | 2435 | — | 9·5 | 10·0 | 3·15 | 3·3 | — | Hollow wood block covered with raw hide | I., 16-pr., M.L., 3 lbs. ... | | | | | | | | | | | | | |
| 9-pr., Mark I. ... ... | 5/6/71 | 2088 | — | 9·0 | 9·75 | 2·4 | 2·6 | — | ,, | I., 9-pr., M.L., 1 lb. 12 oz. | | | | | | | | | | | | | |
| 7-pr., Mark I. ... ... | 27/2/77 | 3093 | — | 5·75 | 6·25 | 2·35 | 2·5 | — | ,, | I., 7-pr., M.L., 12 oz. ... | | | | | | | | | | | | | |

* With paper covers.

For bags containing spare powder, and weight of empty cases, see table p. 220.

# CHAPTER XV.—B.L. SMALL-ARM AMMUNITION.

MARTINI-HENRY AND SNIDER ARMS.—ADAMS' AND COLT'S PISTOLS.—GATLING GUNS.

Martini-Henry bullet.

THE bore of the Martini-Henry rifle is ·45″; the bullet is made of lead hardened with tin,* 1″·27 long, weight 480 grains. Its diameter increases from ·439″ at the shoulder to ·45″ at the base. A small hollow in the base of the bullet tends to slightly expand it, and the great length of the bullet causes it to set up in the bore and fill the grooves. The cannelures allow the cartridge to be secured by choking to the bullet. The latter has two turns of fine white parchment paper wrapped round it from right to left, that is, in a contrary direction to the spin of the bullet, so that the paper untwists in passing through the bore and leaves the bullet free. The paper is lubricated for about half its length and on its base, and its function is to prevent leading. The paper is crimped over the centre of the base into a "rose" which is twisted closely so as not to allow the lubrication to pass through it on to the base of the bullet.

Charge.

The charge is 85 grains R.F.G².

Case.

The case is formed of two turns and a ·5″ overlap of ·004″ brass. It is lined with tissue paper where the powder charge rests to prevent the mutually injurious action which would take place between the powder and brass, if in contact with each other. Inserted between the folds at the base, is a strip of brass (the same as that composing the body of the case) to give additional strength. A small "sighthole" is punched in the case so as to allow the examiners to see that this inner strip is present. The cartridge is made of a bottle-necked shape by crimping in the upper part, so that the whole fits into the shot chamber of the rifle. The earlier rifles were long-chambered and the cartridge the same diameter throughout its length. This made it inconveniently long, and the shooting was not so good as with the bottle-necked cartridge. The base is strengthened by an outer and an inner base cup of brass, and the bottom closed by a blackened wrought-iron disc. Inside, a paper pellet is pressed against the bottom of the cartridge, a brass cap chamber pierced with a fire-hole passes through the disc, cups, and pellet, and rivets the whole together, the top being bulged out over the paper pellet, and the base of the cap chamber being flanged to fit a countersunk recess in the iron disc. The cap chamber contains a brass anvil, on the shoulders of which rests the copper cap which contains cap composition (fulminate of mercury, sulphide of antimony, and chlorate of potash) varnished over. Above the powder is placed a millboard disc, then a wad of beeswax, cupped out to the front to ensure its expanding in cold weather, and then two more millboard discs. On the top of these rests the bullet. The cartridge is

* The proportions are 12 lead to 1 tin. The Sp.Gr. of the mixture is 10·9.

rather more than 3″ long. See plate, p. 362, for dimensions and details of the various parts of the cartridge.

Action. On firing, the cap composition is forced against the point of the anvil, and the flash reaches the charge through the hole in the chamber; the case unwinds, and being pressed firmly against the sides of the bore prevents the gas from escaping in the direction of the breech; after firing, the cartridge contracts to its original proportions, making extraction easy.

The base of the cartridge is very strong owing to its base cups, strengthening strip, and disc, and it is at the base that strength is essential, indeed if the base is perfect, a very faulty cartridge may be fired without throwing any undue strain on the breech.

The cartridge is found to stand rough usage and wet well, what tries it most is a moist hot climate,* it is almost impossible to prevent the moist air from penetrating when aided by great variations of temperature. Experiments have shown that it is impossible to explode these cartridges in a mass, thus firing ¼ lb. of powder along with a number of cartridges in an iron cylinder hardly exploded any rounds.†

§ 2661. Three patterns of Martini-Henry ammunition have been issued, but only that described above as Mark III. is likely to be met with in large quantities.

Mark I. generally resembled Mark III., but the brass was ·003″ thick, and there was no "sight-hole." The bullet had only one cannelure. Mark II. had the brass ·004″ thick, but the strip of brass between the folds of the cartridge was omitted; the outer base cup was slightly longer than in Mark I. The bullet had only one cannelure.

Mark III. is the present service pattern, and was approved on 10/8/73. It has been described above.

Mark IV. (not issued to service) had a 410 grain bullet, 1″·115 long, and a charge of 80 grains R.F.G. The recoil of the rifle with this pattern is reduced to about the same as the Snider, and the alteration was made in consequence of complaints about the heavy recoil experienced with Mark III. Lengthening the butts of the rifles slightly was found to diminish the sensible effect of the recoil, so the manufacture of Mark III. was reverted to in April, 1875. Only about two millions of Mark IV. were made.‡ Space does not admit of any detailed account of the advantages of the Martini-Henry over the Snider.§

* Much information as to the effect of climate on early patterns of Snider cartridges without paper lining will be found in Extracts, Vol. VII, p. 224.

† The experiments were actually made with Snider cartridges, but would hold good with the M.-H. cartridges.

‡ The shooting of Mark IV., at 500 yards, was very nearly as good as that of Mark III. and the trajectories up to that range of both patterns were nearly identical; but the penetration into deal was 10″ with Mark IV. against 12″ with Mark III. (at 500 yards). The cases were identical. Mark IV. cartridge complete was rather shorter (on account of the different bullet) than Mark III., and of course weighed slightly less.

§ Extracts from B.L. Rifles and Ammunition, p. 39. This work by Major Majendie and Capt. Orde Browne contains much valuable detailed information.

At 500 yards range the greatest height of the Martini-Henry trajectory is 8·1 feet; at this range a cavalry soldier will be hit at any point, and an infantry soldier will be hit between 396 and 500 yards. The greatest height of the Snider trajectory at the same range is 11·9 feet; a cavalry soldier will be hit between 400 and 500 yards, and an infantry soldier between 438 and 500 yards.

The Martini penetrated 14½ elm planks ½ inch thick, as against 8½ with the Snider.

The Martini penetrated iron plates over ½ an inch thick at 200 yards, the Snider failed to penetrate.

It has a flatter trajectory, greater accuracy, and penetration, and can be loaded much more rapidly.

The bullet is much less deflected by the wind.

**Packing.** In tens, packed "heads and tails" in fine white paper and brown paper. On the brown paper is printed the nature of the contents and the year of manufacture. One packet weighs very nearly 1 lb. 2 oz. For numbers contained in various packages, see table, p. 80.

**Martini-Henry Cavalry carbine, Mark I. § 3220.** Ammunition for the Martini-Henry cavalry carbine. The case is identical with Mark III., the bullet weighs 410 grains, and the charge is 70 grains R.F.G². The spare length in the case is filled up with a little cotton wool. The cartridges are packed like those for the M.-H. rifle. A packet weighs 1 lb. The cartridges are easily distinguished from Mark III. rifle cartridges by their being shorter. Another pattern is likely to be soon introduced in which the cotton wool is dispensed with, and the space filled up by thickening the paper lining.

*Snider Ammunition.*

The same description of cartridge is used for all the B.L. small arms with Snider action in the service, the same cartridge will fit an Enfield rifle or carbine, or a Lancaster carbine. The diameter of the bore is ·577″, the diameter of the bullet is ·573″, small enough to drop through the bore, and depending on expansion for its fit, the length of bullet is about 1″ (1″·04 in present pattern), and the length of the cartridge is a very little under $2\frac{1}{2}$″. The charge is 70 grs. (very nearly $2\frac{1}{2}$ drs.) of R.F.G. Each packet of ten rounds weighs about 1 lb.

Many patterns of the Boxer* cartridge have been made, they are all serviceable with the exception of Mark I., which may be recognised by having a potêt base, formed by pressing out the rim of the base cup so as to form a beading round the edge instead of having a brass or iron base disc as the other marks have. Mark V. is a weaker† cartridge than the others, it may be recognised by its being the only brown paper covered cartridge without a distinguishing ring or rings marked on it, the brass sheet from which this cartridge is made is ·003″ thick, instead of ·005″ like the later patterns, and its base is not so strong as the earlier ones, the cartridge is not to be issued for foreign service, but is to be expended at home stations.

**Construction.** The sketch illustrates the construction of the present pattern of cartridge, which only differs in minor details from the other service patterns.

The bullet is made from pure lead, weight 480 grs., the hollow in the head is closed by having the lead spun over it, the hollow parts are necessary in order to get the bullet of a sufficient length for good shooting, without unduly increasing its weight, and to get its centre of gravity in the proper place, the hollow in the base is also used to give the expansive action to the bullets, the plug, made of clay, and soaked in beeswax, closes the rear cavity, and on firing expands the bullet, which has three cannelures, the sides of the bullet as far as the front cannelure are coated with beeswax, the cannelures holding a sufficient supply of the lubricant in their recesses; by the expansion of the bullet the lubri-

---

The figure of merit of the Snider is from 15″ to 16″, that of the Martini-Henry from 9″ to 10″. This is taking an average of the shooting throughout the year in all weathers. Under favourable circumstances the Martini has given as low a figure as 5″·25.

Much information as to penetration will be found in Extracts, Vol. VII, pp. 68, 231.

* The name Boxer has been omitted from the nomenclature.

† The case has but one turn and an overlap of ·3″ of sheet brass, previous patterns have $2\frac{1}{2}$ turns of the same sheet brass. § 2436.

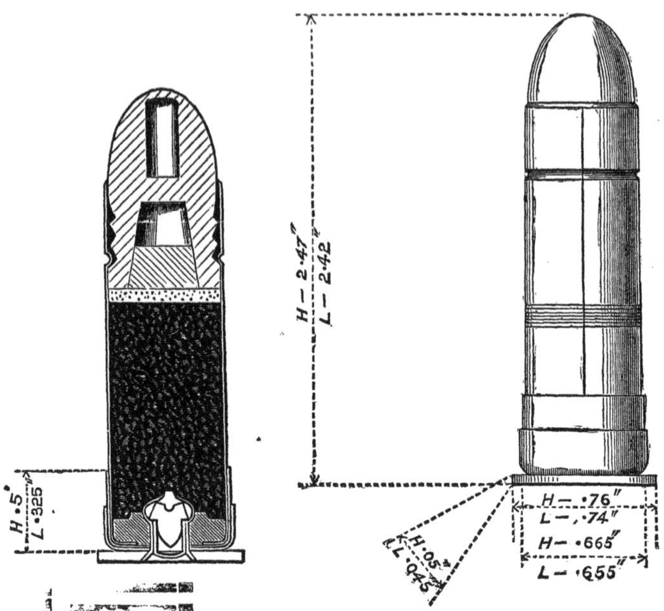


CARTRIDGE, MARK IX, SNIDER ARM.

cant is squeezed out, and the bore is thoroughly cleaned out by the bullets passing through it. As many as 4,000 rounds have been fired without fouling sufficiently to injure the shooting.

The case is formed of sheet brass covered with brown paper. It is lined with shellac and thin white paper to prevent corrosion from the powder. The case overlaps by about $\frac{1}{4}''$, and is cemented together with shellac and glue. The arrangement of the two base cups, iron disc, &c., are identical with those used in the Martini-Henry cartridge, Mark III. There is no strengthening strip in the folds.

Marks. The first four marks had the brass cylinder covered with white paper.

§ 1328. Mark I. has a potêt base, before described.

§ 1448. „ II*. has a brass disc, and heavy bullet 525 grains. 4 cannelures

§ 1449. „ III. has also a brass disc. The bullet is shorter, the distance therefore from choke to top of case is less, but it is hard to distinguish between II. and III. 3 cannelures.

§ 1450. „ IV. is the only cartridge which has white paper and iron disc. Bullet has 3 saw-shaped cannelures.

§ 1496. „ V. is the only brown-paper cover cartridge without a distinguishing ring. Brass ·003″ thick, 4 cannelures.

§ 1703. „ VI. has one black ring. „ ·005″ „ „

§ 1760. „ VII. has one black ring and lead spun over head instead of wood plug. Bullet, 4 cannelures.

§ 1831. „ VIII. has two black rings, lined with tissue paper and shellac to prevent corrosion.†

---

* This cartridge is best suited for the naval rifle, as its long bullet, 525 grains, gives good shooting with the rapid twist of this rifle. A few Mark II. cartridges were made for S.S. with iron discs after the introduction of Mark IV.

† Mark VIII. and IX. cartridges may be met with containing copper instead of brass cap chambers. Mark VIII. up to 4/71 had the bullet lacquered. This was discontinued without causing a change of pattern.

Mark IX. has one red ring, it only differs from VIII. in having a bullet with 3 cannelures of the same size as pattern IV., which was supposed to give better shooting at long ranges. § 2105.

In packets of 10 rounds, bullets all one way, made up in brown paper, see p. 80. Issue.

*Blank Cartridges* are issued in packets of 10 rounds, made up in purple paper. The same blank cartridges fit either the Martini-Henry or Snider rifles, or carbines. Blank cartridges for B.L. rifles. §§ 1552, 2271.

For packing, see p. 80.

Three patterns of blank ammunition may be met with, all are serviceable. Four patterns have been made.

I. Obsolete. It was the same construction as Mark I. ball cartridge. § 1451.

II. The case is made of sheet brass not covered by paper, and contains the powder in the form of a pellet. § 1451.

III. Differed from II. in having a paper case. § 1552.

IV. Differs from III. in having the powder loose, and in some minor details of construction. §§ 2332, 248

Besides the above, a large quantity of blank is converted from condemned ball cartridge, either Snider or Martini-Henry, and is called "Blank. Mark IV. Converted."

MARK IV. BLANK CARTRIDGE.

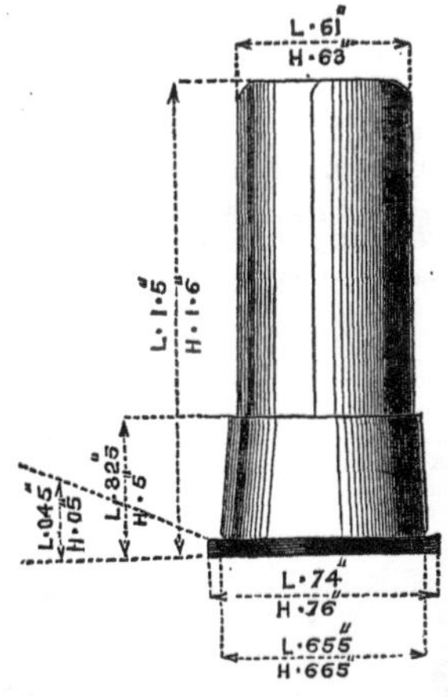


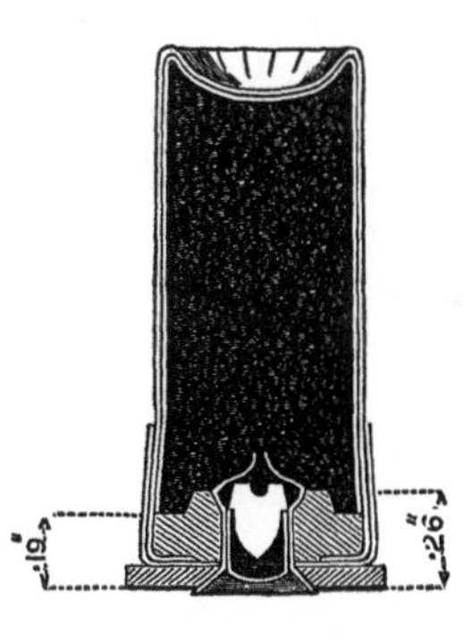


*Ammunition for Gatling Guns.*

There are two calibres of Gatling gun in the service, viz., one with a ·65″ bore, and one with a ·45″ bore. Gatling cartridges. § 2644.

The cartridge for the ·65″ gun has a coiled case made of sheet brass, ·01″ thick. The construction resembles that of the Martini-Henry cartridge, differing only in minor details. The bullet is made of the same alloy as that for the Martini-Henry cartridge, and weighs 1,422 grs. (about $3\frac{1}{4}$ oz.). The charge is 270 grs. R.F.G². A bundle of ten rounds weighs 3 lbs. $7\frac{3}{16}$ oz.*

* A steel-tipped bullet, or a lead-coated steel bullet, have been proposed for service against torpedo boats, with Gatling guns of 1″·0 and 0″·65 bore. The 1″ bore (if the bullet were lead tipped with steel) would have a bullet weighing nearly $\frac{1}{2}$ lb.

The cartridge for the ·45″ bore Gatling is at present procured by contract, and loaded and capped in the R.L. It is a solid brass cylinder, somewhat contracted at the top, and the base is formed by pressing out the material of the case, so as to form the projecting rim necessary for extraction. There is no anvil, but the cap strikes against a small boss or nipple in the chamber, which is pierced with two fire holes. The cases are not lined with paper, but lacquered. The charge, bullet, wads, &c., are the same as those used in the Martini-Henry cartridge. A bundle of ten rounds weighs 1 lb. $2\frac{1}{4}$ oz.

**Army Circulars, June, 1873, Clause 12.**

The rule as to excluding B.L.S.A. cartridges from a gunpowder magazine should be rigidly attended to with this cartridge, as the cap composition is in close proximity to the raised part of the chamber, and seems to be more liable to accidental ignition than the cartridges for B.L. small arms.

Though this cartridge is of the same calibre as the Martini-Henry, the two cartridges are not interchangeable, as the dimensions are different. The Martini-Henry cartridge was found not to answer in the Gatling gun.

**Packing.**

Both calibres of Gatling cartridges are packed, heads and tails, in bundles of 10 in brown paper wrappers. For numbers contained in various boxes, &c., see table, p. 80.

The limit of the effective range of the ·45″ bore is stated to be about 1,200 yards, after that the longitudinal error rapidly increases.—Extracts, Vol. X, p. 51. 100 shots fired at 3 rows of targets, 9″ × 45″, gave 91 hits at 1,000 yards, 76 hits at 1,200, 40 at 1,400, 22 at 1,600, and 13 at 1,800.*

It is stated that the ·65″ bore cannot be used much beyond 2,000 yards without great waste of ammunition.—Extracts, Vol. X, p. 256.

**Buck-shot cartridge, Mark II. § 2546.**

*Buck-shot Cartridge.*—A special cartridge containing buck shot has been issued for convict-guard service, and for service in Ashantee. The case is similar in construction to the service ball cartridge for Snider arms.

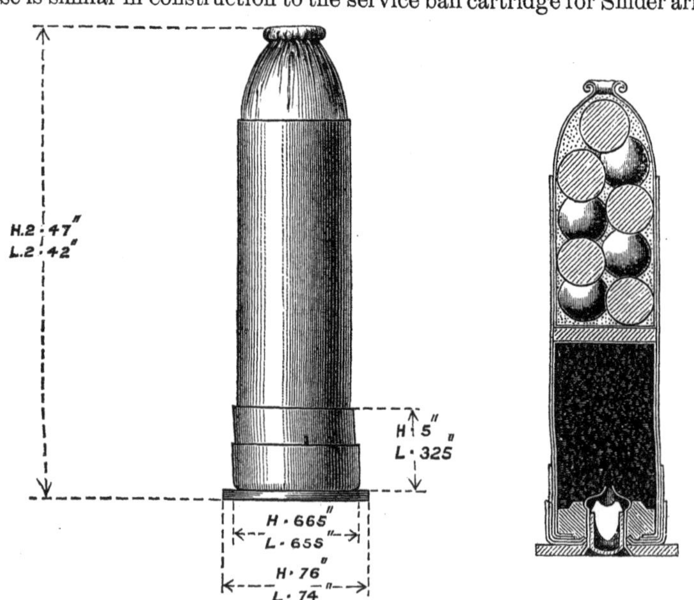


* The ·45″ Gatling at 500 yards gives an average figure of merit of about 20 inches.

The charge is from 52 to 56 grains of R.F.G. powder, on the top of which is a paper disc and a felt wad lubricated with beeswax.

In place of the lead bullet the cartridge contains 13 buck-shot, weighing about 220 to the lb., placed in a paper bag, the interstices between the shot being filled with bone dust; the bag is inserted in the cartridge case and secured with shellac cement. Mark I. had the powder charge pressed into a solid pellet, and plaster of Paris was used instead of bone dust.

§ 1739.
**Mark I.**

*Ammunition for Adams' Revolver.*

The cartridge consists of a small brass cylinder (·01″ thick), with the base attached in the same manner as the rifle cartridges. There are no base cups.

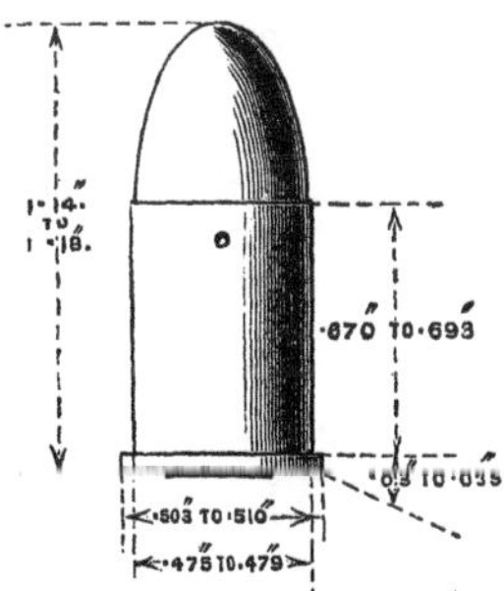

The bullet is of pure lead, and weighs 225 grs. It is lubricated with beeswax. The charge is 13 grs. of pistol powder, or siftings of R.F.G[2]. powder. The bullet is secured in the case by four small indentations between the cannelures.* A small loose coil of paper interposes between the powder and the metal of the cylinder. The base disc is of iron, between ·03″ and ·035″ in thickness.

Mark II. differs from Mark I. only in having the base disc of hard brass instead of iron.

**Mark II.**
§ 3219.

These cartridges are made up in bundles of 12, in brown-paper wrappers. For the number contained in various packages, see table, p. 80.

**Packing.**

*Ammunition for Colt's Pistol.*

The cartridges for Colt's pistol have 13 grs. pistol powder contained in a skin envelope, choked on to the bullet. The latter weighs 135 grs. and is made of pure lead. The lubrication is beeswax. Each cartridge is enclosed in a paper cover, which is torn off by means of a tape before loading.

§ 857.

The cartridges are packed in brown paper in bundles of 18, and each bundle is enclosed in a waterproof bag, consisting of a film of india-rubber between two thicknesses of paper. A bundle of 18 cartridges weighs $6\frac{10}{16}$ oz. and 2,250 rounds are contained in a quarter metal-lined case.

**Packing.**
**§§ 247, 259.**

This ammunition is issued to the Coast Guard.

* In the cut and in the description given in § 1739, ¼ gr. of cotton is placed over the fire-hole of the cap-chamber. This, however, has been discontinued for several years, and it is doubtful if any cartridges so made were issued to the service.

*Proof.*

**Proof.**

*All B.L. ammunition is proved at Woolwich by firing a percentage from a fixed mechanical rest, at a range of 500 yards. It is unnecessary here to describe the rest, its adjustment requires a skilled mechanic; at out-stations sufficiently good results can be obtained by a good marksman firing from a sandbag rest. Care must be taken to use a rifle in perfectly good order. It is advisable to use several rifles, so that defects due to them may be eliminated.

The target is divided into squares by thick lines drawn at intervals of 3 feet, and into smaller squares by thin lines drawn at half a foot interval, and diagrams are drawn and divided so as to correspond to the target.† See plate, p. 363.

The rifle is laid on any desired spot on the target, and the rest once adjusted does not allow the rifle to shift, 20 rounds ‡ are fired, and the position of each shot is marked on the diagram, the position of the point aimed at is also marked.

A point is selected from which to measure the distance of the hits (each hit is numbered in the order of firing), it is found convenient to select a point near to the group, but to the left of and below all the shots; this point is called the origin.

From this point the horizontal distance is measured and placed in the second column.

The vertical distance is measured and placed in the third column. Both columns are added up, and divided by 20 to get the co-ordinates of the point of mean impact.

This point is fixed by measuring the mean horizontal distance from the origin, and then measuring the mean vertical distance. Having thus marked the point of mean impact the distance of each hit from this point is measured by a scale corresponding to that of the diagram, and the distances are placed in the fourth column, and we thus get the deviation of each hit from the point of mean impact. This column is added up, and divided by 20 to get the average deviation from the point of mean impact, or, as it is often called, the figure of merit.§

Though ammunition at out-stations should be examined and its shooting qualities ascertained, it is not to be condemned without the authority of the Surveyor-General.

**A.C. 7/74.**

The velocities of bullets are taken in the R.L. by means of the Bashforth gravity chronograph, of which a full description will be found in the inventor's treatise on the "Motion of Projectiles."

---

* Much information will be found in "Military B.L. Rifles and Ammunition," by Captains Majendie and Browne.

† Half an inch to represent one foot will be found a convenient scale.

‡ The first shot fired from a rifle after cleaning, or when quite cold, is not marked, as there is often a considerable difference between the first shot fired from a cold rifle and the subsequent shots.

§ The strength, direction, and character of the wind; the heights of the barometer and thermometer, and the degree of humidity are noted. These form important factors in judging the shooting. No detailed directions can be given, as practice alone enables one to judge of the proper value to be assigned to each of these factors in particular cases.

As a rule, the Snider ammunition shoots better in cold than in hot weather. Much better shooting will be made with all ammunition when the wind is steady, even though strong, than when it is light and variable. At Woolwich two targets are used close to each other and the rifles fired almost simultaneously so as to eliminate as far as possible uncertainty caused by casual gusts of wind.

Pistol ammunition is generally proved by firing from a table rest at a range of 50 yards. The figure of merit should not exceed 12″, but much depends on the marksman. Adams' B.L. pistol cartridges should pierce on the average three wet ½-inch elm boards, placed ½ inch apart, at a range of 12 or 15 yards.

# CHAPTER XVI.—ROCKETS.

HALES' WAR ROCKETS, 9-PR. AND 24-PR.—SIGNAL ROCKETS.—CALL ROCKETS.—LIFE SAVING ROCKETS.—ROCKET MACHINES.—MISCELLANEOUS STORES.

THE greater part of this chapter has been taken *verbatim* from Ammunition, Part II., by Captain C. O. Browne, late R.A.

## ROCKETS AND ROCKET MACHINES.

**General remarks.**

It is difficult to say at what date signal or sky rockets were first made, and how far small rockets of this description were in old times discharged as projectiles.*

Sir W. Congreve states that rockets of inconsiderable size had been used in India† for war purposes, and that General Desaguliers had attempted to manufacture large rockets in the Royal Laboratory, though without success,† when in 1805 he (Sir W. Congreve) first formed the conception of a systematic construction of projectiles in the form of rockets.

The chief recommendation of such a system appeared to him to be "that projectile force was exerted without any reaction upon the point "from which it was discharged,"‡ or, as he says elsewhere, "on its "fulcrum."§

Consequently, in rockets Sir W. Congreve saw the means of enabling boats' crews and individual men to discharge missiles of equivalent power to those projected by the heaviest cannon of the day. Thus he says, speaking of a rocket,|| "it is ammunition without ordnance"; "it "is the soul of artillery without the body."

* In "Mémoire sur les Fusées de Guerre, par Brussel de Brulard, Ancien Officier Supérieur d'Artillerie," is quoted a statement made by Mailla in his "Histoire de China," tome ix, p. 167, that the Chinese "used rockets and shells against the Tartars in the 13th contury." The same author states, pp. 2 and 3, that rockets were used in Europe in the 15th century. M. De Mongery, Capitaine, Chevalier de St. Louis, &c., makes a similar statement in his "Traité des Fusées de Guerre."

† See a "Concise Account of the origin and progress of the Rocket System," by Sir W. Congreve, p. 1; also conclusion to general view in the same work, p. 20; also same work, p. 14.

Colonel Gerrard, then Adjutant-General of Indian Army, states that he once saw three men killed and four desperately wounded by the same rocket, and states that the British suffered more from the enemy's rockets in the siege of Seringapatam in 1799 than from shells or any other weapon.

Tippoo Sahib undoubtedly taught us the value of rockets as a projectile. *Vide* Concise Account, postscript, p. 14; also Lectures by General Konstantinoff, p. 77.

‡ *Vide* Concise Account, p. 19.

§ " " " p. 1.

|| " " " p. 7, and Lectures by General Konstantinoff, p. 78.

Commencing with the idea of an incendiary projectile which was to be suddenly thrown in large numbers from boats into a seaport town or a harbour crowded with shipping,* he expanded the system so as to comprehend not only shells but also Shrapnel, having for his object both the dispersion of bodies of men, especially cavalry, and the enfilading of trenches; work which he considered could be performed by rockets carried on horses by troops of the lightest description. The idea he had conceived so far became a reality that England obtained some of the advantages he had anticipated by the use of her new weapon.

On October 8th, 1806,† 18 boats discharged 200 rockets into Boulogne and set fire to the town in many places without incurring the slightest opposition or suffering any loss.

At the battle of Copenhagen, in 1807,‡ rockets fired into the town appear to have caused a considerable conflagration; a similar result was produced at Walcheren; and at Leipsic, in 1813, a troop of horse artillery vindicated Sir W. Congreve's opinion as to the possible value of rockets in the field. Lastly, they were successfully used at the river Plate and against the French boats on the Adour.§

It may be well to consider more particularly, before proceeding further, the action of a rocket and the recommendations it possesses.

The cause of motion of a rocket has commonly been briefly described as being due to the difference of pressure on the head and base. Some, however, appear to speak of it as mainly the result of the pressure of the gas generated against the air "as a fulcrum";|| either explanation standing alone is very incomplete.

The following may therefore be added. The smallest quantity of gas generated must have a certain amount of inertia as well as tension, and the force which pushes it away from the point where it was generated must re-act on the rocket; the enormous rapidity at which such gas is formed and accumulates at once gives an atmosphere at a great tension, whether the action commenced in vacuo or in air, thus furnishing a fulcrum against which the nascent gas acts. The resistance of the air to the escaping gas, and also to the rocket, the cooling of the gas, friction, and other causes, further complicate the question.

Thus the rocket itself providing the necessary means of *propulsion*, it follows that means of *direction* only require to be supplied to fire it effectively, and these being very readily provided,¶ rockets may be used in the various ways proposed by Sir W. Congreve.

To expose the surface of composition necessary to obtain the formation of a sufficient quantity of gas for effective propulsion, Sir W. Congreve made each rocket with a large conical hollow in the interior.

The next point to consider is the size of the vents. The propelling power is proved to be inversely as the area of the vent,** but owing to the increased pressure in the interior of the rocket, the strength of the

---

* An attack on Boulogne was his primary object.

† *Vide* Concise Account, p. 4.

‡ " " " p. 15; also Lectures by General Konstantinoff, p. 78.

§ *Vide* MSS. notes on Gen. Gardner's Course for School of Gunnery; also Cust's Wars of the 19th Century, Vol. IV.

|| More frequently implied than distinctly stated.

¶ By the use of tubes, troughs, or by pointing the rocket in the required direction by resting it against a parapet or bank, or even on the ground, where it is to fire along a very level plain, for a sudden rise or broken piece of ground would divert its course, and might even turn it back on those who fired it.

** *Vide* General Boxer's Chapter on Rockets in Treatise on Artillery, pp. 68, 69.

case of the latter must be inversely as the square of the area of the vent.* (Vide Chapter on Congreve Rockets, by General Boxer.)

As an incendiary projectile the rocket possesses great recommendations, and its moral effect is very great, but it labours under certain disadvantages that hardly seem likely to be removed.

1st. Its efficiency and safety depend on the complete contact of a very large service of composition with a thin metal case; hence it is liable to deteriorate from any great variations of temperature, causing expansion of metal; from chemical action causing corrosion; and from vibration in travelling; any of these causes being liable to compromise the safety and efficiency of the rocket.

2nd. The method by which its velocity is gradually imparted to it entails the evil that its flight is very slow, and the rocket is consequently very susceptible to the action of gravity, wind, and accidental causes of deviation.

3rd. The same causes that thus make it peculiarly liable to be acted on by wind and gravity, aggravate the effect of deflection, for, unlike a projectile on which the force of the charge acts entirely in the *desired direction*, a rocket is continually developing velocity in the direction in which it is at the moment pointed; thus, supposing it to be deflected one degree in the first second of time of flight, all the force next added will be applied one degree out of the proper direction, and so on with all impulses in succession; indeed rockets have been occasionally deflected so as to come back, with very great and constantly increasing velocity at the troops who fired them.

4th. In addition to the time afforded for the action of wind or other disturbing causes on a rocket, and also to its own further development of such deviation, Congreve rockets gave a mechanical power to the deflecting action of wind from the lever afforded to it by the stick which they carried attached to them; gravity also had increased power to alter the position of the axis of the rocket, and hence the direction of its impulses.†

5th. From the fact of the composition burning away during flight, the position of the centre of gravity, and consequently the balance of the rocket, is constantly changing; but this is a slight matter compared to those previously mentioned.

In the present day, when firing takes place at extended ranges, and great accuracy is generally requisite, the above objections are grievously felt, but they did not apply with so much force in the time of Sir W. Congreve.

The principle of motion being understood, it is necessary to pass on to the means by which a projectile of such length is kept point first.

Two methods have been employed:—

---

* *Vide* the same, p. 69. For a discussion of the form of vent, *vide* the same, p. 70.

† Captain Browne remarks:—

To my mind there is a strong analogy between the relation borne by gun-shot to rockets, and that of statical to dynamical electricity. In the first-mentioned of each pair compared, the effect is produced by a *sudden impulse*, in the latter by a *constant force*. The former is in each case violent and little liable to be diverted from its course; the latter acting gradually is subject to retardation, and has comparatively little power to force its path in a direct line. Hence the former is specially suited to destructive, the latter to useful work; thus, while lightning shivers the oak and the shot pierces the armour-clad vessel, the galvanic current carries messages round the world, and the life-saving rocket brings the line of communication to the stranded wreck.

1st. The attachment of a stick, fixed on the base or side,* by which the centre of gravity is brought so near the point that pressure of the air acts on the stick end with sufficient leverage to keep the point always towards the direction from which such pressure proceeds; thus on a still day the flight is tolerably direct, but on a windy day it is ever being deflected more and more up the wind, the action of the air being continually to point it in the direction of the resultant of the resistance due to its present line of flight and the pressure of the wind. It is clear, theoretically, that such a tendency of the wind is insatiable, its action being directed to carry the rocket into a line of flight eventually approximating indefinitely closely to the "wind's eye."

Inaccurate as is this methed of obtaining direction, it is peculiarly adapted to a projectile whose great length gives it enormous mechanical advantage if the weight is well forward.

The second method is that of rotation, imparted by some application of either the force of the gas escaping from the rocket, or of the resistance of the air; the great length of the rocket tells against the application of the rotary system of direction, but the slowness of flight is in its favour.

It is unnecessary to dwell long on Sir W. Congreve's rockets. They were first made of paper and then of iron. It was he who established the system of naming a rocket according to its total weight;† he seems never to have achieved success with the 300-pr. rockets‡ which he wished to make, but he dealt largely in 32-prs. With regard to his various projectiles, it should be noticed that the bullets of his Shrapnel§ were projected with a large bursting charge, with a much higher velocity than that of the rocket at the moment of fracture, for he obtained the advantage of keeping his bursting charge behind his bullets constantly.

§ 957.

On the 14th September, 1864, the well-known 24, 12, and 6-pr. Congreve rockets then in the service were provisionally superseded by a rocket of General Boxer's construction, patterns being sealed in August, 1864, with the following improvements:—

1st. The Congreve rocket had the stick attached to a disc which closed up the end of the conical hollow, so that escape holes or vents (five in number) had to be made round it through the base of the composition; the latter was liable to give way on discharge and burst the rocket with the consequent sudden generation of gas; this fault was remedied in the Boxer rocket by slotting three vents in the disc opposite to the base

---

* The stick was originally fixed to the side like those of signal rockets, but Sir W. Congreve altered its position to the centre of the base. It is obvious that a rocket with a stick on one side cannot have its "centre of resistance" opposite to its centre of gravity, and on the principle noticed with reference to shot so fired (*vide* Appendix, p. ), the rocket would tend to describe a spiral path, the longer the stick the less would be the spiral; hence there is a certain advantage in the very long sticks of signal rockets; and the life-saving rocket, with its stick on one side, if fired without the line, carries very badly, as might be expected, for the above-mentioned reason. Also *vide* General Boxer's Chapter on Rockets, pp. 76, 77.

† Vide Concise Account, p. .

‡ M. de Montgéry says, in his "Traité des Fusées de Guerre," that a Capt. Cox saw a rocket in course of manufacture in Burmah which was to contain 10,500 lbs. of powder. He refers to "Journal of a Residence in the Birman Empire," p. 192, published, London, 1821.

§ I have called it "Shrapnel" because in its action it more resembled this projectile than case, and the word "spherical case" being clearly incorrect, "Congreve or rocket Shrapnel" seems the only designation for it. Vide Concise Account, pp. 10, 15; also general view in the same work, p. 7; and detail, p. 7.

of the conical hollow, so as to leave the ring of composition unbroken and strong.

2nd. Greater accuracy was employed by the employment of a stronger composition* which caused the rocket to "jump" with a high velocity from its tube.

There were also minor improvements which need not be noticed. A pattern of Boxer 3-pr. rocket was provisionally approved, 1/10/66. **§ 1336.**

On 24/4/66, the use of war rockets as shells was discontinued. **§ 1245.**

In 1867 Hales' rockets superseded Boxer's improved Congreve rockets, Hales' 3, 6, and 12-pr. being approved on 25/7/67, and Hales' 24-pr. on 31/8/67. It seems necessary to enter into detail in describing these rockets. **§ 1469.**

An account of the Boxer rocket will be found in § 957.

### Hales' Rockets.

**Hales' rockets, 9 and 24-pr.**

† Calibres 9 and 24-prs.

The main difficulties to be contended against in the manufacture of rockets, are the tendency of the case to break up under the heavy pressure of the gas, and the liability of the composition to deteriorate, owing to the tendency of the saltpetre to attack the iron of the case if it is in contact with the metal; the presence of moisture would accelerate this action.

When the metal is corroded by the saltpetre, a spongy porous deposit is formed between the metal and the composition; the result of this is that the flame finds an access to the whole surface of the composition wherever this spongy coating extends, and a large amount of gas is suddenly formed causing an enormous pressure, under which the rocket bursts explosively.

**Construction.**

The head is of cast-iron, plugged with wood and riveted on to the body. The latter is of Atlas metal (a mild steel made by the Bessemer process) lined inside with brown paper and calico, the latter being inside the turns of the former. The object of this lining is to prevent the contact of the metal and the composition. The seam of the body is riveted together and brazed.‡ The base is closed by an iron disc secured to the body by screws, and the disc is tapped to take the tail-piece. It is cupped out in continuation with the cupping out of the tail piece.

The tail-piece is of cast-iron, cupped out inside, and contains three conical vents, the larger part of the cone being towards the interior of the rocket. The vents are cut away on one side; and in consequence of this, the gas issuing from the vents meets with resistance on the side where the vents are prolonged, and, there being no counter-balancing resistance where the vents are cut away, rotation is given to the rocket. The tail-piece and vents are protected from injury, and the interior of the body from damp, by a covering of canvas secured

---

* General Boxer gives his reasons for the use of a stronger composition and the advantages so obtained in his Chapter on Rockets, Treatise on Artillery, pp. 71–76.

† 3, 6, 9, 12 and 24-pr. rockets were originally manufactured, but the designation of the 6-pr. was changed to 9-pr. (§ 1314), and the manufacture of the 3 and 12-prs. discontinued (§ 1940).

‡ Copper was tried in place of spelter, but the heat necessary to melt the copper injured the Bessemer metal.

by twine. Under the canvas is leather, to prevent the sharp edges of the tail-piece from cutting the canvas.*

Composition.

The composition consists of saltpetre, sulphur, and charcoal, mealed and intimately mixed. It is pressed into pellets which are driven one by one† into the rocket by hydraulic pressure. It is separated from the head by a millboard disc, and is bored so as to expose a considerable amount of surface to ignition when the rocket is fired, as otherwise gas would not be formed rapidly enough to start the rocket.

Paint.

§§ 2963, 3138.

The whole of the rocket is painted red, and has its numeral and date stencilled on it in black.‡

The above description applies to the latest patterns of rockets, viz., 24-pr. Mark V., and 9-pr. Mark VI.§

The earlier patterns of rockets are corrugated in three places, to give a good hold to the composition, so as to prevent the case twisting away from it in flight, and there is no paper lining. With the exception of Marks I. and II., the earlier rockets are painted inside, and those made between March, 1874, and the introduction of the present patterns are tinned inside and out as well as painted, to preserve the rockets from deteriorating by corrosion. The various patterns of 9-pr. and 24-pr. rockets are as follows:—

§ 1469.

---

* The paper lining has been found to be unsatisfactory and in all probability will be discontinued. The new patterns of rockets will generally resemble those preceding the patterns with paper lining, but will retain the other improvements introduced in the latter.

† PRESSURES IN DRIVING.

| Nature. | Per sq. in. tons. | cwt. | qrs. | lbs. | On area of rocket. tons. | cwt. | qrs. | lbs. |
|---|---|---|---|---|---|---|---|---|
| 24-pr. .. .. | 10 | 8 | 2 | 8 | 99 | 6 | 0 | 8 |
| 9-pr. .. .. | 7 | 8 | 0 | 0 | 90 | 15 | 3 | 10 |
| Life-saving .. | 8 | 11 | 0 | 1 | 45 | 7 | 3 | 19 |

‡ At first Brunswick black was used as paint. Red lead paint was introduced in 1870, but complaints were received from foreign stations; the paint was found to come off, leaving the iron bare; therefore, in 1873, the ingredients were changed to those given in the table, p. 322.

§ 1515.
§ 1985.
§ 2441.

The numeral and date are also stamped on the base of the rocket. On the case near the head are stamped the letters A.M. (Atlas metal). Each rocket, moreover, has a letter of the alphabet and a number stamped on both head and case. The numbers run up to 1,000, and then the letter is changed.

§ None of the Hales' rockets at present in the service have shells or incendiary composition in their heads, it is highly desirable that they should carry some means of igniting or destroying material.

Sir Samuel Baker, in a lecture on savage warfare, delivered at the U.S. Institution, points out that he would have found rockets much more serviceable if they had carried incendiary composition, the rockets passed through native huts constructed of reeds without firing them; even without shell he points out that he found them most serviceable in searching long grass where the natives used to be in ambush. Attempts are now being made to introduce a carcass rocket, and also rockets carrying gun cotton in their heads. Some satisfactory results have been obtained with 24-prs. carrying 3 lbs. of gun-cotton, and also trials have been made of a rocket 6″ in diameter, weighing about 100 lbs., carrying about 13 lbs. of damp gun-cotton in the head, and having a fuze so arranged as to detonate the gun-cotton with fulminate of mercury on impact. The fuze resembled the R.L. percussion fuze, set in action. The detonators used contained 35 grs. of fulminate of mercury for the 24-pr., and 45 grs. for the 100-pr. A dry gun-cotton primer containing the detonator was next to the fuze. The detonation of such a quantity of gun-cotton against buildings would be no doubt effective.

A range of about 3,000 yards has been obtained from the 6″ rocket at 20° elevation. See Extracts, XII, pp. 62, 150, 225.

For experience in Gold Coast Expedition, see Appendix, p. 301.

Mark I. rocket had a case made of wrought-iron, and the case was greased inside. § 1881.

These rockets were found to be unsafe, as the metal of the case was not strong enough, and they were also not properly protected from corrosion.* They were withdrawn in 1870.

Mark II. and subsequent patterns of rockets are made of Atlas metal,† a mild steel produced by the Bessemer process; the Mark II. is however imperfectly protected against corrosion, being only protected by blackening the cases and by greasing the inside.‡ § 1679.

These rockets stood better than I., but are to be carefully watched, see § 1881, 3 per cent. are to be sent quarterly to Woolwich for examination from stations therein named; details will be found in the Changes of W.S. referred to.

Mark III. differs from II. only in having two coats of paint applied to the interior of the case (see p. 322); since the close of 1873, three coats of paint have been given. § 1940.

As the 9-pr. was found to "puff," a slower burning composition was introduced in 1872. Rockets with this composition are Mark IV.§ § 2211.

It has been found an improvement to increase the length of the conical hole in the composition, and to form a cup-shaped recess in the interior of the tail-piece, this has been found to improve the range and to diminish the tendency of the rocket to "puff." Extracts, Vol. XI., p. 5. §§ 2576, 2662.

24-pr. rockets with the above improvements are Mark IV., and the 9-pr. Mark V.

The length of the cone is increased in the 24-pr. from 11″·875 to 13″, and in the 9-pr. from 7″·6 to 9″·5. The tail-piece is hollowed out ·5″ in the 24-pr., and ·4″ in the 9-pr. § 2.

---

* In manufacture, all the cases are dipped in oil and blackened by holding them over a fire, this to some extent helps to resist rust, it is still carried out with all rockets excepting those with tinned cases, see text above. This is carried out in last pattern of 9-pr. and 24-pr.

† The metal is supplied in sheets of the required thickness; care is taken to cut the sheet so as to have the greatest strength of the fibre in a lateral direction, so as to resist the tendency of the rocket to burst.

‡ This is of little use, as the first pellet pressed in carried down the grease with it.

§ A Mark III. 9-pr. rocket burst within a few yards of the trough at practice at Shoeburyness. Mark IV. having a slower burning composition is safer.—Extracts Vol. XI, p. 141.

The dimensions of the Mark III. rockets are given below.

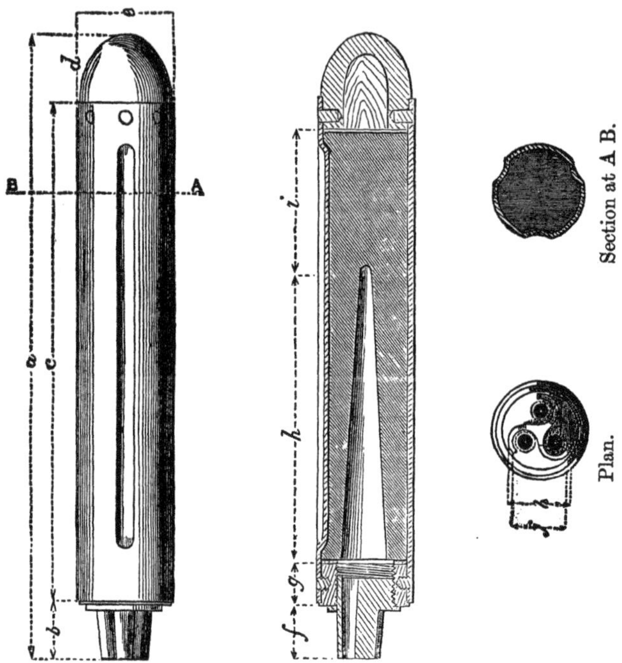


| — | 9-pr. | 24-pr. |
|---|---|---|
| *a* .. .. .. .. .. | 16·25 ins. ± ·025 | 23·18 ins. ± ·025 |
| *b* .. .. .. .. .. | 1·56 in. ± ·01 | 2·43 ins. ± ·01 |
| *c* .. .. .. .. .. | 12·93 ins. ± ·015 | 18·0 ins. ± ·015 |
| *d* .. .. .. .. .. | 1·76 in. ± ·02 | 2·75 ins. ± ·02 |
| *e* .. .. .. .. .. | 2·5 ins. ± ·01 | 3·75 ins. ± ·01 |
| *f* .. .. .. .. .. | 1·48 in. ± ·01 | 2·305 ins. ± ·01 |
| *g* .. .. .. .. .. | ·8 in. ± ·01 | 1·25 in. ± ·01 |
| *h* .. .. .. .. .. | 7·6 ins. | 11·875 ins. |
| *i* .. .. .. .. .. | 3·85 ins. | 3·813 ins. |
| *j* .. .. .. .. .. | 1·4 in. | 2·2 ins. |
| *k* .. .. .. .. .. | 1·5 in. | 2·4 ins. |
| Vents { No. of .. .. .. | 3 | 3 |
| Vents { Diameter of .. .. | 0·4 in. | 0·625 ins. |

24-PR. ROCKET. MARK IV.

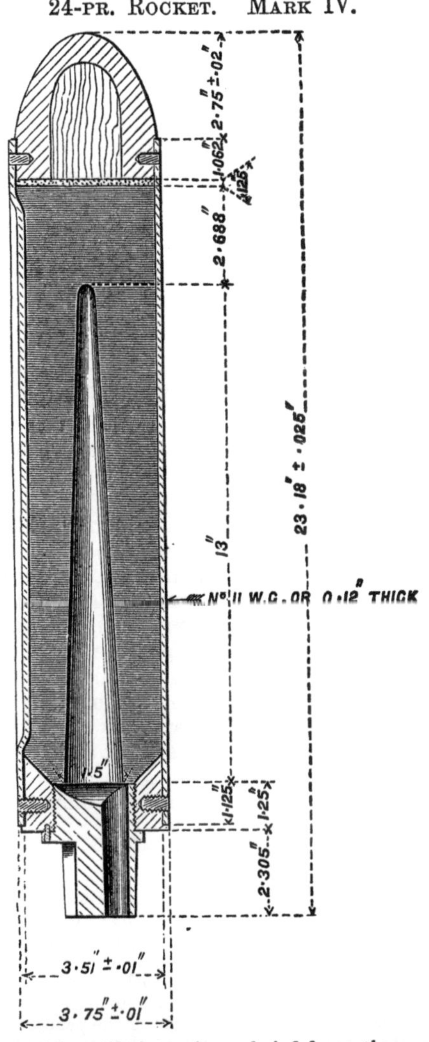


The following table will show in a brief form the general differences of the various patterns of rockets :—

| 24-pr. | 9-pr. | Remarks. |
|---|---|---|
| Mark I. | Mark I. | Paint black. Case of wrought-iron. DANGEROUS. Obsolete. |
| Mark II. | Mark II. | Paint black (unless repaired, when they would be painted red). Case of Atlas metal. Case greased inside. To be treated with caution. |
| Mark III. | Mark III. | Has two or three coats of paint inside. Paint, red. |
| | Mark IV. | Slower burning composition. |
| Mark IV. | Mark V. | Cone lengthened, and tail-piece hollowed out. |
| Mark V. | Mark VI. | No corrugations. Paper lining. No tissue paper protecting vents, but canvas and leather tied over tail-piece. In 24-pr., base-piece fixed to body by 10 screws instead of 8. Rivets stronger. |
| Mark VI. | Mark VII. | |

The following instructions have been issued as to examining, cleaning and repainting rockets :—

Examining, cleaning and repainting Hales rockets.

*Examination.*—All Hales rockets in store will, at frequent intervals, be examined carefully over the surface and particularly along the seam and round the rivets, and those which show the slightest trace of rust or corrosion will be cleaned and repainted.

§ 2441.

*Cleaning.*—The rockets will be placed in a convenient position for scraping, that is, on a couple of rails laid on a table, or a similar arrangement, and the whole of the corroded parts well scraped with a copper knife or scraper, and then rubbed with emery cloth until every trace of rust or corrosion is thoroughly removed (if emery cloth is not available a piece of cloth or serge with fine dry sand will be used); the rockets will then be well washed or rubbed with a piece of serge steeped in spirits of turpentine, and afterwards wiped dry with old linen cloths.

*Repainting.*—A coat of paint will be put over the parts which have been so cleaned, and the rockets then laid to dry; when quite dry they will receive a second coat of paint, and after lying three or four days to become again thoroughly dry, they will be replaced in boxes.

The paint will be composed of the ingredients specified in table, p. 322. Five pounds will paint 100 rockets.

*Marking.*—Rockets when repainted will be re-marked with the original date, &c., as well as with the date and station where they have been repainted; a pencil marking on the red paint, or a piece of paper pasted on, will be sufficient for this purpose.

N.B.—Care must be taken during the above operations not to injure or break through the paper covering the inside of the vents (Mark V. 24-pr. and Mark VI. 9-pr. do not have this paper).

The keeping qualities of rockets are not satisfactory.* They should be stored in as dry a place as possible.

Clause 129, Army Circulars, 1872.

War rockets are withdrawn from the equipment of fortresses, and

---

* Mr. Abel, Chemist, W.D., gives his opinion as follows :—

"The corrosion of the metal at the seam of the case has not been set on foot in the first instance by the borax employed in brazing, as no trace of the existence of borax can be detected upon the metal at the joint. The saline matter scraped from the exterior of the case contains carbonate of potash.

"The deliquescent and alkaline nature of this salt accounts for the collection of moisture on the case and for the destruction of the paint coating.

"This carbonate of potash is a product of the decomposition of the saltpetre from the rocket composition, and it is owing to some imperfection in the brazing that small quantities of saltpetre have been admitted in the operations of pressing that a corrosive action has been established, which has been promoted by gradual access of air and moisture to those points, and by the co-existence of brass and iron in contact with the composition.

"This action of the saltpetre upon the metal appears to have spread in the interior of the case round that part where the brazing extends to a very slight degree, but sufficient to effect a separation between the composition and the case, which are found to be very firmly attached to each other at all other parts of the case.

"The slight symptoms of corrosion round the rivets at the head of the rocket are evidently due to the penetration of minute quantities of saltpetre (forced in by pressure) applied in manufacture to the exterior between the rivets and the holes; the non-existence of brazing at these points renders the action very trifling.

"The employment of brazing in the closing of the rocket cases is evidently a cause of deterioration; the existence of minute imperfections in the joint made by brazing is probably unavoidable, and as the saltpetre must penetrate on pressure, the establishment of corrosion is unavoidable."

will only be employed with siege trains and in the field as circumstances may demand.

Rockets, as before pointed out, are very irregular, both as to range and direction.

They are found to range longer when the wind is from right to left than when it is in a contrary direction.*

### Experience as to firing Hales' War Rockets.

Experience as to firing.

The average range of 25 24-pr. rockets fired successively in proof at Shoeburyness in 1868, with 15° elevation, was 1,895 yards, the maximum being 2,226 yards, and the minimum, which was exceptionally low, 1,546 yards, the average deviation being 49 yards.

The average of 77 24-pr. rockets fired successively in proof in 1870, at Shoeburyness, was 1,460 yards, but the rockets ranged longer with the wind from right to left than from left to right, thus 51 rockets fired with the wind in the former direction had an average range of 1,678 yards, while 26 fired on days with the wind in an opposite direction, ranged only on an average 1,121 yards; the average deviation altogether is 35 yards. The deflection is generally to the left.

### Range of Rockets fired in Proof in 1873.

Range.

The average range of 18 24-pr. Mark III. rockets fired in proof in 1873, elevation 15°, was 1,562 yards, the average deviation was 39 yards.

The maximum range was 2,060, and the minimum 1,117 yards.

The average range of 30 9-pr. Mark IV. rockets, elevation 15°, was 1,536 yards, the average deviation was 38 yards.

The maximum range was 2,300, the minimum was 815 yards.†

Rockets occasionally are found to "puff" in flight; this may be due to the sudden or irregular burning of the composition, or to the vents becoming choked; it is generally injurious to the flight of the rocket; the cupped-out tail-piece not being so liable to have the vents choked with slag, will, it is hoped, diminish the tendency to "puff."

Issue.

Rockets are issued in wood cases, that for the 24-pr. holding 6, and that for the 9-pr. holding 12.‡

---

* The rotation of a rocket is the same as that of a rifled projectile, from left to right; if a cylinder revolving in this direction is pressed on the right side there will be a tendency to run up, and the reverse if pressed on the other side; possibly this may account for the increased range of the rocket.

The times of burning are about the following:—

| | |
|---|---|
| 24-pr. .. .. .. .. | 10 seconds. |
| 9-pr. .. .. .. .. | 8 seconds. |

24-pr. Marks IV. and V., and 9-pr. Marks V. and VI., will burn at rest about half as long as the above.

The life-saving rocket burns about 4·5 seconds.

† Ranges of 9-pr. Marks IV. and V.

| No. fired. | Elevation. | Mean range. | |
|---|---|---|---|
| 5 | 10° | 1,128 yards | Mark IV. When fired at 5° the rockets struck the ground a short way from the trough. |
| 5 | 15° | 1,612 " | |
| 5 | 20° | 2,354 " | |
| 2 | 25° | 2,426 " | |
| 5 | 10 | 1,316 " | Mark V. |
| 5 | 15° | 2,269 " | |
| 5 | 20° | 2,902 " | |

Extracts XII., 1867.

‡ A slight alteration in Mark I. boxes for 24-pr. rocket (§ 1876) was necessitated by the introduction of Mark V. rocket, so a Mark II. box was sealed.

§ 2982.

§ 3058. The box holding six 24-pr. rockets was found inconveniently heavy, and consequently liable to damage in hoisting it in and out of boats; a box, to contain three rockets only, has been sealed to govern future manufacture.

The boxes holding six, already packed in store, will be issued as usual.

## Machine or Trough for Hales' Rockets.*

§ 1637.
§ 1651. On 8/6/68 a trough machine was approved to fire 9-pr. Hale's rockets, and on 10/7/68 a similar trough of larger dimensions was approved for 24-pr.

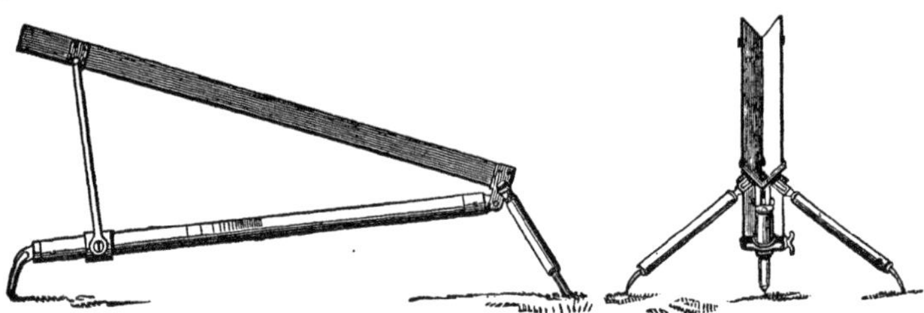

The following general description applies to both :—

Each size consists of a sheet-iron V trough; the sides made at an angle of 80° with each other, supported at rear by three legs made of wrought-iron tubing, two short ones opening right and left, and one long one to the front beneath the trough, each terminating in a prong. On the front one runs a gun-metal ring connected by two bars with a V near the front of the trough, the bars pivot on V and ring; the elevation is given by slipping the ring up and down the front leg, and clamping it with the arrow on the rear edge of the ring at the required line of graduation up to 15° of elevation for 9-pr., and 25° for 24-pr. machine, with reference to the plane on which the machine stands.*

At the back end of the trough is an iron stop preventing the rocket sliding back; it is slotted to form a crutch for copper friction tubes.

In firing the rocket with the friction tube, in order to prevent the machine being disturbed, the lanyard must be pulled smartly or else the foot pressed against the leg of the machine, or again the lanyard may be led under the foot.

---

* Some machines were introduced which have become obsolete. See §§ 1488, 1528.

† A longer trough gave better results, prevented dipping and increased accuracy. Extracts XII., 150.

Letter of Sec. W. to D. of A., 9/2/77, recommends *tube* in two or three pieces, so as to be readily put together. He gives the following results from troughs of various lengths with 24-pr. rockets :—

| Length of Trough. | Mean range in yards. |
|---|---|
| 16′ 4″ | 2947·2 |
| 14′ 4″ | 2798·8 |
| 12′ 4″ | 2691·6 |
| 11′ 4″ | 2618·2 |
| 10′ 4″ | 2800·6 |
| 9′ 4″ | 2676·2 |
| 8′ 4″ | 2703·4 |
| 7′ 4″ | 2659·4 |
| 6′ 4″ | 2297·2 |
| 5′ 4″ | 2143·6 |

The general dimensions of the two sizes are as follows:—

| | 24-pr. Mark II. | 9-pr. Mark II. |
|---|---|---|
| Weight of machine .. .. | 64¾ lbs. .. | 27 lbs. |
| Length of trough .. .. | 5′ 6″ .. | 3′ 5″ |
| Width of sides .. .. | 4″ .. | 2½″ |

Paint, black.

Paint.

*Issue.*

Each machine in a packing case.

Issue.

Sea Service Rocket Machines.*

On 7/6/69 a tube machine, Mark II., proposed by Lieut. Fisher, R.N., was approved. It is described in Changes in War Stores as follows:—

§ 1805.
§ 1860.

"The tube is attached at the centre by a vertically hinged joint to a small iron stanchion (see woodcut) which slips into a tabernacle* that can be fixed at pleasure to the stern, bows, or quarters of the float; in the stem a socket is fitted to receive the stanchion.

"To secure the tabernacle a keep pin is provided, and to prevent the stanchion unshipping from the tabernacle, a similar pin is attached to the heel of the stanchion.

"To prevent the heel of the tube turning into the boat (when firing) a clutch is provided which does not allow of more than a few degrees of lateral motion. This clutch should be shipped when the bearing is roughly on."

"The elevation is given by a straight bar marked in degrees attached to the tube, and worked through a slot in the stanchion. The friction-tube lanyard reeves through a swivel block attached to the rocket tube immediately over the stanchion.

"When it is required to fire abeam or thereabouts, the stanchion should be shipped in the stem or stern according as the wind is from aft or forward, and when it is required to fire nearly ahead or astern, the stanchion should be shipped on the bow or quarter on the *lee* side.

* A description of Mark I. tube machine, now obsolete, will be found in § 1432. An obsolete machine for Congreve and Boxer rockets is given in § 1270.

§ 1860.

† "The word 'tabernacle' has been used for years to designate a short wooden stanchion having iron bands in which the wooden rocket stanchion was supported, the tabernacle itself being attached to the gunwale of the boat. Tabernacles were heretofore supplied by the dockyards, but it has been decided that in future supplies of the machine above described, the metal tabernacle (or socket into which the iron stanchion of the new fitment will fit) shall be supplied by the War Department."

"The copper screen which was used with the rocket machine, pattern I., is not required for the above, and will therefore become obsolete."

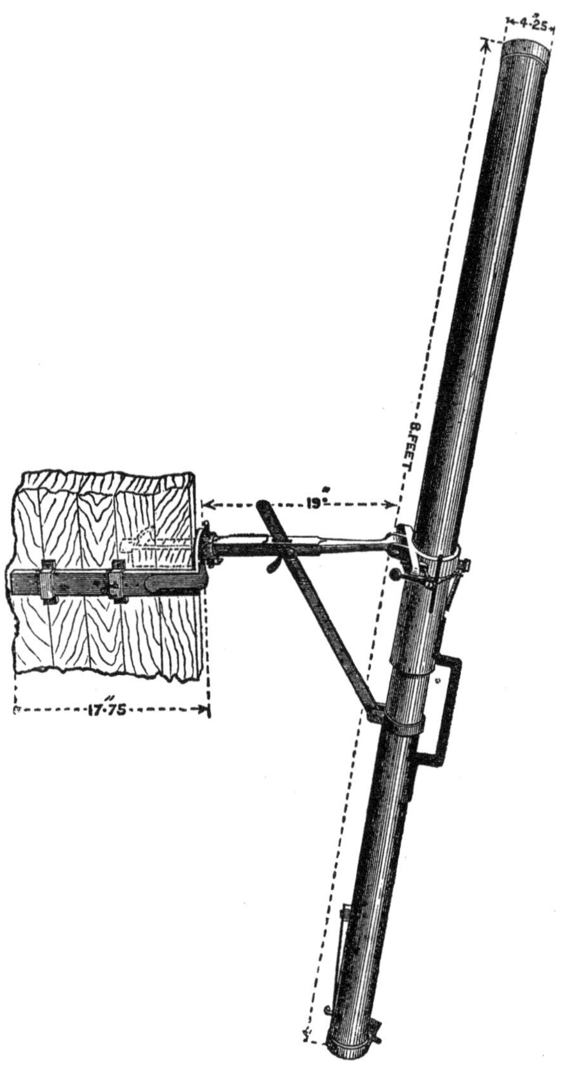


It may be observed that elevation (up to 30°) with this tube is given with reference to the horizontal plane; that is on the supposition that the stanchion is at right angles to the keel of the boat, and that the latter is horizontal.

§ 1931. On 26/5/70 the following store was sealed for use with the above machine, viz., a lanyard with block described in Changes in War Stores, as follows:—

"The arrangement consists of a lanyard rove through two small single blocks so as to form a tackle, to the movable block of which is attached a short lanyard with the hook for the friction tube. The fixed block through which the running end of the lanyard passes is attached to the rocket tube by a swivel immediately over the stanchion. The object is to give additional power so that a violent pull may not be necessary, which was the case with the former arrangement (§ 1805), the consequence being that the friction tube was frequently jerked out without igniting the rocket."

The Mark II. lanyard differs only in the form of the hook, see p. 91. § 3134.

Paint. Black.

*Issue.*

Issue. Each machine in a packing case.

SIGNAL ROCKETS.

1 lb. and ½ lb.*

§ 1709. The case is made of thick brown paper, rolled up into a cylinder, the rocket composition is driven by hand, and the conical hollow is made by a former placed in temporarily.

A light paper case is attached to the head, terminating in a cone, this serves to contain the stars and some mealed powder which serves to open the case and scatter the stars. The star chamber is separated from the rocket composition by some clay driven in at the top of the composition, having a central hole forming a communication; the rocket is choked near the base, and has a priming made up of L.G. powder and isinglass.† The vent is secured by a paper cap.

The stick is attached by means of a copper socket, in Mark I. rocket the edge is fringed and slightly turned in so as to bite the stick. The stick occasionally became detached in flight, this has been remedied by having a small projecting piece at the top of the socket, and by cutting a notch in the stick, so that when the stick is fixed in the rocket the projecting piece can be bent down into the notch. Rockets and sticks with this alteration are Mark II.‡ § 2365.

Composition. The composition is given on p. 319. The dogwood charcoal is used as it gives more sparks, making a brilliant tail to the rocket. The stars consists of small pellets of composition (see p. 320).§

Issue. 60 1-lb. rockets in No. 1 packing case.

84 ½-lb. „ in No. 4 packing case.

Marks and paint. All signal rockets are painted stone colour, and are marked with the numeral and date in blue.

§ 1470. Rockets issued since 20/7/67 have a label of directions pasted on their sides with the words "Before firing remove the paper cap over "the vent of the rocket."||

---

* 2 lbs. and ¼ lb. were made for Abyssinia, 19/11/67. § 151[illegible].

† This occasionally deteriorates while the rest of the rocket is in good order, it can easily be replaced.

‡ A short stick with a rope tail is used for sea service. A Mark II. stick with rope tail has been sealed to suit Mark II. rockets. § 2383.

The sticks for the 1 lb. and ½ lb. rockets are 8 ft. 2 ins. and 6 ft. 6 ins. long.

§ The 1 lb. rocket contains 28 stars and the ½ lb. 20 stars.

|| The signal rockets sent out to the Gold Coast deteriorated rapidly from the imate; special packing in tin-lined cases seems desirable for such a service.

## Rockets Issued to the Board of Trade for purposes of Display.

(Sizes, 1 lb. and ½ lb.)

Owing to the danger involved in firing service signal rockets over the heads of crowds from the metal socket remaining on the rocket stick and causing it to fall swiftly, point first like an arrow, it was decided to make rockets for purposes of display on a different principle ; in these the entire case of the head is made of paper, including the socket to take the stick.* In the small end of the latter (as in the case of the ¼ lb. signal rocket) is a puff of powder communicating by a fire-hole with that part of the rocket composition which burns away just before the flame enters the head and bursts it open, consequently by means of this fire-hole the puff of powder is fired, and the stick ejected so as to float slowly in the air just before the rocket opens.† The sticks are rectangular ; they are not tapered ; their lengths are 5 feet for the 1lb. rocket and 4 feet 2 inches for the ½ lb. rocket.

The 1 lb. and ½ lb. rockets are made of this description, containing red and white, or red, blue, or green stars, the head being painted red, red and white, blue or green, according to the colour of the stars ; the 1 lb. contains in each case 50 stars ; the ½ lb. 28 ; the heads are closed by choking. The puff of powder in the socket of the 1 lb. and ½ lb. rocket is ¾ dram and ½ dram respectively.

The heads are in all cases opened by quickmatch packed in with the stars.

**Issue.**

In metal-lined cases and packing cases,‡ sticks in bundles in numbers to correspond to the numbers of rockets.

**Machine rocket signal.**

Signal rockets may be fired from a T-frame with cleats, from off a nail in a post, or even with the stick end stuck into soft ground ; however, a signal rocket-tube machine, or "gun" for firing 1 lb. and ½ lb. signal rockets from boats, and under circumstances when the back rush of flame might do injury, has long existed in the service, although no pattern was sealed and deposited until July 1866.

This machine consists of an oval tube of sheet iron (2″·8 × 2″·3) to take the rocket with the portion of the stick at its side, a round tube of sheet iron being fixed on to it to take the remainder of the stick in its interior.

The two tubes are joined together by being entered into the opposite ends of a middle piece of gun-metal about 6″·2 long, to which both are riveted.

The larger part of the finished tube being about 1′ 8″ and the smaller 7′ 6″ long.

The metal at the mouth of the finished tube is wire edged, at the opposite end is a ground spike.

A vent is made in the close portion of the base of the oval tube opposite to the vent of the rocket to take a quill tube for firing, which is prevented from falling out when the tube machine is pointed up into the air, by a hinged piece of gun-metal which shuts in behind its head.

---

* The sticks for Board of Trade rockets are much shorter than those for signal rockets of larger size ; the gain in both convenience and strength is great.

† This arrangement is made in red and white rockets only.

‡ Sixteen rockets in a packing case, or, on special demand, 18 in a box fitted with lock and key.

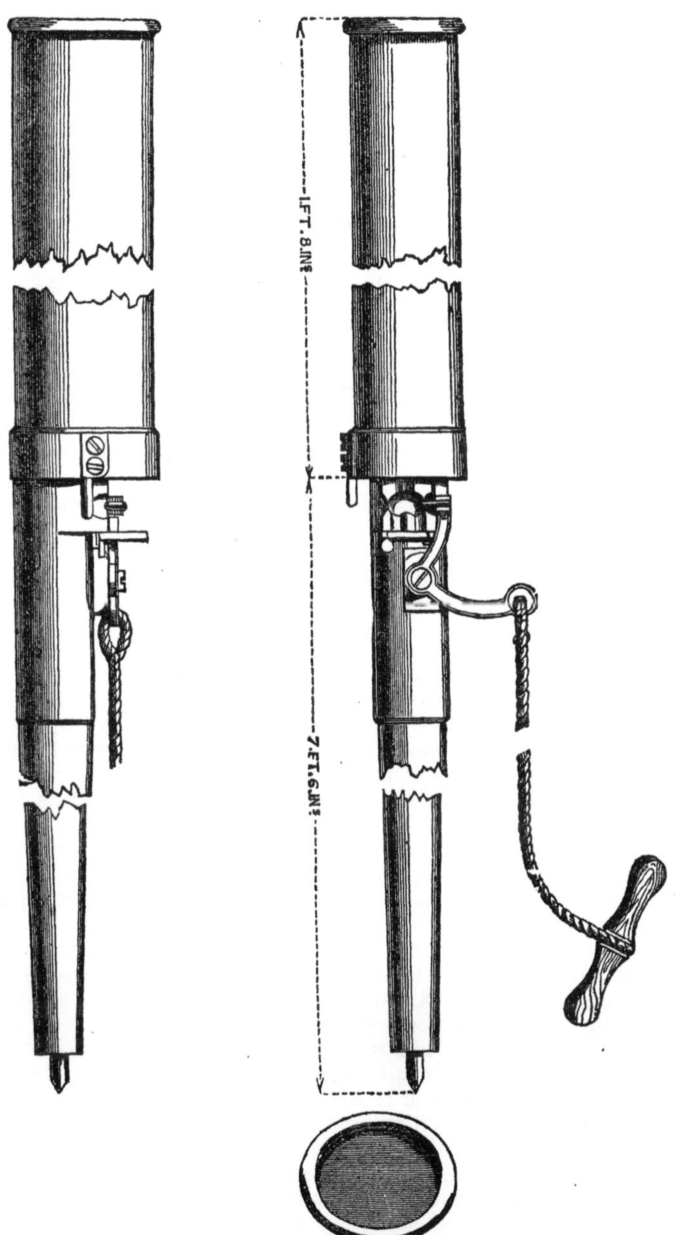


In Mark I. machine the detonating cross-headed tube was used, which was fired by the blow on the side of its head given by a hammer worked on a hinge and made to descend by the pull of a lanyard, in a manner very similar to the hammer formerly fixed on guns fired by detonating cross-headed tubes.

On quill-friction tubes entirely superseding the above, a pattern of this machine was sealed and approved as Mark II. (3/11/66), which **§ 1373.**

differs from Mark I. in having a hook fixed on to the end of a bell-crank lever, which is worked by a lanyard and toggle so as to fire the quill-friction tube, the form of the gun-metal hinged support being altered to suit the dimensions of the tube.

QUICKMATCH LEADER FOR FIRING ROCKETS.

§ 1675. On 8/9/68 a pattern rocket leader, Mark I., was approved, consisting of "10 short pieces of quickmatch, about 4 inches in length, attached at intervals of three feet to a piece of match 31 feet long."

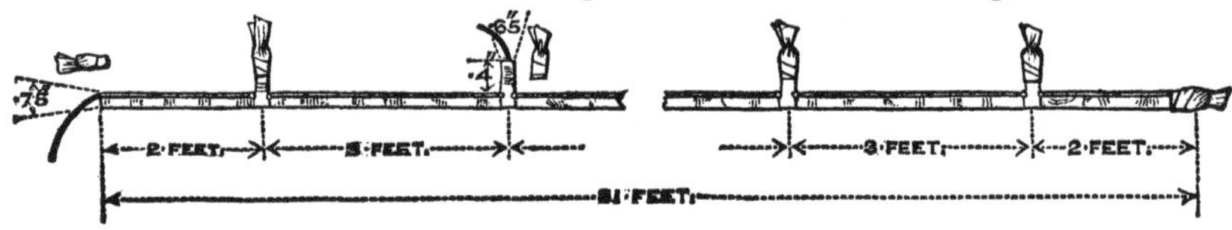


§ 1793. On 22/5/69, a pattern rocket leader was approved; it differs from the last mentioned pattern in having "9 short pieces instead of 10 at intervals of 3 feet; it is 34 feet in length, and has 5 feet of leader clear at each end." None of the previous pattern having been issued, this store is also marked I. The "tubes" are of linen, painted on the exterior to preserve it from damp. The projecting ends of the quick-match are covered with caps of fine white paper.

CALL ROCKET.

§ 2724. *Rocket Call, Mark I. Stick for ditto, Mark I.*—This rocket is arranged to give signals or attract attention by sound as well as light. The rocket is fitted with a stick, and is similar to ordinary signal rockets, excepting that it is made of iron, is much more powerful, and is of stronger construction, so as to enable it to be used in stormy weather and in exposed situations, for the purpose of giving signals from light ships, &c.

The case is made of Atlas metal. It consists of two parts; one, the rocket proper; the other, a cylindrical case or head with ogival top, containing the bursting charge, which consists of three perforated pellets of pressed L.G. powder ($13\frac{1}{4}$ oz.); the perforations in the pellets are filled up with 2 oz. of loose pistol powder, so as to give a very loud report when exploded. The head is fitted on to the rocket, and so arranged by means of a diaphragm and a small blowing charge, that it is blown off from the body of the rocket at the moment of explosion.

A magnesium light enclosed in a copper case is attached to the side of the rocket. This is to be ignited at the bottom end (the paper cap covering being torn off or broken into), and it is so arranged that, after burning a short time, it will communicate by means of a quickmatch leader with the vent of the rocket, which is then fired. If these means fail, the rocket can be fired in the ordinary way with a portfire. The magnesium light is intended to burn during the whole time of ascent of the rocket, and after the bursting of the head, to continue to burn until the case has fallen. Weight of rocket, 7 lbs. 6 oz.; of stick, 1 lb. 3 oz.*

* Experiments are now being carried on with call rockets, consisting of ordinary signal rockets, having a head attached, containing 2, 4, 8 or 12 oz. of dry gun-cotton. The gun-cotton is exploded by a detonator containing fulminate of mercury which is ignited by quickmatch communicating with the rocket composition. The two larger sizes give a very considerable report.

## LIFE-SAVING ROCKET, BOXER, 12-PR.*

Life-saving rocket, Boxer, 12-pr. § 1047.

Dennett's "twin" rockets were superseded by Boxer's on 15/3/65. This consists of two rocket bodies, one being fixed in prolongation of the other, to give great length of burning and flight without any sudden violence which might break the line which it carries,† or irregularity from uneven burning.

Thus it will be seen that "instead of making one cavity in the rocket, " two cavities (*c c'*) are formed, the one behind the other, with a portion " of solid composition (*b*) between them, so that when the solid compo- " sition (*b*) is burnt through, the front cavity (*c'*) is ignited, thereby " imparting to the rocket an additional impulse." The stick (*d d*) is fixed at the side of the rocket. The line (*e e*) is passed through a hollow at each end of the stick, as shown in the annexed cut, and the end of the line is secured by a common overhand knot; 2 india-rubber, and 1 brass washer (*f*) are placed between the knot and the stick, to reduce the effect of the sudden jerk which is given to the line when the rocket

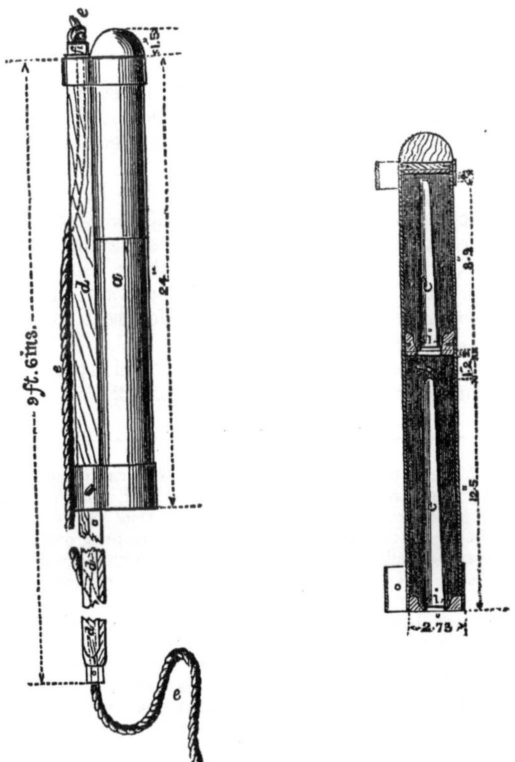


---

* Time of burning, about 4½ seconds.

† General Boxer writes, in letter of 25/5/65, that his object is "the continuance of the propulsion through a much longer period, without any excessive strain upon the line."

Captain Robertson, R.N., writes to Secretary, Marine Department, Board of Trade, is fired. The arrangements for the use of this rocket are the same as those hitherto carried out with Dennett's rockets.

A second knot is usually made in the rope near the hinder end of the stick, in case the line should be burnt through by the flame issuing from the rocket.

§ 2441. N.B.—All Boxer life-saving rocket cases are protected from the action of the composition by an internal coat of anticorrosive paint (see p. 322), being the same as is applied to the interior of Hale's rockets.

All rockets manufactured since 22/9/68 have their cases further protected by blackening by burnt oil.

### Details of Patterns.

§ 1047. The pattern in the woodcut, known as Mark I., was approved 15/3/65. Mark II., approved 9/66,* differs from Mark I. in having no hole to take the keep-pin through the "clip," the pin being passed through the stick in front of the clip, because it was sometimes found troublesome to bring the hole in the stick and "clip" exactly to cover one another.† The sealed pattern is nearly ·5 inch shorter than Mark I., so as to enable the rockets manufactured to conform with it, it having been found that the act of pressing the composition slightly shortens the whole case; hence that of the dummy pattern was longer than the same case would

§ 1679. be after pressing.‡ Mark III.,§ approved 1/9/68, differs from Mark II. in having the case made of Atlas (*i.e.* Bessemer) metal. All manufactured since October, 1870, have the vent covered with paper (instead of

§ 1995. the serge plug). The paper is to be broken before firing.

It is important to distinguish pattern III. clearly from I. and II., the cases of rockets of the latter patterns having been found liable to deteriorate, and even to split, from their being taxed beyond their strength by the pressure of the composition, are ordered to be very carefully examined from time to time for rust spots and indications of

§ 2441. cracks‖ (*vide* p. 310).

Paint: formerly two coats of black varnish; since 5/11/70 two coats of red paint for better protection (see p. 322).

---

9/2/65, that Dennett's rockets "frequently carry away the lines, and sometimes do not ignite. They are also double the expense" of Boxer's rocket.

Inspecting Commander Earle reports on a trial between Boxer's and Dennett's rockets: "Of the three double Dennett's rockets only one wa sany use; two broke their lines and struck the ground. . . . The mean of the five shots with the Boxer rocket gave a range of 370 yards very true, and with much less strain on the line, as it never broke with Boxer's rocket."

Reports from Inspecting Commanders Charles and James, from Yarmouth and Lydd, are confirmatory of this statement, 19/10/65.

At Whitby, on 27/3/66, one of Dennett's rockets, igniting before its twin rocket, came back and struck the Inspecting Commander.

Captain Robertson, in letter 9/2/65, reports that Dennett's rocket attained a greater range than Boxer's.

* Not noticed in Changes in War Stores.

† *Vide* letter by Inspecting Commander Kirby, Sunderland, 28/10/70.

‡ To prevent mistakes arising from comparing an empty pattern with a filled rocket.

§ The numeral marked on the pattern sealed as II. was altered in place of sealing a new pattern.

‖ The crack is generally developed in a longitudinal line running parallel to and within one or two inches of the seam or joint of the rocket.

The 12-pr. life-saving rocket stick* is deal, 9′ 6″ long, square, with corners shaved off; it is the same size from end to end. It is bound at the bottom end with an iron ring, and is plated at the head or front end with plates, which, as well as the stick at the front part, are hollowed to fit close to the rocket. The second or hinder plate is three inches long; it has a flange to rest against the base clip of the rocket.† Over the half of the stick next the rocket is tacked a sheet of tinned iron for a length of 14 inches, to protect the stick from the flame escaping from the rocket. § 1985.

*Iron Pin for Life-saving Rocket Mark I.*

This is an iron pin 1″·2 long, No. 8, Birmingham wire gauge; the end is bent over at a right angle, thus bringing the length down to ·85″. § 1271.

*Brass Washer.*

The brass washer shown in the woodcut of the rocket (p. 279) is 1″ in diameter, with a hole in the centre ·5″ diameter; it is ·15″ thick.

*India-rubber Washer.*

The vulcanized india-rubber washers referred to in the description of the rocket are both alike, each being 1″ in diameter, with a hole in the centre ·5″ in diameter; they are about ·7″ thick.

MACHINE FOR FIRING LIFE-SAVING ROCKET.

The machine for firing the life-saving rocket consists of a bed to hold the rocket, in prolongation of which is fixed a pry pole, and from the rear end of which spring two legs, one opening to the right and one to the left. Both bed and pry pole are made of sheet iron, the former being an open rectangular trough 3·2 inches broad, and 4 inches deep;‡ the latter one, of more rounded form, being 1·65 inch broad at the top, and 1·5 inch deep.

The front end of the pry pole enters the bed for a length of 7 inches the upper edge of the former standing about ·2 inch above those of the latter, so that the bottom of the larger trough is 2·7 inches beneath that of the smaller, to allow for the rocket resting in the bed while the stick lies in the hollow of the pry pole. The two troughs are fixed together by three rivets on each side, the spaces between them on each side, owing to their difference of width, being filled up by a piece of wrought iron through which the rivets pass.

The front edge of the bed trough is iron-strapped, and its remaining edges, as well as those of the pry pole trough, are "wire-edged." With the exception of a strengthening bar running from bed to pry pole, the rear end of the bed trough is left open beneath the front of the pry pole, so as to allow of a free passage to the gas escaping from the rocket base.

Two pieces of wrought-iron, 7 inches long, are riveted along the after

---

* Mark III. stick is strengthened by having the part next the base of the rocket more covered by the tin sheet, which is also passed under and clamped by the iron socket. § 2385.

† It was formerly secured to the stick by two small indentations. On 26/1/77 it was ordered to be secured by two screws.

‡ Interior measurement.

part of the sides of the bed, close to the angles formed with the bottom, their rear ends projecting sufficiently to allow of a bolt secured with a screw washer to pass through them, on which hinges a small flat piece of iron taking two other bolts screwed and nutted, and each long enough to allow of a socket (ending in flanges) which admit the flat iron between them to be hinged on it. Thus the flat iron hinges longitudinally on a bolt transverse to the direction of the troughs, whilst the leg sockets move transversely on hinges longitudinally placed. In each socket is fixed an ash leg with ferrule, having a foot projection and spike, while beneath the pry pole runs a strengthening bar from end to end, which is at the hinder extremity bent down to form a ground spike.

In the right side of the bed is cut an opening to admit of the entrance of a portfire to fire the rocket, and behind this is fixed a brass quadrant plate, on which is hung a plummet and line to give elevation.

On the left side of the bed, protected by a copper cover, is a strong lock of simple construction, with a lever trigger, to which is attached a line, led through one sheave on the left leg socket, and another near the left foot. Near the right foot is fixed by two screws, a strong strap and buckle to enable the two legs and pry pole to be strapped together, for more convenient stowage when not in use.

Mark I. trough or machine has long existed; it was sealed in November, 1865. This pattern has a very small block fixed to a ring near its left foot. It is difficult to pull the trigger line from the right side, owing to the stiff movement of the little block.*

Mark II. was approved 21/10/70; it differs from Mark I. as follows :—

1st. The trigger lever is prolonged to a length of about 4 inches, so as to allow of the lock being worked with a lighter pull.

2nd. The pulley block on the left foot is replaced by a sheave of much larger size fixed through the middle of the wood (which is supported by a band); this pulley enables the machine to be fired from the right side.

3rd. The opening in the right side of the trough is furnished with a sliding cover.

2385. Mark III. machine differs from the previous pattern only in having an arrangement for causing the flash from the detonating tube to strike direct up the axis of the rocket. This is effected by making the vent or channel for the tube in a circular form instead of straight across the machine.

N.B.—A spare spring is ordered to be supplied.

§ 2777. Mark IV. has no lock, and differs in minor details from the above, and is made as free as possible from any projections which might catch the line. The lock is discontinued as the detonating tube is obsolete.

### Stores connected with the Life-saving Rocket.

*Fuze for Life-saving Rocket, Mark I.*

This is 1″·5 long; it is made of paper; it contains an inch of ordinary fuze composition; it is conical in shape, and the sides are covered with kamptulicon, being brought up to fit the vent in the base of the life-saving rocket; it has a paper cap tied on with twine, which need not be removed before firing; it burns for about five seconds, and is required for use with the portfire.

§ 2777. *Mark II.* fuze has 1″·5 of slow burning composition, and burns ten seconds.

---

* Never appeared in "Changes."

*"Light, Long, General Service" (Mark I.).* See p. 105.

This light has been in use in connection with the life-saving rocket, having been approved for that purpose on 15/3/65. It came in for general service in 18/3/68; but it is issued and used with a hollow metal handle in connection with the life-saving apparatus, but for land service with a wooden handle, approved 5/12/68. § 1047. § 1721. § 1726.

For "light illuminating wreck" see p. 106.

*Portfire, Boxer's, for Life-saving Apparatus.*

Differs from a common portfire in being 8 inches long, and in being intended to ignite by means of a detonating primer, in the same way as the G.S. long light, the end being closed by a tin cap and a piece of kamptulicon, and strengthened by a tin band, perforated to take the detonating primer, which enters into a small space beneath the kamptulicon. The composition is primed in the usual method with mealed powder, perforated in the centre.

*Metal Handle for Long Light, General Service (Mark I.), used with Life-saving Apparatus, Mark I.*

Consists of a hollow cylinder of tinned iron, fitting on to a wood end; it is closed at the opposite end by a metal screw cap, to which is hinged on, by means of a brass pin passing through two brass flanges, so as to form a hinge, a copper covered piece of wood, with six transverse cells, each to hold one primer. § 1271.

*Handle for Portfire, used with Life-saving Apparatus, Mark I.*

Consists of a tinned iron cylinder, closed across with tin and red lacquer, so as to form a socket to take the portfire end at one extremity held by a tightening screw. The body is hollow, closed with screw cap and piece of wood, copper covered and recessed, with seven cells to take one detonating primer each.

*Mark II.* has a smaller socket than Mark I., so as to have a steadier hold of the portfire. § 2777.

*Tin Box for Life-saving Rocket Stores, Mark III.*

This is simply a tinned box with a hinged lid. § 2777.

| | |
|---|---|
| Length .. .. .. .. | 6″·1 |
| Breadth .. .. .. .. | 3″·6 |
| Depth .. .. .. .. | 3″·0 |

On the lid is a label giving the contents, viz.:—

12 fuzes.
9 iron pins.
12 india-rubber washers.
6 metal washers.

*Wood Boxes for Lights, &c., for Life-saving Apparatus.*

These are two yellow deal boxes closed with hinged lids, secured with hasps and staples; they have internal fittings to suit the stores. § 2777.

*Box-wood for Lights, Mark II.*, contains 12 lights, long; 2 handles, light; and 15 detonating primers, in a tin cylinder.

*Box-wood for Portfires, Mark I.*, contains 24 portfires, 2 handles, and 30 detonating primers, in a tin cylinder.

## USE OF LIFE-SAVING ROCKET.

Instructions as to the use of the rocket, together with directions as to the formation of volunteer life brigades, the provision of requisite stores, &c., are issued by the Board of Trade, in the form of a pamphlet, entitled "Instructions in respect of the Rocket and Mortar Apparatus for Saving Life from Shipwreck." A short description of the method of using the life-saving apparatus generally adopted is here given, taken partly from this pamphlet and partly from information supplied by Captain Robertson, R.N., also Mr. John Foster Spence, Mr. Gilbert, and members of the Tynemouth Volunteer Life Brigade.

A suitable cart containing the necessary stores* is run down to the best position for action.†

---

* (*b*) "Two or three *rocket lines* laid up loose. One end of the rocket line is to be attached to and launched with the rocket.

(*c*) "*Boxes* fitted with faking pins, in which to stow the rocket lines.

(*d*) "A '*hawser*' of 3″ Manilla right-handed rope, from 40 to 120 fathoms, according to the steepness or flatness of the shore.

(*e*) "A '*whip*' of Manilla line, not exceeding 1½″, rove through a single tailed block. The 'whip' to be made of left-handed rope, the reverse of the hawser, and to be twice as long as the hawser, and the tail of the block to be at least 2 fathoms in length, and the sheave to be brass-bushed. The ends of the 'whip' to be spliced together, so as to convert it into an endless rope.

(*f*) "A '*sling life buoy*,' with petticoat breeches, in which to place the person to be rescued, and haul him ashore.

(*g*) A '*traveller*,' or inverted block, with a brass sheave, to be attached to the '*sling*,' and carry it along the '*hawser*.'

(*h*) "A '*double block tackle purchase*,' for setting taught the hawser, one of the blocks being fitted with two tails to bend on to the hawser, or with luff tackles fitted to put on to the hawser with strop and toggle (like a top-gallant or royal purchase). The blocks to brass-bushed.

(*i*) "Three small *spars* to form a triangle, over which the hawser may be passed, and thereby raised higher above the water. This will be found convenient on parts of the coast where the shore is flat.

"The triangle should be fitted with a swivel snatch block, brass-bushed, instead of standing hooks; the strappings of the block to be of good iron.

(*k*) "An '*anchor*' with one fluke to be buried in the earth, sand, or shingle, to which to set up the hawser by means of the tackle purchase. Or, in some places where the shore is composed of soft shingle or sand, and where an anchor will not hold, a stout plank, 5 or 6 feet long, with a fathom of chain of sufficient strength fastened around it amidships, may be substituted for the anchor. This plank being buried 3 or 4 feet beneath the ground, and the end of the chain, with a ring attached, led to the surface, the hawser may be set up to it by the tackle purchase, in the same manner as to an anchor.

(*l*) "A '*red flag*,' 2 feet by 3 feet, fixed at the end of a staff 5 feet long; and a '*lanthorn*,' with a *red* lens fixed in it; to be used as signals in the manner directed below.

(*m*) "Two or three *spades* or *shovels*, and a *pickaxe*, to be of good quality and suitable for the work; a *selvagee strop*, and a few pieces of *extra rope* to be used as occasion may require.

(*n*) "A light *hand-barrow*, when thought necessary, for carrying portions of the apparatus from the cart to the place where it is to be used.

---

† As the rocket cannot under any circumstances be expected to carry much over 380 yards (vide p. 289), the choice of position must generally be very limited.

The machine is placed to stand as firmly as circumstances will permit for a maximum range the trough should be laid at from 35° to 38°,* the box in which the line is faked being placed from about 6 to 9 feet to the rear, and 6 to 9 feet to leeward,† the top with the pins being taken out and the box slightly tilted with its mouth towards the front with the line lying in it, the end being threaded through the rocket stick and knotted over the washers and also some way along the stick.‡

It is very important, for more than one reason, to effect a communication with as few unsuccessful attempts as possible, not only is precious time wasted, but after the line becomes dirty and wet the chances of success are decreased.

At short ranges it may be desirable to fire the rocket at a lower elevation than 35°, for it is easier to project the rocket between the masts, when the line must of course follow it, than to fire it high in the air with the allowance necessary to cause the line to fall between the masts.§

---

(o) "Three sets of *tally boards*, each set consisting of two boards of hard wood about 9″ long by 5″ wide and ¾″ thick. These boards to have the following words painted on them in white letters on a black ground. English on one side and French the other, viz.:—

"No. 1 tally board to be attached to the whip.

"English:—

"Make the tail of the block fast to the lower mast well up. If the masts are gone, then to the best place you can find. Cast off rocket line, see that the rope in the block runs free, and show signal to the shore.

"French:—

"Fouettez la poulie le plus haut possible sur le bas-mât, ou a l'endroit le plus favorable si les bas-mats sont perdus. Détachez la ligne, voyez que la corde coure facilement dans la poulie, et faites signal au rivage.

"No. 2 tally board to be attached to the hawser.

"English:—

"Make this hawser fast about 2 feet above the tail block. See all clear, and that the rope in the block runs free, and show signal to the shore.

"French:—

"Amarrez cette aussière a deux pieds environ au dessus de la poulie. Voyez que rien n'engage et que la corde coure facilement dans la poulie, puis faites signal au rivage.

(p) "*Long light.* One box of Colonel Boxer's, to be used as occasion may require.

(q) "'*Signal rockets.*' Eighteen, throwing white and red stars.

(r) "Two *heaving sticks* and lines, to be used as occasion may require.

(s) "A *water barrico*, with a large square hinge bung large enough to admit a man's hand, will be supplied if specially demanded.

(t) "A *hawser cutter*, for the purpose of severing a hawser from a wreck.

(u) "A *tarpaulin*, to cover over the apparatus and stores in the cart when the apparatus is not in use, and fitted with beckets and tent pegs, to secure it on the beach or shore for coiling the whip on when the apparatus is in use,

(v) "*Life belts.* Two of Captain Ward's, and two *life lines.*

N.B.—"The whole of the gear and a sufficient supply of rockets, &c., are to be kept in the rocket apparatus cart, IN GOOD ORDER, DRY, AND READY FOR IMMEDIATE USE."

* *Vide* Board of Trade Instructions, § 19.

† The rocket stand may be capsized by the line running out if the line be laid to windward; the coil should be as little out of the line of flight as may be, for it is obvious that the pulling of the line tends to draw the axis of the rocket in the direction of a line passing from the centre of gravity of the rocket to the spot where the rope is coiled. That the position of the coil of rope affected the flight of the rocket considerably was pointed out by Captain Alderson in a proof report on rockets fired at Shoeburyness.

‡ *Vide* page 280.

§ Even at 35° I believe the rocket generally passes between masts. As to flight of rockets, *vide* page 288.

When the crew of the wreck signal that they have the line,* the rocket brigade make fast their "whip" (*e* in note, p. 284) by bending the rocket line round both returns at about 12 feet from the tailed block and signal.†

The wreck's crew then haul in and make fast the tail of the block *about* 18 *inches below the highest secure part of the ship* ‡ (some distance up the mast if possible)§, unbend the rocket line and signal.

While the crew are drawing this "whip" in, it is especially necessary that the brigade on shore should see that the lines are carefully paid out to them, keeping the two parts steadily in hand, at the same time not letting them out faster than the crew on board the wreck can haul in, the men who have charge of the two coils of the whip being specially careful that the lines run out all clear from the coils.

On seeing the ship's signal the brigade attach the hawser 6 or 9 feet from its end to one return of the whip and haul on the other return, so as to carry the hawser to the ship, which the crew make fast 18 inches above the whip (*i.e.*, to the highest safe point), and then disconnect it from the whip and "signal." While those on shore are hauling the hawser on board the ship, it is especially necessary that the men in charge of the whip should keep the returns of the opposite end, if possible, 30 yards or more apart, and the hawser nearest to the hauling part, to prevent the hawser taking turns round the whip, which is very liable to occur even when these precautions are observed, and the wrecked crew should, if possible, ascertain before making the hawser fast that it is all clear.

On this, the brigade having adjusted the block of the breeches buoy to run on the hawser, attach one return of the whip line to it by a clove hitch, and if the motion of the wreck is slight, lead the hawser through the snatch block of the triangle and set it up (*i.e.*, haul it taut) by means of their "double block tackle purchase" (*h* in note, p. 284). This, however, can be paid out or hauled in but slowly if required to follow the motion of the vessel. If, therefore, the sea beats the wreck about violently it will be better not to use the double block tackle, but to keep the hawser taut by manning it with as many hands as can be spared, so as to follow the oscillating motion of the wreck without risk of the communication being broken.

It will be seen in the woodcut that while the whip return by which the buoy is hauled towards shore must be pulled fair along the hawser, the opposite return should throughout be kept wide of it.

The crew may descend one, two, or even three at a time in the breeches buoy.

---

* Either by a wave of hand or flag, a light shown, or a gun fired.

† Generally by red flag by day, and red light by night, *vide* Board of Trade directions.

‡ There are many reasons for this. 1st, The hawser will bend with the weight of any person travelling on it, and perhaps let them into the water. 2nd, If near the water the wash of the sea may twist and foul the ropes. 3rd, The higher the starting point the easier it is to haul a weight to the shore.

§ I have been informed of an instance of a whole crew being drowned by making fast to the knightheads on the deck instead of some point up the mast. I may observe that a brother of my own in travelling experimentally on a low hawser descended into the sea, but it is hardly necessary to enunciate that there is a limit to the distance which a person can be drawn through the surf without drowning. (Note by Captain Browne, R.A.)

In cases of very violent wind the empty breeches buoy has been carried right round over the top of the hawser,* fouling the whip with it; it is therefore well not to let it pause while on a journey, especially when travelling empty back to the wreck.

In urgent cases, such as the threatened immediate break up of the wreck, one or more buoys with lines to them communicating with the shore may be passed to the wreck directly the whip is made fast, or again, the " buoy " may be made fast to one return of the endless line while it travels on the other,† at the same time the hawser should be set up when practicable.‡

## Flight of Life-saving Rocket.

It may be seen that the construction of the life-saving rocket is not such as will enable it to carry truly when fired without its rope. Its stick is fixed on one side of it, hence in flight the resultant of the resistance of the air on its anterior part, acting at a point, termed by General Maievsky its "centre of resistance," will not be opposite to its centre of gravity, and hence a couple tending to deflect the rocket will be established.

On page , the case of a rotating elongated projectile proceeding in a direction not coincident with that of its axis is discussed. The case of the rocket somewhat resembles it, the tendency of the rotation to resist the deflecting couple, being answered by the mechanical action of the stick (described, page ); the velocity of rotation and the length of the stick being the relative "functions" of the steadying force in the two cases.

Now the stick of the life-saving rocket is not only placed on one side, but is also a little curtailed in its length; it may therefore be readily seen that this rocket is constructed on the supposition of its carrying a line, when the pull of the line from the starting point will act to draw the stick and rocket into the production of the line of flight it has taken up to the moment considered; this steadying power (in spite of the wind carrying the middle of the line in a bend to one side) becomes very great indeed after the rocket has proceeded any considerable distance.

From this may be deduced two facts, which it may be vitally important to consider in firing the rocket:—

1st. That the wind will carry the rocket and line with it, because it will not have the power to deflect its axis so as to point the rocket up the wind.

2nd. It is very desirable to start the rocket at a momentary lull. For

* Captain Robertson informs me this has been reported as having occurred.

† The endless line must be cut to effect this; it is best to make fast the ends to the grummets on opposite sides of the life-buoy.

‡ Various methods of escape from a wreck have been devised and some carried out; the crew are generally in a nearly helpless condition with the waves beating over them, the most feasible expedient appears to me to be that of a kite, as there is generally a violent wind blowing from the wreck to the shore, and considering the comparative sizes of the ship and the land it seems reasonable (as proposed by Capt. Nares, R.N., *vide* "Seamanship," by that officer, pp. 220 to 222) to call attention to the possibility of the crew making and getting off a kite when the means on land were insufficient to establish a communication. Once let the kite fly over the land, the sudden paying out of its line would cause it to drop on the shore. Capt. Robertson, R.N., informed me that a man has been known to swim from a ship with a line, assisting himself by a kite; it is here obvious that the kite might have carried a light line by which might have been passed stronger ones till a hawser was at last carried across.

if the first action of the wind carries the rocket to one side it will exert its force afterwards in prolongation of this incorrect direction.

If the rocket machine be brought into action on uneven ground, causing the foot on one side to be lower than that on the other, or if one foot sink deeper than the other, as might occur in yielding sand, the effect will be to cause the rocket to carry towards the lower side.

*Issue.*—Six rockets in a packing case.

**Issue.**

### Experience as to Range and Accuracy.

In 1868,—52 rockets fired in succession, in course of proof at 35° elevation, gave an average range of 378 yards, which may be considered a low one, it certainly includes one or two exceptionally short ranges; the minimum one being 286 yards, the maximum 450 yards. The average deviation from the line on which the rocket was laid was 42 yards.

In 1870,—131 rockets fired successively at proof, gave an average range of 373 yards, the maximum range being 470, and the minimum 330, the mean deviation being about 35 yards.

In calculating for the effects in cases of storm, rather a low range must commonly be expected, the wind generally blowing more or less against the direction which the rocket has to take.

# APPENDIX.

## NOTES ON THE FLIGHT AND PENETRATION OF PROJECTILES.

I 4 Sladen.

Lieut. Sladen, R.A.,* Royal Laboratory, writes with reference to the resistance of the air during flight—

"From recent experiments carried on at Shoeburyness with Professor Bashforth's Chronograph, some important results have been obtained with regard to the resistance of the air to variously shaped projectiles.

The point, I purpose to bring to your notice in this letter, is the experimental proof that the *total* resistance to a projectile moving with a given velocity in the air, is made up of (1) the resistance to the head, (2) and the resistance or "*minus pressure*" on the base.

Now it has been found that the total pressure on a 9-inch spherical shot, moving with a velocity of 1,130 feet-seconds, is about 555 lbs. (A N B M representing the spherical 9-inch shot), and the total pressure on a hemispherical headed elongated shot of the same diameter (represented by A C D B M), and moving with the same velocity, is 487 lbs., thus showing a difference of 68 lbs. in total pressure.

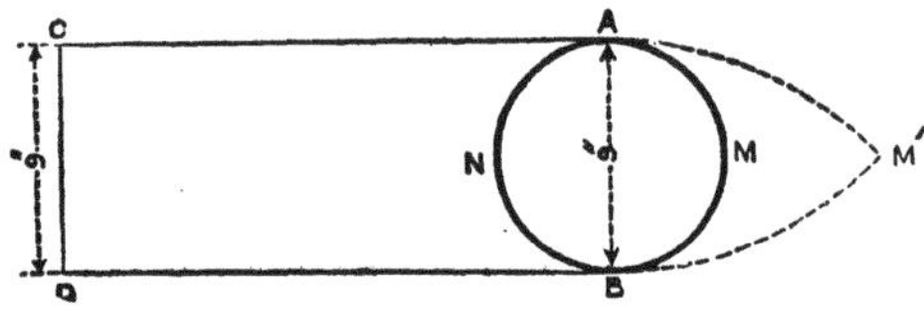


Now supposing the elongated shot to move steadily point first, the pressure on the respective heads A M B must be the same, therefore the difference of total pressure, viz., 68 lbs., must be due to the difference of minus pressure on the bases A N B, A C D B respectively, thus showing conclusively that the form of base of a shot materially influences the total pressure which it meets with when moving through the air at a high velocity.

The total pressure on a service ogival headed shot of 9 inches in diameter (represented by A C D B M′) is only 389 lbs., thus showing the great difference of pressure (viz., 166 lbs.) on an elongated ogival headed shot and a spherical shot of the same diameter when moving at the same velocity through the air. Another advantage which the elongated shot possesses over the spherical is that for the same calibre the momentum of the former is much greater, varying of course in proportion to their respective weights, which would be nearly three to one, depending on the length of the elongated projectile.

* Now Captain Sladen, Professor of Artillery, R.M. Academy.

These two principles taken together constitute the great superiority which elongated rifled projectiles possess over spherical projectiles.

These considerations would indicate the superiority in every respect of rifled over smooth-bore mortars, as, owing to the longer time of flight, these principles have a longer time to act, and therefore would produce more marked results than in the case of horizontal firing.

General Maievsky.

The investigations of General Maievsky, as to the forces acting on projectiles in flight and penetration, are such that they form a study of great magnitude, involving peculiar applications of the highest mathematics. It would be idle to attempt to take up the pursuit of this subject without making it a regular course of study. A few of Maievsky's results quoted by Mr. Mallet in his papers in the "Engineer,"* as well as notes on the subject of penetration, as discussed by Mr. Mallet, may be found valuable, although briefly given in general terms.

1st. As to the irregular spiral path described by rifled projectiles in flight:—

General Maievsky supposes the projectile to be proceeding on a path not absolutely coincident with its axis, and shows that the point where

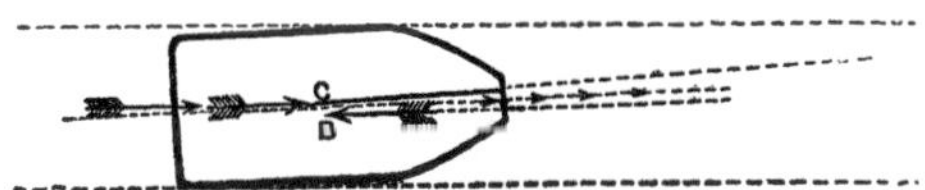


the resultant of the resistance on the interior part of the projectile intersects its axis (which he terms "the centre of resistance"), not being exactly opposite to the centre of gravity, the force of projection and that of resistance act so as to form a couple BC which tends to deflect the head of the shot, this being opposed by the centrifugal force due to rotation and the effect varying continually, the axis of the projectile constantly describes in space a form approaching a cone, the path of its centre of gravity being a helix or spiral.

General Maievsky arrives at the conclusion that the helix becomes wider and wider,† as the projectile proceeds further and further on its path.

When the centre of resistance coincides with the centre of gravity then there is no deflecting couple and no tendency to alter the direction of the axis of the projectile which is merely lifted or pressed downwards the range being increased or decreased by the action of the resisting force, which acts in the plane in which the trajectory and the axis of the projectile are situated.

---

* Also partly taken from "Revue de Technologie Militaire," Vol. V, p. 101.

† Facts seem to contradict this, and though General Maievsky gives reasons in support of his conclusions, it seems likely that the conditions of the question are affected by the velocity of translation (and hence the disturbing cause, viz., the resistance of the air) decreasing much more rapidly than the velocity of rotation.

The small diminution of the velocity of rotation, as compared with that of translation, is a fact that has been frequently overlooked, and the strong reasons there are for believing that the flight of the projectile increases in steadiness suggest the doubt whether General Maievsky has sufficiently considered these disproportionate rates of decrease in his investigations.

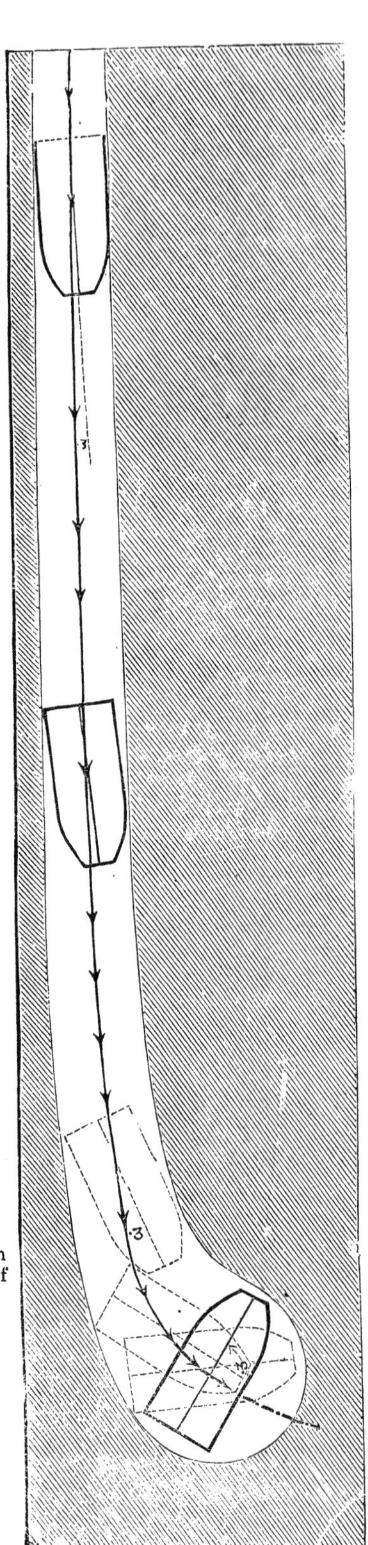

With regard to penetration the writer assumes that friction causes the rotation very shortly to cease, and gravity may be neglected in the investigation of this subject.

Hence, if the projectile entered truly at a normal to the surface of the resisting medium, and its axis was also coincident with the same, the path through the medium penetrated would continue in the same straight line; practically this cannot occur, so that the path of the projectile into the medium may be said to lie in the plane passing through two lines:—

1st. The direction in which the centre of gravity of the projectile is moving at the moment of impact; and,

2nd. The axis of the figure which may be assumed to be at a small angle S with the first.

If the direction of motion is at right angles or at a normal to the surface of the body it penetrates, the resistance of the same tends to increase the angle S made by the axis of the projectile with the path described by its centre of gravity, while it also deflects the path itself towards the same direction; finally the projectile comes to rest with the angle S sometimes as high as 80°, while the deflection of the path itself may be such that the point projectile is towards the rear, that is, pointing towards the surface where it entered, as in fig.

The above result has actually occurred in firing projectiles into earth, and the projectiles when dug out have been found with their points towards the side at which they entered.

Mr. Mallet on penetration of armour.

These effects cannot be fully developed in penetrating such material as armour plates, and Mr. Mallet in his paper next discusses the actual effects which are produced by various forms of projectiles in piercing armour, on the supposition that it is "a homogeneous plate of "parallel thickness of a malleable "material," which is not completely penetrated by a rigid projectile.

The greatest resistance is made by

tough and moderately soft wrought-iron, but even this behaves as a more or less brittle body whenever the velocity of impact reaches about 560 feet per second. The entrance of projectiles into armour, Mr. Mallet considers them as accompanied by a certain amount of direct fracture and a certain amount of lateral displacement, the metal behaving to some extent as a plastic and flexible body. The ogival form of head is specially adapted to perform this work effectually.

Mr. Mallet next illustrates the action of penetration at various angles in a horizontal plane, confining himself to the consideration of effects of distortions in this plane, on the lamina of the shot passing through the axis, supposing that the axis of the projectile coincides with the tangent to the trajectory at the moment of impact.

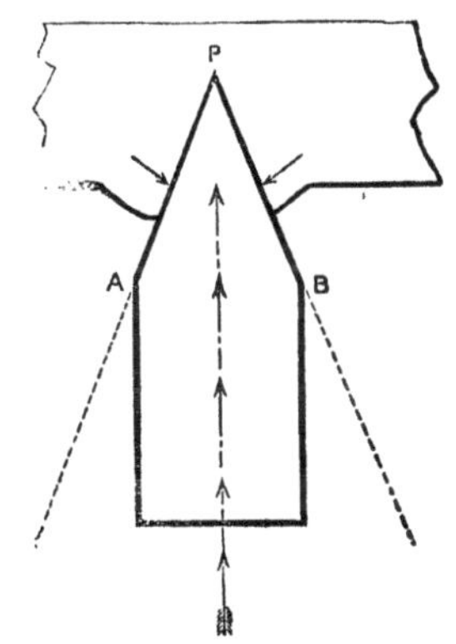


1st Case.—A projectile of cylindro-conic form entering at a true normal to the surface of the plate; here the resistances act at normals to the faces PA and PB; the plastic distortion is symmetrical, and the shot continues its course in the same direction, acting like an isosceles wedge as it enters the plate (as has been noticed this can hardly occur actually).

2nd Case.—Suppose the same shot to enter obliquely, the forces acting on the two sides of the head become unequal for two reasons :—

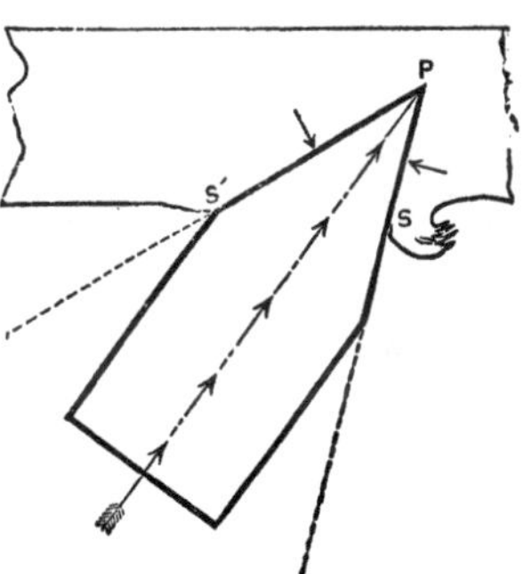


1st, because the surface of resistance PS′ is greater than PS ; and 2nd, because the actual coefficient of resistance (*i.e.*, of friction and compressibility jointly) on the side of PS′ is greater than on the side of PS, because the plate yields and bulges on that side more easily than on the side PS.

Many variations arise even in this simple case; in the angle given by Mr. Mallet the pressure of the shot at a normal to its head, *i.e.*, at right angles to AP in fig. 4, tends to force the plate up in the burr of the illustration, but had the angle of incidence been less oblique the case would have been altered.

Had the line SP been at right angles to the face of the plate, the *coefficient of resistance* on that side would have been at a maximum, and the preponderance of resistance on one side would have depended on the relation of the smaller coefficient acting on the larger surface, to the larger coefficient on the smaller surface.

In this particular case the coefficient is at a maximum, but at all angles of incidence between that and the normal this question of preponderance remains the same.

Taking the resultant of these resistances to act at a point half-way along the faces PS′ and PS, which, as Mr. Mallet notices, is not absolutely true, even considering only the horizontal lamina of the shot in question, it may be seen that there arises a dynamic couple tending to deflect the shot and cause its path to approach more nearly the normal to the plate (*i.e.*, cause the shot to turn in).

Mr. Mallet next considers the forces opposing the direct entrance of a flat-headed cylindrical projectile and of an ogival, and then passes to the important question of their respective powers of penetration at various angles.

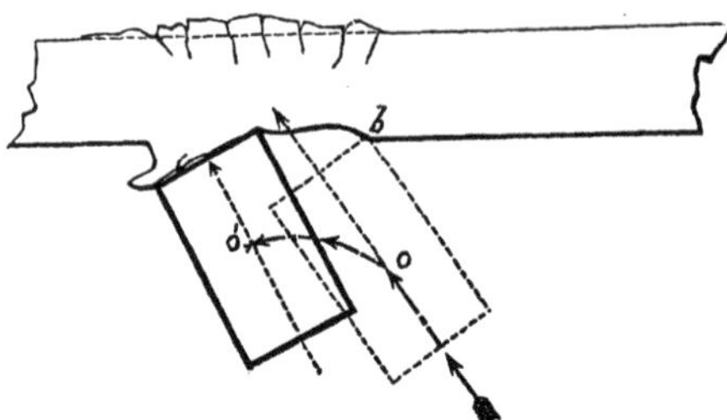


The flat bolt striking at an angle shown in fig. 5 has a tendency at first to rotate on *b* (turning in).

But Mr. Mallet considers that as it proceeds further, on principles which apply to the circumstances which he discussed in the case shown in fig. 4, its centre of gravity *o* advances and slides towards *o'*,* in a direction making a variable angle, greater or less, towards the interior of the plate, the plastic face of which is cut out in a sort of curved form shown in fig, 5, and the material excavated is pushed before the face of the shot, adding to the thickness of the plate to be removed before penetration is effected. Hence Mr. Mallet considers that a flat-headed shot can only penetrate under these circumstances at the expense of great waste of work, and under ordinary circumstances glances off the plate, base first, before it is able to "*immerse itself deep enough to* "*become encastré in its substance.*"

As it glances off it probably continues to whirl rapidly end for end, on a transverse axis,

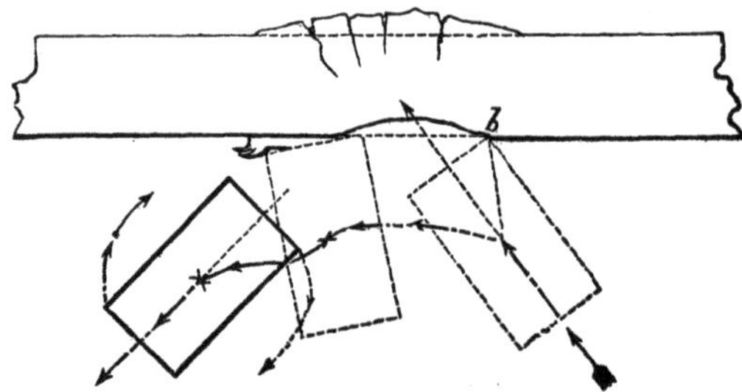


The cylindro-ogival shot, if the angle of incidence be greater than that made with the axis of the shot by a tangent to the curve of the head at the point, digs its point in at once and "the centre of gravity "of the shot at the same time goes forward, turns round more readily "at first than the flat-ended shot upon an equatorial axis, and slides "in the direction of a line making an angle, more or less, towards the

* With regard to the flat-headed shot there is a tendency to slide from the component of the velocity resolved in a direction parallel to the plate and a tendency to turn in from the resistance acting as a normal to the flat head at and near the edge *b*; hence it both slides and turns in rapidly, though it is difficult to say how nearly in the relative proportions shown by Mr. Mallet.

Probably the direct punching of the flat-headed shot compares better with the wedging open of the ogival head if the plates are ***very hard.***

" internal side of the face struck. The forward part of the shot thus
" cuts out and partly pushes before it, normally to the face towards
" which it slides, the plastically distorted part of the iron, and bulges
" or not the opposite face is an *umbo*, whose conditions are such as
" referred to in fig. below."

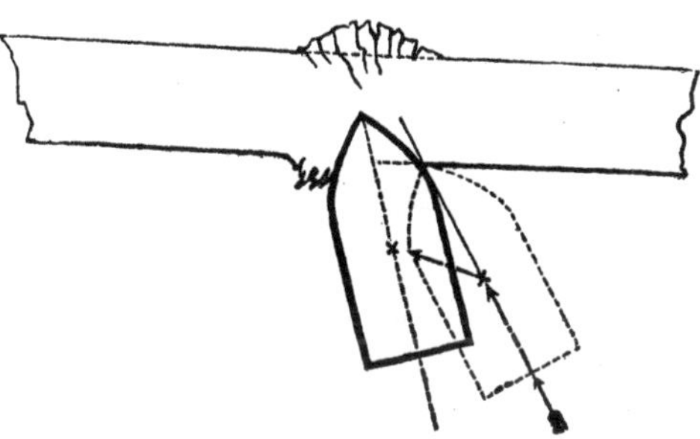

Mr. Mallet, however, considers that the shot of this form soon becomes encastré at its point as regards rotation in the plane of the figure, so that further rotation is prevented by the support at the left side of the head, and also near the point on the right side, the shot finally assuming the position shown in the fig. following.

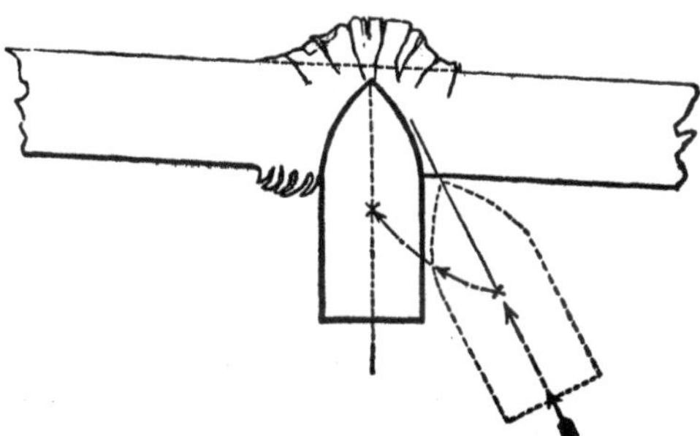

Mr. Mallet further considers that friction may even cause a shot of ogival form to turn, catch its point and penetrate when impinging on a target at an angle of incidence slightly less than that of a tangent to the ogival curve at the axis.

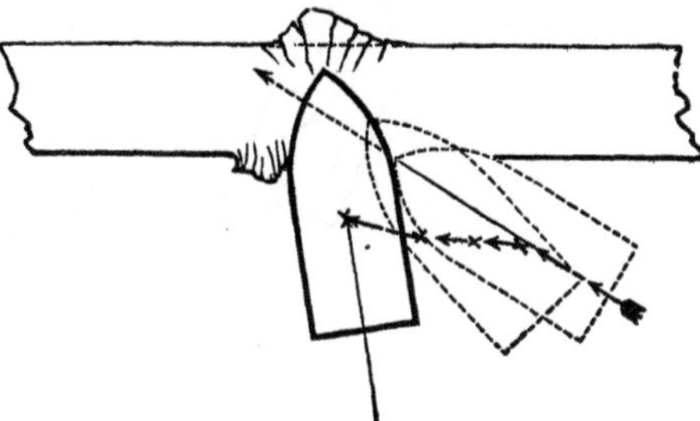

Passing on to the question of still more oblique impact as against convex plates, Mr. Mallet shows that the ogival-pointed shot may glance off *point first*, when there is not enough plastic distortion to hold the point,

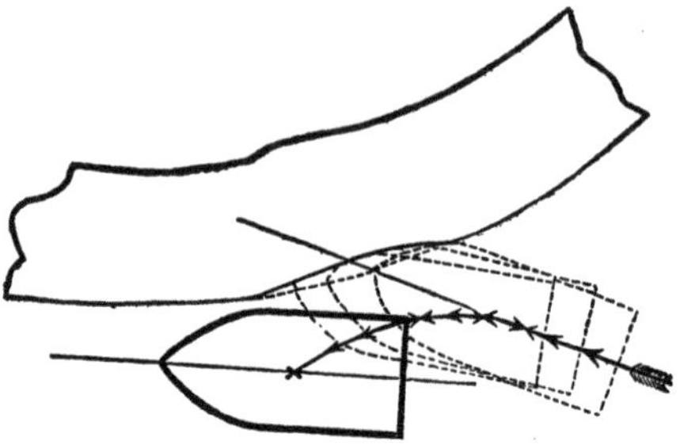

and *base first* when the point is caught but not held.

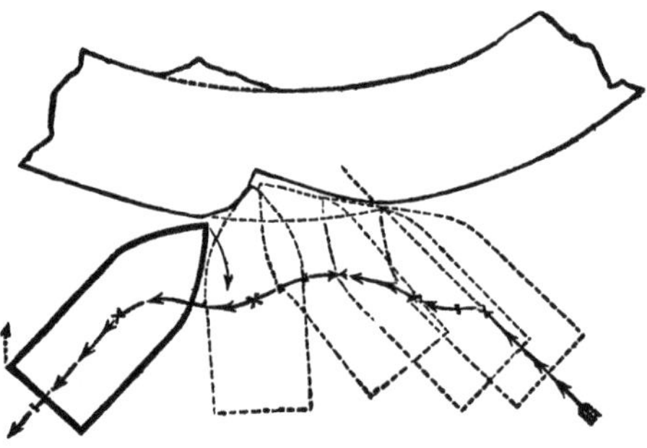

The question of impact against a concave surface is referred to (*vide* fig. below) to illustrate the small power of deflection and the great improbability of a shot glancing off when so fired, and—

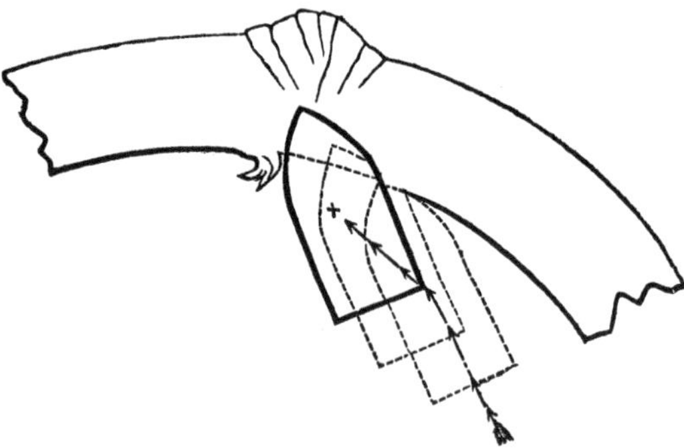

Lastly, Mr. Mallet notices the conditions which cause a shot to ricochet on water, in spite of the slight tendency there is to turn downwards before the head becomes immersed, owing to the inequality of the resistances on the front of it, the coeficient of resistance being much less on the upper side.

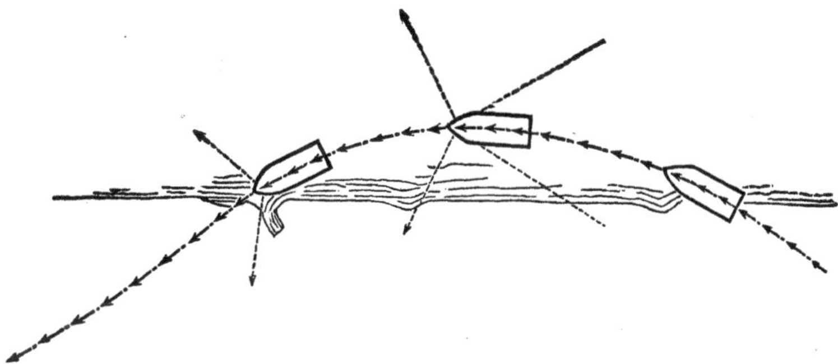

Comparing the wave to that caused by an earthquake (*vide* the *Engineer*, 1867, January 4th, 11th, 18th, and 25th).*

N.B.—It may be observed that the power of turning in instead of glancing off *point first* when striking on armour at an oblique angle increases with the radius of the circle with which the ogival head is described, inasmuch as the limiting angle of penetration is approximate to that made by the tangent to the ogival curve at the apex with the axis of the projectile, *i.e.*, the limiting angle of penetration with ogivals of various radii may be said to be approximately as follows:—

| | | |
|---|---|---|
| For 1 diameter | .. .. .. | 60° |
| „ $1\frac{1}{4}$ „ | .. .. .. | 53° 8′ |
| „ $1\frac{1}{2}$ „ | .. .. .. | 48° 12′ |

## Experience with Segment Shell on Active Service.

Extracts from Report of Capt. Mercer, R.A., to D.A.G., R.A., 6/4/61. See R.A.I. Proceedings, Vol II., page 393:—

"The guns were loaded and laid, and the gunners with lanyard in hand waited for the word from the officer, who was watching until some heads appeared above in that direction, or a puff of smoke revealed their presence, when the gun was instantly fired, and the shell, entering just below the crest of their pits, burst inside.

"The following evidence has been given concerning the action of the Armstrong shell with the concussion fuze (*i.e.*, percussion fuze) only:—Colour-Serjeant J. Morant, Royal Engineers, was at the head of the sap, and saw an Armstrong shell go through a rifle pit, about four feet of earth, and burst inside, and heard the enemy shout as in pain; he also observed that the shell from the Armstrong gun entered the rifle pits as soon or sooner than the report was heard, so that the natives had not time to get out of the way. Bombardier J. Singer, No. 3 battery, 12th brigade, R.A., was at the head of the sap, and in the advance

* Observe the beauty of the effect of the *inelastic* medium water on the lower side pressing against the bottom and superincumbent mass, and on the upper side lifted in a wave, as shown in fig.

I conclude the ricochet will be repeated until the *difference* of the pressures caused by the resistance of the water against the lower half and upper half of the anterior surface of the projectile, which is a function of its velocity, becomes less than the vertical downward component of the shot.

parallel with the Coehorn mortars, when he saw several shell from the Armstrong gun go through the enemy's rifle pits and burst inside. After the cessation of hostilities one of the natives told my serjeant-major that they were sometimes able to get out of the way of the mortar or large shells, but never out of the way of the shell (whether with time and concussion, or concussion fuze only) from the gun 'all the same as the rifle,' meaning the Armstrong guns, as the shell was amongst them as soon as they heard the report. These natives have designated the Armstrong shell 'the quick shell.'

The different statements made both by those who were in the sap as well as by the natives themselves corroborate the observations taken from the battery, viz., that the Armstrong shell with concussion fuze only entered the crest of the enemy's rifle pits and burst inside; whether there were few or many natives in the pit at the time cannot be ascertained."

*Extracts of Evidence given by Captain Seymour, C.B., R.N., before Armstrong and Whitworth Committee, page* 74.

"Saw the Armstrong guns worked in New Zealand the whole time, and almost every shot that was fired from them.

"I have seen a report by Captain Mercer of the Royal Artillery, who commanded the Armstrong battery, and I am only expressing the feelings of the officers and men of the three regiments of the line and of the naval brigade engaged there when I say that we by no means concur in Captain Mercer's statement in his report. We were all most anxious for the success of the Armstrong principle; we had also heard of what it had done in China, but I cannot say that the practice was as satisfactory as we hoped it would have been, I mean from the 12-pr. 8 cwt. guns. Captain Mercer succeeded to a certain extent in driving shell through the earth so as to burst with the concussion fuze inside the rifle pit."

*Extracts of Evidence given by Captain R. Harrison, R.E., before Armstrong and Whitworth Committee, page* 69.

"The shell from the Armstrong guns in China were very destructive. By one shot that was fired at an action that we had near Pekin I think there were as many as 13 killed. It was at a very long range (about 1,500 yards), and there were two or three horses, and, I think, 13 men killed at one single shot. The effect was produced by a segment shell with concussion fuze. It was not hilly ground, but ground dotted over with small jungles and villages comparatively level; there were deeply sunk roads and gullies here and there. You could see a good deal over the country if you got on the top of a little mound or of a house. The shell struck in a roughish field, not hard ground, it was dryish ground. This one shot was very much noted, because we had been fighting with the Tartars all the day, and they had halted in a place which they thought was quite out of range. Then Sir Hope Grant said he wanted to try the effect of one shot at them, and he ordered one gun to be fired, and the effect was very striking. Everybody was surprised, and the Tartars particularly so. There was a large body of 500, I should think, and it completely dispersed them. That was the most remarkable shot I witnessed against men, and we all noted it. The chief effect that I saw with Armstrong guns was against the village of Tang Ku. That was a large fortified village. The guns were all brought up in line, and at a range of from 1,000 to 1,200 yards, opened against the guns of the enemy. There were about 60 enemy's guns firing on the Armstrong batteries and the other batteries which were brought up in line and I noted where the Armstrong guns fired. There was a battery of Armstrong's on the right. There were 16 or 18 gunners of the Tartars killed at one gun. The shells hit a small crenelated wall on the top of the parapet, about 3 feet 6 inches high and 1 foot 3 inches to 18 inches in thickness where they burst, and soon all the Chinese gunners were killed. We counted at one gun from 16 to 18 bodies. The effect was very destructive against those men who were concealed behind this parapet. The parapet was just sufficiently thick to explode the shells. We saw where the shell struck the parapet and broke, that the men were all killed by the Armstrong shell. There were some, of course, in the line that were killed by other shell, but the ones that I noted were killed by the Armstrong shell.

"Certainly the Armstrong guns, we all thought, were terribly effective, and they saved a great deal of loss on our side. Photographs were taken showing how the bodies lay about inside. There were an immense number killed by the Armstrong guns. I am quite sure that the great impression in the army was that almost all the effect thus produced on the fort, as well as on the village of Tang Ku, was by the Armstrong guns. I am sure that the wounds were caused by Armstrong guns. I saw the segments of the Armstrong shells lying about just inside the casemates, and the wounds, you could see, were made by them. Sometimes you found a bit in the wound; you cannot mistake the wound.

*Extract from the late Colonel Milward's Notes on Armstrong* 12-*prs*., *on Service in China, page* 214. *Select Committee on Ordnance*, 1862.

"11th August, 1860. In action at Singho. The battery fired an average of 18 rounds per gun at 1,200 and 450 yards, besides two shells at 2,100 yards, all with most excellent effect. The concussion fuzes acted admirably; the practice was entirely against cavalry in open order, and quite prevented their forming for a charge, which they attempted more than once. The effect of the shells was all that could be desired, the two fired at 2,100 yards dispersing a large body of cavalry instantly.

"13th August. Fired 30 rounds per gun, 25 time fuzes, remainder concussion. Time fuzes did not succeed. The concussion fuzes acted well, but one burst in ramming home. No damage to the gun.

"21st August. Capture of Peiho Forts. Fired 85 rounds per gun, all with concussion fuzes, making excellent practice at various ranges from 1,600 to 400 yards. The practice at one of the south forts at a range of 1,200 yards was most excellent, several shells in succession bursting on the terreplein of the cavalier. The firing at the north fort at 1,600 yards, was also very good and most effective. Some shells have been picked up only partially burst; generally, I hear (I have only seen one) the bottom half of the shell is unbroken. The concussion fuzes are also liable to suffer from damp, but can easily be repaired if time permits.

*Captain (now Colonel) Hay, R.A., B.M. to R.A. in China. Extract from his Report to Brigadier-General Crofton, R.A.*, 6/9/60, *page* 216. *Select Committee on Ordnance.*

"Those (Armstrong shell) that struck the parapets in places where more than three feet thick, appeared to be blind, whilst the others passed through, scattering their segments on the other side.

"On our first advance from Pehtang I assisted in laying the gun on bodies of cavalry, at distances varying from 1,600 yards and upwards, and found that those that burst as desired were most efficient. The same day I had also an opportunity of seeing the gun fired at very long ranges, viz., about 2,700 yards; this was at an enemy retreating at Sinho, along a narrow causeway; only one shell that I observed burst at that range; they grazed well. I observed in many instances that the shells had not broken into segments, but had broken in two, or the head or bottom was merely blown out; they (the segments) certainly at present do their work, as, although I found many dead, I never found a man wounded by a segment; they appear to pass right through, stopped by nothing.

"Concussion fuzes, a great number were blind. * * * I also observed some premature explosions with them, which must have done some mischief to our troops."

*Extract from remarks by Lieut. Pickard, R.A.. Vol. IV., Proceedings of R.A. Institute, page* 371.

"From these plans and descriptions two sections were constructed, one of which, similar to the pahs constructed in the northern part of the island, consisted of two rows of young trees, or strong spars, about 18 inches in diameter, placed upright side by side from 8 to 10 feet high in the clear, the rows being about 3 feet apart.

"The other description of pah consisted of strong spars about 18 inches in diameter, placed upright in the ground about 8 feet apart; cross bars connected them, and on these were hung smaller spars from 4 inches to 6 inches in diameter; these were placed close to one another, and their ends were kept about a foot above the ground. No nails were used, but, as is customary with the natives, flax secured everything. The line of stockading was again double. The natives used to fire out of pits dug out behind the second row, and made deep enough for them to fire under the stockades.

"These latter pahs were used by the southern natives in the war of 1860–61, and on one occasion one 8-inch gun and two 24-pr. howitzers, besides a 9-pr. gun, failed to make a practicable breach in a pah formed as above stated, after two hours' firing, at a distance of 200 yards.

"The battery of six 12-pr. Armstrong's took up a position 900 yards from the stockade, and after rather more than an hour's firing made a breach in the section constructed after the northern fashion large enough to allow a section of men to go through abreast.

"The southern description of pah was more difficult to breach, as it gave more to the shell on bursting; and the débris, supported by the flax, was most difficult to clear away. With both descriptions of pahs, salvoes were found most effective, all the guns being laid on the same part of the stockade, until the posts which seemed to afford most support were destroyed.

"The great accuracy with which these guns could be fired, and the tearing damage done by the shells on bursting contributed greatly to effect a breach in the latter description of pah. About two hours were employed in making a practicable breach, and concussion fuzes were chiefly used.

"The guns at Whangamarino then annoyed the Maories at Mere-Mere as much as possible. It was found that the best way to damage canoes at great distances was to fire shells with concussion fuzes, to strike rather short of the canoes, the shell burst on striking the water, and the pieces ricocheted forward among the canoes. It was found also that by firing with time fuzes fixed to burst at the extreme range of the fuze, the pieces of the shell all went over 2,600 yards if the gun was laid with about half a degree more elevation than the actual range required. The shells thus fired burst high in the air, and the pieces being propelled forward and downward went to a great distance.

"When the assault was ordered the rapidity of fire was increased; shells had been prepared in expectation of the assault, and the guns were loaded as fast as they were fired.

"The shells burst beautifully and prevented the Maories taking any aim at the advancing troops, but, when from the near approach of the storming party to the works, the shelling was stopped, the soldiers suffered severely.

"When a long range or great precision is required, the Armstrong shell is most effective, but it would fail if used as a substitute for a howitzer shell in breaching field parapets, and in ricochet fire at short distances."

## Extracts from Reports on Ammunition used in the Expedition to Coomassie.

From the report of Major Rait, R.A., C.B., it appears that the common shell and case were very good; the latter appeared to break up in all instances, and were most effective. The time and percussion fuzes acted well, but the latter were little used, as, firing through the bush, the shell would be prematurely exploded. The friction tubes acted well, and the cylinders in which the fuzes and tubes were enclosed protected them from deterioration.

The star shell were not used on service; some fired experimentally did not appear to act well.

Major Rait recommends a quarter metal lined case suitable for mule transport for the cartridges, as they suffered from climate in the leather cases.

The rockets last received from England were in very good condition, but the composition is subject to rapid deterioration if exposed to the climatic effects of the west coast.

Although their precision could not be compared to guns, yet, owing to their portability and moral effect, Major Rait would not be inclined to advocate their disuse.

They were liable to deflect greatly in the bush, and might be occasionally dangerous to those using them.

Both Major Rait and Lieut. Knox recommended alterations in the rocket trough, the former stating that the elevating bar is too weak, and if bent by rough usage there is a difficulty in sliding the elevating socket up or down; the latter states that the trough should be higher, as the rockets dip and may have their direction altered by uneven ground.

Lieut. Knox observed that the rockets burst explosively on striking a solid object such as a tree.

Lieut. Allen, R.M.A., reports favourably of the rockets and recommends that they should be made explosive.

Some rockets which were taken out of store at Elmina and appeared useless, judging by the exterior, were well up to range and did not explode in flight.

The rockets employed were 9-pr. Hale's rockets, and were packed in metal lined cases, from which, however, they were removed in transport.

## Saluting and Exercising Cartridges of Silk Cloth, for Muzzle-Loading Iron Guns, Land Service.

1. Adverting to § 1780 in List of Changes, June 1869, it must be fully understood that the object of the introduction of silk cloth cartridges is to lessen the risk of accident consequent on rapid firing from the same gun or guns.

2. The use of these cartridges will therefore be restricted, *first* to the firing of salutes at such stations as are named in the Queen's Regulations, provided the number of guns available for firing salutes is less than the number of rounds to be fired; *secondly* to garrison guns fired at reviews (such as are specially allowed to be held at Dover and Portsmouth), when authorised by the Secretary of State for War; and, *lastly*, to the guns used in the "dismissal" of recruits of Garrison Brigades.

3. On receipt of this order, the nature of the guns to be used, in future, for salutes (at those stations where salutes are authorised to be fired), and for the dismissal of recruits, will be decided upon by the general or other officer commanding the troops at the station, and notified in General Orders, and a copy of the order transmitted for the confirmation of the Secretary of State.

4. In the event of the gun or guns selected not being of the smallest available calibre, the reasons for the selection made will be given.

5. At those stations, where morning, noon, evening, and signal guns are authorised by the Secretary of State to be fired, cartridges of the present pattern and charge will be made use of, and also for all other purposes not involving risk from rapid firing.

6. Estimates of the quantities of silk cloth cartridges required for the different calibres for two years' future expenditure will be prepared in accordance with the foregoing conditions, by the officer commanding the Royal Artillery at each station, and will be based on the actual expenditure of the last two years, with such modifications as may be probable in the ensuing year.

7. The estimated quantities required for salutes, for reviews, and for the dismissal of recruits, will be shown separately; the estimate will be submitted for the approval of the general officer commanding, and be forwarded to the district controller.

8. On receipt of these estimates the district controller or commissary in charge, after carefully checking the same, will consolidate the whole into one demand, which will be forwarded to the Controller, Royal Arsenal, Woolwich, for supply, on the approval of the Controller-in-Chief being obtained.

9. On receipt of the cartridges by the store officer of the station, they will be held in charge until required by the Officer commanding the Royal Artillery, who will from time to time make such demand as the expenditure renders necessary.

10. The cartridges required for the usual salutes will be supplied with the sanction of the general or other officer commanding on authority conveyed by the district controller.

11. The cartridges for reviews will be issued on the authority of the order of the Secretary of State for War for the holding of the review, and for the guns to be used, being given.

12. The cartridges for the dismissal of recruits, namely, six rounds, will be issued on the authority of the Regulations for the Supply of Ammunition, dated February 1869, page 5, section 4.

13. Before supplies are made, on the authority of the foregoing three paragraphs, the controller will be responsible that the conditions of the supply, as well as the quantities, are correctly stated.

14. Clauses 2, 6, and 7 of § 1780 of List of Changes, June 1869, will stand; the remaining clauses are hereby cancelled.

INSTRUCTIONS FOR PROOF OF TUBES, PRIMERS, DETONATORS OF FUZES, &c., TO ACCOMPANY THE APPARATUS FOR PROOF OF TUBES, PRIMERS, AND DETONATORS OF FUZES. MARK II.

1. The apparatus consists of two stands with the necessary vents, anvils, and weights for proof of all descriptions of tubes, primers, and detonators for fuzes.

*Proof of Tubes and Primers.*

2. The stand for proof of tubes and primers is a cast-iron table fitted with two vents, one for proof of "tubes, friction, copper," and one for proof of "tubes, friction, quill," in connexion with which is a "pin friction tube" and a "guide plate," the same as fitted on guns for Naval Service; the stand is also fitted for a special vent for proof of primers for Shrapnel shell. Two iron spanners for fixing and unfixing vents, &c., are furnished with the stand.

3. For proof of "tubes, friction, copper," a brass vent A in two parts, upper and lower, and a receiver A for a powder puff, are screwed upon the proper vent provided for that purpose: the upper part alone with the receiver represents a vent of 14 inches, with a clear space of about half an inch to the top of the powder puff; the addition of the lower part represents a vent of 30 inches. All long tubes are to be tested in the 30-inch vent; tubes for general service in the 14-inch vent.

4. For proof of "tubes, friction, quill," a brass vent B in two parts, and a receiver B for a powder puff, are provided, and fitted on the proper vent, as described for the vent A for copper friction tubes.

5. For proof of "tubes, friction, copper, 7-pr. R.M.L. gun," a brass vent with receiver for a powder puff C 3 inches long is provided; it is to be screwed on the vent for quill friction tubes in place of the brass vent B.

6. For proof of "primers, vent piece," a special brass vent D, and a receiver D for a powder puff 4 inches long, are provided; this vent is also screwed on the vent for quill friction tubes; the primer is to be inserted in the horizontal end of the vent, and the receiver then screwed on. The distance between the end of the primer and the puff to be 4 inches. The primer is to be fired by means of an ordinary friction or other tube, placed in the vertical portion of the vent. Tubes known to be good should be used for this purpose.

7. For proof of "primers, brass, Shrapnel, shell," a special brass vent with receiver E 30 inches long, is provided, the upper end of this vent is fitted internally to receive the primer, it is passed up and screwed through the top of the stand, and through a steel boss which is fixed on the top of the stand; the boss is screwed and fitted to receive a brass cap for covering and enclosing the top of the vent when fitted with the primer; the brass cap has a small hole bored through the side; as a means of communication for a piece of quickmatch for firing the primer. About 2 drams of loose powder is to be placed in the receiver at the bottom of the vent.

8. The powder puff contains 2 drams of F.G. powder, and is made with two thicknesses of serge. A special puff of one thickness of serge is used for the tubes for 7-pr. R.M.L. gun; and a puff of one thickness of serge and one of waterproof material is used for the copper friction tube for waterproof cartridges.

9. A failure of 3 per cent. in tubes that have never been issued for

service, or of 5 per cent. of tubes that have been issued and returned will warrant *suspicion* that the tubes are unserviceable.

10. Cylinders of tubes thus suspected, being set aside, one or more will be tested by firing every tube in them.

11. No report of the unserviceability of friction tubes is to be made, or percentage of failures deduced, upon a less trial than 100 tubes.

12. Inspectors of warlike stores will keep up their knowledge of the condition of the tubes in their district by obtaining from Commanding Officers of Artillery sample cylinders of tubes manufactured at different dates. Should the tubes be otherwise packed than in the tin cylinders closed with a soldered tin band, the inspecting officer will use his judgment as to the proportion to be tested. If these samples are unsatisfactory, more cylinders must be opened in order to form a judgment upon theirc ondition.

31. Tubes, friction, quill, *without* loops, should be returned (if in good order) as repairable (to tubes with loops),

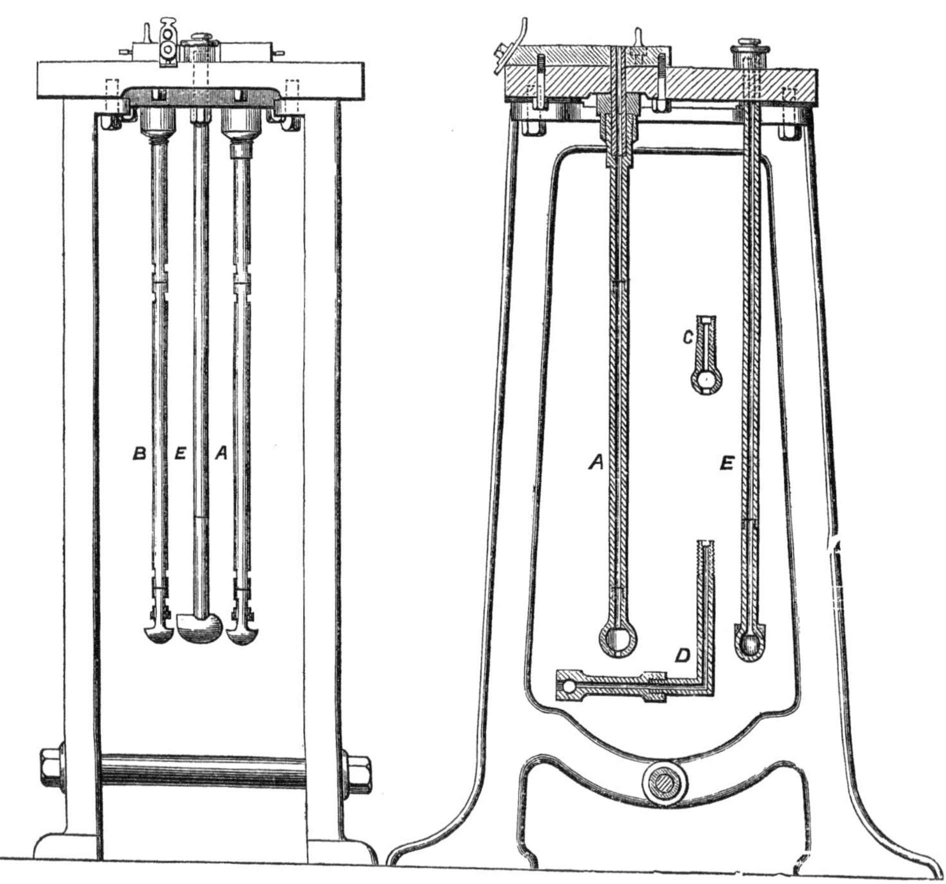

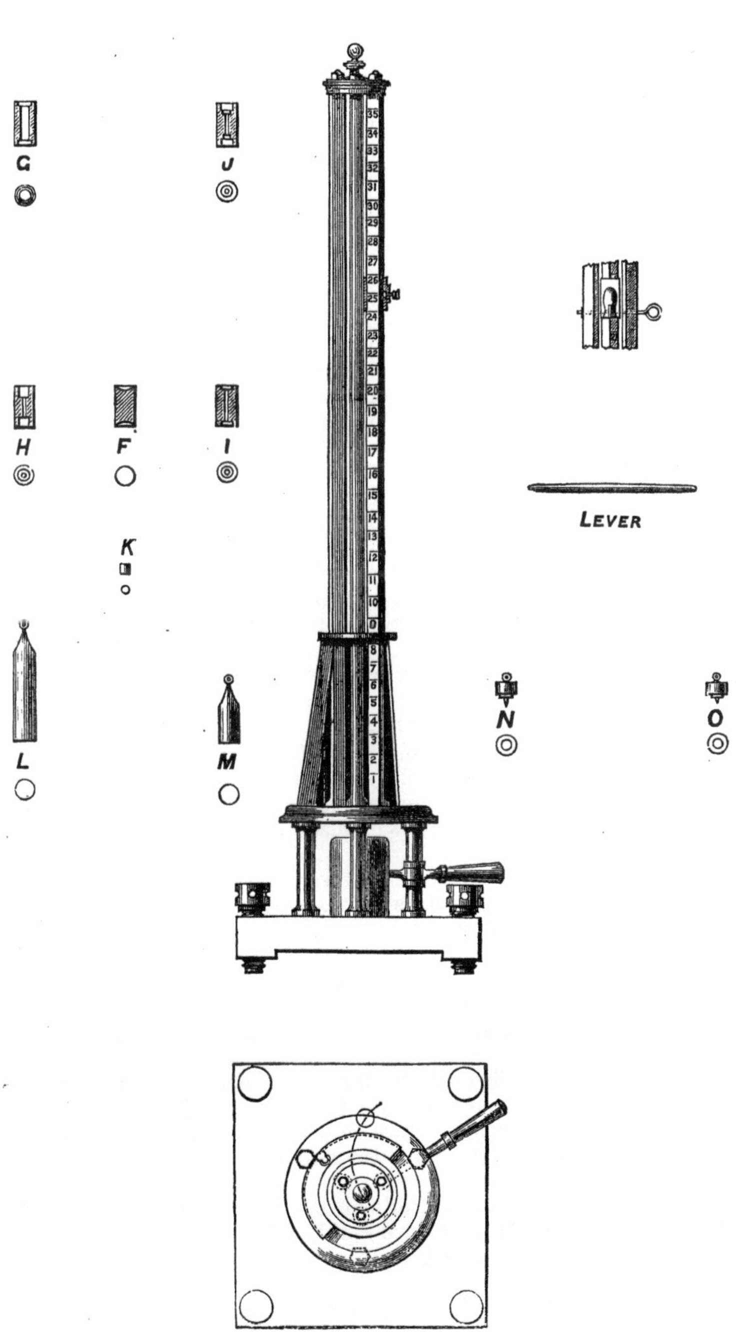


*Proof of Detonators of Fuzes.*

14. The stand for proof of detonators of fuzes consists of a cast-iron base plate fitted on the top with 3 wrought iron pillars, carrying a brass plate, on which is erected 3 steel guide bars of such a height as to admit

of a weight, placed within the bars, being dropped a clear distance of 25 inches on to the top of a detonator on an anvil placed on an anvil bed beneath the guide bars: one of the guide bars is graduated in inches, and is fitted with a suspender or sliding bracket with fixing screw for carrying the weights; the weights are attached to the suspender by a brass pin; when all is prepared for releasing the weight, the pin is to be drawn out with a *sharp* pull, so as to allow the weight to drop freely. The anvil bed of wrought iron is attached to one of the wrought iron pillars by a handle, which acts as a lever with the pillar as a pivot, so as to allow the anvil bed to be moved from the centre of the stand whilst the required weight and anvil are being inserted or removed from within the guide bars at the bottom.

To protect the operator from the effect of the explosion of the detonator, a shield of sheet brass, in the form of a half frustum of a cone, is fitted on the brass plate so as to revolve round the guide bars, the open side of the shield being brought next the operator when placing the detonator, weight, and anvil, and the shield reversed before the weight is released.

This stand is intended to rest on the top of the stand for proof of tubes when in use, and, in order to adjust it perpendicularly, the base plate is fitted with four adjusting screws; a "Lever, steel, for adjusting screws," is provided for this purpose. Before using the stand care should be taken that the fall is perfectly true; this may be ascertained by hanging any small weight or plummet by a thread to the pin through the eyes of the suspender.

15. The detonators are to be proved as described in the following paragraphs, care being taken in each operation to insert the proper anvils and weights, as undermentioned, viz.:—

**F.** Anvil, steel, for detonating balls of Pettman fuzes.
**G.** Anvil, steel, for steady plugs of Pettman fuzes.
**H.** Anvil, steel, for detonators of B.L. wood time fuzes.
**I.** Anvil, steel, for detonators of fuzes, percussion, R. L., Mark I., and B. L. plain.
**J.** Anvil, steel, for detonators of fuzes, percussion, R.L., Mark II.
**K.** Plug, steel, for detonators of B.L. wood time fuzes.
**L.** Weight, steel, 18-oz.
**M.** Weight, steel, 7½-oz.
**N.** Weight, brass, and steel-pointed, 1-oz. for fuzes, percussion, R.L., Mark I., and B.L. plain.
**O.** Weight, brass, and steel-pointed, 1-oz., for fuzes, percussion, R.L., Mark II.

16. The detonating balls of Pettman general service percussion fuzes are to be proved by allowing the 18-oz. steel weight **L** to fall on them through a height of 25 inches. Care must be taken that the face of the anvil **F** on which the ball rests, and the striking-face of the weight, are cleaned between each explosion.

17. The detonating balls of Pettman land service percussion fuzes are to be proved with the 18-oz. weight **L** falling through a height of 22 inches, the faces of the anvil **F** and the weight being cleaned between each explosion.

18. The steady plugs of Pettman general service percussion fuzes are to be proved by allowing the 18-oz. weight **L** to fall through 25 inches on to the brass ball of the fuze placed over the ring of composition in the steady plug, on the anvil **G.**

19. The brass detonators of Boxer breech-loading wood time fuzes are to be removed from the fuzes, the brass plug unscrewed from the top,

and then (without removing the suspending wire) placed on the anvil **H** and proved by allowing the 7½-oz. weight **M** to fall through a distance of 22 inches. To enable the blow to be communicated to the detonator, a small steel plug **K** is provlded, which is to be placed upon the hammer of the detonator.

20. The detonators of the B.L. percusssion fuzes, plain, and the R.L. percussion fuzes, are to be proved by placing the pellets containing them on an anvil and allowing a pointed weight of 1 oz. to fall on them through a distance of 10 inches, the powder in the pellets is not to be removed:—there are two anvils and two weights provided for the proof of these detonators, the anvil **I** and the weight **N** for R.L. fuzes of Mark I. and for B.L. plain fuzes, and the anvil **J** and the weight **O**, with a shorter needle point, for R.L. fuzes of Mark II.

21. The following table gives the details of the different proofs:—

| | | Weight. | Height. |
|---|---|---|---|
| Detonating ball .. | Pettman G.S. fuzes .. .. | 18 oz. | 25 inches. |
| | " L.S. " .. .. | 18 " | 22 " |
| Steady plug .. | " G.S. " .. .. | 18 " | 25 " |
| Detonator B.L. or R.L. percussion fuzes .. .. | | 1 " | 10 " |
| Detonator Boxer B.L. wood time fuzes .. .. | | 7½ " | 22 " |

22. If the detonator fails the first time, it should be tried a second and third time; but the fact of a detonator requiring more than one blow should excite suspicion.

23. Detonators of fuzes should not be condemned on the trial of a less number than 3 per cent.; a failure of more than 5 per cent. of this number is sufficient to condemn them. Care should be taken in reporting the result of the examination to quote all the marks on the fuzes and the cylinders, amd to state if the latter have ever been opened. Rule 12 for tubes should be applied also for fuzes.

### Proof of Wood Time Fuzes for B.L. and M.L. Ordnance.

24. Judgment of the state of these fuzes will be formed from the average time of burning of not less than 20; and selections for proof should be taken at the rate of about 2 per cent., from those manufactured about the same date.

25. In proving B.L. fuzes, the detonators will be removed, and quick-match inserted, by which the fuzes can be ignited. (For proof of detonators, see above.)

26. Fuzes will be condemned if their time of burning is not within the following limits, viz.:—

| | | | |
|---|---|---|---|
| The 5 secs. fuze should not reach | 5·5 secs., | nor burn less than | 4·8 secs. |
| 9 " " " | 11·0 " | " " | 9·7 " |
| 20 " " " | 22·0 " | " " | 19·4 " |

*Approved*, 22/12/76

57
24
8142

# HINTS ON THE EXAMINATION OF AMMUNITION.

---

The method of examining powder and its classification will be found in the regulations for gunpowder magazines. The "flashing" test is a ready way to ascertain whether powder is of good quality and in good condition. About eight drams of powder are poured on a glass plate so as to form a conical heap and "flashed" by applying a hot iron; no residue should be left, only a few smoke marks should be seen on the plate.

If powder has been damaged by damp it will be "caked," and a close inspection will generally detect a white appearance, due to the saltpetre having been dissolved and deposited in crystals on the surface.

*Examination of Cartridges.*

The condition of the powder must be examined as above given. Owing to pressure, cartridges which have been tightly packed sometimes feel hard as if the powder was caked, in this case the powder will crumble into its proper condition when handled, and so cannot be confounded with powder which is "caked" from damp.

The condition of the serge should be closely looked to, and any cartridges having holes or traces of being moth-eaten should be put aside for repair or condemnation, according to the amount of damage sustained. Silk cartridges should be examined in a similar manner; they are said be much less liable to the attack of insects than serge.

The cartridges should be gauged, the choking and hooping should be looked to; the directions as to these operations have been given in the notes. Specially see that the silk cartridges are choked and hooped with silk, and that blank cartridges for B.L. guns are choked with worsted, and service cartridges for B.L. guns with twine. The knots of cartridges for rifled guns require careful examination, as often a slip knot is made instead of a fast one.

*Proof of Friction Tubes, &c.*

See instructions for proof of tubes, &c., given in the Appendix, p. 303.

Care should be taken to keep the vents clear, and to ensure their being free from damp a tube should be fired before commencing to test.

The proof of primers for vent-pieces and of fuzes, both time and percussion, will be found in the instructions. It is well to gauge the time fuzes, as the wood sometimes alters its form; in the case of fuzes of the common gauge having powder channels this is important, as sometimes they are so much enlarged as to bring the side holes above the fuze-hole. This can hardly happen with the fuzes of G.S. gauge.

Lights, portfires, &c., can be readily examined by burning, and ascertaining that they burn about the time laid down; if they burn well there is no harm in their burning long.

The primers for lights can be tested at the same time.

*Examination of Projectiles.*

All projectiles are examined by gauging. S.B. projectiles are so simple, and are so well known in the service, that it is not necessary to give any rules.

Lead coated projectiles should be carefully examined as to the attachment of the lead coatings; in very bad cases the eye will detect a loose coat; and in doubtful cases, tapping the shell with a hammer will detect a loose place, as a peculiar dull sound is given out. The high ring gauge should be passed over the shells, and the lead, if set up, can be filed down.

If blisters appear on the surface they should be pricked, and the lead hammered down.

The fuze-hole of shells of garrison calibres should be examined, and any having the Moorsom gauge must be converted by using a G.S. adapter, if the shell are likely to be required for use.

The Moorsom gauge is readily known by the large plug with a shoulder and cylindrical body.

The adapter is screwed into its place by the "key, fuze, and plug, G.S.," when screwed home, it fits well down in the socket, about ·2″ below the top of the fuze-hole. The space between the side of the adapter and the iron of the shell is filled by a composition of rosin, 12 lbs., Spanish brown, 2 lbs., plaster of Paris, 1 lb., turpentine, half a pint. The composition is poured in hot, the adapter being closed with a wooden plug.

§ 1583.

In examining Shrapnel shells, R.M.L. or B.L., special attention should be paid to the junction between the head and the body; loose heads may be met with, especially in F.S. Shapnel of early patterns. A loose head renders the shell unserviceable; also early patterns should be examined to see whether any rosin has worked up into the socket, which might prevent the flash of the fuze from igniting the primer.

Studded projectiles should be examined by passing the cylinder (or ring, body and studs) gauge over them. As the gauge is slightly smaller than the calibre of the gun, a shot which passes the gauge is certain to load easily.

In examining common shells, M.L. or B.L., the condition of the lacquer should be looked to, and also as to whether any loose iron filings may be present in the shell. B.L. shells, with black lacquer, may occasionally be found and prematures may occur when using them. Any loose matter may be detected by "upending" the shell.

Extracts, Vol. III, pp. 143–240.

Palliser projectiles are sometimes damaged in transit, if the point is broken off the shell becomes unserviceable. They can, however, be utilized at practice.* Any Palliser shot having the base closed with a wedge of wrought-iron must be returned for repair, as directed in § 2040. All made prior to 1870 require alteration.

See § 1872 for the patterns which do not require alteration.

Case shot are sometimes damaged in transit, or by the jolting motion of a limber. They can generally be repaired by a tinsmith, a little solder is often all that is wanting.

Shells that have been stored in the open air are sometimes found to have admitted water; they should be carefully dried.†

Cl. 143, paras. 32, 33, 34. A. C. /69.

Nothing requires more care than the examination of shells returned into store as empty; frequent accidents have happened from the presence of powder in shells so returned; they are therefore received into store

---

* See page 202, for regulations on this point.

† If Shrapnel shell once get wet inside it is most difficult to dry them.

as *doubtful* and carefully examined. At out-stations shell so examined are marked with an E in red.*

When it is necessary to break up old shells it may be done by placing an iron wedge in the fuze-hole and striking it with a sledge hammer. Shell should be washed out with water before this is done.

It is important to remember that projectiles which are not fit for service may often be used for practice; thus shells may sometimes be found so damaged about the bush as to be unfit to use with a fuze, but they can be fired at practice as plugged shell.

B.L.S. arm ammunition. Examination.

*B.L.S.A. Ammunition.*—By opening the cartridge the state of the powder and the condition of the brass case can be ascertained.

In some of the early patterns, especially in Mark V., the brass may be found to be corroded by the action of the saltpetre on the metal

The condition of the bullets as to dents and corrosion is rarely important, a bullet may be much knocked about, and still will be found to shoot well.

Firing some targets from a rest will determine whether the cartridges are serviceable.

The Snider cartridge should be condemned† if it gives a figure of merit over 20 inches when fired at 500 yards in fine calm weather from a fixed rest: where no mechanical rests are provided, a good marksman using a sand bag will be able to fire with sufficient accuracy to test the ammunition.

The Martini-Henry should shoot about five inches better than the Snider.

Missfires are a most important defect and should be reported, all details as to date of ammunition, &c., being given if possible.

Breaking up small arm ammunition.

Breaking up B.L.S.A. ammunition is an operation requiring great care, the cartridge should be opened with a copper tool, and the powder at once placed in water; on no account should any accumulation of loose powder be allowed on the table.

It is necessary to stir the powder to prevent its floating on the top of the water. In order to prevent risk in transit, the empty cases should be boiled to destroy the cap composition and the powder which is apt to stick to the cases.

Hales' and life-saving rockets. Examination.

*Hales' Rockets* should be examined carefully and frequently for rust, especially along the seam and rivets, if the slightest trace is found the rockets are to be repainted, full directions will be found in § 2441, and p. 270. Similar instructions apply to the life-saving rockets. It is important to distinguish between rust due to a case itself and rust caused by contact with another rusty case. A magnifying glass is useful to detect any flaw in the iron. Rockets with flaws or cracks are unsafe.

If rockets are very rusty they are probably dangerous, and should be returned to Woolwich. As before pointed out Mark I. is unserviceable.

Rockets of all kinds may be tested as to soundness of composition by entering them head first down the bore of the gun, firing them with a quickmatch leader, and noting the time of burning of the composition,

---

* At Woolwich they are first marked E in *yellow*, and then re-examined; if empty the E is then marked in *white*.

† B.L.S.A. ammunition is not to be condemned without the sanction of the Surveyor-General.

and whether any sudden puffs are heard before the end. Times of burning are about the following: 24-pr. Hale, 10 seconds; 9-pr., 8 seconds; Boxer life-saving, 4½ seconds.*

## General Instructions for the Guidance of the Royal Artillery in Charge of Magazines or Ammunition Stores.

1. No one will pass the barrier at the entrance of a magazine or ammunition store except in the presence of the officer, master gunner, or non-commissioned officer, in local charge of the building, who will be responsible that all persons entering comply with the necessary precautions, and that they have no articles of a combustible nature in their possession.

2. All persons employed in magazines, cartridge stores, and in shell stores when connected with cartridge stores, will, before entering the same, change their own clothes and boots for magazine clothing and slippers. This will be effected in the place appointed for shifting, where the boots and clothing taken off will be deposited.

3. Smoking is strictly prohibited near any magazine or ammunition store, and any soldier entering them with a pipe or lucifer match in his possession will be made a prisoner.

4. Only the hand magazine lantern will be used within the magazine or ammunition stores, and then only in the presence of the person in actual charge.

5. Laboratory operations will not be carried on in any magazine, cartridge, or shell store, or in any of the passages connected therewith, but only in the building, or tent, specially provided for the purpose (*vide* Regulations for Laboratories).

6. Every favourable opportunity will be taken for airing the magazines on the principles given in the memorandum attached. Common thermometers will be issued to all magazines containing 100 barrels and upwards of loose powder.

7. Magazines will never be left open unguarded, and sentries will be particularly attentive to the earliest appearance of a storm, however distant, and upon hearing thunder, or seeing a flash of lightning, they will give the necessary notice, in order that the doors and ventilators may be immediately closed.

8. The floor of a magazine or ammunition store will be kept scrupulously clean and free from loose grains of powder. The passages will be covered with hides, wadmiltilts, or hair cloths, when powder in bulk is being moved; these coverings should be frequently lifted and dusted.

9. Barrels, cylinders, and cases will be placed so that the air can circulate freely round them. They should be at least six inches from the masonry of the building.

10. No packing or shifting of cartridges, or issue of powder from cases or barrels, will be permitted within the block containing the magazine or cartridge stores. These operations will only be carried on in the Laboratory.

11. No friction, detonating or common tubes, fuzes, quick or slow

---

* As previously mentioned in the text, the 24-pr., Marks IV. and V., will burn about 5 seconds, the 9-pr., Marks V. and VI., about 4 seconds.

match, signal lights, rockets or primers will be kept in any magazine or cartridge store, or admitted within the enclosure of a magazine where gunpowder alone is stored. Tubes and fuzes may be kept in the shell stores.

12. Small-arm ammunition, which contains its own means of ignition, will not be stowed in the same chamber of a magazine with gunpowder, whether the latter is loose or in filled cartridges.

13. Oiled rags, cotton waste, oakum, or cloths for cleaning, are not to be kept in magazines, ammunition stores, or their passages.

14. All boxes, cases, and barrels placed in magazines or ammunition stores will be labelled, and no empty boxes, cases, or barrels will be allowed to remain in them. Barrels containing powder will never be rolled along the floors of magazines or passages, but will be carefully transported from one place to another.

15. Officers, master gunners, and non-commissioned officers in charge will at all times be particularly careful with everything in or about the magazines, and will take immediate notice of any irregularity they may observe. They will also be very prompt in reporting any defects or repairs necessary either to the interior or exterior of the buildings.

16. An inventory board, showing the contents of the magazine or ammunition store, will be hung up in the lobby or passage leading thereto.

17. The keys of the magazines and ammunition stores will be labelled, and when not in use deposited in a secure place.

18. A copy of these instructions attached to a board will be hung up on the inside of outer doors and on the wall of the entrance to the magazines. Copies for this purpose, printed on foolscap paper, can be obtained on demand.

19. W. O. Form 939 (Standing Orders for Artillery Magazines, dated Horse Guards, 1st December, 1865) has been cancelled.

## Memorandum respecting the Ventilation of Magazines.

1. The dampness complained of in buildings will frequently be found to arise from condensation of the watery vapour of the air which enters the building. Buildings with thick walls and vaulted roofs, and especially those covered with earth, are particularly liable to dampness from this cause.

2. Air always contains some proportion of watery vapour. When the proportion is small the air is said to be dry, and when large the air is said to be damp; when the proportion is the greatest that can be diffused through air at a given temperature, the air is said to be saturated at that temperature.

3. The proportion of watery vapour which saturated air contains varies with the temperature, being greater for high than for low temperatures. Air containing a particular proportion of moisture is rendered less capable of depositing moisture by its temperature being raised and the reverse when it is lowered.

4. Air may be brought to a state of saturation by reducing its temperature. If the air contains but little moisture, the reduction of

temperature must be considerable; but if it contain much a slight reduction will bring it to a state of saturation.

5. If air be cooled below the degree of temperature at which it will be in a state of saturation, a portion of the watery vapour contained therein will be deposited on any cold substance with which it may come in contact. The degree of temperature at which air will thus begin to deposit moisture is called its *dew-point.*

6. When warm air enters a comparatively cold building the temperature of the air is reduced by coming in contact with the interior walls and other cold surfaces: and if its temperature be thus reduced below the *dew-point*, condensation will take place. In the latter case it is obvious that the admission of fresh air will not tend to dry a building, but to render it damp.

7. If a magazine, 40 feet by 24 feet by 12 feet, the temperature of whose internal walls, &c., is 45 degrees, were to be filled with saturated air having a temperature of 50 degrees, and the magazines were then closed, nearly a pint of moisture would be deposited during the cooling of the fresh air to the temperature of the walls. The pint of moisture would result from the quantity of air sufficient merely to fill the magazine; but if the ventilators were open, the air might be renewed many times in the course of a day, and very much more than a pint of moisture be deposited.

8. Air entering a building, whose temperature is higher than its own, becomes capable of absorbing moisture from damp surfaces.

9. The efficiency of the ventilation of a magazine will depend upon the degree of dryness which the fresh air admitted into it possesses, and the rapidity of the current of dry air passing through the building.

10. The dryness of air is indicated by the number of degrees by which its temperature exceeds its dew-point.

11. The ventilators of magazines should, in all cases, be constructed so as to exclude or admit the external air at discretion, *and the instructions for their use should be framed with a view to the exclusion of the external air, when the temperature of its dew-point is above that of the interior of the building, and the admission of the air when its dew-point is below the temperature of the interior of the building.*

12. For the foregoing reasons, the common practice by which, under Art. 491, Ordnance Regulations, 1855, magazines are open for purpose of ventilation on "every fine day," is considerably modified.

13. The interior of a bombproof magazine with thick walls and a vaulted roof is commonly colder than the outside air in summer and warmer in winter. Winter is therefore the most favourable season for ventilation; but in the climate of England the exceptions to this rule are numerous, owing to the prevalence during winter of warm damp winds from the south and west, and during summer of cold dry winds from the north and east.

TABLE showing the Dew-point of the Air at different degrees of temperature, when the reading of the Wet Bulb of the Thermometer is from 1 to 10 degrees below that of the Dry Bulb.

| Temperature (Fahrenheit). | Dew-point when the Wet Bulb stands from 1° to 10° lower than the Dry Bulb. | | | | | | | | | |
|---|---|---|---|---|---|---|---|---|---|---|
| | 1° Lower. | 2° Lower. | 3° Lower. | 4° Lower. | 5° Lower. | 6° Lower. | 7° Lower. | 8° Lower. | 9° Lower. | 10° Lower. |
| 34 | 31¼ | 28½ | | | | | | | | |
| 36 | 33½ | 31 | 28½ | | | | | | | |
| 38 | 35½ | 33¼ | 30¾ | 28½ | | | | | | |
| 40 | 37¾ | 35½ | 33½ | 30¾ | 28½ | | | | | |
| 42 | 39¾ | 37½ | 35½ | 33¼ | 31 | 28¾ | | | | |
| 44 | 41¾ | 39½ | 37½ | 35¼ | 33 | 30¾ | 28½ | | | |
| 46 | 44 | 41¾ | 39¾ | 37½ | 35½ | 33½ | 31¼ | 29¼ | | |
| 48 | 46 | 43¾ | 41¾ | 39½ | 37½ | 35½ | 33¼ | 31¼ | 29 | |
| 50 | 48 | 45¾ | 43¾ | 41½ | 39½ | 37½ | 35¼ | 33¼ | 31 | 29 |
| 52 | 50 | 48 | 46 | 44 | 42 | 40 | 38 | 36 | 34 | 32 |
| 54 | 52 | 50 | 48 | 46 | 44 | 42 | 40 | 38 | 36 | 34 |
| 56 | 54 | 52 | 50 | 48 | 46 | 44 | 42 | 40 | 38 | 36 |
| 58 | 56 | 54¼ | 52¼ | 50½ | 48½ | 46½ | 44¾ | 42¾ | 41 | 39 |
| 60 | 58 | 56¼ | 54¼ | 52½ | 50½ | 48½ | 46¾ | 44¾ | 43 | 41 |
| 62 | 60 | 58¼ | 56¼ | 54½ | 52½ | 50½ | 48¾ | 46¾ | 45 | 43 |
| 64 | 62 | 60¼ | 58¼ | 56½ | 54½ | 52½ | 50¾ | 48¾ | 47 | 45 |
| 66 | 64¼ | 62½ | 60½ | 58¾ | 57 | 55¼ | 53½ | 51½ | 49¾ | 48 |
| 68 | 66¼ | 64½ | 62½ | 60¾ | 59 | 57¼ | 55½ | 53½ | 51¾ | 50 |
| 70 | 68¼ | 66½ | 64½, | 62¾ | 61 | 59¼ | 57½ | 55½ | 53¾ | 52 |
| 72 | 70¼ | 68½ | 66½ | 64¾ | 63 | 61¼ | 59½ | 57½ | 55¾ | 54 |
| 74 | 72¼ | 70½ | 69¼ | 67¼ | 65½ | 63¾ | 62 | 60½ | 58¾ | 57 |
| 76 | 74¼ | 72½ | 71 | 69¼ | 67½ | 65¾ | 64 | 62½ | 60¾ | 59 |
| 78 | 76¼ | 74½ | 73 | 71¼ | 69½ | 67¾ | 66 | 64½ | 62¾ | 61 |
| 80 | 78¼ | 76½ | 75 | 73¼ | 71½ | 69¾ | 68 | 66½ | 64¾ | 63 |
| 82 | 80¼ | 78½ | 77 | 75¼ | 73½ | 71¾ | 70 | 68½ | 66¾ | 65 |
| 84 | 82¼ | 80½ | 79 | 77¼ | 75½ | 73¾ | 72 | 70½ | 68¾ | 67 |
| 86 | 84¼ | 82½ | 81 | 79¼ | 77½ | 75¾ | 74 | 72½ | 70¾ | 69 |
| 88 | 86½ | 84¾ | 83¼ | 81½ | 80 | 78½ | 76¾ | 75¼ | 73½ | 72 |
| 90 | 88½ | 86¾ | 85¼ | 83½ | 82 | 80½ | 78¾ | 77¼ | 75½ | 74 |

[*Issued with Army Circulars, dated June*, 1873.]

REGULATIONS TO BE OBSERVED IN MAKING UP CARTRIDGES, FILLING SHELLS, AND EXAMINING AMMUNITION IN LABORATORIES IN ARTILLERY CHARGE.

1. By the term "Laboratory" is meant the block of buildings (with the passages and ways leading thereto) in which the examination of all ammunition will take place, cartridges made up, and shells filled. In most works suitable buildings for the purpose have been erected, consisting of a lobby with barrier at the entrance, and filling room for shells or cartridges, with hatches or openings for the admission and delivery of powder or filled cartridges and shells. Where no laboratory building exists a tent will be used.

2. Laboratory operations will be carried on under the superintendence of an officer, who must satisfy himself that the several men, as detailed

in § 3, understand the duties entrusted to them. The presence of an officer will not, however, be necessary at small detached forts or batteries in charge of master gunners where ammunition is made up for the auxiliary artillery.

3. The party for laboratory operations will be detailed as follows:—

Two men as magazine men, to issue powder in barrels from the magazine, and receive and stow cases or cylinders containing made-up cartridges.

Four men for conveying the powder barrels and cases with cartridges or filled shell to and from the laboratory. Two men will be sufficient if only one barrel of powder, or less, be required.

Eight men for the cartridge or shell-filling room, or less, if a small quantity of ammunition is to be made up. These men will be detailed to unhead the barrels, weigh out charges, make up cartridges or fill shells, as may be required.

Two men will be required at the entrance or receiving hatch, to pass in empty shells.

4. The men engaged in the laboratory will exchange their clothing and boots for laboratory clothing and slippers, in the place provided at the entrance, and will on no account re-pass the barrier without again changing their clothing.

5. Previous to being told off, the men must be warned to lay aside any knives, pipes, matches, or combustibles they may have about them. Any infringement of this rule will be dealt with as *disobedience of orders.*

6. The operations of filling shells and making up cartridges will never be carried on at the same time in the same room or tent.

7. The greatest attention will be paid to cleanliness in all parts of the laboratory and ways leading thereto; also in the wagons and barrows used for the conveyance of the powder or ammunition to or from the laboratory. Any loose grains of powder, dust, or grit will at once be swept up.

8. No barrow, tool, or tackle used outside will be admitted within the barrier at the entrance of the laboratory.

9. Nothing will be kept inside the barrier of the laboratory but the authorised articles for a laboratory, a proportion of clothing, also a supply of zinc cylinders for cartridges, which must be thoroughly examined previous to admission.

10. The shells to be filled will be piled on old shot or stone, outside the entrance to the laboratory or hatch, lettered "For admission of "shell," if there be one. The shells will be thoroughly cleaned and brushed externally before being passed into the laboratory—planks being laid down when the heavier shells are filled, in order to save the floor.

11. All shells, previous to filling, will be carefully searched internally, and all loose filings or pieces of lacquer removed.

12. All shells, up to the 9-inch inclusive, can be *up-ended* by hand on the blocks, for examination and filling. Heavier shells require the tackle and strap.

13. Made-up cartridges or filled shells will on no account be issued by the same door or hatch through which the loose powder or empty shells are passed in.

14. Powder barrels will be conveyed to the laboratory, and zinc cylinders containing filled cartridges to the magazine or cartridge stores, in barrows, in order to keep them free from dirt and grit.

15. The floor of the chamber in the laboratory, appropriated as a filling room, will be covered with hides when in use.

16. Empty powder barrels will be passed out at once, and stored in a clean place; and in the event of a barrel not being emptied, it will be re-headed and returned to the magazine.

17. Not more than the equivalent to two barrels of powder will be in the laboratory, or in transit between the magazine and the laboratory, at the same time.

18. A copy of these instructions attached to a board will be hung up in the entrance to the laboratory. Copies for this purpose, printed on foolscap, can be obtained on demand.

---

[*Issued with Army Circulars, dated June* 1873.]

REGULATIONS TO BE OBSERVED IN THE EMPLOYMENT OF LAMPS FOR LIGHTING MAGAZINES, AMMUNITION STORES, LABORATORIES, AND THEIR PASSAGES.

1. On no account will any but the authorized lamps be used for the purpose of lighting magazines, ammunition stores, laboratories, and their passages.

2. Such lamps only will be lighted from time to time as the officer commanding may direct.

3. A magazine copper lantern will be used for the inspection of the ammunition stores and underground passages.

4. One or more men, as may be required, will be specially detailed as "lampmen" *for each work*, to attend to all the lighting arrangements and stores connected therewith.

5. As all passage and wall lamps required for lighting the ammunition stores can be placed in position from the "light" passages, the lampmen will on no account pass beyond the junction of the "light" passages with the general passages.

6. When it is impossible to clean the glass of the light recess from the lamp passage, such glass must be cleaned by one of the magazine men from the inside. This in some cases may necessitate the unscrewing and removing of the frame; if so, care will be taken that it is properly replaced. This operation should be effected in the presence of the officer or non-commissioned officer in charge.

A.C. /77.
Cl. 162.

6*a*. Lamp barrows and trays are provided for the carriage of the lamps to and from the lamp room. Care must be taken that they are always used, and that the lamps are not placed on the ground or floor, as the glasses are thus broken.

7. All lamps, when not in use, will be kept in the lamp room.

8. A copy of these instructions attached to a board will be hung up in each lamp room. Copies for this purpose, printed on foolscap paper, can be obtained on demand.

9. Should any special instructions be required in any particular work for the guidance of the lampmen in the management of any peculiar lamp recesses, they should be added in manuscript.

## Instructions, Scraping, Gauging, Painting, and Piling Shot and Shell (A.C. 1868, Cl. 81 & 115).

1. Shot runs and trestles are to be supplied to all working parties employed in cleaning shot and shell. The shot or shell is to be placed on the run, and scraped with the swords until quite free from all paint and rust.

2. The plugs or corks of the shells are to be removed, and the interiors examined; those of spherical shells are to be scraped with a copper scraper, and freed from rust; as the interiors of rifled shell are lacquered, they are not to be scraped, but simply inverted to ensure their being free from water.

3. The collars should be examined, and, if necessary, removed and replaced by new ones.

4. The rivet holes should be examined, and, if necessary, cleaned out and re-waxed.

5. The threads of the screw plug are to be smeared with a solution of beeswax and rangoon oil,* and the plug screwed home into the metal bouche.

6. Previous to painting, the shot and shell are to be gauged with a high gauge by the non-commissioned officer in charge of the party, who is to put aside all that do not gauge properly, and report the circumstance to his Commanding Officer.

7. The shot or shell are to be painted on the runs with two coats of paint; the second coat is not to be given until the first coat has thoroughly set. Tarpaulins are to be supplied for placing over the projectiles after they have received the first coat.

8. The paint is to be applied over the whole of the exterior iron surface, including the portion inside the fuze-hole as far as the metal bouche, the metal plug being for the time removed; in the case of lead-coated projectiles the painting should extend at least half an-inch over the lead at either end of the projectile.

9. The bases for all piles should be firm and level, and formed of stone, concrete, or other hard material. Unserviceable shot or shell may be used.† Shot and shell should never be piled on sand, mud or loose shingle, even when garlands are used. Spherical shell are to be piled with their fuze-holes downwards. Projectiles for rifled ordnance should be piled on their sides, especial care being taken not to injure the lead-coating or the studs.

It is important to remember that mixed paint deteriorates by keeping; if kept too long it dries up and becomes useless. Cl. 173 A. C. 1872 states, "The paint and composition will be supplied from Woolwich on demand, prepared ready for use." In order, however, to ensure as far as possible that these articles may always be in good condition, they must not be demanded in larger quantities than will suffice for the requirements of each district for 12 months.

---

The chief constituents of the most important paints are spirits of turpentine, boiled oil, and some metallic body which is not liable to be oxodised or altered when exposed to the air.

---

* A mixture of equal parts of finely powdered chalk and cocoa nut oil is now used for this purpose.

† All spherical projectiles, from 32-pr. calibre upwards, make good bases. Wood is *not* suitable.

Spirits of turpentine possess the property of dissolving fatty and resinous substances. On exposure to the air it absorbs oxygen and hardens; this quality renders it suitable as an ingredient for paints.

Linseed oil takes up oxygen on exposure to air, and hardens; from this quality it is known as a drying oil. Its drying powers are much increased by boiling along with litharge, red lead, or binoxide of manganese; after this process it is known as boiled oil.

From the above properties it will be seen that paint when mixed will soon become unserviceable when exposed to the air. It is not desirable to mix paints long before they are wanted for use; even in closed tins they are apt to spoil.

An oxide of iron is the metallic body employed in the paint for projectiles; it is termed "Pulford's magnetic paint," (called "magnetic" from the property of being attracted by a magnet). Pulford's paint is obtained by contract.

Oxides of metal are suitable for the body of paint, as they are cheap, easily obtained in a state of fine division, and not subject to oxidation on exposure to the air. Other metallic bodies, such as white lead, &c., are largely used for paints, but are more expensive.*

---

* Much information on the subject of paints will be found in a lecture by Mr. Dent, Chemical Department, R.L., contained in Lectures on Building Materials, Chatham, 1871.

# Table of Combustible Compositions for Laboratory Stores.

**Carcass Composition :—**

| | lb. | oz. |
|---|---|---|
| Saltpetre, ground .. | 6 | 4 |
| Sulphur, ground.. .. | 2 | 8 |
| Rosin, pounded .. .. | 1 | 14 |
| Antimony, Sulphide of .. | 0 | 10 |
| Tallow, Russian .. .. | 0 | 10 |
| Turpentine, Venice .. | 0 | 10 |

**Smoke Ball :—**

| | lb. | oz. |
|---|---|---|
| Powder, L.G., bruised .. | 5 | 0 |
| Saltpetre, ground .. | 1 | 0 |
| Coal, Sea, pounded .. | 1 | 8 |
| Pitch, Swedish .. .. | 2 | 0 |
| Tallow, Russian .. .. | 0 | 8 |

## Portfires.

**Common Portfire :—**

| | lb. | oz. |
|---|---|---|
| Saltpetre, ground .. | 6 | 0 |
| Sulphur, ground .. | 2 | 0 |
| Powder, mealed, cylinder | 1 | 4 |

**Blue or Slow Portfire :—**

| | lb. | oz. |
|---|---|---|
| Water, distilled, from one to two quarts, according to the nature of the paper | | |
| Saltpetre, ground .. | 0 | 3 |

**Slow-burning Composition for Life-buoy Portfire :—**

| | lb. | oz. |
|---|---|---|
| Saltpetre, ground .. | 8 | 0 |
| Sulphur, ground .. | 4 | 0 |
| Powder, mealed, cylinder | 1 | 0 |

**Quick-burning Composition for Life-buoy Portfire :—**

| | lb. | oz. |
|---|---|---|
| Saltpetre, ground .. | 3 | 0 |
| Sulphur, ground .. | 2 | 0 |
| Powder, mealed, pit. .. | 1 | 0 |

## Lights.

**Ground Light Ball :—**

| | lb. | oz. | dr. |
|---|---|---|---|
| Saltpetre, ground .. | 6 | 4 | 0 |
| Sulphur, ground .. | 2 | 8 | 0 |
| Rosin, pounded .. | 1 | 14 | 0 |
| Oil, Linseed, boiled .. | 0 | 7 | 8 |

**Parachute Light Ball :—**

| | lb. | oz. |
|---|---|---|
| Saltpetre, ground .. | 7 | 0 |
| Sulphur, ground .. | 1 | 2 |
| Orpiment, red .. .. | 0 | 11 |

**Stars, 7-pr. Star Shell :—**

| | lb. | oz. |
|---|---|---|
| Nitrate of Baryta .. | 6 | 12 |
| Chlorate of Potash .. | 4 | 8 |
| Magnesium Powder .. | 6 | 0 |
| Boiled Oil, 3 per cent. | | |

**Long and Coast Guard Lights, and Lights for Wrecks :—**

| | lb. | oz. |
|---|---|---|
| Salpetre, ground .. | 7 | 0 |
| Sulphur, ground .. | 1 | 12 |
| Orpiment, red* .. .. | 0 | 8 |

**Signal Light, Magnesium :—**

| | oz. |
|---|---|
| Saltpetre, ground .. .. | 14 |
| Sulphur, ground .. .. | 3½ |
| Orpiment, red .. .. | 1 |
| Magnesium, containing 25 per cent. of Paraffin .. | 0½ |

## Fuze Composition.

**All Wood Fuzes except 5 seconds :—**

| | lb. | oz. |
|---|---|---|
| Saltpetre, ground .. | 3 | 4 |
| Sulphur, ground .. | 1 | 0 |
| Powder, mealed, pit .. | 2 | 12 |

**5 Seconds Fuzes :—**

| | lb. | oz. |
|---|---|---|
| Saltpetre, ground† .. | 0 | 1 |
| Powder, mealed, pit .. | 1 | 0 |

## Rockets.

**Hale Rocket :—**

| | 24-pr. Parts. | 9-pr. Parts. |
|---|---|---|
| Saltpetre, ground.. | 70 | 8·75 |
| Sulphur, ground .. | 16 | 2·0 |
| Charcoal, Alder, ground .. | 23 | 2·5 |

**Signal Rocket Composition :—**

| | lb. | oz. |
|---|---|---|
| Saltpetre, ground .. | 8 | 0 |
| Sulphur, ground .. | 2 | 0 |
| Charcoal, Dogwood .. | 3 | 0 |

* 1 oz. more Orpiment is added for Lights for Wrecks.

† Saltpetre only necessary at times, to regulate time of burning.

N.B. Previous to 1877 sublimed sulphur was used in many compositions. It is liable to contain free acid, which acts injuriously on the compositions. Ground sulphur is now used.

## Table of Combustible Compositions for Laboratory Stores—*continued*.

Composition for Stars of Signal Rocket:—

| | lb. | oz. |
|---|---|---|
| Saltpetre, ground .. | 8 | 0 |
| Sulphur, ground.. .. | 2 | 0 |
| Antimony, Sulphide of .. | 2 | 0 |
| Isinglass .. .. .. | 0 | 3¼ |
| Spirits, methylated .. | pint | 1 |
| Vinegar .. .. . | quart | 1 |
| Powder, L.G. mealed for priming .. | lb. 1 | oz. 0 |

### Match.

Slow Match:—

| | |
|---|---|
| Hemp, Yarn, pure, Russian .. .. .. | lb. 100 |
| Ashes, Wood .. | bushel 1 |
| Water .. .. | gals. 50 |

Quick Match:—

| | 4 threads. lb. | oz. | 6 threads. lb. | oz. | 10 threads. lb. | oz. |
|---|---|---|---|---|---|---|
| Cotton Wick | 1 | 10 | 2 | 2 | 2 | 7 |
| Gum Arabic | 0 | 8 | 0 | 9 | 0 | 10 |
| Powder, mealed, cylinder | 20 | 0 | 20 | 0 | 24 | 0 |
| Water, distilled | 8 pts. | | 9 pts. | | 10 pts. | |

### Coloured Fires.

| | | lb. | oz. | dr. |
|---|---|---|---|---|
| Blue | Chlorate of potash and copper .. | 4 | 8 | 0 |
| | Calomel .. .. | 2 | 10 | 0 |
| | Shellac, pale .. | 1 | 5 | 0 |
| | Chlorate of Potash | 4 | 14 | 0 |
| Red | Nitrate of Strontia | 2 | 8 | 0 |
| | Calomel .. .. | 1 | 9 | 8 |
| | Dextrine .. .. | 0 | 11 | 0 |
| | Shellac, pale .. | 0 | 9 | 0 |
| | Sulphide of Copper | 0 | 2 | 0 |
| | Chlorate of Potash | 2 | 7 | 6 |
| Green | Nitrate of Baryta . | 5 | 0 | 0 |
| | Calomel .. .. | 1 | 0 | 0 |
| | Shellac, pale .. | 0 | 2 | 0 |
| | Charcoal (120 mesh) | 0 | 3 | 0 |
| | Sulphur, sublimed | 1 | 8 | 0 |
| | Chlorate of Potash | 2 | 0 | 0 |

### Detonating Compositions.

#### Fuzes.

Detonating Composition for Boxer 5, 9, and 20 seconds Time-Fuze, B.L.:—

| | Parts. |
|---|---|
| Potash, Chlorate of .. | 6 |
| Antimony, Sulphide of .. | 4 |
| Mercury, Fulminate of .. | 4 |

Damped with varnish, of spirits, methylated, 1 pint, shellac 645 grains, in the proportion of 24 minims to 100 grains of composition.

Composition for Percussion Caps, E Time, R.L. and B.L. plain, and Sensitive Percussion Fuzes:—

| | lb. | oz. |
|---|---|---|
| Mercury, Fulminate of .. | 0 | 6 |
| Potash, Chlorate of .. | 0 | 6 |
| Antimony, Sulphide of .. | 0 | 4 |

Detonating Composition for Pettman Percussion Fuzes:—

| | Parts. |
|---|---|
| Potash, Chlorate of .. | 12 |
| Antimony, Sulphide of .. | 12 |
| Sulphur, ground .. | 1 |
| Powder, mealed, L.G. .. | 1 |

Damped with varnish, of spirits, methylated, 1 pint, shellac 112 grains, in the proportion of 40 minims to 100 grains of composition.

#### Tubes.

Detonating Composition for Copper Friction Tube:—

| | lb. | oz. |
|---|---|---|
| Potash, Chlorate of .. | 0 | 6 |
| Antimony, Sulphide of . | 0 | 6 |
| Sulphur, ground .. | 0 | 0½ |

Damped with spirits, methylated, 1 quart, shellac 824 grains, in the proportion of 200 minims to 1,000 grains of composition.

Detonating Composition for Quill Friction Tube:—*

| | lb. | oz. |
|---|---|---|
| Potash, Chlorate of .. | 0 | 6 |
| Antimony, Sulphide of .. | 0 | 6 |
| Sulphur, ground .. | 0 | 0½ |
| Powder, mealed.. .. | 0 | 0½ |
| Glass, ground .. .. | 0 | 0½ |

Damped with spirits, methylated, 1 quart, shellac 448 grains, in the proportion of 200 minims to 1,000 grains of composition.

For the Detonating Compositions used with Electric Tubes and Detonators, see p. .

* Previous to 6/75 there was 1 oz. mealed powder and no ground glass in this composition (§ 2785).

N.B. Previous to 1877 sublimed sulphur was used in many compositions. It is liable to contain free acid, which acts injuriously on the compositions. Ground sulphur is now used.

## Table of Paints and other Non-combustible Compositions for Laboratory Stores.

**1.—Pulford's Magnetic Black Paint for all Shot and Shell:—**

| | lb. | oz. |
|---|---|---|
| Pulford's .. .. | 112 | 0 |
| Litharge .. .. | 10 | 0 |
| Oil, boiled .. .. | qrts. 20 | |
| Turpentine, Spirits | „ | 12 |

**2.—White Paint for Tips of Palliser Shell:—**

| | | |
|---|---|---|
| Lead, white, ground | lbs. | 112 |
| Driers, patent .. | „ | 10½ |
| Oil, linseed, raw .. | pints | 8 |
| Turpentine .. .. | „ | 8 |

**3.—Red Paint for Heads of F.S. Shrapnel Shell:—**

| | | |
|---|---|---|
| Lead, red, dry .. | lbs. | 112 |
| „ white, ground | „ | 14 |
| Litharge .. .. | „ | 14 |
| Oil, linseed, boiled . | pints | 18 |
| Turpentine .. .. | „ | 4 |

**4.—Slate Colour Paint for Ground Light Ball:—**

| | lb. | oz. |
|---|---|---|
| Lead, white, ground | 28 | 0 |
| Black, Lamp .. | 2 | 0 |
| Litharge .. .. | 0 | 12 |
| Oil, Linseed, boiled | qrts. 2 | |
| Turpentine, Spirits .. | „ | 1 |

**5.—Yellow Paint for Smoke Ball:—**

| | lb. | oz. |
|---|---|---|
| Lead, white, ground | 5 | 0 |
| Lead, Sugar of .. | 0 | 4 |
| Chrome Yellow .. | 0 | 0¼ |
| Turpentine, Spirits .. | pint 1 | |
| Oil, Linseed, raw .. | „ | 0½ |

**6.—Luting:—**

| | |
|---|---|
| Tallow .. .. | parts 6 |
| Beeswax .. .. | „ 6 |

**7.—Kit Composition:—**

| | lb. | oz. |
|---|---|---|
| Pitch, Swedish .. | 6 | 14 |
| Tallow, Russian .. | 1 | 14 |
| Beeswax .. .. | 6 | 14 |
| Rosin .. .. .. | 7 | 8 |

**8.—Stone Colour for Brass Pentagon, Plain and Corrugated Rectangular Powder Cases, and Zinc Cylinders, § 2437:—**

| | lb. | oz. |
|---|---|---|
| Lead, white .. | 28 | 0 |
| Copperas .. .. | 0 | 12 |
| Umber .. .. | 0 | 8 |
| Varnish, Copal .. | qrts. 3 | |
| Gold Size .. .. | „ | 2 |
| Turpentine, Spirits | „ | 2 |

**9.—Black for B.L. plain Percussion Fuze, Bodies of Quill, and Copper Friction Tubes:—**

| | lb. | oz. |
|---|---|---|
| Black, Vegetable .. | 0 | 8 |
| Litharge .. .. | 0 | 8 |
| Varnish, Copal .. | qrts. 3 | |
| Turpentine, Spirits .. | „ | 2 |
| Gold Size .. .. | pints 1½ | |

**10.—Thick Brown Varnish:—**

| | lb. | oz. |
|---|---|---|
| Shellac, Gum .. .. | 16 | 0 |
| Spirits, methylated | gals. 2 | |

**11.—Thin Brown Varnish:—**

| | lb. | oz. |
|---|---|---|
| Shellac, Gum .. | 8 | 0 |
| Spirits, methylated .. | gals. 2 | |

**12.—Shellac Putty:—**

| | lb. | oz. |
|---|---|---|
| Whiting .. .. | 6 | 0 |
| Shellac, Gum .. | 2 | 0 |
| Spirits, methylated .. | qrt. 1 | |

**13.—Paste:—**

| | lb. | oz. |
|---|---|---|
| Flour .. .. .. | 2 | 0 |
| Alum, pounded .. | 0 | 1 |
| Water .. .. | gal. 1 | |

**14.—Flesh Colour for Common Portfire, &c.:—**

| | lb. | oz. |
|---|---|---|
| Lead, white, ground | 20 | 0 |
| Lead, red, dry .. | 0 | 4 |
| Shellac, Gum .. | 10 | 0 |
| Spirits, methylated | gals. 3 | |

**15.—Black for Plates of Life-buoy Portfire, &c.:—**

| | lb. | oz. |
|---|---|---|
| Black, Lamp .. .. | 1 | 0 |
| Litharge .. .. | 0 | 2 |
| Oil, Linseed, boiled .. | qrt. 1 | |
| Turpentine, Spirits .. | pint 1 | |

**16.—Drab Colour for Long, Signal, and Coast Guard Lights, and 1lb. Signal Coloured Rockets:—**

| | lb. | oz. |
|---|---|---|
| Lead, white, ground | 20 | 0 |
| Shellac, Gum .. | 10 | 0 |
| Spirits, methylated | gals. 3 | |

**17.—Black for Life-buoy Portfire and Body of Wood Fuze:—**

| | lb. | oz. |
|---|---|---|
| Black, Lamp .. | 2 | 0 |
| Shellac, Gum .. | 10 | 0 |
| Spirits, methylated | gals. 3 | |

## TABLE OF PAINTS AND OTHER NON-COMBUSTIBLE COMPOSITIONS FOR LABORATORY STORES—*continued.*

18.—Thick Black Paint for Ring round Cap of Wood Fuze:—

| | lb. | oz. |
|---|---|---|
| Shellac, Gum .. | 1 | 0 |
| Black, Lamp .. | 0 | 8 |
| Spirits, methylated .. | pints $1\frac{1}{2}$ | |

19.—List of Ingredients used for the Paint of M.L. and B.L. Wood Time Fuzes:—

| | | | lbs. |
|---|---|---|---|
| Body of Fuze | | Lead, white .. | 10 |
| | | Shellac .. | 5 |
| | | Spirits, methylated .. | gal. $1\frac{1}{2}$ |
| | | | lbs. |
| Distinguishing Colour, 5 sec. M.L. and B.L. | For body | Vermilion .. | 3 |
| | | Shellac .. | 4 |
| | | Spirits, methylated | gal. 1 |
| | For figures | Blue Ultramarine | oz. 4 |
| | | Oil, raw | pints 1 |
| | | Turpentine | gill $\frac{1}{2}$ |

20.—Blue for Parachute Fuze:—

| | lb. | oz. |
|---|---|---|
| Ultramarine .. | 0 | 4 |
| Lead, white, ground | 3 | 0 |
| Shellac, Gum .. | 1 | 0 |
| Spirits, methylated .. | gal. $\frac{1}{2}$ | |

21.—Varnish for the Gun and Silk over Detonating Ball of Pettman fuzes:—

| | lb. | oz. |
|---|---|---|
| Shellac, Gum .. | 0 | $8\frac{1}{2}$ |
| Spirits, methylated .. | qrt. 1 | |

22.—Lacquer for Brass Fuzes and other Brass Work:—

| | lb. | oz. |
|---|---|---|
| Seed lac .. .. | 5 | 0 |
| Turmeric .. .. | 2 | 8 |
| Spirits, methylated .. | gals. 5 | |

23.—Fuze Grease:—(See also 29.)

| | lb. | oz. |
|---|---|---|
| Tallow, Russian .. | 3 | 0 |
| Beeswax .. .. | 3 | 4 |
| Oil, Sweet .. .. | qrts. 2 | |

24.—Black Paint for Iron Burster:—

| | lb. | oz. |
|---|---|---|
| Black, Lamp .. | 3 | 0 |
| Litharge .. .. | 1 | 0 |
| Oil, Linseed, boiled .. | pints 3 | |
| Turpentine, Spirits .. | ,, 3 | |

The second coat is Brunswick Black.

25.—Lacquer for Inside of 100-pr. S.B. Shell and Shells for Rifled Guns:—

| | lb. | oz. |
|---|---|---|
| Rosin .. .. | 12 | 0 |
| Brown, Spanish .. | 2 | 0 |
| Plaster of Paris .. | 1 | 0 |
| Turpentine, Spirits | pint $\frac{1}{2}$ | |

26. Lubricant for Boxer Lubricating Wad, for Armstrong B.L. Gun:—

| | lb. | oz. |
|---|---|---|
| Tallow, Russian .. | 6 | 0 |
| Oil, Linseed, raw .. | 6 | 0 |

27.—Waterproof Varnish for Percussion caps:—

| | lb. | oz. |
|---|---|---|
| Shellac, Gum .. | 2 | 2 |
| Spirits, methylated .. | gal. 1 | |

28.—Red for Outside of Case of Hale's War, and Boxer Life-Saving Rockets, § 2441:—

| | lb. | oz. |
|---|---|---|
| Lead, Red .. .. | 3 | 0 |
| Lead, White .. | 1 | 0 |
| Litharge .. .. | 0 | 4 |
| Copperas .. .. | 0 | 2 |
| Oil, Linseed, boiled .. | pint 1 | |

29.—Anti-corrosive for Inside of Hale's War, and Boxer Life-Saving Rocket Cases, § 1940:—

| | |
|---|---|
| Copal Varnish .. | pint $\frac{1}{4}$ |
| Gold Size .. | ,, 1 |
| Turpentine, Spirits | ,, $1\frac{1}{4}$ |
| Lead, white, dry | lbs. 7 |

30.—White, for Case of Signal Rocket:—

| | lb. | oz. |
|---|---|---|
| Lead, white, ground | 12 | 0 |
| Blue, Prussian .. | 0 | $0\frac{1}{2}$ |
| Lead, Sugar of .. | 0 | 12 |
| Oil, Linseed, boiled | qrts. 2 | |
| Turpentine, Spirits | ,, 1 | |

31.—Composition for Lubricating Fuzes, Shell Plugs, &c., equal parts of:—*

Cocoa-nut Oil.
Finely Powdered Chalk.

32.—Cement, securing, Pettman G.S. Fuze against damp:—

| | lb. | oz. |
|---|---|---|
| Shellac .. .. | 7 | 8 |
| Spirits, methylated .. | gal. 1 | |
| Tar, Stockholm .. | ,, $\frac{1}{2}$ | |

To the above add equal parts, by weight, of Venetian Red.

* This composition is also used for the bright parts of B.L.R. guns, and it supersedes fuze grease.

## RING GAUGES FOR PROJECTILES,* § 1314.

### SMOOTH BORE.

CALIBRE OF 13-INCH MORTAR, 13·0 INCHES.

| | |
|---|---|
| H.G. 13-inch .. .. .. | 12·88 |
| L.G. 13-inch .. .. .. | 12·8 |

CALIBRE OF 10-INCH GUN, HOWITZER, AND MORTAR, 10·0 INCHES.

| | |
|---|---|
| H.G. 10-inch .. .. .. | 9·88 |
| L.G. 10-inch .. .. .. | 9·82 |

CALIBRE OF 100-PR. GUN, 9·0 INCHES.

| | |
|---|---|
| H.G. 100-pr. .. .. .. | 8·92 |
| L.G. 100-pr. .. .. .. | 8·88 |

CALIBRE OF 68-PR. GUN, 8·12 INCHES; 8-INCH GUN, 8·05 INCHES; 8-INCH HOWITZER AND MORTAR, 8·0 INCHES.

| | |
|---|---|
| H.G. 68-pr. .. .. .. | 7·95 |
| L.G. 68-pr. .. .. .. | 7·82 |

CALIBRE OF 56-PR. GUN, 7·65 INCHES.

| | |
|---|---|
| H.G. 56-pr. .. .. .. | 7·51 |
| L.G. 56-pr. .. .. .. | 7·45 |

CALIBRE OF 42-PR. GUN, 6·97 INCHES.

| | |
|---|---|
| H.G. 42-pr. .. .. .. | 6·795 |
| L.G. 42-pr. .. .. .. | 6·735 |

CALIBRE OF 32-PR. GUN, 6·41, 6·375, 6·35, AND 6·3 INCHES; 32-PR. HOWITZER, 6·3 INCHES.

| | |
|---|---|
| H.G. 32-pr. .. .. .. | 6·207 |
| L.G. 32-pr. .. .. .. | 6·147 |

CALIBRE OF 24-PR. GUN, 5·823; HOWITZER, 5·72; 5½-INCH HOWITZER, 5·68 5½-INCH MORTAR, 5·62 INCHES.

| | |
|---|---|
| H.G. 24-pr. .. .. .. | 5·639† |
| L.G. 24-pr. .. .. .. | 5·57 |

CALIBRE OF 18-PR. GUN, 5·292 AND 5·17 INCHES.

| | |
|---|---|
| H.G. 18-pr. .. .. .. | 5·124 |
| L.G. 18-pr. .. .. .. | 5·074 |

CALIBRE OF 12-PR. GUN, 4·623; 12-PR. HOWITZER, 4·58; 4⅖-INCH OR COEHORN HOWITZER AND MORTAR, 4·52 INCHES.

| | |
|---|---|
| H.G. 12-pr. .. .. .. | 4·54 |
| L.G. 12-pr. .. .. .. | 4·432 |

CALIBRE OF 9-PR. GUN, 4·2 INCHES.

| | |
|---|---|
| H.G. 9-pr. .. .. .. | 4·117 |
| L.G. 9-pr. .. .. .. | 4·06 |

CALIBRE OF 6-PR. GUN, 3·668 INCHES.

| | |
|---|---|
| H.G. 6-pr. .. .. .. | 3·585 |
| L.G. 6-pr. .. .. .. | 3·532 |

SEA SERVICE HAND GRENADE.

| | |
|---|---|
| H.G. hand grenade, S.S. .. | 3·496 |
| L.G. hand grenade, S.S... .. | 3·456 |

LAND SERVICE HAND GRENADE.

| | |
|---|---|
| H.G. hand grenade, L.S. .. | 2·778 |
| L.G. hand grenade, L.S... .. | 2·738 |

* "The issue of low gauges is to be restricted to stations of inspection." § 1314. "It is not intended to interfere with the limits of manufacture at present allowed in spherical projectiles, but merely to simplify the regulations for their inspection at out-stations." W.O.O. 31st Dec., 1866, being errata on Circular 11 (New Series).

† For future manufacture this gauge will be 5″·64,—W.O. Letter, 29/9/66, 75/12/2914.

### RING GAUGES FOR SAND SHOT.

| | |
|---|---|
| H.G. 4 lb. sand shot .. .. | 3·1 |
| L.G. 4 lb. sand shot.. .. .. | 3·05 |
| H.G. 3 lb. sand shot .. .. | 2·81 |
| L.G. 3 lb. sand shot.. .. .. | 2·77 |
| H.G. 2 lb sand shot.. .. .. | 2·46 |
| L.G. 2 lb. sand shot.. .. .. | 2·42 |
| H.G. 1½ lb. sand shot .. .. | 2·23 |
| L.G. 1½ lb. sand shot .. .. | 2·2 |
| H.G. 1 lb. sand shot .. .. | 1·95 |
| L.G. 1 lb. sand shot.. .. .. | 1·92 |
| H.G. 13⅛ oz. sand shot .. .. | 1·83 |
| L.G. 13⅛ oz. sand shot .. .. | 1·8 |
| H.G. 8 oz. sand shot .. .. | 1·55 |
| L.G. 8 oz. sand shot .. .. | 1·52 |
| H.G. 6½ oz. sand shot .. .. | 1·48 |
| L.G. 6½ oz. sand shot .. .. | 1·45 |
| H.G. 6 oz. sand shot .. .. | 1·4 |
| L.G. 6 oz. sand shot .. .. | 1·37 |
| H.G. 5 oz. sand shot .. .. | 1·34 |
| L.G. 5 oz. sand shot .. .. | 1·31 |
| H.G. 4 oz. sand shot .. .. | 1·23 |
| L.G. 4 oz. sand shot .. .. | 1·2 |
| H.G. 3¼ oz. sand shot .. .. | 1·16 |
| L.G. 3¼ oz. sand shot .. .. | 1·13 |
| H.G. 3 oz. sand shot .. .. | 1·12 |
| L.G. 3 oz. sand shot .. .. | 1·1 |
| H.G. 2 oz. sand shot .. .. | 0·98 |
| L.G. 2 oz. sand shot .. .. | 0·96 |
| H.G. 1½ oz. sand shot .. .. | 0·88 |
| L.G. 1½ oz. sand shot .. .. | 0·86 |

## GAUGES, SHELL AND SHOT, RIFLED, MUZZLE-LOADING.

| Nature. | For Stations of Inspection. | | | For Military Store Department, Garrison Brigades, and H.M. Ships in Commission. | | For Manufacture and Examination. | | | | | | | |
|---|---|---|---|---|---|---|---|---|---|---|---|---|---|
| | Cylinder, (Cast-Iron). | | Ring, Low over Studs (Wrought Iron, with wood handle). | Ring, Body and Studs (Cast-Iron). | | Cylinder, (Cast-Iron). | | Ring (Wrought-Iron). Diameter over Body. | | | | Ring (Wrought-Iron). Diameter over Studs. | |
| | Diameter over | | Diameter over | Diameter over | | Diameter over | | High Gauges. | | Low Gauges. | | | |
| | Body. | Studs. | Studs. | Body. | Studs. | Body. | Studs. | Palliser. | Common Shrapnel or Case Shot. | Palliser, Common or Shrapnel. | Case Shot. | High Gauge. | Low Gauge. |
| | Inches. | Inches. | Inches. | Inches. | Inches. | Inches. | Inches. | Inches. | Inches. | Inches. | Inches. | Inches. | Inches. |
| 12"·5 .. | 12·455 | 12·865 | .... | 12·455 | 12·865 | 12·455 | 12·865 | 12·44 | 12·43 | 12·41 | 12·33 | 12·855 | 12·845 |
| 12", 35-ton .. | 11·955 | 12·365 * | 12·345 | 11·955 | 12·365 | 11·955 | 12·365 | 11·94 | 11·93 | 11·91 | 11·83 | 12·355 | 12·345 |
| 12", 25-ton .. | 11·955 | 12·365 | | 11·955 | 12·365 | 11·955 | 12·365 | | | | | | |
| 11" .. .. | 10·955 | 11·365 | 11·345 | 10·955 | 11·365 | 10·955 | 11·365 | 10·94 | 10·93 | 10·91 | 10·83 | 11·355 | 11·345 |
| 10" .. .. | 9·955 | 10·365 * | 10·345 | 9·955 | 10·365 | 9·955 | 10·365 | 9·94 | 9·93 | 9·91 | 9·83 | 10·355 | 10·345 |
| 10" Howitzer .. | .. | .. | | .. | .. | .. | .. | — | | | | | |
| 9" .. .. | 8·955 | 9·325 | 9·305 | 8·955 | 9·325 | 8·955 | 9·325 | 8·94 | 8·93 | 8·91 | 8·84 | 9·315 | 9·305 |
| 8" .. .. | 7·955 | 8·325 * | 8·305 | 7·955 | 8·325 | 7·955 | 8·325 | 7·94 | 7·93 | 7·91 | 7·85 | 8·315 | 8·305 |
| 8" Howitzer .. | 7·955 | 8·325 | | 7·955 | 8·325 | 7·955 | 8·325 | — | | | | | |
| 7" .. .. | 6·955 | 7·325 | 7·305 | 6·955 | 7·325 | 6·955 | 7·325 | 6·94 | 6·93 | 6·91 | 6·86 | 7·315<br>7·115‡ | 7·305<br>7·105‡ |
| 80-pr... .. | 6·255 | 6·545 | 6·525 | 6·255 | 6·545 | 6·255 | 6·545 | — | 6·23 | 6·21 | 6·17 | 6·535 | 6·525 |
| 64-pr... .. | 6·255 | 6·485 | 6·465 | 6·255 | 6·485 | 6·255 | 6·485 | — | | | | 6·475 | 6·465 |
| 40-pr... .. | 4·73 | 4·942 | 4·925 | 4·73 | 4·942 | 4·7† | 4·935 | — | 4·7 | 4·68 | 4·66 | 4·935 | 4·925 |
| 25-pr... .. | 3·98 | 4·192 | 4·175 | 3·98 | 4·192 | 3·95† | 4·185 | — | 3·95 | 3·93 | 3·92 | 4·185 | 4·175 |
| 16-pr.. .. | 3·58 | 3·812 | 3·795 | 3·58 | 3·812 | 3·55† | 3·805 | — | 3·55 | 3·53 | 3·52 | 3·805 | 3·795 |
| 9-pr.. .. | 2·98 | 3·212 | 3·195 | 2·98 | 3·212 | 2·95† | 3·205 | — | 2·95 | 2·93 | 2·92 | 3·205 | 3·195 |
| 7-pr.. .. | 2·98 | 3·182 | 3·165 | 2·98 | 3·182 | 2·95† | 3·175 | — | | | | 3·175 | 3·165 |

* When two gauges are shown the same in diameter, they differ as regards rifling. † Plus 0"·01 for Asst.-Superintendent's gauges. ‡ Case shot.

## Table of Filled Cannon Cartridges.—Smooth-Bore.

| Nature. | Charge.—L.G. Powder. | Diameter of Cartridge Body. | Diameter of Ring Gauge | Length of Cartridge. | | Number of Hoops. | | Number Packed and Weight of Package. | | | | | | | | | | | | | | | | | | Remarks. |
|---|---|---|---|---|---|---|---|---|---|---|---|---|---|---|---|---|---|---|---|---|---|---|---|---|---|---|
| | | | | | | | | Barrel, Whole. | | Ammunition Box. | | Case, Powder, Copper-lined. | | | | | | Case, Powder, Brass. | | | | | | | | |
| | | | | | | | | | | | | Whole. | | Half. | | Quarter. | | Pentagon. | | | | Rectangular. | | | | |
| | | | | | | | | | | | | | | | | | | Whole. | | Sectional. | | Plain. | | Corrugated "A" | | |
| | | | | From | To | Braid. | Worsted. | Number. | Weight. | Number. | Weight. | Number. | Weight. | Number. | Weight. | Number. | Weight. | Number. | Weight. | Number. | Weight. | Number. | Weight. | Number. | Weight. | |
| | lb. oz. | in. | in. | in. | in. | | | | lb. | | lb. | | lb. | | lb. | | lb. | | lb. | | lb. | | lb. | | lb. | |
| 100-pr. ... | 25 0 | 8·25 | 8·25 | 13·5 | 14·5 | 3 | — | 3 | 109 | 1 | 47 | 3 | 124 | — | — | — | — | 3 | 137 | — | — | 4 | 171 | 4 | 151 | |
| | 20 0 | 8·25 | 8·25 | 12·0 | 12·5 | 2 | — | 4 | 115 | 2 | 62 | 4 | 129 | — | — | — | — | 4 | 142 | — | — | 5 | 171 | 6 | 171 | |
| | 12 0 | 8·25 | 8·25 | 8·0 | 8·5 | 1 | — | 7 | 119 | 4 | 70 | 9 | 158 | — | — | — | — | 8 | 159 | — | — | 8 | 168 | 9 | 160 | |
| 10″ ... | 12 0 | 7·75 | 7·76 | 8·5 | — | — | 3 | 7 | 119 | 4 | 70 | 9 | 158 | — | — | — | — | 9 | 171 | — | — | 10 | 192 | 12 | 196 | |
| | 8 0 | 7·4 | 7·76 | 7·0 | — | — | 1 | 12 | 132 | 5 | 63 | 14 | 162 | 6 | 78 | — | — | 14 | 175 | — | — | 15 | 192 | 17 | 188 | |
| 8″ ... ... | 10 0 | 7·0 | 7·2 | 10·2 | — | — | 3 | 9 | 125 | 4 | 62 | 11 | 160 | 4 | 70 | — | — | 11 | 172 | — | — | 12 | 192 | 14 | 192 | |
| 68-pr. ... | 18 0 | 7·75 | 7·76 | 12·8 | — | — | 3 | 5 | 125 | 2 | 58 | 6 | 158 | — | — | — | — | 6 | 171 | — | — | 6 | 180 | 7 | 177 | |
| | 16 0 | 7·55 | 7·76 | 11·5 | — | — | 3 | 6 | 131 | 3 | 70 | 6 | 145 | 2 | 62 | — | — | 6 | 158 | — | — | 7 | 183 | 8 | 179 | |
| | 12 0 | 7·5 | 7·76 | 9·5 | — | — | 2 | 8 | 132 | 4 | 70 | 9 | 158 | 4 | 78 | — | — | 9 | 171 | — | — | 10 | 192 | 12 | 196 | |
| | 8 0 | 7·3 | 7·76 | 7·4 | — | — | 1 | 12 | 132 | 5 | 63 | 14 | 162 | 6 | 78 | — | — | 14 | 175 | — | — | 16 | 200 | 17 | 188 | |
| 56-pr. ... ... | 14 0 | 7·0 | 7·2 | 11·6 | — | — | 3 | 6 | 119 | 3 | 65 | 8 | 161 | 3 | 71 | — | — | 8 | 174 | — | — | 9 | 198 | — | — | |
| 42-pr. ... | 14 0 | 6·6 | 6·67 | 12·3 | — | — | 3 | 6 | 119 | 3 | 65 | 8 | 161 | 3 | 71 | — | — | 8 | 174 | — | — | 9 | 198 | — | — | |
| | 12 0 | 6·6 | 6·67 | 11·2 | — | — | 3 | 8 | 132 | 4 | 70 | 9 | 158 | 4 | 78 | — | — | 9 | 171 | — | — | 10 | 192 | — | — | |
| | 10 0 | 6·6 | 6·67 | 10·2 | — | — | 3 | 9 | 130 | 4 | 64 | 11 | 164 | 4 | 72 | — | — | 11 | 177 | — | — | 12 | 198 | — | — | |

## Table of Filled Cannon Cartridges.—Smooth-Bore—*continued.*

| Nature. | Charge.—L.G. Powder. | Diameter of Cartridge Body. | Diameter of Ring Gauge | Length of Cartridges. | | Number of Hoops. | | Number Packed and Weight of Package. | | | | | | | | | | | | | | | | | | Remarks. |
|---|---|---|---|---|---|---|---|---|---|---|---|---|---|---|---|---|---|---|---|---|---|---|---|---|---|---|
| | | | | | | | | Barrel, Whole. | | Ammunition Box. | | Case, Powder, Copper-lined. | | | | | | Case, Powder, Brass. | | | | | | | | |
| | | | | | | | | | | | | Whole. | | Half. | | Quarter. | | Pentagon. | | | | Rectangular. | | | | |
| | | | | | | | | | | | | | | | | | | Whole. | | Sectional. | | Plain. | | Corrugated "A" | | |
| | | | | From | To | Braid. | Worsted. | Number. | Weight. | Number. | Weight. | Number. | Weight. | Number. | Weight. | Number. | Weight. | Number. | Weight. | Number. | Weight. | Number. | Weight. | Number. | Weight. | |
| | lb. oz. | in. | in. | in. | in. | | | | lb. | | lb. | | lb. | | lb. | | lb. | | lb. | | lb. | | lb. | | lb. | |
| 32-pr. ... | 10 0 | 6·0 | 6·09 | 10·75 | — | — | 3 | 9 | 125 | 4 | 62 | 11 | 160 | 4 | 70 | — | — | 11 | 172 | — | — | 12 | 192 | 14 | 192 | |
| | 8 0 | 6·0 | 6·09 | 9·2 | — | — | 2 | 12 | 132 | 5 | 63 | 14 | 162 | 6 | 78 | — | — | 14 | 175 | — | — | 16 | 200 | 18 | 196 | |
| | 7 0 | 6·0 | 6·09 | 8·0 | — | — | 2 | 13 | 126 | 7 | 72 | 16 | 162 | 7 | 78 | — | — | 16 | 175 | — | — | 18 | 198 | 21 | 199 | |
| | 6 0 | 5·9 | 6·09 | 7·5 | — | — | 2 | 16 | 131 | 8 | 71 | 19 | 164 | 8 | 78 | — | — | 19 | 177 | — | — | 21 | 198 | 24 | 196 | |
| | 5 0 | 5·8 | 6·09 | 6·5 | — | — | 2 | 19 | 130 | 10 | 73 | 22 | 160 | 10 | 80 | — | — | 22 | 173 | — | — | 26 | 202 | 30 | 202 | |
| | 4 0 | 5·8 | 6·09 | 5·75 | — | — | 1 | 25 | 136 | 12 | 71 | 27 | 158 | 13 | 82 | — | — | 27 | 171 | — | — | 32 | 201 | 36 | 196 | |
| 24-pr. ... | 8 0 | 5·5 | 5·52 | 10·5 | — | — | 3 | 12 | 131 | 5 | 63 | 14 | 162 | 6 | 78 | — | — | 14 | 175 | — | — | 16 | 200 | — | — | |
| | 6 0 | 5·5 | 5·52 | 8·3 | — | — | 2 | 16 | 131 | 8 | 71 | 19 | 164 | 8 | 78 | — | — | 19 | 177 | — | — | 21 | 198 | — | — | |
| | 5 0 | 5·5 | 5·52 | 7·5 | — | — | 2 | 19 | 130 | 10 | 73 | 22 | 160 | 10 | 80 | — | — | 22 | 173 | — | — | 26 | 202 | — | — | |
| | 4 0 | 5·4 | 5·52 | 6·5 | — | — | 2 | 25 | 136 | 12 | 71 | 27 | 158 | 13 | 82 | — | — | 27 | 171 | — | — | 32 | 201 | — | — | |
| | 3 0 | 5·2 | 5·52 | 5·2 | — | — | 1 | 32 | 132 | 18 | 77 | 37 | 161 | 17 | 81 | — | — | 37 | 174 | — | — | 43 | 202 | — | — | |
| 18-pr. ... | 6 0 | 5·0 | 5·02 | 9·0 | — | — | 3 | 16 | 131 | 8 | 71 | 19 | 164 | 8 | 78 | — | — | 19 | 177 | 7 | 85 | 21 | 198 | 25 | 202 | |
| | 3 0 | 4·9 | 5·02 | 5·75 | — | — | 1 | 32 | 132 | 18 | 77 | 37 | 161 | 17 | 81 | — | — | 37 | 174 | 17 | 94 | 43 | 202 | 52 | 202 | |
| 12-pr. ... | 4 0 | 4·35 | 4·39 | 8·6 | — | — | 3 | 25 | 136 | 12 | 71 | 27 | 158 | 13 | 82 | 5 | 39 | 27 | 171 | 12 | 91 | 32 | 201 | — | — | |
| | 2 0 | 4·3 | 4·39 | 6·0 | — | — | 2 | 37 | 130 | 18 | 69 | 44 | 161 | 20 | 81 | 8 | 39 | 44 | 174 | 20 | 94 | 51 | 201 | — | — | |

TABLE OF FILLED CANNON CARTRIDGES.—SMOOTH-BORE—*continued.*

| Nature. | Charge. — L.G. Powder. | Diameter of Cartridge Body. | Diameter of Ring Gauge | Length of Cartridge. | | Number of Hoops. | | Number Packed and Weight of Package. | | | | | | | | | | | | | | | | | | Remarks. |
|---|---|---|---|---|---|---|---|---|---|---|---|---|---|---|---|---|---|---|---|---|---|---|---|---|---|---|
| | | | | | | | | Barrel, Whole. | | Ammunition Box. | | Case, Powder, Copper-lined. | | | | | | Case, Powder, Brass. | | | | | | | | |
| | | | | | | | | | | | | Whole. | | Half | | Quarter. | | Pentagon. | | | | Rectangular. | | | | |
| | | | | | | | | | | | | | | | | | | Whole. | | Sectional. | | Plain. | | Corrugated "A" | | |
| | | | | From | To | Braid. | Worsted. | Number. | Weight. | Number. | Weight. | Number. | Weight. | Number. | Weight. | Number. | Weight. | Number. | Weight. | Number. | Weight. | Number. | Weight. | Number. | Weight. | |
| | lb. oz. | in. | in. | in. | in. | | | | lb. | | lb. | | lb. | | lb. | | lb. | | lb. | | lb. | | lb. | | lb. | |
| 9-pr. ... | 3 0 | 3·9 | 4·02 | 8·2 | — | — | 3 | 32 | 132 | 18 | 77 | 37 | 161 | 17 | 81 | 6 | 37 | 37 | 174 | 17 | 94 | 43 | 202 | — | — | |
| | 2 8 | 3·9 | 4·02 | 6·7 | — | — | 2 | 38 | 132 | 18 | 69 | 44 | 161 | 20 | 81 | 8 | 39 | 44 | 174 | 20 | 94 | 51 | 201 | — | — | |
| | 1 8 | 3·8 | 4·02 | 4·9 | — | — | 1 | 58 | 124 | 30 | 69 | 74 | 162 | 35 | 83 | 12 | 37 | 74 | 175 | 35 | 96 | 84 | 199 | — | — | |
| 6-pr. ... | 1 8 | 3·35 | 3·5 | 5·5 | — | — | 2 | 58 | 124 | 30 | 69 | 74 | 162 | 35 | 83 | 12 | 37 | 74 | 175 | 35 | 96 | 84 | 199 | — | — | |
| | 1 0 | 3·35 | 3·5 | 4·4 | — | — | 1 | 94 | 134 | 42 | 65 | 110 | 162 | 50 | 81 | 20 | 39 | 110 | 175 | 50 | 94 | 125 | 199 | — | — | |
| 13″ Mortar | 20 0 | 8·35 | — | 12·75 | — | — | 3 | 4 | 115 | 2 | 62 | 5 | 150 | — | — | — | — | 4 | 142 | — | — | 5 | 172 | — | — | |
| | 16 0 | 8·2 | — | 12·0 | — | — | 3 | 5 | 115 | 3 | 70 | 6 | 145 | — | — | — | — | 6 | 158 | — | — | 7 | 183 | — | — | |
| | 9 0 | 7·85 | — | 7·3 | — | — | 1 | 10 | 125 | 5 | 67 | 12 | 158 | — | — | — | — | 11 | 162 | — | — | 12 | 180 | — | — | |
| 10″ ,, | 9 8 | 6·5 | — | 9·7 | — | — | 3 | 10 | 131 | 4 | 60 | 12 | 164 | 5 | 78 | — | — | 11 | 167 | — | — | 12 | 186 | — | — | |
| | 4 0 | 6·0 | — | 5·5 | — | — | 1 | 25 | 136 | 12 | 71 | 27 | 158 | 13 | 82 | — | — | 27 | 171 | — | — | 32 | 201 | — | — | |
| 8″ ,, ... | 2 0 | 4·9 | — | 5·0 | — | — | 1 | 48 | 133 | 24 | 72 | 55 | 161 | 25 | 81 | 10 | 39 | 55 | 175 | 25 | 94 | 64 | 202 | — | — | |
| 5½″ ,, ... | 0 7 | 3·1 | — | 3·1 | — | — | 1 | 220 | 135 | 120 | 77 | 240 | 157 | 120 | 84 | 50 | 40 | 240 | 170 | 120 | 96 | 280 | 197 | — | — | |
| 4⅖″ ,, ... | 0 5 | 2·6 | — | 2·9 | — | — | 1 | 260 | 122 | 160 | 76 | 300 | 147 | 160 | 82 | 70 | 41 | 300 | 160 | 150 | 91 | 350 | 184 | — | — | |
| 10″ Howitzer | 7 0 | 7·15 | 7·76 | 6·3 | — | — | 1 | 13 | 126 | 7 | 72 | 16 | 162 | 7 | 78 | — | — | 16 | 175 | — | — | 18 | 198 | — | — | |
| | 4 0 | 6·25 | 7·76 | 5·3 | — | — | 1 | 25 | 136 | 12 | 71 | 27 | 158 | 13 | 82 | — | — | 27 | 171 | — | — | 32 | 201 | — | — | |
| 8″ ,, | 4 0 | 5·75 | 6·67 | 6·0 | — | — | 1 | 25 | 136 | 12 | 71 | 27 | 158 | 13 | 82 | — | — | 27 | 171 | — | — | 33 | 201 | — | — | |
| | 3 0 | 5·2 | 6·67 | 5·1 | — | — | 1 | 32 | 132 | 18 | | 37 | 161 | 17 | 81 | — | — | 37 | 174 | — | — | 43 | 202 | — | — | |

## TABLE OF FILLED CANNON CARTRIDGES.—SMOOTH-BORE—*continued.*

| Nature. | Charge. — L.G. Powder. | Diameter of Cartridge Body. | Diameter of Ring Gauge | Length of Cartridge. | | Number of Hoops. | | Number Packed and Weight of Package. | | | | | | | | | | | | | | | | | | Remarks. |
|---|---|---|---|---|---|---|---|---|---|---|---|---|---|---|---|---|---|---|---|---|---|---|---|---|---|---|
| | | | | | | | | Barrel, Whole. | | Ammunition Box. | | Case, Powder, Copper-lined. | | | | | | Case, Powder, Brass. | | | | | | | | |
| | | | | | | | | | | | | Whole. | | Half. | | Quarter. | | Pentagon. | | | | Rectangular. | | | | |
| | | | | | | | | | | | | | | | | | | Whole. | | Sectional. | | Plain. | | Corrugated "A" | | |
| | | | | From | To | Braid. | Worsted. | Number. | Weight. | Number. | Weight. | Number. | Weight. | Number. | Weight. | Number. | Weight. | Number. | Weight. | Number. | Weight. | Number. | Weight. | Number. | Weight. | |
| | lb. oz. | in. | in. | in. | in. | | | | lb. | | lb. | | lb. | | lb. | | lb. | | lb. | | lb. | | lb. | | lb. | |
| 5½″ Howitzer | 2 0 | 4·15 | 5·1 | 5·2 | — | — | 1 | 48 | 133 | 24 | 72 | 55 | 161 | 25 | 81 | 10 | 39 | 55 | 175 | 25 | 94 | 64 | 202 | — | — | |
| 4⅖″ ,, | 0 8 | 2·2 | 2·2 | 4·6 | — | — | 1 | 200 | 136 | 110 | 78 | 230 | 167 | 110 | 86 | 40 | 39 | — | — | — | — | — | — | — | — | |
| | 0 4 | 2·2 | 2·2 | 3·5 | — | — | 1 | 350 | 126 | 180 | 70 | 400 | 154 | 200 | 83 | 70 | 37 | — | — | — | — | — | — | — | — | |
| 32-pr. ,, | 3 0 | 4·3 | 4·39 | 6·8 | — | — | 2 | 32 | 132 | 18 | 77 | 37 | 161 | 17 | 81 | 7 | 40 | 37 | 174 | 17 | 94 | 43 | 202 | — | — | |
| | 2 0 | 4·3 | 4·39 | 5·4 | — | — | 1 | 48 | 133 | 24 | 72 | 55 | 161 | 25 | 81 | 10 | 39 | 55 | 175 | 25 | 94 | 64 | 202 | — | — | |
| 24-pr. ,, | 2 8 | 4·75 | 5·1 | 5·6 | — | — | 1 | 37 | 130 | 13 | 69 | 44 | 161 | 20 | 81 | 8 | 39 | 44 | 174 | 20 | 94 | 51 | 201 | 60 | 202 | A conical papier mâché gauge is used in the Royal Laboratory. |
| | 1 8 | 4·2 | 5·1 | 4·6 | — | — | 1 | 58 | 124 | 30 | 69 | 74 | 162 | 35 | 83 | 12 | 37 | 74 | 175 | 35 | 96 | 84 | 199 | 96 | 197 | |
| 12-pr. ,, | 2 0 | 3·9 | 4·2 | 6·6 | — | — | 2 | 48 | 133 | 24 | 72 | 55 | 161 | 25 | 81 | 10 | 39 | 55 | 175 | 25 | 94 | 64 | 202 | — | — | |
| | 1 4 | 3·5 | 4·2 | 5·3 | — | — | 1 | 77 | 134 | 38 | 71 | 88 | 161 | 40 | 81 | 16 | 39 | 88 | 175 | 40 | 94 | 102 | 201 | — | — | |
| | 1 0 | 3·4 | 4·2 | 4·5 | — | — | 1 | 94 | 134 | 42 | 65 | 110 | 162 | 50 | 81 | 20 | 39 | 110 | 175 | 50 | 94 | 125 | 199 | — | — | |
| 68-pr. Carronade | 5 0 | 6·3 | — | 6·3 | — | — | 1 | 19 | 130 | 10 | 73 | 22 | 160 | 10 | 80 | — | — | 22 | 173 | — | — | 26 | 202 | — | — | |
| 42-pr. ,, | 3 8 | 5·7 | — | 5·3 | — | — | 1 | 27 | 131 | 13 | 69 | 32 | 163 | 14 | 79 | — | — | 32 | 176 | — | — | 36 | 199 | — | — | |
| 32-pr. ,, | 2 11 | 5·15 | — | 5·2 | — | — | 1 | 33 | 126 | 18 | 72 | 42 | 164 | 18 | 79 | — | — | 42 | 177 | — | — | 45 | 195 | — | — | |
| 24-pr. ,, | 2 0 | 4·7 | — | 4·6 | — | — | 1 | 48 | 133 | 24 | 72 | 55 | 161 | 25 | 81 | 10 | 39 | 55 | 175 | 25 | 94 | 64 | 202 | — | — | |
| 18-pr. ,, | 1 8 | 4·1 | — | 4·4 | — | — | 1 | 58 | 124 | 30 | 69 | 74 | 162 | 35 | 83 | 12 | 37 | 74 | 175 | 35 | 96 | 84 | 199 | — | — | |
| 12-pr. ,, | 1 0 | 3·8 | — | 3·4 | — | — | 1 | 94 | 134 | 42 | 65 | 110 | 162 | 50 | 81 | 20 | 39 | 110 | 175 | 50 | 94 | 125 | 199 | — | — | |
| 6-pr. ,, | 0 10 | 3·1 | — | 3·4 | — | — | 1 | 130 | 121 | 72 | 69 | 160 | 153 | 72 | 76 | 30 | 30 | 160 | 166 | 70 | 88 | 190 | 195 | — | — | |

TABLES showing quantities of Braid, Silk, Thread, and Worsted required to make up 100 of each of the undermentioned Cartridges.

**Muzzle-loading.**

| Description. | | Serge Cartridges. Making, Empty. Braid, Worsted. Broad. | Serge Cartridges. Making, Empty. Braid, Worsted. Narrow. | Serge Cartridges. Making, Empty. Worsted, White, No. 20. | Serge Cartridges. Filling and Completing. Worsted White, No. 14. | Silk Cloth Cartridges. Making, Empty. Braid, Silk Cloth. Broad. | Silk Cloth Cartridges. Making, Empty. Braid, Silk Cloth. Narrow. | Silk Cloth Cartridges. Making, Empty. Silk Twist. | Silk Cloth Cartridges. Making, Empty. Silk, Sewing. | Silk Cloth Cartridges. Filling and Completing. Silk Twist. |
|---|---|---|---|---|---|---|---|---|---|---|
| | | yards. | yards. | oz. | oz. | yards. | yards. | oz. | oz. | oz. drs. |
| 12″ (35-ton) ... | 110 lb. P. | 2460 | ... | 9 | $5\frac{1}{2}$ | 2460 | ... | $7\frac{1}{2}$ | ... | 1 4 |
| 12″ (35 or 25-ton) ... | 85 ,, P. or 67 lb. R.L.G. | 1800 | ... | 7 | $5\frac{1}{2}$ | 1800 | ... | $5\frac{3}{4}$ | ... | 1 4 |
| 12″ (25-ton) ... ... | 55 ,, P. or 50 lb. R.L.G. | 1050 | ... | 6 | $5\frac{1}{2}$ | 1050 | ... | $5\frac{1}{4}$ | ... | 1 4 |
| 11″ ... ... ... | 85 ,, P. or 70 lb. R.L.G. | 1940 | ... | 7 | $4\frac{1}{2}$ | 1940 | ... | $6\frac{3}{4}$ | ... | 1 4 |
| | 60 ,, P. or 50 lb. R.L.G. | 1400 | ... | $6\frac{1}{2}$ | $4\frac{1}{2}$ | 1400 | ... | $5\frac{1}{2}$ | ... | 1 4 |
| 10″ ... ... ... | 70 ,, P. or 60 lb. R.L.G. | 1680 | ... | $6\frac{1}{2}$ | $4\frac{1}{2}$ | 1680 | ... | $5\frac{1}{2}$ | ... | 1 4 |
| | 44 ,, P. or 40 lb. R.L.G. | 1025 | ... | 6 | $4\frac{1}{2}$ | 1025 | ... | 5 | ... | 1 4 |
| 9″ ... ... | 50 ,, P. or 43 lb. R.L.G. | 1475 | ... | 6 | 4 | 1475 | ... | $5\frac{1}{4}$ | ... | 1 3 |
| | 30 ,, R.L.G.* | ... | 820 | $4\frac{1}{2}$ | 4 | ... | 820 | $4\frac{1}{4}$ | ... | 1 3 |
| | 15 ,, Blank, R.L.G. or L.G. | ... | 370 | 3 | 4 | ... | 370 | 3 | ... | 1 1 |
| 8″ ... ... ... | 35 ,, P. or 30 lb. R.L.G. | ... | 1415 | $5\frac{1}{2}$ | 4 | ... | 1415 | $5\frac{1}{4}$ | ... | 1 3 |
| | 20 ,, R.L.G.* | ... | 570 | $4\frac{1}{2}$ | 4 | ... | 570 | $4\frac{1}{4}$ | ... | 1 1 |
| | 12 ,, Blank, R.L.G. or L.G. | ... | 450 | 3 | 4 | ... | 450 | 3 | ... | 1 1 |
| 8″ Howitzer... ... | 10 ,, R.L.G.* | ... | 320 | $4\frac{1}{2}$ | $3\frac{1}{2}$ | ... | ... | ... | ... | ... |
| | 5 ,, R.L.G.* | ... | 120 | 4 | $3\frac{1}{2}$ | ... | ... | ... | ... | ... |
| | $2\frac{1}{2}$,, R.L.G.* | ... | ... | 3 | $3\frac{1}{2}$ | ... | ... | ... | ... | ... |
| 7″ ... ... ... | 30 ,, P. or 22 lb. R.L.G. | ... | 1200 | 5 | $3\frac{1}{2}$ | ... | 1200 | $4\frac{3}{4}$ | ... | 1 3 |
| | 14 ,, R.L.G.* | ... | 538 | $3\frac{1}{2}$ | $3\frac{1}{2}$ | ... | 538 | $3\frac{1}{4}$ | ... | 1 1 |
| 7″ or 32-pr. ... ... | 10 ,, Blank, R.L.G. or L.G. | ... | ... | 3 | $16\frac{1}{2}$ | ... | ... | ... | ... | ... |
| 80-pr. ... ... ... | 10 ,, L.G. | ... | 360 | 3 | $3\frac{1}{2}$ | ... | ... | ... | ... | ... |
| 64-pr. ... ... .. | 12 ,, L.G. | ... | 450 | 4 | $3\frac{1}{2}$ | ... | ... | ... | ... | ... |
| | 10 ,, L.G. | ... | 360 | 3 | $3\frac{1}{2}$ | ... | ... | ... | ... | ... |
| | 8 ,, L.G. | ... | 278 | $2\frac{1}{2}$ | $3\frac{1}{2}$ | ... | ... | ... | ... | ... |
| 64-pr. or 32-pr. ... | 6 ,, L.G. | ... | ... | 2 | $8\frac{13}{16}$ | ... | ... | ... | ... | ... |
| 64-pr., 80-pr. or 32-pr. | 5 ,, Blank, R.L.G. or L.G. | ... | ... | ... | ... | ... | ... | ... | $\frac{3}{4}$ | 2 8 |
| 40-pr. ... ... ... | 7 ,, L.G. | ... | 500 | 4 | $3\frac{1}{2}$ | ... | ... | ... | ... | ... |
| 25 pr. .. ... ... | 4 ,, L.G.‡ | ... | 400 | $3\frac{1}{2}$ | $2\frac{1}{2}$ | ... | ... | ... | $1\frac{1}{4}$ | 3 0 |
| 16-pr. ... ... ... | 3 ,, L.G.‡ † | ... | 320 | 3 | $2\frac{1}{2}$ | ... | ... | ... | $\frac{14}{16}$ | 3 6 |
| | $1\frac{1}{2}$,, Blank, R.L.G. or L.G. | ... | ... | ... | ... | ... | ... | ... | $\frac{3}{4}$ | 1 10 |
| 9-pr. ... ... ... | $1\frac{3}{4}$,, L.G.‡ | ... | 255 | 2 | 2 | ... | ... | ... | ... | ... |
| | $1\frac{1}{2}$,, L.G.‡ | ... | 255 | 2 | 2 | ... | ... | ... | ... | ... |
| | 1 ,, Blank, R.L.G. or L.G. | ... | ... | ... | ... | ... | ... | ... | $\frac{10}{16}$ | 1 4 |
| 7-pr. ... ... ... | 12 oz. F.G.‖ | ... | 150 | $1\frac{1}{2}$ | $1\frac{1}{4}$ | ... | ... | ... | ... | ... |
| | 8 ,, F.G.‖ | ... | 50 | 1 | $1\frac{1}{4}$ | ... | ... | ... | ... | ... |
| | 6 ,, F.G.‖ | ... | 50 | 1 | $1\frac{1}{4}$ | ... | ... | ... | ... | ... |
| | 4 ,, F.G.‖ (Shalloon) | ... | 50 | $\frac{3}{4}$ | $1\frac{1}{4}$ | ... | ... | ... | ... | ... |

**Breech-loading.**

| Description. | | Serge Cartridges. Making, Empty. Braid, Worsted, Blue, Narrow. | Serge Cartridges. Making, Empty. Worsted, White, No. 20. | Serge Cartridges. Making, Empty. Worsted, White, No. 14. | Serge Cartridges. Making, Empty. Silk, Sewing. | Serge Cartridges. Making, Empty. Thread, Pack, Kitted. | Serge Cartridges. Filling and Completing. Worsted, White, No. 14. | Serge Cartridges. Filling and Completing. Thread, 3 Cord. |
|---|---|---|---|---|---|---|---|---|
| | | yds. | oz. | oz. | oz | oz. | oz. | oz |
| 7″ | 11 lb. L.G. | 450 | $4\frac{1}{2}$ | ... | ... | 7 | ... | ... |
| | 10 ,, L.G. | 450 | $4\frac{1}{2}$ | ... | ... | 7 | ... | ... |
| | 7 ,, Blank, R.L.G. or L.G. | 200 | 3 | ... | ... | ... | $3\frac{1}{2}$ | ... |
| 40-pr. | 5 ,, S.S. L.G. | 365 | 4 | ... | ... | 6 | ... | ... |
| | 5 ,, L.S. L.G. | 438 | 4 | ... | ... | ... | ... | $1\frac{1}{2}$ |
| | 3 ,, Blank, R.L.G. or L.G. | 142 | $2\frac{1}{2}$ | ... | ... | ... | $2\frac{1}{2}$ | ... |
| 20-pr. | $2\frac{1}{2}$ ,, L.G. | 340 | 3 | ... | ... | ... | ... | $1\frac{1}{2}$ |
| | $1\frac{1}{2}$ ,, Blank, R.L.G. or L.G. | 128 | $2\frac{1}{2}$ | ... | ... | ... | 2 | ... |
| 12-pr. | $1\frac{1}{2}$ ,, L.G. | 215 | 2 | ... | ... | ... | ... | $1\frac{1}{2}$ |
| 12 or 9-pr. | 1 ,, Blank, R.L.G. or L.G. | 320 | 1 | 2 | $\frac{6}{16}$ | ... | 4 | ... |
| 9-pr. | $1\frac{2}{16}$ lb. L.G. | 160 | 2 | ... | ... | ... | ... | $1\frac{1}{2}$ |
| 6-pr. | 12 oz. L.G. | 144 | 2 | ... | ... | ... | ... | 1 |
| | 10 ,, Blank, R.L.G. or L.G. | 156 | 1 | ... | $\frac{6}{16}$ | ... | 2 | ... |

* L.G. is to be used for land service.

† Serge to be used until the present supply is exhausted.

‖ R.F.G. may be substituted for F.G.

‡ R.L.G. when issued to R.A. or R.M.A. and S. of G.

## SMOOTH-BORE ORDNANCE.

### CHARGES, BURSTING,—APPROXIMATE.*

| Nature of Shell. | Description of Shell. | | | | | | Number of Calico and Paper Bags. |
|---|---|---|---|---|---|---|---|
| | Common. | Naval. | Mortar. | Shrapnel Diaphragm. | Hand Grenades. Sea Service | Hand Grenades. Land Service | Diaphragm. |
| | lb. oz. dr. | lb. oz. dr. | lb. oz. dr. | drs. | oz. | oz. | No. |
| 13-inch .. .. | — | — | 10 15 0 | — | — | — | — |
| 10-inch .. .. | 6 12 0 | 6 5 0 | 5 4 0 | — | — | — | — |
| 8-inch, or 68 pr... | 2 9 0 | 2 9 0 | 2 9 0 | 80 | — | — | — |
| 100-pr. .. .. | — | 3 13 0 | — | 96 | — | — | — |
| 56-pr. .. .. | 2 7 0 | — | — | 70 | — | — | — |
| 42-pr. .. .. | 1 12 0 | — | — | 60 | — | — | — |
| 32-pr. .. .. | 1 5 0 | 1 5 0 | — | 50 | — | — | — |
| 24-pr., or 5½-inch.. | 1 0 0 | — | 1 0 0 | 40 | — | — | 3 |
| 18-pr. .. .. | 0 12 0 | — | — | 30 | — | — | 2 |
| 12-pr., or 4⅖-inch.. | 0 7 0 | — | 0 7 0 | 24 | — | — | 2 |
| 9-pr. .. .. | — | — | — | 13 | — | — | 1 |
| 6-pr. .. .. | — | — | — | 10 | 5 | — | 1 |
| 3-pr. Hand Grenade .. | — | — | — | — | — | 3 | — |

* All smooth-bore shells are now filled by capacity instead of by weight, and the charges here given, except in the case of Shrapnel, are taken from W.O.C. 927, the shells being filled in accordance with § 954; also W.O.C. 884; and Royal Artillery Circular Memo., 13th December, 1864, paragraph 3; "the shell being tapped with a mallet during the process."

The approximate amount of powder required was determined by an experiment in the Royal Laboratory in 1865, when ten shells of each nature were filled, and the quantities given above are slightly in excess of the mean result of this experiment, to give even weights; and "it is assumed that as shells of small and large capacity are supplied in about equal proportions, the powder saved in the one case will suffice to make up the deficiency in the others."—W.O.C. 927.

An allowance must be made for displacement by the fuze.—See also § 1116.

Thus the 13-inch mortar shell will take approximately 10 lb. 15 oz. of powder without a fuze, and 10 lb. 14 oz. 8 dr. with a fuze. In the case of the Shrapnel shell the charge is the minimum sufficient to open them, and is either weighed or measured. For field service the weighed or measured charge is issued in paper and calico bags as above.

## Nature and Number of Empty Cannon Cartridges, Packed in a Bale.

### R.M.L. Ordnance.

| Nature of Cartridge. | Flannel. Number in bale. | Silk Cloth. Number in bale. | Nature of Cartridge. | Flannel. Number in bale. | Silk Cloth. Number in bale. |
|---|---|---|---|---|---|
| 12″ 110 lbs. | 100 | 100 | | | |
| 12″ 85 or 67 lbs. | 100 | 100 | | | |
| 12″ 55 or 50 ,, | 150 | 150 | 6″·3 Howr. 80 or 64-pr., 10 lbs. | 400 | 400 |
| 11″ 85 or 70 ,, | 100 | 100 | 64-pr. 12 lbs. .. | 400 | 400 |
| 11″ 60 or 50 ,, | 150 | 100 | 64-pr. 8 ,, .. | 600 | 600 |
| 10″ 70 or 60 ,, | 150 | 100 | 64-pr. 6 ,, .. | 600 | 600 |
| 10″ 44 or 40 ,, | 200 | 150 | 40-pr. 7 ,, .. | 600 | 600 |
| 9″ 50 or 43 ,, | 150 | 150 | 25-pr. 4 ,, .. | 800 | 800 |
| 9″ 30 .. ,, | 250 | 250 | 16-pr. 3 ,, .. | 800 | 800 |
| 9″ 15 .. ,, | 400 | 400 | 16-pr. 1½ ,, .. | — | 1,000 |
| 8″ 35 or 30 ,, | 200 | 200 | 9-pr. 1¾ ,, .. | 1,000 | 1,000 |
| 8″ 20 .. ,, | 300 | 300 | 9-pr. 1½ ,, .. | 1,000 | 1,000 |
| 8″ 12 .. ,, | 400 | 400 | 9-pr. 1 ,, .. | — | 1,000 |
| 7″ 30 or 22 ,, | 250 | 250 | 7-pr. 12 oz. .. | 2,000 | 2,000 |
| 7″ 14 .. ,, | 400 | 300 | 7-pr. 8 ,, .. | 2,000 | 2,000 |
| 7″ 10 .. ,, | 400 | 400 | 7-pr. 6 ,, .. | 2,000 | 2,000 |
| 10″ Howr. | | | 7-pr. 4 ,, .. | 2,000 | 2,000 |
| 8″ Howr. 10 lbs. .. | — | 600 | | | |
| 8″ Howr. 5 ,, .. | — | 800 | | | |
| 8″ Howr. 2½ ,, .. | — | 800 | | | |

### R.B.L. Ordnance.

| Nature of Cartridge. | Flannel. Number in bale. | Silk Cloth. Number in bale. | Nature of Cartridge. | Flannel. Number in bale. | Silk Cloth. Number in bale. |
|---|---|---|---|---|---|
| 7″ 11 lbs. .. | 500 | — | 12-pr. 1½ lbs. .. | 1,000 | — |
| 7″ 10 ,, .. | 500 | — | 12 or 9-pr. 1 ,, .. | 800 | — |
| 7″ 7 ,, .. | 500 | — | 9-pr. $1\frac{2}{16}$ ,, .. | 1,000 | — |
| 40-pr. 5 ,, .. | 600 | — | 6-pr. 12 oz. .. | 1,500 | — |
| 40-pr. 3 ,, .. | 800 | — | 6-pr. 10 ,, .. | 1,000 | — |
| 20-pr. 2½ ,, .. | 800 | — | | | |
| 20-pr. 1½ ,, .. | 1,000 | — | | | |

### S.B. Ordnance.

| Nature of Cartridge. | Flannel. Number in bale. | Silk Cloth. Number in bale. | Nature of Cartridge. | Flannel. Number in bale. | Silk Cloth. Number in bale. |
|---|---|---|---|---|---|
| To contain 25 lbs. .. | 250 | — | To contain 5 lbs. .. | 800 | 800 |
| ,, 20 ,, .. | 300 | — | ,, 4 ,, .. | 800 | — |
| ,, 18 ,, .. | 400 | — | ,, 3 ,, .. | 800 | 800 |
| ,, 16 ,, .. | 400 | — | ,, 2½ ,, .. | 1,000 | — |
| ,, 14 ,, .. | 400 | — | ,, 2 ,, .. | 1,000 | 1,000 |
| ,, 12 ,, .. | 400 | — | ,, 1½ ,, .. | 1,000 | 1,000 |
| ,, 10½ ,, .. | 600 | — | ,, 1¼ ,, .. | 1,500 | — |
| ,, 10 ,, .. | 600 | — | ,, 1 ,, .. | 1,500 | 1,000 |
| ,, 8 ,, .. | 700 | 700 | ,, 0¼ ,, .. | 3,500 | — |
| ,, 7 ,, .. | 700 | — | Spare powder bags to contain 15 lbs. | 300 | — |
| ,, 6 ,, .. | 700 | 700 | | | |

**PROPORTION per Cent. of PROJECTILES and FUZES issued for each nature of Rifled Gun, Garrison Service, Army Equipment, 1876.**

| Nature of Gun. | | Projectiles. Shell. Common. | Projectiles. Shell. Double. | Projectiles. Shell. Shrapnel. | Projectiles. Shell. Segment. | Projectiles. Shell. Palliser. | Projectiles. Shot. Palliser. | Projectiles. Shot. Case. | Fuzes.* Percussion. Pettman's G.S. Sea Fronts only. | Fuzes.* Percussion. R.L. II. Land Fronts only. | Fuzes.* Time. 5 seconds. Land Fronts only. | Fuzes.* Time. 9 seconds. Sea Fronts. | Fuzes.* Time. 9 seconds. Land Fronts. | Fuzes.* Time. 20 seconds. Sea Fronts. | Fuzes.* Time. 20 seconds. Land Fronts. |
|---|---|---|---|---|---|---|---|---|---|---|---|---|---|---|---|
| R.M.L. | 12″·5 | — | — | — | — | — | — | — | — | — | — | — | — | — | — |
| | 12″ 35 tons | 20 | — | — | — | 77 | — | 3 | 20 | — | — | — | — | — | — |
| | 12″ 25 ,, | 20 | — | — | — | 42 | 35 | 3 | 20 | — | — | — | — | — | — |
| | 11″ | 20 | — | — | — | 77 | — | 3 | 20 | — | — | — | — | — | — |
| | 10″ | 20 | — | — | — | 42 | 35 | 3 | 20 | — | — | — | — | — | — |
| | 9″ | 31 | — | — | — | 32 | 32 | 5 | 31 | — | — | — | — | — | — |
| | 7″ | 26 | 5 | — | — | 32 | 32 | 5 | 31 | — | — | — | — | — | — |
| | 80-pr. | 60 | — | 30 | — | — | — | 10 | †60 | †60 | ‡ | ‡ | ‡ | — | — |
| | 64-pr. | 60 | — | 30 | — | — | — | 10 | †60 | †60 | ‡ | ‡ | ‡ | — | — |
| R.B.L. | 7″ | 60 | — | — | 30 | — | — | 10 | 60 | 60 | — | 30 | 20 | 20 | 20 |
| | 40-pr. | 60 | — | — | 30 | — | — | 10 | 60 | 60 | — | 30 | 20 | 20 | 20 |

* 1 cylinder only of each nature to be kept with each gun in time of peace.
† For common shell.
‡ 1 per Shrapnel, and 10 per cent. spare, ⅓ being 5 sec., the remainder 9 sec. fuzes.

## ANALYSIS OF METAL.

| Proportions according to Specifications. | |
|---|---|
| No. 1.—Copper .. .. | 90·96 |
| Tin .. .. .. | 8·27 |
| Zinc .. .. .. | 0·77 |
| *Pettman L.S.*, body, top, and bottom plug. | |
| *Pettman G.S.*, body and top plug. | |
| No. 2.—Copper .. .. | 87·5 |
| Tin .. .. .. | 12·5 |
| *Pettman L.S. and G.S.*, steady, and cone plugs and ball. | |
| No. 3.—Pure Lead .. .. | 100·0 |
| *Pettman L.S. and G.S. cups.* | |
| No. 4.—Copper .. .. | 77·11 |
| Tin .. .. .. | 1·20 |
| Zinc .. .. .. | 19·27 |
| Lead .. .. .. | 2·40 |
| *E, Time fuze*, body, cap, screw plug, pellet (II. and III.) and nut (II. and III.) | |
| No. 5.—Lead .. .. .. | 50·0 |
| Tin .. .. .. | 50·0 |
| Pellet of B.L. plain, and R.L. percussion fuzes. | |
| No. 6.—Copper .. .. | 90·90 |
| Tin .. .. .. | 9·09 |
| Studs, R.M.L. projectiles, *16, 40 and 25-pr. and 7″ and over. | |

| Proportions according to Specifications. | |
|---|---|
| No. 7.—Copper .. .. | 9·786 |
| Zinc .. .. | ·214 |
| Studs, 80-pr. R.M.L. | |
| No. 8.—Studs for Shrapnel shell for 25-pr. and 16-pr., and studs for 9 and 7-pr. projectiles— | |
| Copper .. .. | 100·0 |
| No. 9.—Copper .. .. | 5·77 |
| Tin .. .. .. | 93·30 |
| Zinc .. .. | 1·93 |
| Detonator, B.L. fuzes, Plug, M.L. fuzes. | |
| No. 10.—Copper .. .. | 86·96 |
| Tin .. .. | 3·62 |
| Zinc .. .. | 5·80 |
| Lead .. .. | 3·62 |
| B.L. plain fuze, body, bottom plug, and guard. | |
| No. 11.—Copper .. .. | 84·51 |
| Tin .. .. | 4·93 |
| Zinc .. .. | 5·63 |
| Lead .. .. | 4·93 |
| R.L. percussion fuze, body, bottom plug, guard. | |
| Also for Sensitive fuze, body, bottom, pellets, and detonating cap. | |

* The studs for the 64-pr. R.M.L. are stamped from rods of pure copper. See also No. 8.

Cylinders, Tin, for Fuzes, Tubes, &c., &c., arranged in consecutive order according to the Numbers on their Bases; and showing the purposes for which the different Cylinders are used.

| No. on Base. | Description of Articles issued packed in them. |
|---|---|
| 1 | Caps, percussion. |
| 2 | ,, ,, |
| 3 | ,, ,, |
| 4 | ,, ,, |
| 5 | Fuzes, time, Boxer, R.B.L., 5 and 9 seconds. |
| 6 | ,, ,, ,, ,, 20 ,, |
| 7 | ,, ,, ,, common; and Primers, vent piece. |
| 8 | ,, ,, ,, diaphragm. |
| 9 | ,, electric, Abel (No. 1). |
| 10 | ,, time, Boxer, R.M.L., 5 and 9 seconds. |
| 11 | ,, ,, ,, ,, 20 ,, |
| 12 | Lights; Coast Guard, and their Primers. |
| 13 | ,, long, general service; and Portfires, life-saving apparatus |
| 14 | Caps, percussion; and Lights, signal, magnesium. |
| 15 | Tubes, friction, copper, long. |
| 16 | ,, ,, ,, service. |
| 17 | ,, ,, 7-pr. |
| 18 | Fuzes, Bickford; and Tubes, friction, quill, long, with loops. |
| 19 | Tubes, friction, quill, short, with loops. |
| 20 | Fuzes, Pettman; and wax, bees', for Harvey torpedo. |
| 21 | Discs, paper, tubes, electric, drill (No. 11); and Primers, gun-cotton. |
| 22 | Fuzes, electric, platinum wire (No. 3); and Submarine (No. 2).* |
| 23 | Detonators, electric, Abel (No. 5). Obsolete, see No. 32. |
| 24 | ,, ,, platinum wire (No. 7); and Submarine (No. 6). |
| 25 | ,, Bickford fuze (No. 8). |
| 26 | Caps, percussion; and Lights, long, general service. |
| 27 | Tubes, friction, copper, long, with wire loops. |
| 28 | ,, quill, common and friction, for time guns. |
| 29 | Sensitive fuzes. |
| 30 | Washers, leather, for plugs, base and fuze; and for disconnectors. |

* These fuzes being now obsolete, the cylinder is to contain fuzes, electric, No. 14.

## Cylinders, Tin, for Fuzes, Tubes, &c., &c.—*continued.*

| No on Base. | Description of Articles issued packed in them. |
|---|---|
| 31 | Detonators (service and special), exploding bolt of Harvey torpedo. |
| , | Primers, for Lights, Portfires, and Shrapnel shell. |
| 32 | Tubes, electric, Abel (No. 4); Detonator, electric (No, 5), |
| 33 | Fuzes, metal, percussion, R.L. |
| 34 | Tubes, electric, gun, low tension, naval service (No. 10). |
| 35 | Primers exploding bolt of Harvey torpedo. |
| 36 | Packing ,, ,, ,, ,, |
| 37 | Washers, leather, exploding bolt and loading hole of Harvey torpedo. |
| 38 | ,, ,, ,, ,, of Harvey torpedo |
| 39 | Tubing, india-rubber, $\frac{3}{16}$ and $\frac{1}{4}$ inch |
| 40 | ,, ,, ,, $\frac{4}{16}$ and $\frac{3}{8}$ ,, |
| 41 | ,, ,, ,, $\frac{1}{2}$ inch |
| 42 | Detonators, electric, naval service (No. 9). |
| 43 | Tubes, electric, (No. 11), dummy, drill. |
| 44 | Cards, with platinum-silver wire, for test coil. |
| 45 | Tape, india-rubber, one 8 oz. coil. |
| 46 | Detonators for Pistol, Whitehead torpedo |
| 47 | Detonators, electric, low tension (Nos. 12 and 13). |
| 48 | Solution, india-rubber, 8 oz. |
| 49 | Detonators (No. 8), with Bickford fuze for Cavalry Pioneers. |
| 50 | Primers, gun-cotton, Whitehead torpedo (1). |
| 51 | Primers, gun-cotton, 2 oz. (5). |
| 52 | ,, ,, ,, 1 oz. (5). |
| 53 | Discs, gun-cotton, 1 oz. in waterproof bags for Cavalry Pioneers. |
| 54 | Detonators (No. 8), with Bickford fuze, for Cavalry Pioneers (25). |
| 55 | Tallow for jointers, 8 oz. |
| 56 | Tubing, india-rubber, for jointers. |
| 57 | Tape ,, ,, ,, ,, |

N.B.—This table has been corrected up to date, but is liable to frequent modification and additions. It is useful to refer to in the event of the printed labels being defaced or missing,

# INDEX.

## A.



## B.



* This sign shows that the store in question is illustrated by a wood cut.
† This sign shows that the store in question is illustrated by a plate at end of book.

## C.



* This sign shows that the store in question is illustrated by a wood cut.
† This sign shows that the store in question is illustrated by a plate at end of book.

* This sign shows that the store in question is illustrated by a wood cut.
† This sign shows that the store in question is illustrated by a plate at end of book.

## D.



## E.



* This sign shows that the store in question is illustrated by a wood cut.
† This sign shows that the store in question is illustrated by a plate at end of book.

## F.



* This sign shows that the store in question is illustrated by a wood cut.
† This sign shows that the store in question is illustrated by a plate at end of book.

## G.



* This sign shows that the store in question is illustrated by a wood cut.
† This sign shows that the store in question is illustrated by a plate at end of book,

* This sign shows that the store in question is illustrated by a wood cut.
† This sign shows that the store in question is illustrated by a plate at end of book.

## H.



## I.



## K.



* This sign shows that the store in question is illustrated by a wood cut.
† This sign shows that the store in question is illustrated by a plate at end of book.

---

* This sign shows that the store in question is illustrated by a wood cut.
† This sign shows that the store in question is illustrated by a plate at end of book.

## P.



* This sign shows that the store in question is illustrated by a wood cut.
† This sign shows that the store in question is illustrated by a plate at end of book.

* This sign shows that the store in question is illustrated by a wood cut.
† This sign shows that the store in question is illustrated by a plate at end of book.

## R.



* This sign shows that the store in question is illustrated by a wood cut.
† This sign shows that the store in question is illustrated by a plate at end of book.

* This sign shows that the store in question is illustrated by a wood cut.
† This sign shows that the store in question is illustrated by a plate at end of book.

## T.



* This sign shows that the store in question is illustrated by a wood cut.
† This sign shows that the store in question is illustrated by a plate at end of book.

## U.



## V.



## W.



* This sign shows that the store in question is illustrated by a wood cut.
† This sign shows that the store in question is illustrated by a plate at end of book.

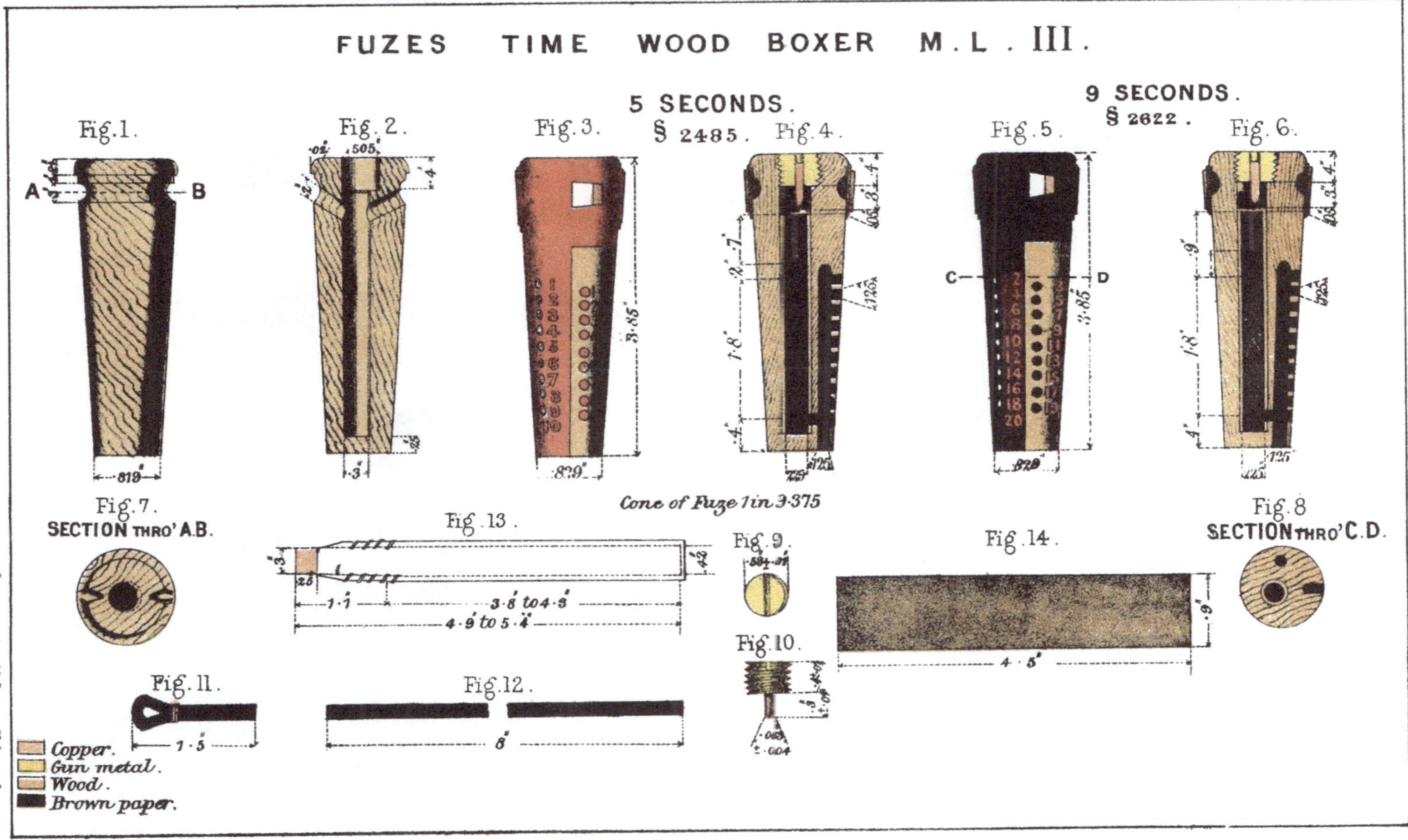

FUZES TIME WOOD BOXER M.L. III.
5 SECONDS.
§ 2485.
9 SECONDS.
§ 2622.
Fig. 1.
Fig. 2.
Fig. 3.
Fig. 4.
Fig. 5.
Fig. 6.
A
B
C
D
Cone of Fuze 1 in 3·375
Fig. 7.
SECTION THRO' A.B.
Fig. 13.
Fig. 9.
Fig. 14.
Fig. 8
SECTION THRO' C.D.
Fig. 10.
Fig. 11.
Fig. 12.
3·8 to 4·8
4·9 to 5·4
1·5
8
4·5
Copper.
Gun metal.
Wood.
Brown paper.
Dangerfield Lith. 22. Bedford St. Covent Garden

# FUZE PERCUSSION PETTMAN GENERAL SERVICE

## I

## §§ 1235 & 1634.

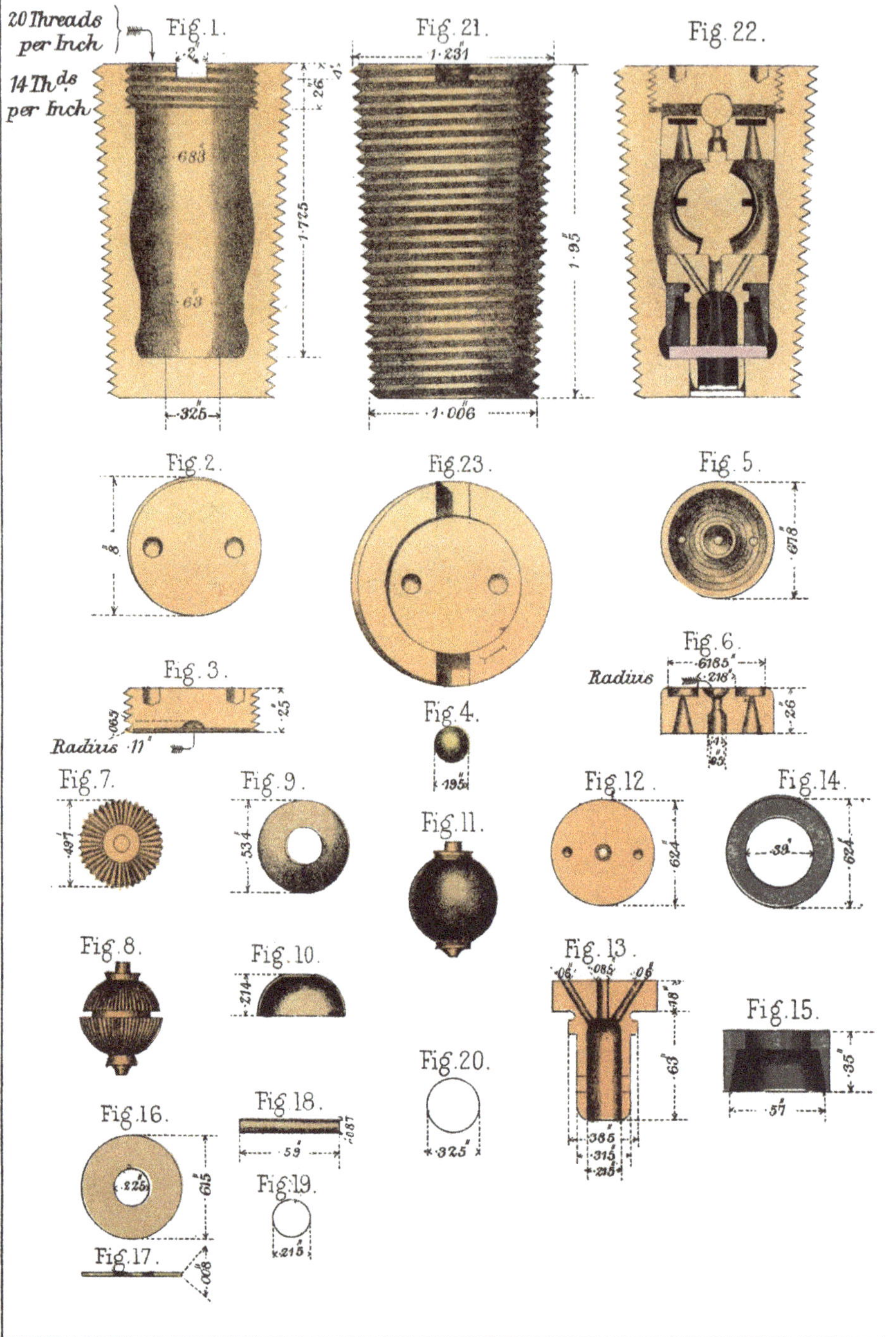


Dangerfield Lith. 22, Bedford St Covent Garden

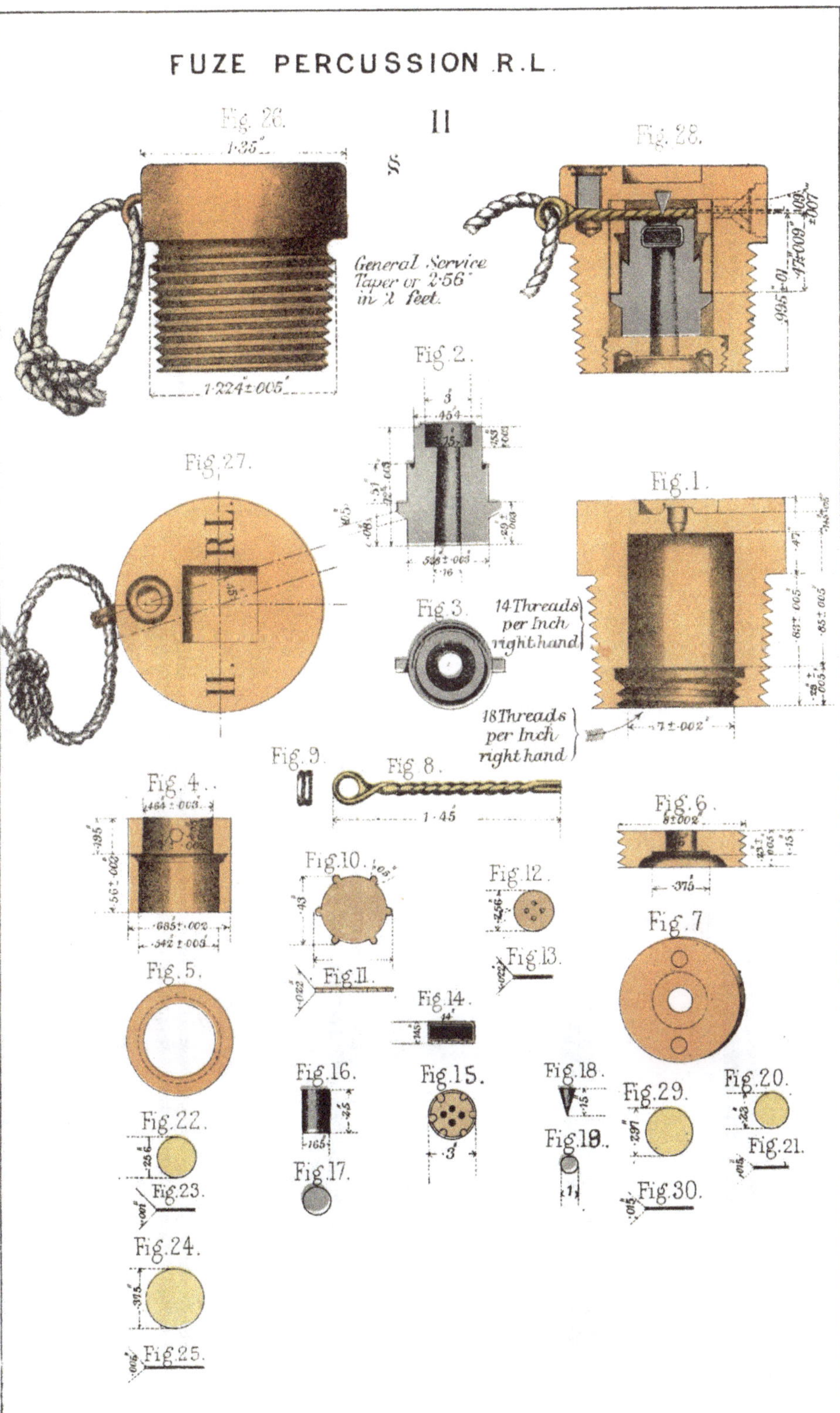


Dangerfield Lith. 22. Bedford St. Covent Garden.

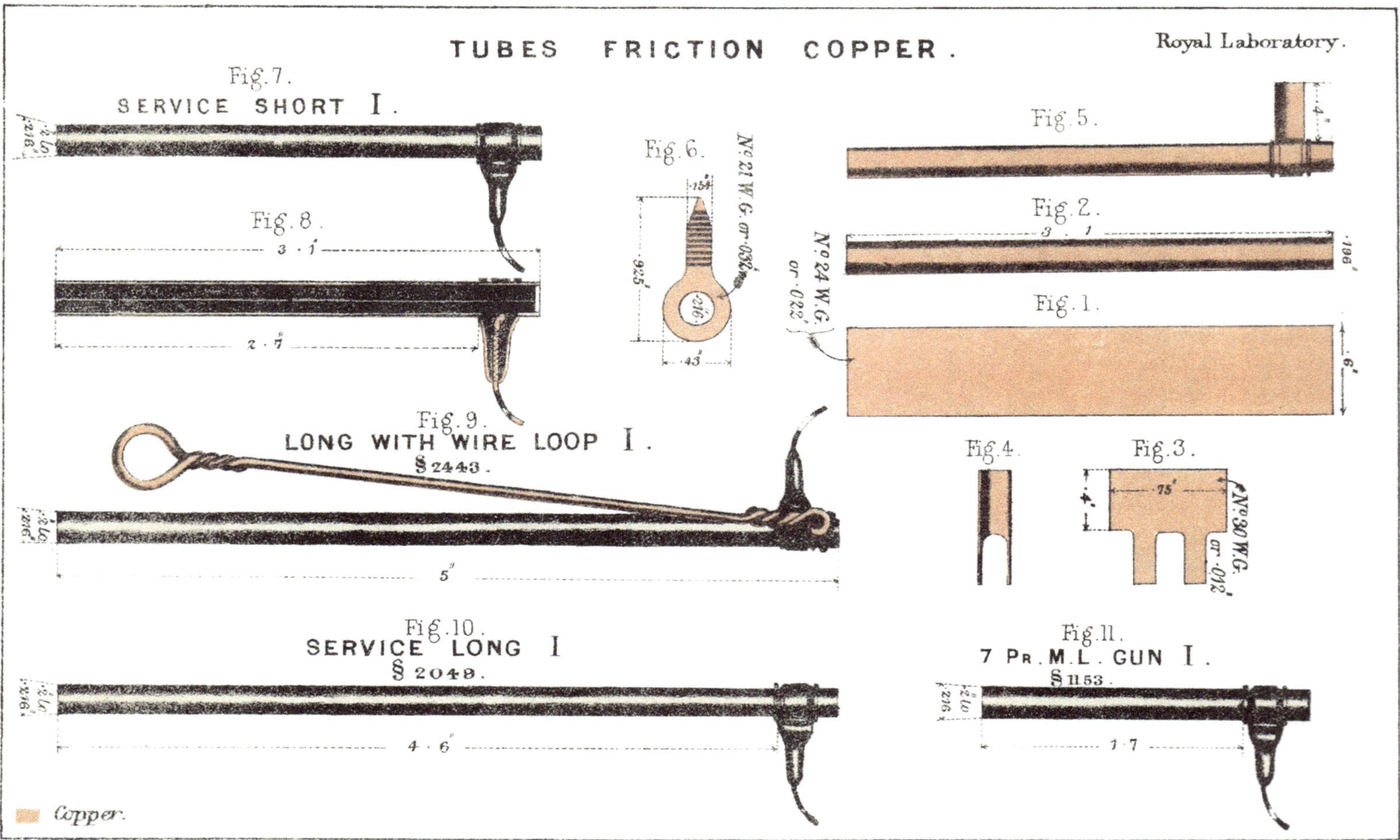

Dangerfield Lith. 22, Bedford St. Covent Garden

## SHELL RIFLED BREECH LOADING SEGMENT 12 PR

I.

SCALE ½.

*Weight 10.lbs 8.ozs ± 4.ozs*

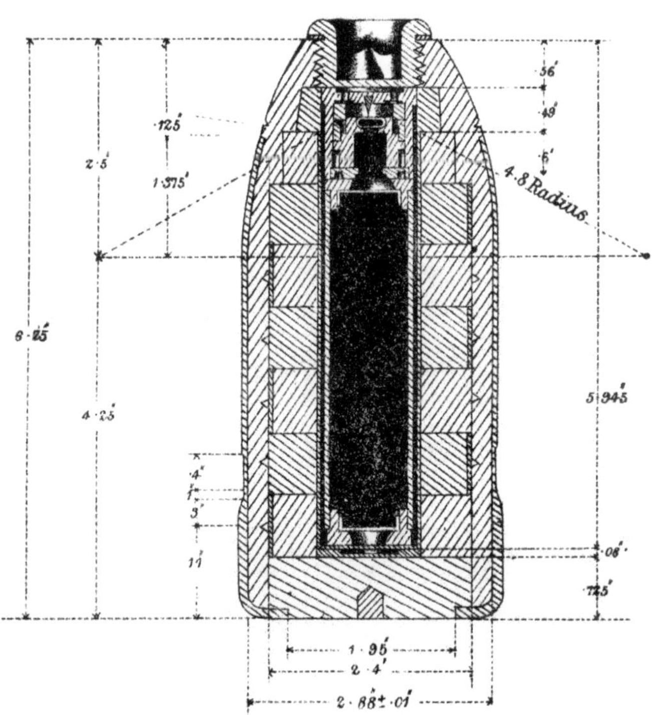


*Diameters High Low*

*Back End....3·074....3·067"*

*Body..........3·034...3·024"*

*Commt. of taper 3·015"*

*is not to pass on the Body.*

Dangerfield Lith. 22 Bedford St Covent Garden

SHELL RIFLED BREECH LOADING COMMON 12 PR.

III

*Scale ½*

23 · 2 · 74

| WEIGHT | LBS | OZS |
|---|---|---|
| CAST IRON | 10 - | 12 |
| POWDER | | 8 |
| TOTAL | 11 - | 4±.4 OZS |

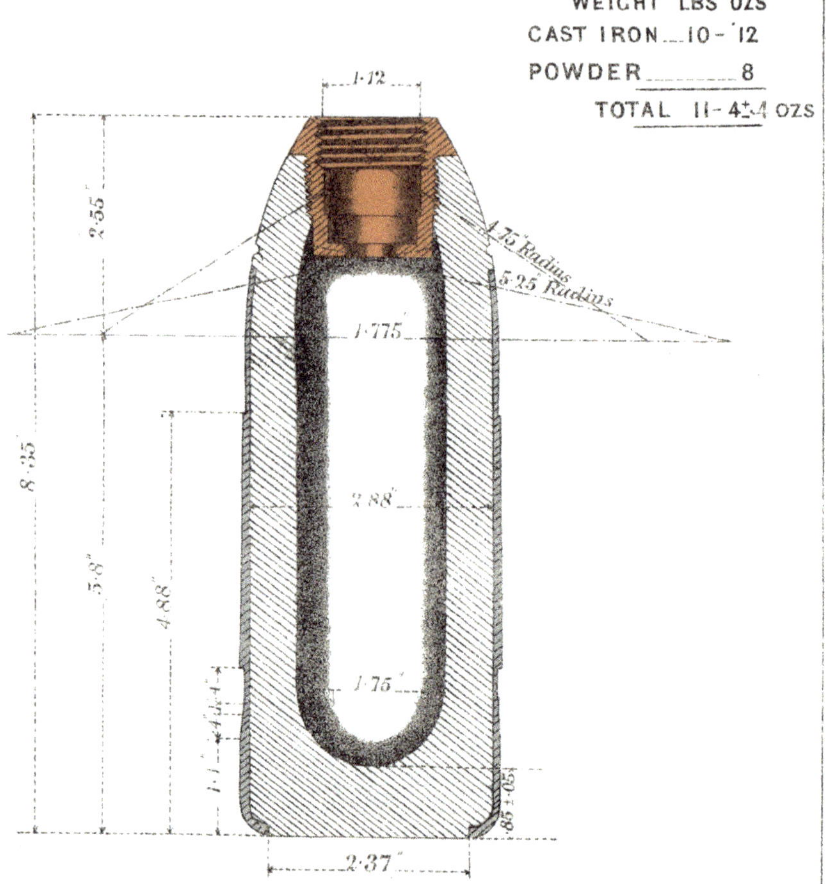


| DIAMETERS | HIGH | LOW |
|---|---|---|
| BACK END | 3·074" | 3·067" |
| BODY | 3·034" | 3·024" |

COMMT. OF TAPER = 3 · 015"
IS NOT TO PASS ON TO THE BODY

DANGERFIELD. LITH. 22. BEDFORD ST. COVENT GARDEN

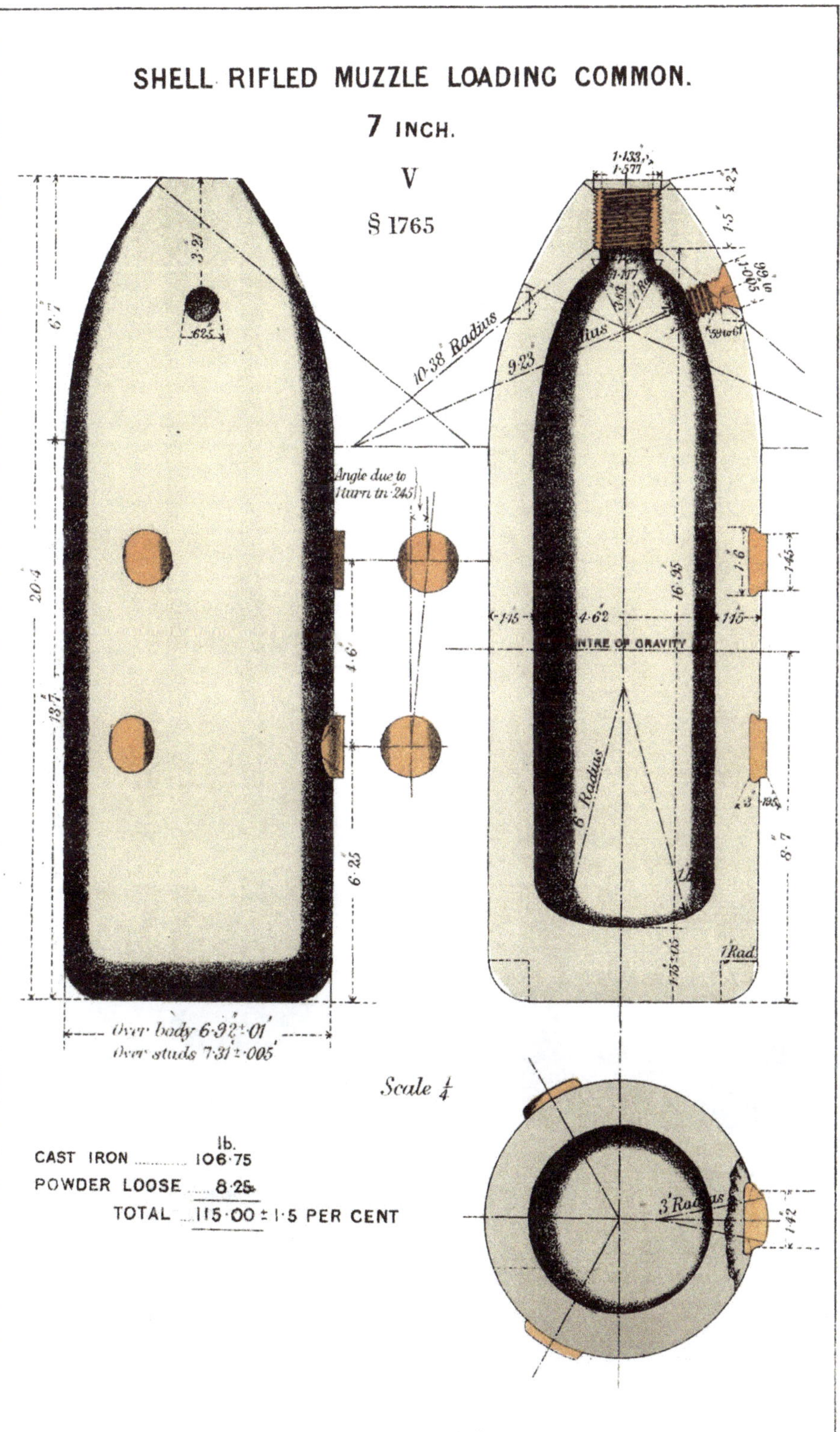


Dangerfield. Lith. 22. Bedford St Covent Garden

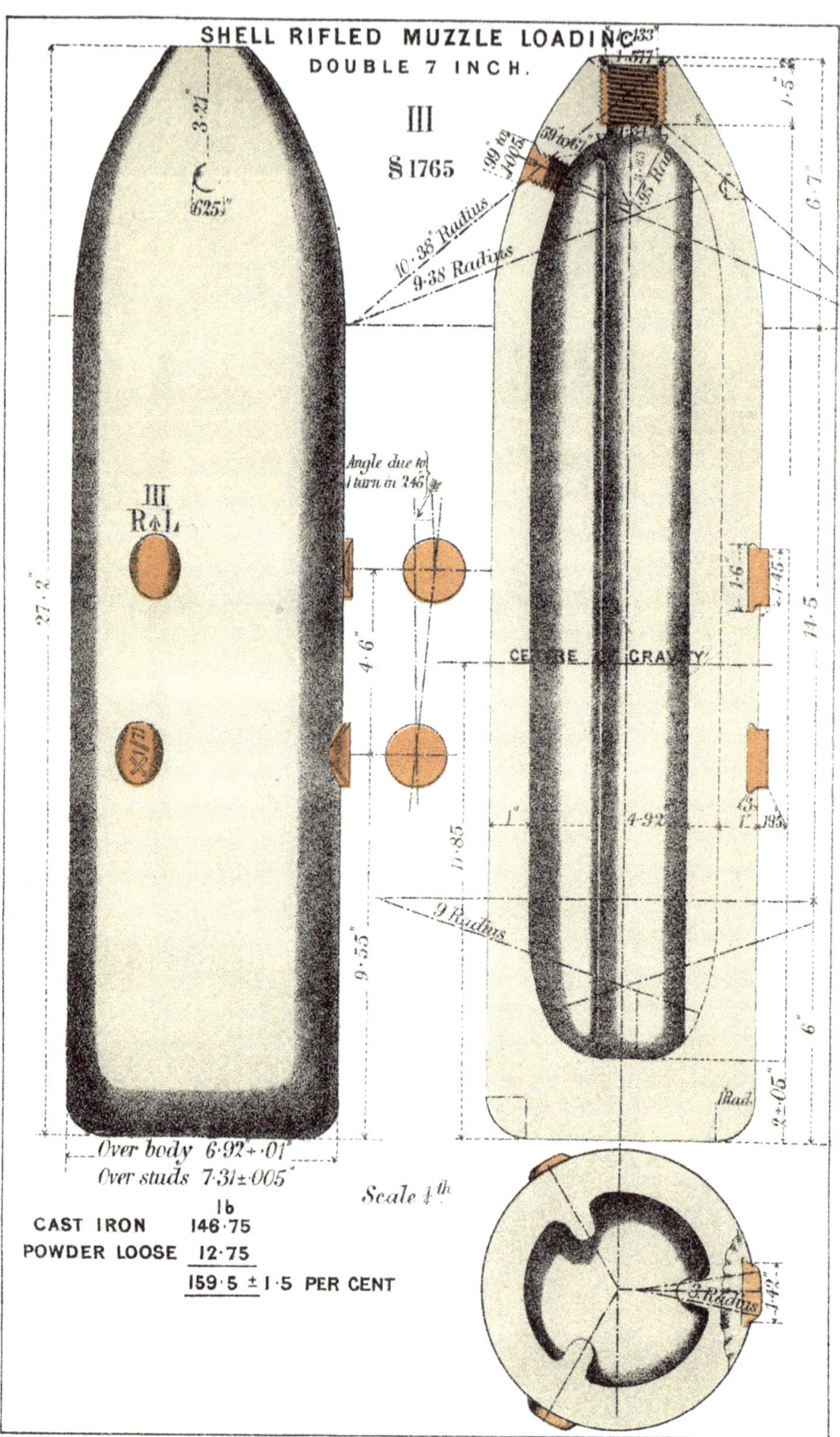


DANGERFIELD. LITH. 22. BEDFORD ST. COVENT GARDEN.

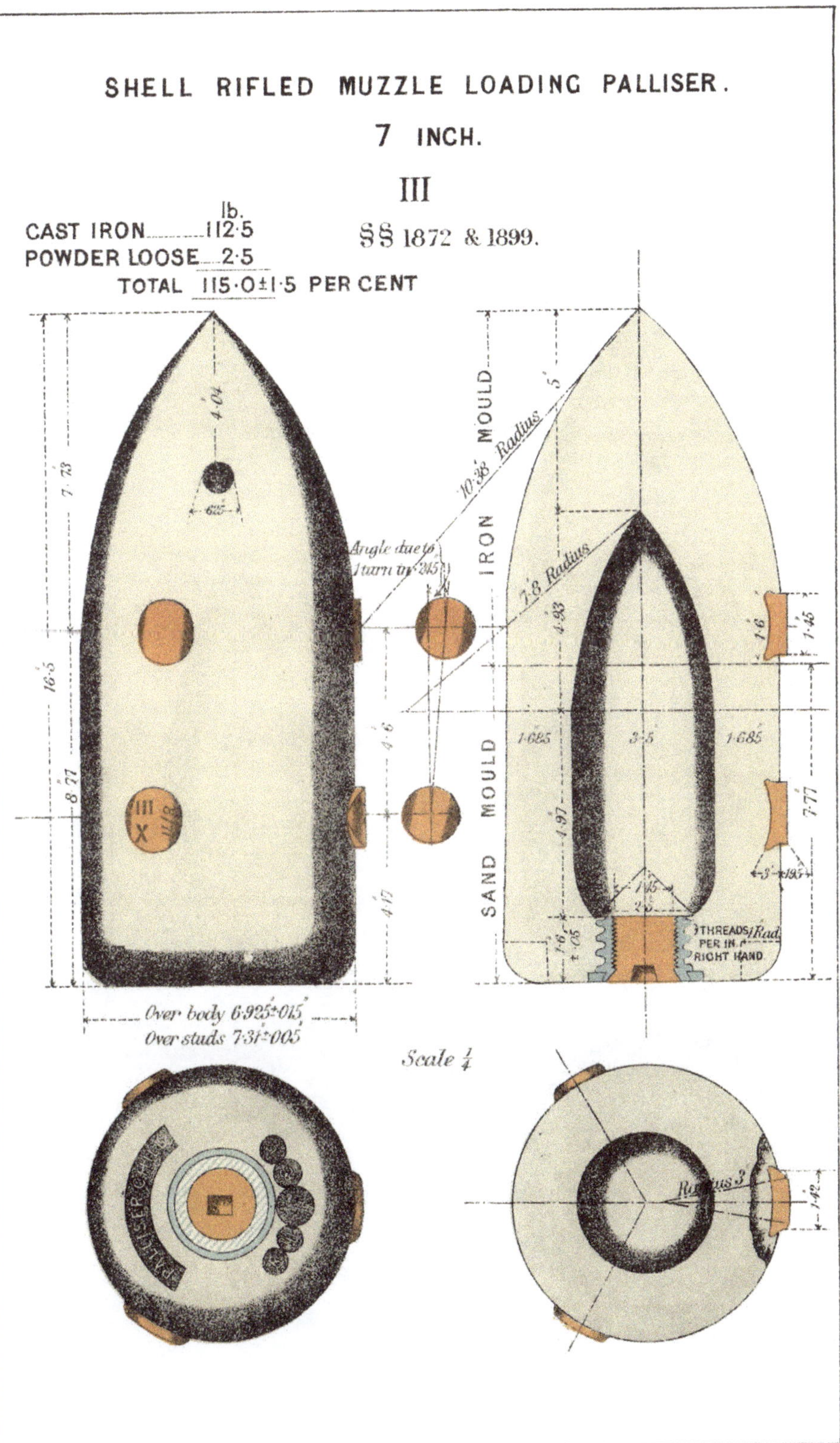


Dangerfield Lith, 22, Bedford St Covent Garden

# SHOT RIFLED MUZZLE LOADING PALLISER.

## 7 INCH

## VI

## § 2222

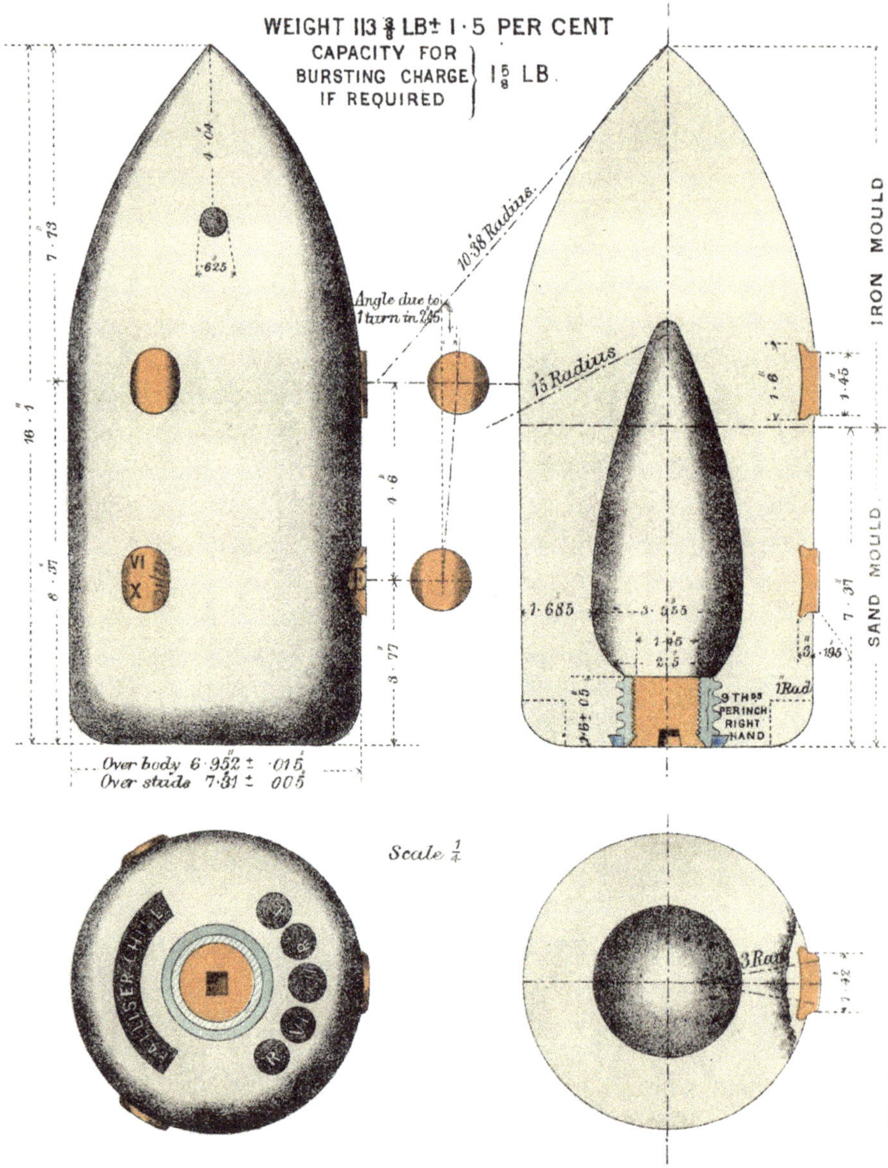

WEIGHT 113 $\frac{3}{8}$ LB ± 1·5 PER CENT

CAPACITY FOR BURSTING CHARGE IF REQUIRED } 1$\frac{5}{8}$ LB.

IRON MOULD

SAND MOULD

10·38 Radius.

1·5 Radius

Angle due to 1 turn in 245

9 THDS PER INCH RIGHT HAND

Over body 6·952 ± ·015

Over studs 7·31 ± ·005

Scale $\frac{1}{4}$

PALLISER CHILL

Dangerfield Lith. 22 Bedford St. Covent Garden

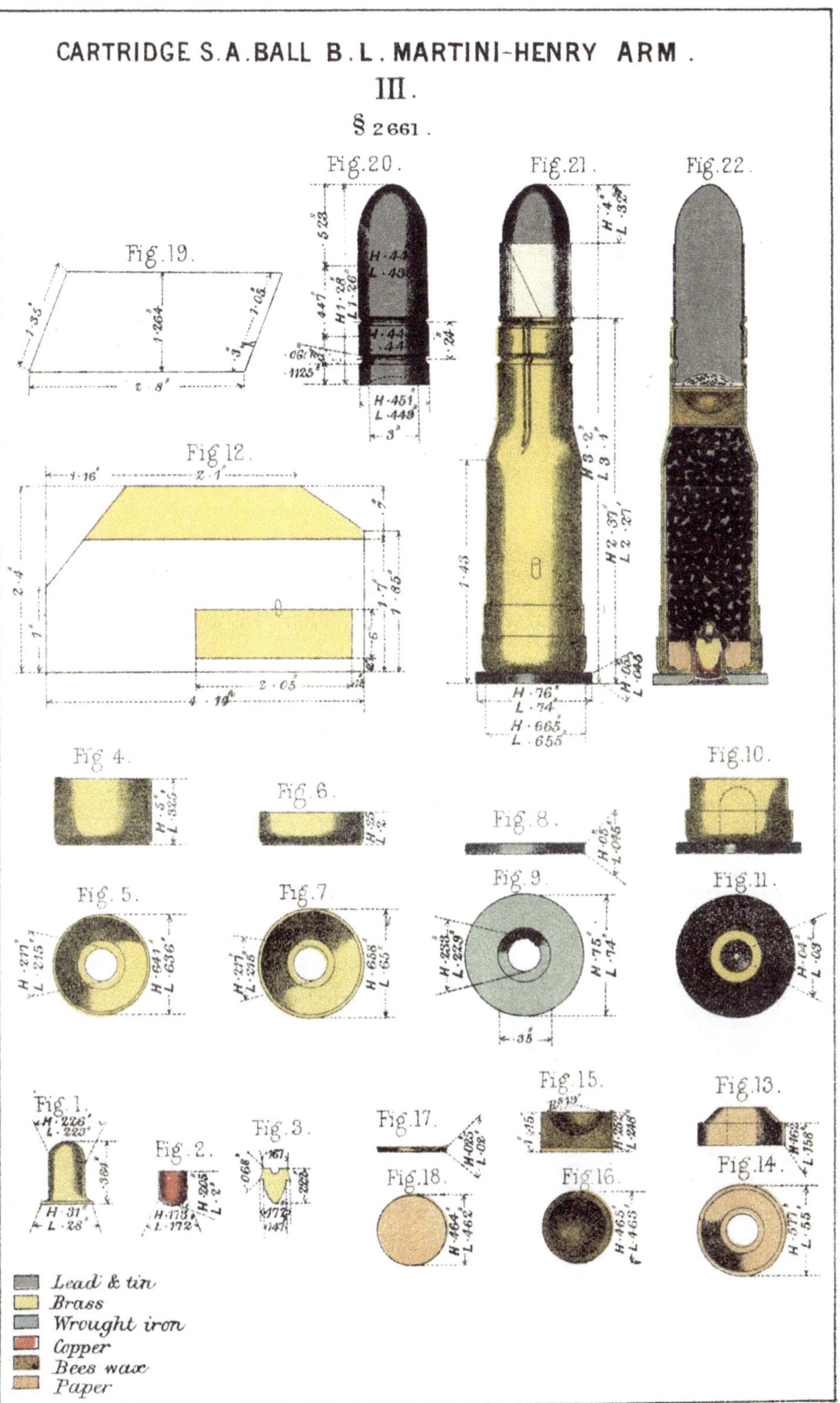


DANGERFIELD LITH. 22. BEDFORD ST COVENT GARDEN.

## Nº 8. Diagram

**Left TARGET.**

**DATA.**

Rifle Nº 3 B description of Snider Breech Loading Enfield Rifle
Powder 70 Grains R.F.G Lot 1237.
Bullets 573 Clay Plug
Lubrication Wax
Cartridge Boxer Amⁿ VI.
Fired from Fixed Rest
Hits 20
Missed 0
Mean Absolute Deviation 12 Inches
Number of Shots 21 to 40
Range in Yards 500
Elevation 1°45'
Point Aimed at ○

Direction of Range N.N.E.
Wind: Direction of S.W.; Strength ½ to 1½; Character Gusty
Thermometer 47
Barometer 29·464
Degree of Humidity 93

# PROOF OF AMMUNITION.

RECORD OF 20 SHOTS FROM ⅜ GAUGER *Work of the 23d*

1 2 3 4 5 6 7 8 9 10 11 12 13 14 15 16 17 18 19 20 21 22 23 24

23 22 21 20 19 18 17 16 15 14 13 12 11 10 9 8 7 6 5 4 3 2 1

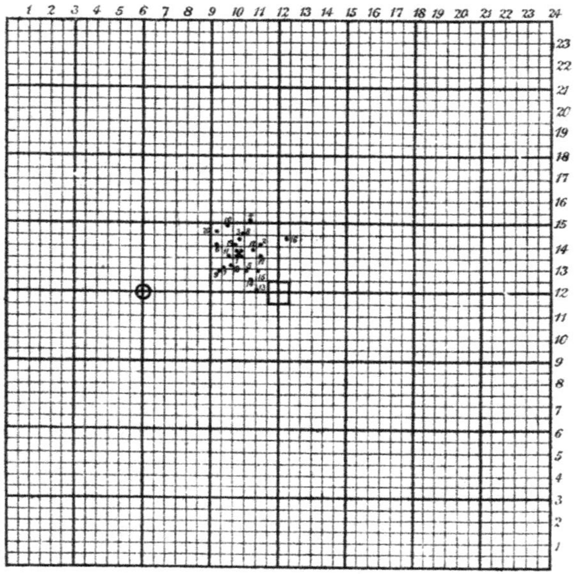

## ROYAL LABORATORY WOOLWICH.

23 DEC. 1868.

*Showing Deviation of each Shot.*

| Nº OF SHOT | HORIZONTAL MEASUREMENT | VERTICAL MEASUREMENT | ABSOLUTE DEVIATION FROM POINT OF MEAN IMPACT |
|---|---|---|---|
| 1 | 1 - 1 | 1 - 7 | 0 - 4 |
| 2 | 2 - 1 | 2 - 1 | 0 - 10 |
| 3 | 1 - 1 | 2 - 3 | 0 - 7 |
| 4 | 1 - 10 | 3 - 1 | 1 - 6 |
| 5 | 1 - 6 | 1 - 0 | 0 - 9 |
| 6 | 0 - 5 | 2 - 0 | 1 - 0 |
| 7 | 0 - 6 | 1 - 0 | 1 - 1 |
| 8 | 1 - 3 | 2 - 6 | 0 - 9 |
| 9 | 0 - 5 | 1 - 0 | 1 - 3 |
| 10 | 0 - 6 | 3 - 0 | 1 - 6 |
| 11 | 0 - 11 | 1 - 7 | 0 - 5 |
| 12 | 1 - 10 | 1 - 11 | 0 - 6 |
| 13 | 2 - 0 | 0 - 2 | 1 - 8 |
| 14 | 1 - 8 | 0 - 7 | 1 - 2 |
| 15 | 2 - 1 | 0 - 11 | 1 - 1 |
| 16 | 3 - 5 | 2 - 5 | 2 - 2 |
| 17 | 2 - 1 | 1 - 9 | 0 - 8 |
| 18 | 0 - 10 | 1 - 1 | 0 - 10 |
| 19 | 1 - 0 | 2 - 0 | 0 - 6 |
| 20 | 0 - 3 | 2 - 9 | 1 - 6 |
|  | 26 - 9 | 34 - 8 | 20 - 1 |
|  | 1 - 4 | 1 - 8¾ | 1 - 0 |

F. Dangerfield Lith. 22 Bedford St Covent Garden London

# SHOT RIFLED MUZZLE LOADING PALLISER.

7 INCH

VI

§ 2222

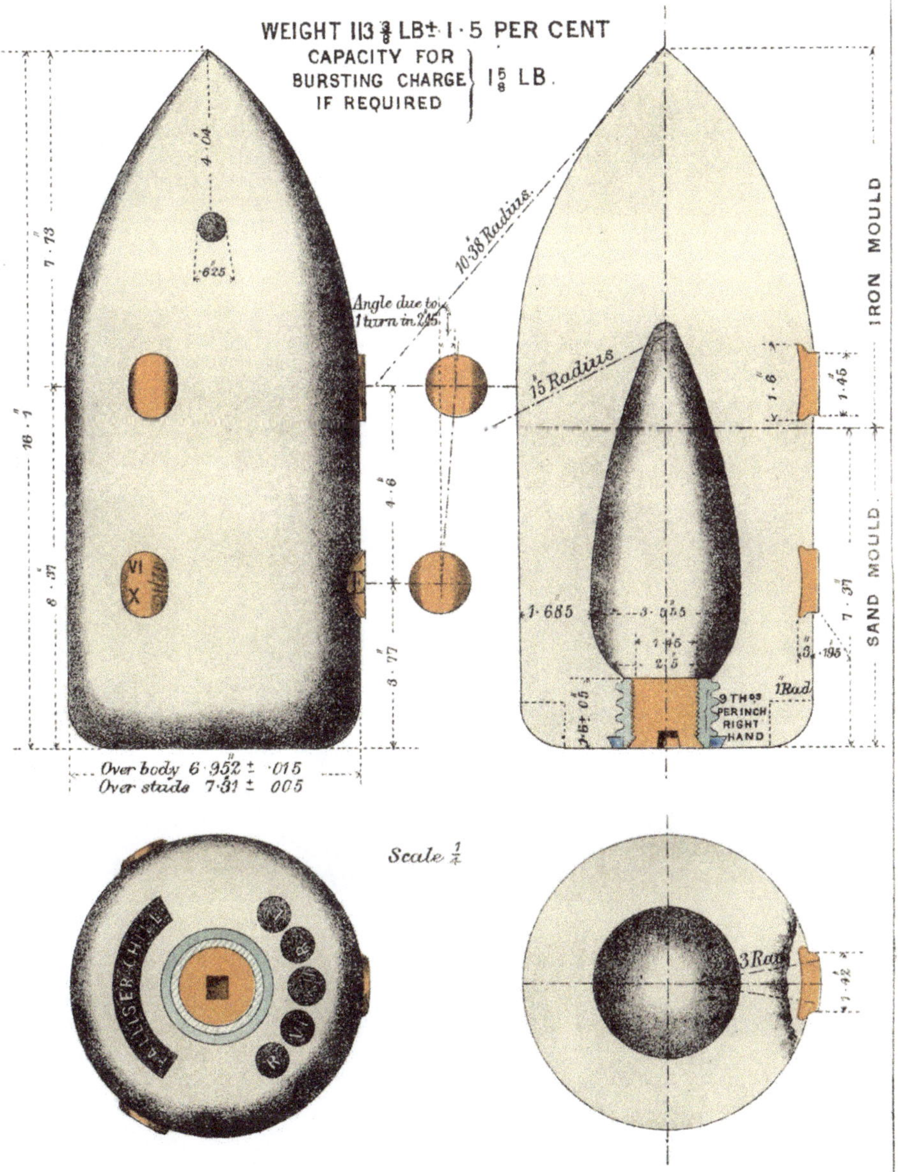

WEIGHT 113 3/8 LB ± 1·5 PER CENT

CAPACITY FOR BURSTING CHARGE IF REQUIRED } 1 5/8 LB.

4·04

·625

7·73

16·1

8·37

3·77

4·6

Angle due to 1 turn in 245

10·38 Radius.

15 Radius

IRON MOULD

SAND MOULD

1·6

1·45

7·37

1·685

3·555

1·95

2·5

3·195

1 Rad

9 THDS PER INCH RIGHT HAND

Over body 6·952 ± ·015

Over studs 7·31 ± ·005

Scale 1/4

PALLISER CHILL

3 Rad

1·42

DANGERFIELD, LITH, 22, BEDFORD ST COVENT GARDEN

www.ingramcontent.com/pod-product-compliance
Ingram Content Group UK Ltd.
Pitfield, Milton Keynes, MK11 3LW, UK
UKHW021840270726
14058UKWH00002B/251